Utilisation of Bioactive Compounds from Agricultural and Food Waste

Utilisation of Bioactive Compounds from Agricultural and Food Waste

Editor

Quan V. Vuong

School of Environmental and Life Sciences
University of Newcastle, Australia

CRC Press
Taylor & Francis Group
Boca Raton   London   New York

CRC Press is an imprint of the
Taylor & Francis Group, an **informa** business

A SCIENCE PUBLISHERS BOOK

CRC Press
Taylor & Francis Group
6000 Broken Sound Parkway NW, Suite 300
Boca Raton, FL 33487-2742

First issued in paperback 2020

ISBN-13: 978-1-4987-4131-6 (hbk)
ISBN-13: 978-0-367-78195-8 (pbk)

Library of Congress Cataloging-in-Publication Data

Names: Vuong, Quan V., editor.
Title: Utilisation of bioactive compounds from agricultural and food waste / editor: Quan V. Vuong.
Description: Boca Raton, FL : CRC Press, Taylor & Francis Group, 2017.
| Includes bibliographical references and index.
Identifiers: LCCN 2017014543| ISBN 9781498741316 (hardback : alk. paper) | ISBN 9781498741323 (e-book)
Subjects: LCSH: Agricultural wastes--Recycling. | Bioactive compounds.
Classification: LCC TD930 .U85 2017 | DDC 628/.746--dc23
LC record available at https://lccn.loc.gov/2017014543

Visit the Taylor & Francis Web site at
http://www.taylorandfrancis.com

and the CRC Press Web site at
http://www.crcpress.com

Preface

World food production has increased significantly over the last century largely due to the rise in population and consumer demand for food variation. Increase in food production has resulted in generation of high quantities of waste from the food production chain. Approximately 50 per cent of food from 'farm to table' is wasted. This is both a national and global problem because of the numerous risks caused by food waste to humans, animals and the environment.

Waste generated from agriculture and food production is considered as general waste because of its limited utilisation and low economic value. The large quantity of waste generated annually from agricultural and food production requires billions of dollars to be spent on agricultural and food waste treatment. Numerous studies undertaken on food waste reveal that it is a rich source of bioactive compounds, which can be extracted and isolated for further utilisation in the food, cosmetic and pharmaceutical industries.

Utilisation of bioactive compounds isolated from food waste not only reduces the risks and costs of waste treatment, but also adds more value to agricultural and food production. Information on different aspects of waste bioactive compounds proves useful for students, academics, researchers and professionals engaged in food science and the food industry. This book was developed with the aim of providing comprehensive information related to extraction and isolation of bioactive compounds from agricultural and food waste for utilisation in the food, cosmetic and pharmaceutical industries.

The topics range from bioactive compounds and potential health benefits, bioactive compounds in waste, techniques used to analyze, extract, isolate and encapsulate these compounds to several specific examples for potential utilisation of waste generated by the agricultural and food industry, such as rice, oil, wine and juice production. This book also discusses the potential of bioactives isolated from waste for re-use in important applications. It may be noted that the book covers only the main aspects of utilisation of bioactive compounds derived from plant waste materials, not from animal or marine materials. In addition, utilisation of by-products of agriculture and food produce is a complex issue. Although this book cannot cover the entire spectrum of utilisation of food waste, it is expected that the readers will find the information useful for their related works.

This book is an excellent compilation of knowledge gleaned by world experts, working on food waste and bioactive compounds. I would like to acknowledge the efforts of the authors in making invaluable contribution to this book. I am grateful to the reviewers for their commitment to improving of the quality of this book. A word of thanks go to my wife, son and daughter for their encouragement. Finally, I would like to thank the CRC Press for publishing this book.

Quan V. Vuong
University of Newcastle Brush Road,
Ourimbah, NSW 2258 Australia
E-mail: vanquan.vuong@newcastle.edu.au

Content

Preface   v

1. **Bioactive Compounds in Agricultural and Food Production Waste**   1
   Nenad Naumovski, Senaka Ranadheera, Jackson Thomas,
   Ekavi Georgousopoulou and *Duane Mellor*

2. **Phenolic Compounds: Potential Health Benefits and Toxicity**   27
   Deep Jyoti Bhuyan and *Amrita Basu*

3. **Alkaloids: Potential Health Benefits and Toxicity**   60
   Renée A. Street, Gerhard Prinsloo and *Lyndy J. McGaw*

4. **Analytic Methods for the Bioactive Compounds in Waste**   86
   Mark Tarleton

5. **Extraction and Utilisation of Bioactive Compounds from**   127
   Agricultural Waste
   Shamina Azeez, C.K. Narayana and *H.S. Oberoi*

6. **Isolation, Purification and Encapsulation Techniques for Bioactive** 159
   Compounds from Agricultural and Food Production Waste
   Viktor A. Nedović, Fani Th Mantzouridou, Verica B. Đorđević,
   Ana M. Kaluševič, Nikolaos Nenadis and *Branko Bugarski*

7. **Extraction, Isolation and Utilisation of Bioactive Compounds**   195
   from Rice Waste
   Binh T. Ho and *Khang N. Tran*

8. **Extraction, Characterization and Utilisation of Bioactive**   213
   Compounds from Wine Industry Waste
   Ariel R. Fontana, Andrea Antoniolli and *Rubén Bottini*

9. **Extraction, Isolation and Utilisation of Bioactive Compounds**   230
   from Waste Generated by the Olive Oil Industry
   J. Lozano-Sánchez, I. Cea Pavez, E. González-Cáceres,
   H. Núñez Kalasic, P. Robert Canales and *A. Segura Carretero*

10. **Extraction, Isolation and Utilisation of Bioactive Compounds**   252
    from Fresh Fruit and Vegetable Waste
    Narashans Alok Sagar, Sunil Sharma and *Sunil Pareek*

11. **Extraction, Isolation and Utilisation of Bioactive Compounds**   272
    from Fruit Juice Industry Waste
    Suwimol Chockchaisawasdee and *Costas E. Stathopoulos*

12. **Valorization of Waste and By-products from the Agrofood**   314
    Industry using Fermentation Processes and Enzyme Treatments
    Phuong Nguyen Nhat Minh, Thien Trung Le, John Van Camp and
    Katleen Raes

13. **Utilisation of Bioactive Compounds Derived from Waste in the**   342
    Food Industry
    Quan V. Vuong and *Mirella A. Atherton*

14. **Potential Application of Bioactive Compounds from Agroindustrial**   358
    Waste in the Cosmetic Industry
    Francisca Rodrigues, Ana F. Vinha, M. Antónia Nunes and
    M. Beatriz P. P. Oliveira

15. **Potential Use of Bioactive Compounds from Waste in the**   383
    Pharmaceutical Industry
    Antonietta Baiano

Index   403

CHAPTER 1

Bioactive Compounds in Agricultural and Food Production Waste

Nenad Naumovski,[1,*] *Senaka Ranadheera,*[2] *Jackson Thomas,*[3] *Ekavi Georgousopoulou*[1] and *Duane Mellor*[1]

Introduction

The use of natural products from plants and foodstuffs as medicines or as a part of the medicinal approach has been recorded since ancient times. These medicines constituted key components of *traditional medicines* in the form of extracts, powders, potions and oils. However, they relied on combination of very complex and matrix-imbedded combinations of compounds, rather than on *pure* and individual compounds themselves. Today the pure components of these compounds are products predominantly identified as *secondary plant metabolites*. Interest in these compounds as potential drug-leads could be incorporated in new aspects of pharmaceutical design and as such pose significant new areas of development. This is mainly due to their very diverse structure when compared to the standard combinatorial chemistry, which allows the discovery of predominately low-molecular-weight lead compounds. From this perspective, current *modern-medicine* can be identified as having its own foundation based on the more traditional medicine approach (Naumovski 2015).

Today's society is described as one of an ageing population, predominantly due to the very large proportion of baby boomers (born after the Second World War) that are now approaching their *third age* of life, with the highest life expectancy ever. These trends are also seen globally in the developed and some developing countries. It is therefore expected that the new structure of the society will mainly be composed

[1] Discipline of Nutrition and Dietetics, Faculty of Health, University of Canberra, Locked Bag 1, Canberra, ACT, 2601, Australia. E-mail: Ekavi.Georgousopoulou@canberra.edu.au; duane.mellor@canberra.edu.au

[2] Advanced Food Systems Research Unit, College of Health & Biomedicine, Victoria University Werribee, VIC 3030, Australia. E-mail: Senaka.Ranadheera@vu.edu.au

[3] Jackson Thomas, Discipline of Pharmacy, Faculty of Health, University of Canberra, Locked Bag 1, Canberra, ACT, 2601, Australia. E-mail: Jackson.thomas@canberra.edu.au

* Corresponding author: nenad.naumovski@canberra.edu.au

of retired individuals or close to their retirement. Although the initial stage of the *third age* is commonly accepted to be relatively healthy and active, the last two-thirds are commonly burdened with increased incidence of illnesses, such as cardiovascular disease (WHO 2014), cognitive decline (Kumar et al. 2015) and various types of cancer (Cragg and Newman 2005). Demographic changes based on the age of the individual are responsible for development and re-design of new foods and functional foods to meet the new health and lifestyle challenges. Therefore, it is not surprising that the search for potential *functional foods* and *natural supplements* that can potentially delay the onset of these diseases (and associated ones) is on the increase both from the viewpoint of the consumer and the manufacturer (Covolo et al. 2013). Importantly, dependence on the use of these natural products (and their derivatives) can not only be seen from the specific age time-frame, but rather be viewed independent of the cross-cultural and geographical needs.

In addition to the increased proportion of society comprising an ageing population, there is a steady and consistent increase in the world population overall. This leaves humanity facing formidable challenges in securing adequate food sources for the well-being of the overall population (Sutovsky et al. 2016). Food waste reduction and reutilisation can been seen as the single and most easily approachable method to address issues of food security and health. The lowering of the traditionally seen food waste, such as usable but unused component of the food source, can effectively increase food usability (Godfray et al. 2011, Tilman et al. 2011). In addition, the use of non-usable food products commonly seen as the leftover of the primary food production, can also prove a significant source of bioactive compounds and as such, potentially reduce the burden on the primary food product itself. In 2011, the Food and Agriculture Organisation (FAO) identified that every year one-third of the world's food produced for human consumption is wasted (FAO 2011). This report specifically focuses on the *grown-but-not-eaten* foods and although it emphasizes the importance of food losses to combat hunger, raise income and improve the food security in some of the poorest countries, it points to the missed opportunity to improve global food output and utilisation.

Currently, the increased global need is for reduction of food waste from the socio-economic perspective and from all aspects of extraction and re-utilisation into the food system. This must be seen from the perspective of the single identifiable compound and also from the cocktail of compounds that can target increased health response. Therefore, the main aim of this chapter is to provide introduction to the current sources of food waste from agricultural production and to identify some of the most important classes of bioactive compounds found in the food waste. In addition, the potential significance of these compounds in use in today's nutraceutical industry will also be discussed briefly.

Sources of Agricultural and Food Production Waste

In general, *food waste* includes products that are not used, but are directly related to human consumption (FAO 2011). As such, the division can be made in five very broad systems boundaries (Table 1). Although not included in the strict definition of food waste, the use of *leftovers* of agricultural food production must be taken into account. Therefore, from the food industry perspective, waste can be derived from

Table 1. Identifiable 'Five system boundaries' of food waste (Adapted from FAO 2011).

System boundary	Examples of potential food losses
Agricultural production	Mechanical damage, spillage during harvesting, fruit picking
Post-harvest handling and storage	Spillage and degradation during handling, storage and transport
Processing	Industrial or domestic processing
Distribution	Market systems
Consumption	During consumption at the household level

raw vegetable and animal material processing during edible food material production (Baiano 2014). Interestingly, the food industry itself produces a significant amount of food waste, with the highest amounts being produced during the handling of fruit and vegetables, followed by milk, meat, fish and wine food productions (FAO 2011, Baiano 2014).

A large majority of the food industry's waste products contain potentially marketable bioactive compounds that are normally present in foods and associated products. Although interest in these compounds was reported since several decades, it is only recently that significant attention has been channelled towards utilisation and extraction of these bioactives from industry waste. Therefore, the new emerging aim of the food industry is to completely exploit the high value components, such as macronutrients (proteins and carbohydrates) and secondary plant metabolites, such as phytochemicals that contain potential nutraceutical-related properties.

Bioactive Compounds from Food Waste

It is well established that food, nutrition and pharmaceutical industries have a significant number of overlapping interests and areas of research. Since the rise of the term *nutraceuticals*, the search for natural compounds in the form of plant extract or as single compounds with beneficial health effects has risen exponentially and become a lead topic of many research laboratories around the globe (Naumovski 2015).

The term *bioactive compounds* is loosely used in today's literature despite the consensus in terminology defining it as the majority of non-nutritive compounds present in foods with the strong potential to improve human health (Biesalski et al. 2009). Interestingly, the use of non-nutritive compounds raises a query that is arguable, as some of these compounds can play a major role in nutrition and human health (Bernhoft 2010). In addition, minerals and vitamins are active components of plants and can induce beneficial and toxicological effects when ingested in relatively large quantities and over prolonged periods of time. However, when discussing the bioactives as a separate entity of compounds, these compounds must be produced as secondary plant metabolites (Cragg and Newman 2005, Biesalski et al. 2009).

Secondary plant metabolites (Fig. 1) are exceptionally diverse compounds, both structurally and chemically, and are quite often localised in specialised plant cells. These compounds are not directly required for the plant metabolism (such as photosynthetic or respiratory metabolism) but rather, their importance is sought in the

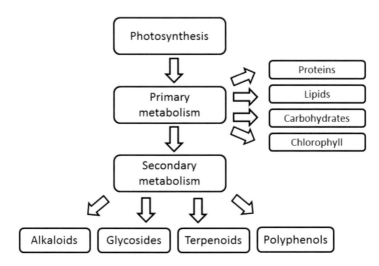

Figure 1. Brief outline of primary and secondary plant metabolites.

plant's survival in the environment. As such, their accumulation is associated with the developmental stages of the plant itself and can provide defensive (against parasites or competing plants) signalling (attraction to pollinating insects) and overall protective support to the plant (oxidant protection). Additionally, the pattern of distribution of these compounds is very diverse between the plant organs and organelles and also between the individual and different plant populations (such as same plant species, but from different geographical positions) (Lattanzio 2013).

The classification of plant secondary metabolites is commonly based on the biosynthetic route or structural features (Lattanzio 2013). However, these metabolites can also be viewed from the basis of clinical function related to their pharmacological effect and botanical approach considering their families (Cragg and Newman 2013, Naumovski 2015). Alkaloids and glycosides are amongst the important categories that are currently attracting a significant amount of research in relation to their recovery from industrial food waste sources—terpenoids and polyphenols (for nutraceutical development).

Structure, Classification and Occurrence of Bioactives in Food Sources

Alkaloids

Alkaloids are natural nitrogen-containing organic compounds with tremendous biological activities (Shi et al. 2014). These compounds are efficiently biosynthesized from amino acids, such as tyrosine (Evans and Mitch 1982) or from amination and transamination reactions (Aniszewski 2015). The boundary between alkaloids and other nitrogen-containing natural compounds is yet to be defined. However, in addition to carbon, hydrogen and nitrogen, alkaloids may also contain oxygen, sulphur and rarely other elements, such as chlorine, bromine and phosphorus (Kabera et al. 2014). The majority of alkaloids are found in higher plants as secondary metabolites, especially in dicots while a few exist in the lower plants (Shi et al. 2014). Certain fungi, such as *Claviceps*, *Penicillium*, *Rhizopus* and *Aspergillus* (Dembitsky 2014) and bacteria including genus *Streptomyces* (Zotchev 2013) also contain natural alkaloids. Naturally-occurring plant alkaloids, in particular the alkaloids present in herbal medicinal plants, have been used in therapeutic application for centuries due to their wide range of pharmacological activities, including potential anti-inflammatory effect, antibacterial and antiviral effects, antimalarial, anticancer activity, hypoglycaemic effects, with positive effects on the central nervous system (Shi et al. 2014, Pereira et al. 2016) and adrenergic (stimulant) activity to promote weight loss (Fugh-Berman and Myers 2004). Thus alkaloids may also contribute to increased perception of flavor, taste and quality of fruits and vegetables (He et al. 2011).

At present more than 12,000 different natural compounds and their derivatives are recognized as alkaloids and their classification is challenging due to their great structural diversity (Hesse 2002, Chowański et al. 2016). Alkaloids can be classified broadly in terms of (1) biological and ecological activity, (2) relation to chemical and technological innovations, (3) chemical structure, and (4) biosynthetic pathway (Aniszewski 2015). There are some other classifications of alkaloids based on their natural sources and the similarity in the carbon skeleton. According to the position of the N-atom in the main structural element, alkaloids can be divided into five different groups: (1) Heterocyclic alkaloids, (2) Alkaloids with N-atoms in exocyclic position, including aliphatic amines, (3) Putrescence, spermidine and spermine alkaloids, (4) Peptide alkaloids and (5) Terpene and storied alkaloids (Hesse 2002).

Alkaloids are usually distributed non-homogeneously over plant tissues and the maximum alkaloid concentration in a plant can be made into any of its tissues, including leaves, fruits or seeds, root or bark. Different tissues of the same plants may possibly contain different alkaloids as well (Grinkevich and Safronich 1983, Hesse 2002). Some well-known alkaloids and their plant sources are listed in Table 2. Chemical structures of two alkaloids which can potentially be recovered from food production by-products, such as potato and tomato wastes, are represented in Fig. 2.

There is no common or recommended method for isolation of alkaloids due to their higher diversity. Alkaloids, which are slightly soluble in water, are soluble in ethanol, benzene, ether, and chloroform (Aniszewski 2015). Based on these characteristics, a number of extraction techniques for alkaloids are available, such as ultrasound-assisted extraction, pressurized liquid extraction, microwave assisted extraction, enzyme assisted extraction, Soxhlet extraction and solid–liquid extraction (Hossain et al. 2015).

Glycosides

Glycosides are organic compounds usually found in plants and may contain phenol, alcohol or sulphur compounds within their structure. They are characterized by a sugar portion or moiety attached by a special bond (mostly a glycosidic bond) to one or non-sugar functional compounds. Many plants store chemicals in the form of inactive glycosides, which can be activated by enzyme hydrolysis, causing the sugar part to be detached and making the chemical available for use (Brito-Arias 2007, Kabera et al. 2014). These plant glycosides are extremely diverse and mostly

Table 2. Some well-known alkaloids and their sources (Adapted from Kabera et al. 2014).

Alkaloid name	Source
Atropine	*Atropa belladonna, Darura stramonium, Mandragora officinarum*
Berberine	*Berberis* species, *Hydrastis, Canadensis, Xanthorhiza simplicissima, Phellodendron amurense, Coptis chinensis, Tinospora cordifolia, Argemone mexicana* and *Eschscholzia californica*
Codeine	*Papaver somniferum*
Coniine	*Conium macularum, Sarracenia flave*
Cytisine (baptitoxine, sophorine)	*Labum* and *Cytisus* of *Fabaceae* family, most extracted from seeds of *Cytisus laborinum*
Morphine	*Papaver somniferum* and poppy derivatives
Nicotine	Solanaceae plant family
Quinine	*Cinchona succirubra, C. calisya, C. ledgeriana*, plants of Rubiaceae family
Solanine	*Solanum tuberosum, S. lycopersiam, S. igrum*, plants of Solanaceae family
Strychnine	*Strychnos nux-vomica*, Loganiaceae plants family
Thebaine (paramorphine)	*Papaver bracteatum*
Tomatine	Green parts of tomato plants

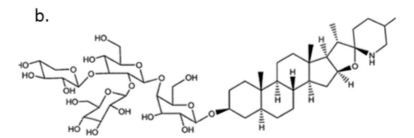

Figure 2. Chemical structures of; (a) solanine; and (b) α-tomatine (Adapted from Lee et al. 2013, Manrique-Moreno et al. 2014).

categorized as prodrugs since they remain inactive until they are hydrolyzed in the large intestine, leading to the release of the aglycone or the non-sugar portion, which is the right active constituent (Kabera et al. 2014). Glycosides are heterogeneous in structure and their classification is based on the nature of aglycone, which can be a wide range of molecular types, including phenols, quinines, terpenes and steroids (Kabera et al. 2014). Certain classifications are based on the therapeutic properties of glycosides, which can be further categorized into different groups, such as cardiac/steroidal, flavonoid, saponin, anthraquinone, cyanophore, isothiocyanate, alcohol, lactone, phenolic, coumarin, chromone and iridoid glycosides. Glycosides can also be categorized into four major groups based on the atoms involved in glycosidic linkage or the type of glycosidic bond: C-glycoside (when sugar moiety is lined with carbon atom), O-glycoside (with oxygen), S-glycoside (with sulphur) and N-glycoside (with nitrogen) (Brito-Arias 2007).

Glycosides contribute to the flavor, taste, color and quality of a number of plant-derived foods. For example, the most widespread anthocyanin which contributes color to fruits is cyanidin-3-glucoside (Fig. 3) (Kong et al. 2003, Shahidi and Ambigaipalan 2015). Additionally, therapeutic properties of glycosides include anticancer effects (Newman et al. 2008), expectorant, sedative, antidepressant, anticonvulsant activities and effect on the central nervous system (Fernández et al. 2006, Kabera et al. 2014, Wang et al. 2016). Glycosides can also be recovered from similar techniques used in isolation of alkaloids with water, methanol or ethanol used as the solvent for extraction (Brito-Arias 2007).

Phenolic compounds

Phenolic compounds are the most ubiquitous, diverse and widely distributed secondary plant metabolites lacking presence in bacteria, fungi and algae. To date, there are over

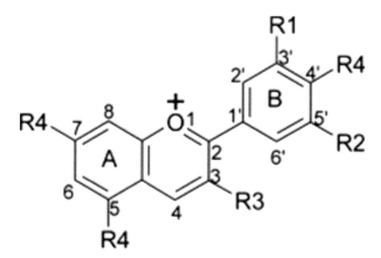

Figure 3. The flavylium cation. R1 and R2 are H, OH, or OCH3; R3 is a glycosyl or H; and R4 is OH or a glycosyl (Adapted from Kong et al. 2003).

8,000 different types of compounds and more than 4,000 flavonoids that have been classified (Tsao 2010). The *phenol*, as a broad term, refers to the chemical structure that defines the phenyl ring bearing one or more of the hydroxyl substituents. Therefore, the commonly used terminology referred to as polyphenols only further extends the structural component of the substance where multiple (at least two) phenol rings are present and contain at least one (or more) of the hydroxyl groups. This classification of phenolic compounds in literature is loosely used and is not necessarily related to the hydroxyl groups but also includes their functional derivatives (such as glycosides). Glycosides, discussed earlier in this chapter, actually contain the majority of plant polyphenols with different sugar units and acylated sugars at different positions of the polyphenolic skeleton. Interestingly, the term polyphenol could also include compounds, such as gossypol and oestrogene (Lattanzio 2013). Therefore, the classification of polyphenols in this chapter will be fundamentally based on different groups as a function of the number of phenol rings and on the structural elements that bind these rings together. Therefore, distinctions can be made from the flavonoids, phenolic acids, stilbenes and lignans (Manach et al. 2004, Tsao 2010, Naumovski 2015).

Flavonoids: Flavonoids are most predominant polyphenols and secondary metabolites with over 4,500 different compounds identified to date. In their structure, the most common feature that flavonoids share is the 2 aromatic rings (A and B) bound together by 3 carbon atoms, effectively forming an oxygenated heterocycle ring (C) (Fig. 4). The diversity of the type of heterocyclic rings is the primary driver of the functional classification of the flavonoids subgroups. Therefore, based on the functional properties, the flavonoids can be divided into flavonols, flavones, isoflavones, flavanones, anthocyanidins and flavanols (Manach et al. 2004, Huber and Rupasinghe 2009, Bernhoft 2010, Naumovski 2015).

Flavones: Flavones (Fig. 4b), are most commonly found in the skins of fruits and vegetables as well as in some culinary herbs. Two of the most common flavones in the edible plants are apigenin and luteolin. The apigenin is present in abundance in food sources, such as celery, onion (0.05 mg/100 g) and also in the culinary herbs, such as fresh sage and dried marjoram (4.4 mg/100 g). On the other hand, the luteolin is predominately present in fruit and vegetables, such as celery, broccoli, carrots, cabbages and apple skins at levels up to 60 mg/100 g (Neveu et al. 2010, Naumovski 2015).

Flavonols: Flavonols (Fig. 4c), are regarded as the most ubiquitous flavonoids present in foods and predominantly found as diverse glycosides with sugar moiety bound to the C-3 position. These compounds are found in different fruits and vegetables, such as onions, apples, grapes and some food products, such as ciders, wine and tea (Manach et al. 2004, Naumovski 2015). The main compounds—quercetin, myricetin and kaempferol—are the lead structures representing this group and although they are the most distributed compounds, their concentration levels in food are relatively low. For a majority of food sources, flavonol levels are less than 30 mg/kg of fresh weight for individual compounds (Manach et al. 2004). However, some food sources do contain relatively high levels of these compounds, such as cranberries, onions (with quercetin levels above 15 mg/100 g) and kale (with myricetin levels above 25 mg/100 g) (Bhagwat et al. 2013, Naumovski 2015). Interestingly, myricetin levels are

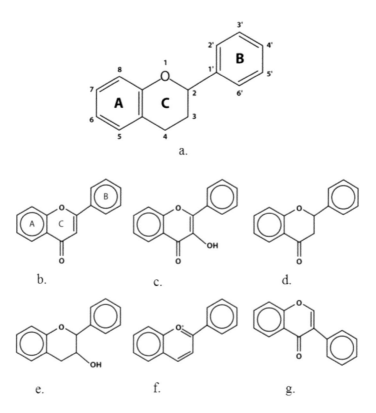

Figure 4. Common structure of flavonoids: a) containing 2 aromatic rings (A and B) and oxygenated heterocyclic ring (C); b) flavone; c) flavonol; d) flavanone; e) flavanol; f) anthocyanidin; and g) isoflavone.

related to the ripening stages of berries, with increased levels detected at fruit ripening (Ong and Khoo 1997).

Flavanones: Flavanones (Fig. 4d) are polyphenolic compounds that are rather unique for their presence in foods, as these compounds are predominately found in citrus fruits (Manach et al. 2003). From the food production perspective, orange juice is one of the most commonly represented foods that contains very high levels (90 per cent) of hesperidin (hesperidin-7-rutinoside) and much smaller percentage (10 per cent) being assigned to narirutin (naringenin-7-rutinoside) (Coelho et al. 2013, Naumovski 2015).

The food sources of flavanols (Fig. 4e) were reported to be in many fruits and vegetables, tea and a variety of different legumes, herbs and spices. Although these compounds exist in both monomer (catechins) and polymer forms (proanthocyanidins), catechins as flavanols have aroused significant interest in the latest literature due to their potential beneficial health effects (Mellor et al. 2010, Sathyapalan et al. 2010, Mellor et al. 2013, Lau et al. 2016). Catechins are found in fruits like apricots (250 mg/kg) but also present in food products, such as red wine (300 mg/L) (Manach et al.

2004, Naumovski 2015). However, green tea (Coelho et al. 2013) and chocolate (Gu et al. 2006) are two of the most catechin-dense foods available on the current market. Green tea in particular contains much higher levels of total catechin content while black tea has significantly lesser amounts due to the fermentation processes associated with black tea production (Vuong et al. 2010, Vuong et al. 2011). Proanthocyanidins are also referred to as tanins which are responsible for the astringent characteristics of fruits like apples, grapes and pears but also for the perceived bitterness of chocolate. Interestingly, these characteristics of foods diminish as the fruit ripens (Manach et al. 2004, Naumovski 2015).

Anthocyanidins: Anthocyanidins (Fig. 4f) are compounds commonly associated with the presence of different and vibrant colours (mainly blue and red) in numerous fruits, vegetables and flowers. These compounds are often bound to sugar groups and structurally, anthocyanidins are glycosylated polyhydroxy and polymethoxy derivatives of the flavium salts (Wallace 2011, Naumovski 2015). Despite the relatively large number of anthocyanidins already identified (over 630), only six (cyaniding, delphinidin, malvidin, pelargonidin, peonidin and pertuindin) form over 90 per cent of the anthocyanidins found in food products. The highest levels of anthocyanidins are reported in black grapes (up to 39.23 mg/100 g) and black currants (86.68 mg/100 g) (Neveu et al. 2010, Naumovski 2015).

Isoflavones: Isoflavones (Fig. 4g) are diphenolic compounds present in legumes, such as soybean and common black beans. These compounds have a structure very similar to mammalian estrogen and their consumption was reported to induce estrogenic and non-estrogenic effects. Two of the most predominant isoflavanones are genistein and daidzein with levels reaching 3 mg/g in various types of soybeans (Manach et al. 2004, Neveu et al. 2010, Naumovski 2015). Additionally, these compounds can also be found in different forms of conjugations, such as isoflavone glucosides (sugar conjugated) and acetyl- and malonyl glucosides.

Phenolic acids: Phenolic acids can be divided into two main categories, derivatives of benzoic acid and derivatives of cinnamic acid (Fig. 5) and these compounds are found free and as conjugates in several different foods. The cinammic acid derivatives commonly occur in foods than benzoic acid derivatives and predominately consist of coumaric (Fig. 5b1), caffeic (Fig. 5b2) and ferullic acids (Fig. 5b3). In Nature, phenolic acids are most commonly found in bound glycosylated derivatives of esters of quinic, shikimic and tartaric acids, except in the processed foods that has undergone freezing, sterilization or fermentation (Manach et al. 2004, Naumovski 2015).

Caffeic acid is the most abundant phenolic acid with over 75 per cent acid content of most fruits with highest concentrations seen in the outside layer of the fruit. The concentration to decreased during fruit ripening, but the overall quantities are proportional to the increase in fruit's size.

Ferullic acid, however, is predominantly found in seeds and cereal grains and may be responsible for up to 90 per cent of total polyphenolic content. Similarly to caffeic acid, the frullic acid is chiefly found in the outermost layer of wheat endosperm (aleurone and pericarp layers) and potential losses of this phenolic acid during flour production can be quite significant (Manach et al. 2004, Naumovski 2015).

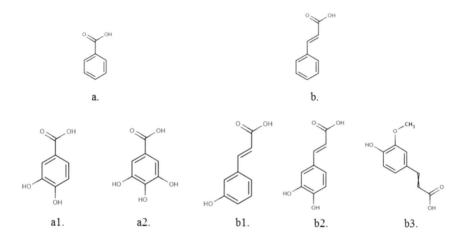

Figure 5. Structures of two classes of phenolic acids: a) bezoic acid; and b) Cinnamic acid and their derivatives; Protocafechuric acid–a1; Gallic acid–a2; Coumeric acid–b1; Caffeic acid–b2; and Ferullic acid–b3 (Adapted from Kim et al. 2016).

The benzoic acid content of fruit and vegetables is generally low, except in foods such as onions and some red fruits with values reported to be a few tens of mg/kg. Interestingly, tea is reported to have the most important source of gallic acid with concentrations reaching 4.5 g/kg (Neveu et al. 2010, Naumovski 2015, Naumovski et al. 2015).

Stilbenes: Stilbenes are one of the only polyphenols that are found in very small concentrations in the human diet due to very small quantities occurring naturally in foods (Neveu et al. 2010, Naumovski 2015). In this class of polyphenols, resveratrol (Fig. 6) is the most abundant stilbene found in a small number of edible plants. Resveratrol (trans-3,4',5-trihydroxystilbene) is a polyphenol commonly found in grapes and has been associated with several beneficial health effects chiefly driven via the anti-inflammatory metabolic regulation in humans (Christenson et al. 2016). Resveratrol is found in strawberries (0.35 mg/100 g) and red currants (1.57 mg/100 g) (Ehala et al. 2005). However, the most investigated and *advertised* food product of resveratrol occurrence is wine. It is found in the skin and seeds of grapes, but larger quantities are found in red wine (0.19 mg/100 ml). It is important to note that resveratrol is found in white and rose wines as well, but to a much smaller extent (around tenfold less). Furthermore, the quantity of resveratrol found in different varieties of red wine is mainly dependent on the grape variety and is also regulated by different external factors, such as geographical region and climatic factors that grapes are exposed to (Gambuti et al. 2007, Neveu et al. 2010, Naumovski 2015).

Lignans: The basic lignan structure is composed of two cinnamic acid residues (or their biogenetic equivalents) (Ayers and Lokie 1990) and as such, belong to a class of a very large group of pharmacologically active compounds (Teponno et al. 2016). Most of the lignans commonly occur freely in a variety of the human food sources. However a relatively small proportion also occurs as glycosides in wood and resin

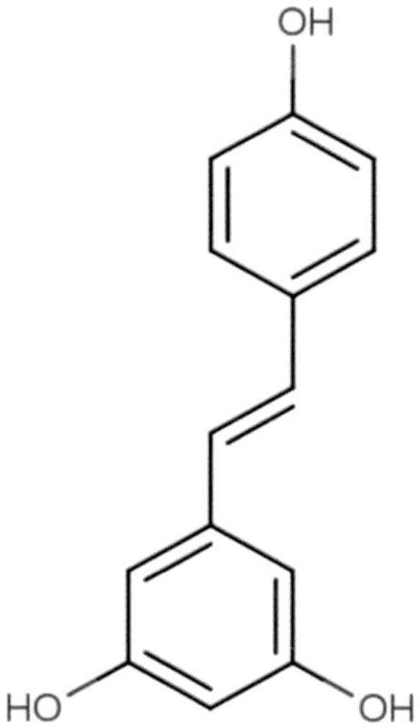

Figure 6. Chemical structure of resveratrol (*trans*-3,4',5-trihydroxystilbene) (Adapted from Kim et al. 2016).

of some plants. These compounds are also referred to as dimers possessing a fairly complex chemical structure. Further, a relatively few trimers and tetramers are also of relatively low molecular weight found in nearly all morphological structures of the plant itself. According to their structure, ligans can be classified into five main categories, namely, lignans, neolignans, norlignans, hybrid lignans and oligomeric lignans, which are further classified into their own respective subgroups (Landete 2012, Zhang et al. 2014, Teponno et al. 2016).

These fiber-related polyphenols are very common in foods with significant amounts of fiber, such as whole-grain products (Landete 2012). Additionally, lignans are also found in other food sources, such as nuts and oilseeds, cereals and breads as well some fruits with varying levels of concentrations (Manach et al. 2004, Landete 2012). A food source with the highest amount of lignans is flaxseed (around 300 mg/100 g) followed by sesame seeds (up to 30 mg/100 g), sunflower seeds (0.89 mg/100 g) and cashew nuts (0.63 mg/kg). Cereals also present a significant source of lignans with highest amounts being detected in rye (8.6 mg/100 g), wheat (3.2 mg/100 g) and oat (2.3 mg/100 g). Vegetables have high levels of lignans (0.19–2.3 mg/100 g) in sources such as cabbage, Brussel sprouts and kale, while the range of lignans in fruits is reported to range from 0 (banana) to 0.45 mg/100 g for apricot (Landete 2012).

Terpenoids

Bioactive plant constituents have been actively studied for their health-enhancing properties since numerous years and as described above, key bioactive plant-derived constituents include flavonoids, phenolic acids, carotenoids, tocopherols, alkaloids, lignans, tannins, salicylates, glucosinolates and triterpenoids (Han and Bakovic 2015).

Triterpenoids are Nature-derived compounds present in free form as well as in the form of numerous glycosides. The complex structures of triterpenoids mainly involve squalene derivatives, lanostanes, cycloartanes, dammaranes, euphanes, tirucallanes, tetranortriterpenoids, quassinoids, lupanes, oleananes, friedelanes, ursanes, hopanes, serratanes and other less known groups (Lesellier et al. 2012, Han and Bakovic 2015). The special triterpenic skeletons allow performance of chemical modifications in order to obtain many new derivatives with improved pharmacological activity (Lesellier et al. 2012). All forms of triterpenoids are widely distributed in edible and medicinal plants. Consequently they form an integral part of the human diet (Han and Bakovic 2015).

The three key triterpene groups include oleane, ursane and lupane triterpenes (Hill and Connolly 2011). The main triterpenoids found in the oleane category include oleanolic acid, erythrodiol and β-amyrin, whereas ursane and lupane families contain ursolic acid, uvaol; and lupeol, betulin and betulinic acid respectively (Han and Bakovic 2015). Oleanolic acid, maslinic acid, and β-amyrin are the chief compounds present in the oleanane triterpene family. These compounds are predominantly found in the skin of grapes, olives and tomatoes. Ursolic acid and uvaol are the chief compounds present in the ursane family. They are mainly present in the cuticle of apples (Han and Bakovic 2015). The chief compounds of lupine family, namely lupeol, betulin and betulinic acid are mainly found in the cuticle of mangoes and grape berries. Triterpenoids, in their free and esterified form are found to have low polarity. Therefore they are found in abundance in plant parts, such as surface cuticle waxes and stem bark (Szakiel et al. 2012). Therefore, fruit peels serve as a promising and highly available source material for such bioactive components (Jager et al. 2009).

Compared with plant vegetative organs, especially leaves, far less information is available regarding the triterpenoid content of fruit cuticular waxes (Szakiel et al. 2012). Fruit peels are a waste product of juice and canned products. Their utilisation to generate a value-added, therapeutically active, food ingredient can be economically beneficial (Wolfe and Liu 2003, Djilas et al. 2009). Triterpenoids, both in their natural form and as templates for synthetic modification, are of considerable interest to academia and industry. As previously indicated, food industries produce large volumes of wastes and the disposal of it may cause significant environmental pollution and burden. Millions of tonnes of apple, grape berry, olive, tomato, orange and other fruit peels are generated each year as agro-industrial waste. This could be utilized to formulate various useful pharmaceutical and/or nutraceutical applications (Mintz-Oron et al. 2008, Szakiel et al. 2012).

Bioactives from Agricultural and Food Production Wastes

Although the total alkaloid content in plants is usually low, the plant-based agricultural by-products in particular food production waste can be considered as a potential source for alkaloids. Various alkaloids have been successfully isolated from agricultural and food wastes, such as potato peels (Hossain et al. 2015), potato tubers (Petersson et al. 2013), green tea leaves (Bermejo et al. 2015), tomato roots (Nagaoka et al. 1993), tomato (fruit) (Yahara et al. 2004), citrus and orange peel (epicarp and mesocarp) (He et al. 2011), dried pepper fruits (*Capsicum frutescens*) (Santos et al. 2016), lupin grains (Przybylak et al. 2005, Sujak et al. 2006), barley (*Hordeum vulgare*), rye (*Secale cereal*) (Aniszewski 2015), common wheat (*Triticum aestivum*), Triticale (Scott and Lawrence Guillaume 1982), Maté (*Ilex paraguariensis*) leaf (Clifford and Ramirez-Martinez 1990), cocoa beans (*Theobroma cacao*) (Nazaruddin et al. 2006), castor leaves (*Ricinus communis*) (Kang et al. 1985), coffee beans (Mehari et al. 2016) and Kolanut (*Cola nitida*) (Muhammad and Fatima 2014).

When plants are under stress due to improper storage conditions, mechanical damages, insect or pest attack, the injured plant tissues instigate synthesis of higher concentrations of alkaloids and other compounds (Chowański et al. 2016). For example, potatoes that have been exposed to light in the open or during storage may become green due to accumulation of chlorophyll, which may affect the surface (peel) or may penetrate into the flesh. Such tubers and damaged potato tubers usually possess higher levels of alkaloids. Therefore, isolation of alkaloid compounds from agricultural and food production waste is a feasible approach for economic development (Petersson et al. 2013, Chowański et al. 2016).

Since most common phenolic acids and flavonoids in plants are present in conjugated forms as esters or glycosides (Lu et al. 2011), agricultural and plant-based food wastes are among the best sources for recovering glycosides and its derivatives. Apple pomace, one of the major global waste products generated primarily during apple juice production, is rich in a vast array of polyphenolic compounds, including quercetin and phloretin glycosides with notable functional properties. For example, quercetin glycosides isolated from apple pomace showed excellent antioxidant properties compared to other phenolic compounds (Lu and Foo 2000). Apple peel skins (Lommen et al. 2000, Huber and Rupasinghe 2009) and seeds (Lu and Foo 1998) are rich in glycosides. Grape pomace as a by-product is approximately 20 per cent of the harvested grapes. Flavonol glycoside is one of the principal phenolic constituents of grape (*Vitis vinifera* cultivars) pomace (Schieber et al. 2001) and has been isolated from both grape fruit skins and seeds (Williams et al. 1983, Ruberto et al. 2007). Grape stem, the other major waste product in the wine industry, also possesses glycosides (Souquet et al. 2000, Spatafora et al. 2013). Glycosides have been identified and characterized from numerous fruits and fruit-based by-products, such as mango peel and puree (Schiebe et al. 2000, Schieber et al. 2001), kiwifruit pulp (Dawes and Keene 1999), bitter apricot seeds (*Prunus armeniaca* L., Rosaceae) (Tunçel et al. 1995), citrus fruits including lime, lemon, orange, grapefruit and tangerine peel (Mouly et al. 1994, Kanaze et al. 2004, Lu et al. 2011, Shahidi and Ambigaipalan 2015), black currant seeds (Lu and Yeap Foo 2003, Shahidi and Ambigaipalan 2015) and skins and seeds of blueberries and blueberry processing waste (Lee and Wrolstad 2004).

Vegetable-based food commodities, such as onion (*Allium cepa* L., Alliaceae) waste, which includes brown skin, the outer two fleshy leaves and the top and bottom bulbs (Schieber et al. 2001, Turner et al. 2006), potato peel (Reyes et al. 2005), lettuce and collard leaves (Young et al. 2005), cabbage (*Brassica*) leaves (Nielsen et al. 1993), pumpkin varieties, such as *Curcubita pepo* (Iheanacho and Udebuani 2009) and pericarp of red pepper fruit (*Capsicum annuum* L.) (Materska and Perucka 2005) have also been used in recovering glycosides. Various other agricultural and food production wastes including cottonseed as a by-product (Piccinelli et al. 2007), defatted soybean (Nemitz et al. 2015), tea leaves (Wan et al. 2009), almond hulls (*Prunus amygdalus*) (Sang et al. 2002), industrial horse chestnut (*Aesculus hippocastanum*) waste (Kapusta et al. 2007), green barley (*Hordeum vulgare* L.) leaves (Kitta et al. 1992) and seed coat of lentils (Dueñas et al. 2002) and beans (Madhujith et al. 2004, Shahidi and Ambigaipalan 2015) are recognized as natural sources of glucosides.

Vegetative parts of certain culinary herbs and spices, such as ginger (*Zingiber officinale* Rosc.), thyme (*Thymus vulgaris* L.), parsley (*Petroselinum crispum* Mill.), curry leaves (*Murraya koenigii* L. Spreng), peppermint (*Mentha piperita* L.), turmeric (*Curcuma longa* L.), green onion scallion (*Allium fistulosum* L.), coriander (*Coriandrum sativum* L.) and Chinese star anise (*Illicium verum* Hook) also contain cardiac glycosides (Wang et al. 2011, Ramkissoon et al. 2016). Hence, portions of these herbs and spices which are unsuitable for culinary purposes can be potential sources of extracting glycosides.

Individual Sources and Potential Uses of Some of the Bioactives from Agricultural Food Waste

Apples and apple peel

Regular consumption of apples has been associated with prevention of chronic diseases, such as lung cancer, cardiovascular disease, symptoms of chronic obstructive pulmonary disease and the risk of thrombotic stroke (He and Liu 2007, Szakiel et al. 2012). Apple peel is found to contain greater antioxidant and antiproliferative activity than the flesh (Wolfe and Liu 2003), thus suggesting that the peel contains a major share of the bioactive phytochemicals (Cefarelli et al. 2006). The triterpenoid composition of apple peel and its noted pharmacological properties are summarised in Tables 3 and 4. Previous reports suggest that the method of cultivation and post-harvest conditions, including cold storage of fruit, can influence the triterpenoid composition of the cuticular wax (Szakiel et al. 2012).

Grape berry

Grapevine (*Vitis vinifera* L.) consumption is associated with reduction of chronic illnesses, such as cancer, cardiovascular diseases, ischemic stroke, neurodegenerative disorders and aging (Yadav et al. 2009, Ali et al. 2010). Grapevine-derived products (fresh berries, raisins, juice, wine) are well known for their antioxidant content and therefore, grape extracts are widely incorporated in various cosmetic formulas. Bioactive components, including polyphenols, resveratrol, hydroxytyrosol and melatonin, may provide health benefits associated with regular consumption of grape products (Leifert and Abeywardena 2008, Ali et al. 2010). The triterpenoid composition of grape berry skin and its noted pharmacological properties are summarised in Tables 5 and 6. In addition to oleanolic acid, other triterpenoids have been identified in grape berry skin, including oleanolic aldehyde, erythrodiol and a group of phytosterols and their derivatives (e.g., β-sitosterol, campesterol, stigmasterol, and lanosterol) (Dagna et al. 1982, Orban et al. 2009, Zhang et al. 2014).

Olives

The olive (*Olea europaea* L.) is a fruit of substantial agricultural importance in the Mediterranean region. Olive tree cultivation started approximately 6,000 years ago

Table 3. Summary of the triterpenoid composition of apple peel (Adapted from Szakiel et al. 2012).

Fruit	Composition	Amount	Main compounds	Reference
Apple (*Malus pumila* Mill.)	Euscaphic acid; 2a,3a-dihydroxy-olean-12-en-28-oic acid; 2a,3a-dihydroxy-urs-12-en-28-oic acid; 2a-hydroxyursolic acid; ursolic acid, uvaol	77% of the peel extract	Ursolic acid (98% of triterpenoid mixture)	(Ma et al. 2005)
Apple (*Malus domestica* Borkh.) cv. Holsteiner Cox	Oleanolic acid; ursolic acid; uvaol	0.34–0.42% of the peel extract	Ursolic acid (0.28–0.34% of the peel extract)	(Ellgardt 2006)
Apple (*M. pumila* Mill.) cv. Red Delicious	3b-cis-p-coumaroyloxy-2a-hydroxyolean-12-en-28-oic acid; 3b-cis-p-coumaroyloxy-2a- hydroxyurs-12-en-28-oic acid; 3b,28-dihydroxy-12-ursene; 3b,13b-dihydroxyurs-11-en-28-oic acid; 2a-hydroxyursolic acid; maslinic acid; 3b-trans-cinnamoyloxy-2a-hydroxyurs-12-en-28-oic acid; 3b-trans-p-cinnamoyloxy-2a-hydroxyolean-12-en-28-oic acid; 3b-trans-p-coumaroyloxy-2a-hydroxyolean-12-en-28-oic acid; 3b-trans-p-coumaroyloxy-2a-hydroxyurs-12-en-28-oic acid; 2a,3b,13b-trihydroxyurs-11-en-28-oic acid; ursolic acid	0.15–19.5% of the peel extract	Ursolic acid (0.15% of the mass of fresh peels, 18% of the peel extract)	(He and Liu 2007)

Table 4. Summary of the pharmacological properties of triterpenoids found in apple peel seen in *in vitro* models (Adapted from Szakiel et al. 2012).

Pharmacological properties	Reference
Antitumor activity, with inhibitory activity against four tumour cell lines, HL-60, BGC, Bel-7402, and HeLa, with ED50 values ranging from 45 to 72 µg/ml	(Ma et al. 2005)
Antiproliferative activities against human cell lines of HepG2 liver cancer, MCF-7 breast cancer and Caco-2 colon cancer	(Yamaguchi et al. 2008)
Antitumorigenic effects—suppression of p65 phosphorylation, TNF-induced expression of cyclin D1, cyclooxygenase 2 (COX-2) and matrix metalloproteinase 9, which are involved in the initiation, promotion and metastasis of tumours	(He and Liu 2007)

Table 5. A summary of the triterpenoid and phytosterol composition of grape berry skin (Adapted from Szakiel et al. 2012).

Fruit	Composition	Amount	Main compounds	Reference
Grape berry (*Vitis vinifera* L.) cv. Cabernet Sauvignon	Oleanolic acid; oleanolic aldehyde; β-sitosterol; β-sitosterol 3-O-β–D–glucoside; β-sitosterol-60-linolenoyl-3-O-β-D-glucopyranoside	0.075% of fresh skin mass	Oleanolic acid (86% of triterpenoid mixture)	(Zhang et al. 2004)
Grape berry (*V. vinifera* L.)	Oleanolic acid; β-sitosterol; β-sitosterol-3-O-β–D–glucoside	0.027% of fresh berry mass	Oleanolic acid (0.003–0.016% of fresh berry mass)	(Orban et al. 2009)

Table 6. Summary of the reported pharmacological properties of triterpenoids found in grape berry skin (Adapted from Szakiel et al. 2012).

Pharmacological properties	Reference
Anticancer, anti-inflammatory, antidiabetogenic, antimicrobial, hepato- and cardioprotective, anti-HIV and anti-multiple sclerosis effects	(Liu 2005, Martin et al. 2010)
Regulation of insulin secretion, activity against type-2 diabetes and inflammation	(Zhang et al. 2004)
Hepato-protective activity, prevention and treatment of liver disorders, to treat diarrhoea, hepatitis and stomach aches	(Liu et al. 2010)
Antimicrobial activity against Streptococcus mutans and Porphyromonas gingivalis, potential benefits to oral health and disease prevention	(Wu 2009)
Serum cholesterol lowering, cancer preventive, antimutagenic and anti-inflammatory activities	(Awad and Fink 2000, Piironen et al. 2000, Villasenor et al. 2002)
Antiprotozoal and chemo preventive activities	(Gallo and Sarachine 2009)

and thus the olive is the oldest cultivated tree (Szakiel et al. 2012). Olive oil is one of the key components in the traditional Mediterranean diet and has significant nutritional and medicinal properties. These include reduction of the risk of coronary heart disease and atherosclerosis, prevention of several types of cancer and modification of the immune and inflammatory responses (Ortega 2006). The potential of triterpenoids and some other polyphenols present in olive oil has not been explored to its full potential (Stiti and Hartmann 2012). The triterpenoid composition of olive oil and its pharmacological properties are summarized in Tables 7 and 8. The triterpenoids composition in olives that are sold in the market is influenced by cultivar or the stage of fruit ripeness, and also the method of processing (Romero et al. 2010).

Tomato

Tomatoes (*Solanum lycopersicum* L. or *Lycopersicon esculentum* L.) are one of the most important fruit crops globally. They are cultivated practically in every country in the world in outdoor fields, glass houses and net houses (Szakiel et al. 2012) with the triterpenoid composition summarised in Table 9. The triterpenoid levels in the fruit vary considerably during various stages of fruit development. For example, the levels of the most abundant triterpenols (α-, β- and δ-amyrins) increase significantly 25–42 days after flower anthesis (Mintz-Oron et al. 2008). The potential pharmacological properties of bioactives identified from tomato skin include prevention of certain types of cancer, liver disorders, heart disease, osteoporosis, cataracts, anti-allergic, antidepressant, anti-inflammatory, anti-nociceptive, antipruritic, anxiolytic, gastro protective and hepato-protective activities (Soldi et al. 2008, Melo et al. 2010, Ching et al. 2011).

Table 7. Summary of the triterpenoid composition of olive skin (Adapted from Szakiel et al. 2012).

Fruit	Composition	Amount	Main compounds	Reference
Olive (*Olea europaea* L.) cv. Coratina	α-amyrin; β-amyrin; betulinic acid; erythrodiol; maslinic acid; oleanolic acid; β-sitosterol, stigmasterol, uvaol	0.075% of fresh skin mass	Oleanolic acid (86% of triterpenoid mixture)	(Bianchi et al. 1992)
Olive (*O. europaea* L.) cv. Arbequina	Maslinic acid; oleanolic acid	0.23 and 0.19% of Fruit, in green and black fruit, respectively	Oleanolic acid (0.003–0.016% of fresh berry mass)	(Guinda et al. 2010)

Table 8. Summary of the pharmacological properties of triterpenoids found in olive skin (Adapted from Szakiel et al. 2012).

Pharmacological properties	Reference
Anticancer, antihyperglycemic, and antiparasitic activities	(De Pablos et al. 2010, Moneriz et al. 2011)
Antiproliferative effect on HT-29 cells (EC50 of 160 μmol/l)	(Juan et al. 2008)
Prevention and treatment of hyperlipidemia	(Liu et al. 2011)
Prevention and treatment of brain cancers, potent inhibitory activity against human 1321N1 astrocytoma cells	(Wu 2009)
Antiparasitic properties with activity against *Plasmodium falciparum*, *Eimeria tenella*, and *Toxoplasma gondii*	(De Pablos et al. 2010, Moneriz et al. 2011)

Table 9. Summary of the triterpenoid composition of tomato skin (Adapted from Szakiel et al. 2012).

Fruit	Composition	Amount	Main compounds	Reference
Tomato (*L. Esculentum* L.)	α-amyrin; β-amyrin; δ-amyrin; bauerenol; cycloartenol; germanicol; lupeol; multiflorenol; β-sitosterol; stigmasterol; taraxasterol; ω-taraxasterol; taraxerol	13.7% of the total wax extract (average from 26 cultivars)	δ-amyrin (5.6% of wax extract, i.e., 41.2% of total triterpenoids); β-amyrin (3.2% of wax extract); α-amyrin (3% of wax extract)	(Bauer et al. 2004)
Tomato (*L. esculentum* L.) cv. MicroTom (wild-type)	α-amyrin; β-amyrin; β-amyrin derivative; δ-amyrin; sterol; lanosterol; lupeol derivative I; multiflorenol; β-sitosterol; stigmasterol; taraxasterol; ω-taraxasterol; taraxerol	21% of the total wax extract (in mature fruit)	α-, β-, δ-amyrins (76–91% of total triterpenoids)	(Leide et al. 2007)

Conclusion

Bioactive compounds have aroused significant interest in the recent past, predominately due to their antioxidant properties, which suppress and inhibit the production of common free radicals, such as superoxide anion and hydroxyl radical. Although these properties are well described in several *in vitro* studies, human and animal data still remain controversial (Lau et al. 2016, Mellor and Naumovski 2016). From the perspective of human consumption, it is important to note that though these compounds may be present in very high doses in some food products, their intake and absorption is regulated by the food intake and combination of different foods independently, whether they are consumed as a pure compound (Naumovski et al. 2015) or as a part of the polyphenolic matrix (Chow et al. 2005, Stalmach et al. 2008). However, formulations based on nanotechnology are proposed to replace the conventional dosage forms to address this concern (Chen et al. 2011). Much of the data supporting the concept of beneficial effects of these compounds is based on *in vitro* and animal model studies, with conflicting data often coming from clinical trials in humans. In some cases, such as with resveratrol, the dietary consumption of the compound is well below the threshold required to produce any significant physiological effect. Therefore, extraction, isolation and purification of individual compounds from the food matrix has already aroused significant levels of interest in the food and pharmaceutical industry. This search for pure compounds has also created a requirement for the production of foods that are primary carriers of bioactives where the utilisation of agricultural food waste can be seen as a perfect fit to this part of the food industry. The main reason for this is due to the fact that agricultural food waste can contain significant levels of individual bioactives and rather than waste being re-used in the form of a landfill, it can be directed to re-utilisation of bioactives. Nevertheless, the presence of bioactives in foods is of crucial importance, not only from the perspective of potential beneficial health effects in humans, but also from the protective role that these compounds play in the overall plant metabolism.

References

Ali, K., F. Maltese, Y. H. Choi and R. Verpoorte. 2010. Metabolic constituents of grapevine and grape-derived products. Phytochem. Rev. 9(3): 357–378.

Aniszewski, T. 2015. Definition, typology, and occurrence of alkaloids. Alkaloids. Boston, Elsevier: 1–97.

Awad, A. B. and C. S. Fink. 2000. Phytosterols as anticancer dietary components: Evidence and mechanism of action. J. Nutr. 130(9): 2127–2130.

Ayers, D. C. and J. D. Lokie. 1990. Lignans: Chemical, Biological and Clinical Properties. Cambridge, United Kingdom, Cambridge University Press.

Baiano, A. 2014. Recovery of biomolecules from food wastes—a review. Molecules 19(9): 14821–14842.

Bauer, S., E. Schulte and H. P. Their. 2004. Composition of the surface wax from tomatoes—I. Identification of the components by GC/MS. European Food Research and Technology 219(3): 223–228.

Bermejo, D. V., J. A. Mendiola, E. Ibáñez, G. Reglero and T. Fornari. 2015. Pressurized liquid extraction of caffeine and catechins from green tea leaves using ethyl lactate, water and ethyl lactate + water mixtures. Food and Bioproducts Processing 96: 106–112.

Bernhoft, A. 2010. A brief review of bioactive compounds in plants. Bioactive Compounds in Plants—Benefits and Risks for Man and Animals. Oslo, Norway, Norw. Acad. Sc. and Lett.

Bhagwat, S., D. B. Haytowitz and J. M. Holden. 2013. Database for the Flavonoid Content of Selected Foods. Nutrient Data Laboratory, USDA Department of Agriculture.

Bianchi, G., C. Murelli and G. Vlahov. 1992. Surface waxes from olive fruits. Phytochemistry 31(10): 3503–3506.

Biesalski, H. K., L. O. Dragsted, I. Elmadfa, R. Grossklaus, M. Muller, D. Schrenk, P. Walter and P. Weber. 2009. Bioactive compounds: Definition and assessment of activity. Nutrition 25(11-12): 1202–1205.

Biesalski, H. K., L. O. Dragsted, I. Elmadfa, R. Grossklaus, M. Muller, D. Schrenk, P. Walter and P. Weber. 2009. Bioactive compounds: Safety and efficacy. Nutrition 25(11-12): 1206–1211.

Brito-Arias, M. 2007. Synthesis and Characterization of Glycosides. New York, USA, Springer Science+Business Media, LLC.

Cefarelli, G., B. D'Abrosca, A. Fiorentino, A. Izzo, C. Mastellone, S. Pacifico and V. Piscopo. 2006. Free-radical-scavenging and antioxidant activities of secondary metabolites from reddened cv. Annurca apple fruits. J. Agric. Food Chem. 54(3): 803–809.

Chen, M., Z. Zhong, W. Tan, S. Wang and Y. Wang. 2011. Recent advances in nanoparticle formulation of oleanolic acid. Chin. Med. 6(1): 20.

Ching, J., H. S. Lin, C. H. Tan and H. L. Koh. 2011. Quantification of alpha- and beta-amyrin in rat plasma by gas chromatography-mass spectrometry: Application to preclinical pharmacokinetic study. Journal of Mass Spectrometry 46(5): 457–464.

Chow, H. H., I. A. Hakim, D. R. Vining, J. A. Crowell, J. Ranger-Moore, W. M. Chew, C. A. Celaya, S. R. Rodney, Y. Hara and D. S. Alberts. 2005. Effects of dosing condition on the oral bioavailability of green tea catechins after single-dose administration of Polyphenon E in healthy individuals. Clin. Cancer Res. 11(12): 4627–4633.

Chowański, S., Z. Adamski, P. Marciniak, G. Rosiński, E. Büyükgüzel, K. Büyükgüzel, P. Falabella, L. Scrano, E. Ventrella, F. Lelario and S. A. Bufo. 2016. A review of bioinsecticidal activity of Solanaceae alkaloids. Toxins 8(3): 60.

Christenson, J., S. J. Whitby, D. D. Mellor, J. Thomas, A. McKune, P. D. Roach and N. Naumovski. 2016. The effects of Resveratrol supplementation in overweight and obese humans—A systematic review of randomised trials. Metab. Syndr. Relat. Disord. In Press.

Clifford, M. N. and J. R. Ramirez-Martinez. 1990. Chlorogenic acids and purine alkaloids contents of Maté (*Ilex paraguariensis*) leaf and beverage. Food Chemistry 35(1): 13–21.

Coelho, R. C., H. H. Hermsdorff and J. Bressan. 2013. Anti-inflammatory properties of orange juice: Possible favorable molecular and metabolic effects. Plant Foods Hum. Nutr. 68(1): 1–10.

Covolo, L., M. Capelli, E. Ceretti, D. Feretti, L. Caimi and U. Gelatti. 2013. Nutritional supplements for diabetes sold on the internet: business or health promotion? BMC Public Health 13: 777.

Cragg, G. M. and D. J. Newman. 2005. Plants as a source of anti-cancer agents. J. Ethnopharmacol. 100(1-2): 72–79.

Cragg, G. M. and D. J. Newman. 2013. Natural products: A continuing source of novel drug leads. Biochim. Biophys. Acta 1830(6): 3670–3695.

Dagna, L., G. Gasparini, M. L. Icardi and E. Sesia. 1982. Study of some components of the unsaponifiable fraction in the skin of grapes. American Journal of Enology and Viticulture 33(4): 201–206.

Dawes, H. M. and J. B. Keene. 1999. Phenolic composition of kiwifruit juice. Journal of Agricultural and Food Chemistry 47(6): 2398–2403.

De Pablos, L. M., M. F. dos Santos, E. Montero, A. Garcia-Granados, A. Parra and A. Osuna. 2010. Anticoccidial activity of maslinic acid against infection with Eimeria tenella in chickens. Parasitol. Res. 107(3): 601–604.

Dembitsky, V. M. 2014. Naturally-occurring bioactive Cyclobutane-containing (CBC) alkaloids in fungi, fungal endophytes and plants. Phytomedicine 21(12): 1559–1581.

Djilas, S., J. Canadanovic-Brunet and G. Cetkovic. 2009. By-products of fruits processing as a source of phytochemicals. Chemical Industry & Chemical Engineering Quarterly 15(4): 191–202.

Dueñas, M., T. Hernández and I. Estrella. 2002. Phenolic composition of the cotyledon and the seed coat of lentils (*Lens culinaris* L.). European Food Research and Technology 215(6): 478–483.

Ehala, S., M. Vaher and M. Kaljurand. 2005. Characterization of phenolic profiles of Northern European berries by capillary electrophoresis and determination of their antioxidant activity. J. Agric. Food Chem. 53(16): 6484–6490.

Ellgardt, K. 2006. Triterpenes in Apple Cuticle of Organically and IP Cultivated Apples. Bachelor project in the Dannish-Swedish Horticulture Programme, Sveriges Lantbruksuniversitet.

Evans, D. A. and C. H. Mitch. 1982. Studies directed towards the total synthesis of morphine alkaloids. Tetrahedron Lett. 23(3): 285–288.

FAO. 2011. Global Food Losses and Waste. Extent, Causes and Prevention. Rome, Italy, Food and Agriculture Organization of the United Nations.

Fernández, S. P., C. Wasowski, L. M. Loscalzo, R. E. Granger, G. A. R. Johnston, A. C. Paladini and M. Marder. 2006. Central nervous system depressant action of flavonoid glycosides. Eur. J. Pharmacol. 539(3): 168 176.

Fugh-Berman, A. and A. Myers. 2004. Citrus aurantium, an ingredient of dietary supplements marketed for weight loss: Current status of clinical and basic research. Exp. Biol. and Med. 229(8): 698–704.

Gallo, M. B. C. and M. J. Sarachine. 2009. Biological activities of Lupeol. Int. J. Biomed. Pharm. Sci. 3(1): 46–66.

Gambuti, A., D. Strollo, A. Erbaggio, L. Lecce and L. Moio. 2007. Effect of winemaking practices on color indexes and selected bioactive phenolics of Aglianico wine. J. Food Sci. 72(9): S623–628.

Godfray, H. C., J. Pretty, S. M. Thomas, E. J. Warham and J. R. Beddington. 2011. Global food supply. Linking policy on climate and food. Science 331(6020): 1013–1014.

Grinkevich, N. I. and L. N. Safronich. 1983. Chemical Analysis of Medicinal Plants. Moscow, Vysshaya Shkola.

Gu, L., S. E. House, X. Wu, B. Ou and R. L. Prior. 2006. Procyanidin and catechin contents and antioxidant capacity of cocoa and chocolate products. J. Agric. Food Chem. 54(11): 4057–4061.

Guinda, A., M. Rada, T. Delgado, P. Gutierrez-Adanez and J. M. Castellano. 2010. Pentacyclic triterpenoids from olive fruit and leaf. J. Agric. Food Chem. 58(17): 9685–9691.

Han, N. and M. Bakovic. 2015. Biologically active triterpenoids and their cardioprotective and anti-inflammatory effects. J. Bioanal. Biomed. S12(005): 11.

He, D., Y. Shan, Y. Wu, G. Liu, B. Chen and S. Yao. 2011. Simultaneous determination of flavanones, hydroxycinnamic acids and alkaloids in citrus fruits by HPLC-DAD–ESI/MS. Food Chem. 127(2): 880–885.

He, X. and R. H. Liu. 2007. Triterpenoids isolated from apple peels have potent antiproliferative activity and may be partially responsible for apple's anticancer activity. J. Agric. Food Chem. 55(11): 4366–4370.

Hesse, M. 2002. Alkaloids: Nature's Curse or Blessing? Zurich, John Wiley & Sons.

Hill, R. A. and J. D. Connolly. 2011. Triterpenoids. Nat. Prod. Rep. 28(6): 1087–1117.

Hossain, M. B., I. Aguiló-Aguayo, J. G. Lyng, N. P. Brunton and D. K. Rai. 2015. Effect of pulsed electric field and pulsed light pre-treatment on the extraction of steroidal alkaloids from potato peels. Innov. Food Sc. Eme. Tech. 29: 9–14.

Huber, G. M. and H. P. V. Rupasinghe. 2009. Phenolic profiles and antioxidant properties of apple skin extracts. Journal of Food Science 74(9): C693–C700.

Iheanacho, K. M. and A. C. Udebuani. 2009. Nutritional composition of some leafy vegetables consumed in Imo state, Nigeria. Journal of Applied Sciences and Environmental Management 13: 3.

Jager, S., H. Trojan, T. Kopp, M. N. Laszczyk and A. Scheffler. 2009. Pentacyclic triterpene distribution in various plants—Rich sources for a new group of multi-potent plant extracts. Molecules 14(6): 2016–2031.

Juan, M. E., J. M. Planas, V. Ruiz-Gutierrez, H. Daniel and U. Wenzel. 2008. Antiproliferative and apoptosis-inducing effects of maslinic and oleanolic acids, two pentacyclic triterpenes from olives, on HT-29 colon cancer cells. Br. J. Nutr. 100(1): 36–43.

Kabera, J. N., E. Semana, A. R. Mussa and X. He. 2014. Plant secondary metabolites: Biosynthesis, classification, function and pharmacological properties. J. Pharm. Pharmacol. 2(377–392).

Kanaze, F. I., E. Kokkalou, M. Georgarakis and I. Niopas. 2004. Validated high-performance liquid chromatographic method utilizing solid-phase extraction for the simultaneous determination of naringenin and hesperetin in human plasma. Journal of Chromatography B 801(2): 363–367.

Kang, S. S., G. A. Cordell, D. D. Soejarto and H. H. S. Fong. 1985. Alkaloids and flavonoids from Ricinus communis. Journal of Natural Products 48(1): 155–156.

Kapusta, I., B. Janda, B. Szajwaj, A. Stochmal, S. Piacente, C. Pizza, F. Franceschi, C. Franz and W. Oleszek. 2007. Flavonoids in horse chestnut (*Aesculus hippocastanum*) seeds and powdered waste water byproducts. Journal of Agricultural and Food Chemistry 55(21): 8485–8490.

Kim, S., P. A. Thiessen, E. E. Bolton, J. Chen, G. Fu, A. Gindulyte, L. Han, J. He, S. He, B. A. Shoemaker, J. Wang, B. Yu, J. Zhang and S. H. Bryant. 2016. PubChem substance and compound databases. Nucleic Acids Res. 44(D1): D1202–1213.

Kitta, K., Y. Hagiwara and T. Shibamoto. 1992. Antioxidative activity of an isoflavonoid, 2-O-glycosylisovitexin isolated from green barley leaves. Journal of Agricultural and Food Chemistry 40(10): 1843–1845.

Kong, J. -M., L. -S. Chia, N. -K. Goh, T. -F. Chia and R. Brouillard. 2003. Analysis and biological activities of anthocyanins. Phytochemistry 64(5): 923–933.

Kumar, A., A. Singh and E. Ekavali. 2015. A review on Alzheimer's disease pathophysiology and its management: An update. Pharmacol. Rep. 67(2): 195–203.

Landete, J. M. 2012. Plant and mammalian lignans: A review of source, intake, metabolism, intestinal bacteria and health. Food Res. Int. 46(1): 410–424.

Lattanzio, V. 2013. Phenolic Compounds: Introduction. Natural Products. K. G. Ramawat and J. M. Merillon. Berlin, Springer-Verlag.

Lau, S. O., E. N. Georgousopoulou, J. Kellett, J. Thomas, A. McKune, D. D. Mellor, P. D. Roach and N. Naumovski. 2016. The effect of dietary supplementation of green tea catechins on cardiovascular disease risk markers in humans: A systematic review of clinical trials. Beverages 2(2): 16.

Lee, J. and R. E. Wrolstad. 2004. Extraction of anthocyanins and polyphenolics from blueberry processing waste. Journal of Food Science 69(7): 564–573.

Lee, S. -T., P. -F. Wong, J. D. Hooper and M. R. Mustafa. 2013. Alpha-tomatine synergises with paclitaxel to enhance apoptosis of androgen-independent human prostate cancer PC-3 cells *in vitro* and *in vivo*. Phytomedicine 20(14): 1297–1305.

Leide, J., U. Hildebrandt, K. Reussing, M. Riederer and G. Vogg. 2007. The developmental pattern of tomato fruit wax accumulation and its impact on cuticular transpiration barrier properties:

Effects of a deficiency in a beta-ketoacyl-coenzyme A synthase (LeCER6). Plant Physiol. 144(3): 1667–1679.

Leifert, W. R. and M. Y. Abeywardena. 2008. Cardioprotective actions of grape polyphenols. Nutr. Res. 28(11): 729–737.

Lesellier, E., E. Destandau, C. Grigoras, L. Fougere and C. Elfakir. 2012. Fast separation of triterpenoids by supercritical fluid chromatography/evaporative light scattering detector. J. Chromatogr. A 1268: 157–165.

Liu, J. 2005. Oleanolic acid and ursolic acid: Research perspectives. J. Ethnopharmacol. 100(1-2): 92–94.

Liu, J., H. Sun, J. Shang, Y. Yong and L. Zhang. 2011. Effect of olive pomace extracts on hyperlipidaemia. Nat. Prod. Res. 25(12): 1190–1194.

Liu, T., J. Zhao, H. Li and L. Ma. 2010. Evaluation on anti-hepatitis viral activity of *Vitis vinifer* L. Molecules 15(10): 7415–7422.

Lommen, A., M. Godejohann, D. P. Venema, P. C. H. Hollman and M. Spraul. 2000. Application of directly coupled HPLC−NMR−MS to the identification and confirmation of quercetin glycosides and phloretin glycosides in apple peel. Analytical Chemistry 72(8): 1793–1797.

Lu, L. and L. Y. Foo. 2000. Antioxidant and radical scavenging activities of polyphenols from apple pomace. Food Chemistry 68(1): 81–85.

Lu, M., B. Yuan, M. Zeng and J. Chen. 2011. Antioxidant capacity and major phenolic compounds of spices commonly consumed in China. Food Research International 44(2): 530–536.

Lu, Y. and L. Y. Foo. 1998. Constitution of some chemical components of apple seed. Food Chemistry 61(1-2): 29–33.

Lu, Y. and L. Yeap Foo. 2003. Polyphenolic constituents of blackcurrant seed residue. Food Chemistry 80(1): 71–76.

Ma, C. M., S. Q. Cai, J. R. Cui, R. Q. Wang, P. F. Tu, M. Hattori and M. Daneshtalab. 2005. The cytotoxic activity of ursolic acid derivatives. Eur. J. Med. Chem. 40(6): 582–589.

Madhujith, T., M. Naczk and F. Shahidi. 2004. Antioxidant activity of common beans (*Phaseolus vulgaris* L.). Journal of Food Lipids 11(3): 220–233.

Manach, C., C. Morand, A. Gil-Izquierdo, C. Bouteloup-Demange and C. Remesy. 2003. Bioavailability in humans of the flavanones hesperidin and narirutin after the ingestion of two doses of orange juice. Eur. J. Clin. Nutr. 57(2): 235–242.

Manach, C., A. Scalbert, C. Morand, C. Remesy and L. Jimenez. 2004. Polyphenols: Food sources and bioavailability. Am. J. Clin. Nutr. 79(5): 727–747.

Manrique-Moreno, M., J. Londoño-Londoño, M. Jemioła-Rzemińska, K. Strzałka, F. Villena, M. Avello and M. Suwalsky. 2014. Structural effects of the Solanum steroids solasodine, diosgenin and solanine on human erythrocytes and molecular models of eukaryotic membranes. BBA Biomembranes 1838(1, Part B): 266–277.

Martin, R., J. Carvalho-Tavares, M. Hernandez, M. Arnes, V. Ruiz-Gutierrez and M. L. Nieto. 2010. Beneficial actions of oleanolic acid in an experimental model of multiple sclerosis. A potential therapeutic role. Biochem. Pharmacol. 79(2): 198–208.

Materska, M. and I. Perucka. 2005. Antioxidant activity of the main phenolic compounds isolated from hot pepper fruit (*Capsicum annuum* L.). Journal of Agricultural and Food Chemistry 53(5): 1750–1756.

Mehari, B., M. Redi-Abshiro, B. S. Chandravanshi, M. Atlabachew, S. Combrinck and R. McCrindle. 2016. Simultaneous determination of alkaloids in green coffee beans from ethiopia: Chemometric evaluation of geographical origin. Food Analytical Methods 9(6): 1627–1637.

Mellor, D. D., T. Sathyapalan, E. S. Kilpatrick, S. Beckett and S. L. Atkin. 2010. High-cocoa polyphenol-rich chocolate improves HDL cholesterol in Type 2 diabetes patients. Diabet. Med. 27(11): 1318–1321.

Mellor, D. D., L. A. Madden, K. A. Smith, E. S. Kilpatrick and S. L. Atkin. 2013. High-polyphenol chocolate reduces endothelial dysfunction and oxidative stress during acute transient hyperglycaemia in Type 2 diabetes: A pilot randomized controlled trial. Diabet. Med. 30(4): 478–483.

Mellor, D. D. and N. Naumovski. 2016. Effect of cocoa in diabetes: The potential of the pancreas and liver as key target organs, more than an antioxidant effect? International Journal of Food Science and Technology 51(4): 829–841.

Melo, C. M., K. M. M. B. Carvalho, L. C. D. S. Neves, T. C. Morais, V. S. Rao, F. A. Santos, G. A. D. Brito and M. H. Chaves. 2010. Alpha, beta-amyrin, a natural triterpenoid ameliorates L-arginine-induced acute pancreatitis in rats. World Journal of Gastroenterology 16(34): 4272–4280.

Mintz-Oron, S., T. Mandel, I. Rogachev, L. Feldberg, O. Lotan, M. Yativ, Z. Wang, R. Jetter, I. Venger, A. Adato and A. Aharoni. 2008. Gene expression and metabolism in tomato fruit surface tissues. Plant Physiol. 147(2): 823–851.

Moneriz, C., J. Mestres, J. M. Bautista, A. Diez and A. Puyet. 2011. Multi-targeted activity of maslinic acid as an antimalarial natural compound. FEBS J. 278(16): 2951–2961.

Mouly, P. P., C. R. Arzouyan, E. M. Gaydou and J. M. Estienne. 1994. Differentiation of citrus juices by factorial discriminant analysis using liquid chromatography of flavanone glycosides. Journal of Agricultural and Food Chemistry 42(1): 70–79.

Muhammad, S. and A. Fatima. 2014. Studies on phytochemical evaluation and antibacterial properties of two varieties of kolanut (*Cola nitida*) in Nigeria. Journal of Biosciences and Medicines 2(2): 37.

Nagaoka, T., T. Yoshihara, J. Ohra and S. Sakamura. 1993. The international journal of plant biochemistry steroidal alkaloids from roots of tomato stock. Phytochemistry 34(4): 1153–1157.

Naumovski, N. 2015. Bioactive Composition of Plants and Plant Foods. Plant Bioactive Compounds for Pancreatic Cancer Prevention and Treatment. C. J. Scarlett and Q. V. Vuong. New York, Nova Publishers: 322.

Naumovski, N., B. L. Blades and P. D. Roach. 2015. Food inhibits the oral bioavailability of the major green tea antioxidant epigallocatechin gallate in humans. Antioxidants 4: 373–393.

Nazaruddin, R., L. K. Seng, O. Hassan and M. Said. 2006. Effect of pulp preconditioning on the content of polyphenols in cocoa beans (*Theobroma cacao*) during fermentation. Industrial Crops and Products 24(1): 87–94.

Nemitz, M. C., F. K. J. Yatsu, J. Bidone, L. S. Koester, V. L. Bassani, C. V. Garcia, A. S. L. Mendez, G. L. von Poser and H. F. Teixeira. 2015. A versatile, stability-indicating and high-throughput ultra-fast liquid chromatography method for the determination of isoflavone aglycones in soybeans, topical formulations, and permeation assays. Talanta 134: 183–193.

Neveu, V., J. Perez-Jimenez, F. Vos, V. Crespy, L. du Chaffaut, L. Mennen, C. Knox, R. Eisner, J. Cruz, D. Wishart and A. Scalbert. 2010. Phenol-explorer: An online comprehensive database on polyphenol contents in foods. Database (Oxford) 2010: bap024.

Newman, R. A., P. Yang, A. D. Pawlus and K. I. Block. 2008. Cardiac glycosides as novel cancer therapeutic agents. Mol. Interv. 8(1): 36.

Nielsen, J. K., C. E. Olsen and M. K. Petersen. 1993. Acylated flavonol glycosides from cabbage leaves. Phytochemistry 34(2): 539–544.

Ong, K. C. and H. E. Khoo. 1997. Biological effects of myricetin. Gen. Pharmacol. 29(2): 121–126.

Orban, N., I. O. Kozak, M. Dravucz and A. Kiss. 2009. LC-MS method development to evaluate major triterpenes in skins and cuticular waxes of grape berries. International Journal of Food Science and Technology 44(4): 869–873.

Ortega, R. 2006. Importance of functional foods in the Mediterranean diet. Public Health Nutr. 9(8A): 1136–1140.

Pereira, R. M., G. Á. Ferreira-Silva, M. Pivatto, L. d. Á. Santos, V. d. S. Bolzani, D. A. Chagas de Paula, J. C. d. Oliveira, C. Viegas Júnior and M. Ionta. 2016. Alkaloids derived from flowers of Senna spectabilis, (−)-cassine and (−)-spectaline, have antiproliferative activity on HepG2 cells for inducing cell cycle arrest in G1/S transition through ERK inactivation and downregulation of cyclin D1 expression. Toxicol. In Vitro 31: 86–92.

Petersson, E. V., U. Arif, V. Schulzova, V. Krtková, J. Hajšlová, J. Meijer, H. C. Andersson, L. Jonsson and F. Sitbon. 2013. Glycoalkaloid and calystegine levels in table potato cultivars subjected to wounding, light, and heat treatments. Journal of Agricultural and Food Chemistry 61(24): 5893–5902.

Piccinelli, A. L., A. Veneziano, S. Passi, F. D. Simone and L. Rastrelli. 2007. Flavonol glycosides from whole cottonseed by-product. Food Chemistry 100(1): 344–349.

Piironen, V., D. G. Lindsay, T. A. Miettinen, J. Toivo and A. M. Lampi. 2000. Plant sterols: Biosynthesis, biological function and their importance to human nutrition. Journal of the Science of Food and Agriculture 80(7): 939–966.

Przybylak, J. K., D. Ciesiołka, W. Wysocka, P. M. García-López, M. A. Ruiz-López, W. Wysocki and K. Gulewicz. 2005. Alkaloid profiles of Mexican wild lupin and an effect of alkaloid preparation from Lupinus exaltatus seeds on growth and yield of paprika (*Capsicum annuum* L.). Industrial Crops and Products 21(1): 1–7.

Ramkissoon, J. S., M. F. Mahomoodally, A. H. Subratty and N. Ahmed. 2016. Inhibition of glucose- and fructose-mediated protein glycation by infusions and ethanolic extracts of ten culinary herbs and spices. Asian Pacific Journal of Tropical Biomedicine 6(6): 492–500.

Reyes, L. F., J. C. Miller and L. Cisneros-Zevallos. 2005. Antioxidant capacity, anthocyanins and total phenolics in purple-and red-fleshed potato (*Solanum tuberosum* L.) genotypes. American Journal of Potato Research 82(4): 271–277.

Romero, C., A. Garcia, E. Medina, M. V. Ruiz-Mendez, A. de Castro and M. Brenes. 2010. Triterpenic acids in table olives. Food Chemistry 118(3): 670–674.

Ruberto, G., A. Renda, C. Daquino, V. Amico, C. Spatafora, C. Tringali and N. D. Tommasi. 2007. Polyphenol constituents and antioxidant activity of grape pomace extracts from five Sicilian red grape cultivars. Food Chemistry 100(1): 203–210.

Sang, S., K. Lapsley, R. T. Rosen and C. -T. Ho. 2002. New prenylated benzoic acid and other constituents from almond hulls (*Prunus amygdalus* batsch). Journal of Agricultural and Food Chemistry 50(3): 607–609.

Santos, P. L., L. N. S. Santos, S. P. M. Ventura, R. L. d. Souza, J. A. P. Coutinho, C. M. F. Soares and Á. S. Lima. 2016. Recovery of capsaicin from *Capsicum frutescens* by applying aqueous two-phase systems based on acetonitrile and cholinium-based ionic liquids. Chemical Engineering Research and Design In press.

Sathyapalan, T., S. Beckett, A. S. Rigby, D. D. Mellor and S. L. Atkin. 2010. High cocoa polyphenol rich chocolate may reduce the burden of the symptoms in chronic fatigue syndrome. Nutr. J. 9: 55.

Schieber, A., F. C. Stintzing and R. Carle. 2001. By-products of plant food processing as a source of functional compounds—Recent developments. Trends in Food Science & Technology 12(11): 401–413.

Schieber, A., W. Ullrich and R. Carle. 2000. Characterization of polyphenols in mango puree concentrate by HPLC with diode array and mass spectrometric detection. Innovative Food Science & Emerging Technologies 1(2): 161–166.

Scott, P. M. and A. Lawrence Guillaume. 1982. Losses of ergot alkaloids during making of bread and pancakes. Journal of Agricultural and Food Chemistry 30(3): 445–450.

Shahidi, F. and P. Ambigaipalan. 2015. Phenolics and polyphenolics in foods, beverages and spices: Antioxidant activity and health effects—A review. J. Funct. Food 18, Part B: 820–897.

Shi, Q., S. Hui, Z. Ai-Hua, X. Hong-Ying, Y. Guang-Li, Y. Ying and W. Xi-Jun. 2014. Natural alkaloids: Basic aspects, biological roles, and future perspectives. Chinese J. Nat. Med. 12(6): 401–406.

Soldi, C., M. G. Pizzolatti, A. P. Luiz, R. Marcon, F. C. Meotti, L. A. Mioto and A. R. Santos. 2008. Synthetic derivatives of the alpha- and beta-amyrin triterpenes and their antinociceptive properties. Bioorg. Med. Chem. 16(6): 3377–3386.

Souquet, J. -M., B. Labarbe, C. Le Guernevé, V. Cheynier and M. Moutounet. 2000. Phenolic composition of grape stems. Journal of Agricultural and Food Chemistry 48(4): 1076–1080.

Spatafora, C., E. Barbagallo, V. Amico and C. Tringali. 2013. Grape stems from Sicilian *Vitis vinifera* cultivars as a source of polyphenol-enriched fractions with enhanced antioxidant activity. LWT-Food Science and Technology 54(2): 542–548.

Stalmach, A., S. Troufflard, M. Serafini and A. Crozier. 2008. Absorption, metabolism and excretion of Choladi green tea flavan-3-ols by humans. Mol. Nutr. Food Res.

Stiti, N. and M. A. Hartmann. 2012. Nonsterol triterpenoids as major constituents of Olea europaea. J. Lipids 2012: 476595.

Sujak, A., A. Kotlarz and W. Strobel. 2006. Compositional and nutritional evaluation of several lupin seeds. Food Chemistry 98(4): 711–719.

Sutovsky, P., A. S. Cupp, W. Thompson and M. Baker. 2016. Reproductive systems biology tackles global issues of population growth, food safety and reproductive health. Cell Tissue Res. 363(1): 1–5.

Szakiel, A., C. Paczkowski, F. Pensec and C. Bertsch. 2012. Fruit cuticular waxes as a source of biologically active triterpenoids. Phytochem. Rev. 11(2-3): 263–284.

Teponno, R. B., S. Kusari and M. Spiteller. 2016. Recent advances in research on lignans and neolignans. Nat. Prod. Rep.

Tilman, D., C. Balzer, J. Hill and B. L. Befort. 2011. Global food demand and the sustainable intensification of agriculture. Proc. Natl. Acad. Sci. USA 108(50): 20260–20264.

Tsao, R. 2010. Chemistry and biochemistry of dietary polyphenols. Nutrients 2(12): 1231–1246.

Tunçel, G., M. J. R. Nout and L. Brimer. 1995. The effects of grinding, soaking and cooking on the degradation of amygdalin of bitter apricot seeds. Food Chemistry 53(4): 447–451.

Turner, C., P. Turner, G. Jacobson, K. Almgren, M. Waldebäck, P. Sjöberg, E. N. Karlsson and K. E. Markides. 2006. Subcritical water extraction and β-glucosidase-catalyzed hydrolysis of quercetin glycosides in onion waste. Green Chemistry 8(11): 949–959.

Villasenor, I. M., J. Angelada, A. P. Canlas and D. Echegoyen. 2002. Bioactivity studies on beta-sitosterol and its glucoside. Phytotherapy Research 16(5): 417–421.

Vuong, Q. V., J. B. Golding, M. Nguyen and P. D. Roach. 2010. Extraction and isolation of catechins from tea. J. Sep. Sci. 33(21): 3415–3428.

Vuong, Q. V., J. B. Golding, C. E. Stathopoulos, M. H. Nguyen and P. D. Roach. 2011. Optimizing conditions for the extraction of catechins from green tea using hot water. J. Sep. Sci. 34(21): 3099–3106.

Wallace, T. C. 2011. Anthocyanins in cardiovascular disease. Adv. Nutr. 2(1): 1–7.

Wan, X., D. Li and Z. Zhang. 2009. Antioxidant properties and mechanisms of tea polyphenols. pp. 131–160. *In*: T. C. Ho, J. K. Lin and F. Shahidi (eds.). Tea and Tea Products: Chemistry and Health Promoting Properties. Boca Raton, FL, CRC Press.

Wang, G. -W., W. -T. Hu, B. -K. Huang and L. -P. Qin. 2011. *Illicium verum*: A review on its botany, traditional use, chemistry and pharmacology. Journal of Ethnopharmacology 136(1): 10–20.

Wang, K., A. Li, R. Zhang, S. Li, F. Zhang, L. Zhao, Z. Zhao, S. Li, J. Cai and J. Cao. 2016. Five new steroidal glycosides from the roots of *Cynanchum stauntonii*. Phytochem. Lett. 16: 178–184.

WHO. 2014. Global status report on noncommunicable diseases 2014. 2015, from http://apps.who.int/iris/bitstream/10665/148114/1/9789241564854_eng.pdf?ua=1.

Williams, P. J., C. R. Strauss, B. Wilson and R. A. Massy-Westropp. 1983. Glycosides of 2-phenylethanol and benzyl alcohol in *Vitis Vinifera* grapes. Phytochemistry 22(9): 2039–2041.

Wolfe, K. L. and R. H. Liu. 2003. Apple peels as a value-added food ingredient. J. Agric. Food Chem. 51(6): 1676–1683.

Wu, C. D. 2009. Grape products and oral health. J. Nutr. 139(9): 1818S–1823S.

Yadav, M., S. Jain, A. Bhardwaj, R. Nagpal, M. Puniya, R. Tomar, V. Singh, O. Parkash, G. B. K. S. Prasad, F. Marotta and H. Yadav. 2009. Biological and medicinal properties of grapes and their bioactive constituents: An update. Journal of Medicinal Food 12(3): 473–484.

Yahara, S., N. Uda, E. Yoshio and E. Yae. 2004. Steroidal alkaloid glycosides from tomato (*Lycopersicon esculentum*). Journal of Natural Products 67(3): 500–502.

Yamaguchi, H., T. Noshita, Y. Kidachi, H. Umetsu, M. Hayashi, K. Komiyama, S. Funayama and K. Ryoyama. 2008. Isolation of ursolic acid from apple peels and its specific efficacy as a potent antitumor agent. Journal of Health Science 54(6): 654–660.

Young, J. E., X. Zhao, E. E. Carey, R. Welti, S. -S. Yang and W. Wang. 2005. Phytochemical phenolics in organically grown vegetables. Molecular Nutrition & Food Research 49(12): 1136–1142.

Zhang, J., J. Chen, Z. Liang and C. Zhao. 2014. New lignans and their biological activities. Chem. Biodivers. 11(1): 1–54.

Zhang, Y. J., B. Jayaprakasam, N. P. Seeram, L. K. Olson, D. DeWitt and M. G. Nair. 2004. Insulin secretion and cyclooxygenase enzyme inhibition by Cabernet Sauvignon grape skin compounds. Journal of Agricultural and Food Chemistry 52(2): 228–233.

Zotchev, S. B. 2013. Alkaloids from Marine Bacteria. Advances in Botanical Research. G. -G. Nathalie, Academic Press 68: 301–333.

CHAPTER 2

Phenolic Compounds
Potential Health Benefits and Toxicity

Deep Jyoti Bhuyan[1],* and *Amrita Basu*[2]

Introduction

Phenolic compounds are probably the most explored natural compounds due to their potential health benefits as demonstrated in a number of studies (Del Rio et al. 2010). Generic terms 'phenolic compounds', 'phenolics' or 'polyphenolics' refer to more than 8,000 compounds found in the plant kingdom and possessing at least an aromatic ring with one or more hydroxyl substituents, including functional derivatives like esters, methyl ethers, glycosides, etc. (Ho 1992, Cartea et al. 2011). These are plant secondary metabolites produced via shikimic acid pathway (Cartea et al. 2011, Talapatra and Talapatra 2015). Phenolic compounds regulate the various metabolic functions including structure and growth, pigmentation and are resistant to different pathogens in plants (Naumovski 2015). The organoleptic properties of the plant food (fruits, vegetables, cereals, legumes, etc.) and beverages (tea, coffee, beer, wine, etc.) are also partially ascribed to phenolic compounds (Dai and Mumper 2010). For instance, the interactions between phenolic compounds (such as procyanidins) and the glycoprotein present in our saliva contribute to the bitterness and astringency of fruit and juices (Dai and Mumper 2010). These phenolics have varied chemical structures ranging from simple molecules (i.e., phenolic acids) to more complex polymerized compounds (i.e., proanthocyanidins) (Galleano et al. 2010). They also help in defense against ultraviolet radiation, insects and predators (Dai and Mumper 2010). Phenolics derived from various natural sources are linked to antioxidant, anti-inflammatory, anti-allergic, anti-carcinogenic, antihypertensive, cardioprotective, anti-arthritic and antimicrobial activities (Rauha et al. 2000, Penna et al. 2001, Puupponen-Pimia et al. 2001, Wang and Mazza 2002, Liu et al. 2004, Dai and Mumper 2010). Studies

[1] Researcher and Casual Academic, Pancreatic Cancer Research Group, School of Environmental & Life Sciences, University of Newcastle, 10 Chittaway Rd., Ourimbah, NSW 2258, Australia.
[2] Research Centre for Toxic Compounds in the Environment, Masaryk University, Brno, Czech Republic.
* Corresponding author: deepjyoti.bhuyan@uon.edu.au

on natural antioxidants has developed significantly in the last few years due to restrictions on the use of synthetic antioxidants and enhanced public awareness of health-related issues (Vázquez et al. 2012). Because of their potential health benefits, natural antioxidants are considered to be a better alternative than the synthetic ones (Fu et al. 2010). Hence, the identification of novel antioxidants from natural sources is one of the main research focuses in natural product development these days. Various studies validate the positive correlation between phenolic content and the antioxidant activity (Dimitrios 2006, Galleano et al. 2010, Bhuyan et al. 2015). Free radicals play an important role in the development of cancer, diabetes, neurodegenerative, ageing-related and cardiovascular diseases. Therefore, antioxidants, such as flavonoids and other phenolics have gained more attention in recent years as potential agents for preventing and treating a number of oxidative stress-related and chronic diseases (Rice-Evans et al. 1996, Stanner et al. 2004, Dimitrios 2006, Fu et al. 2010, Galleano et al. 2010, Gharekhani et al. 2012). The antioxidant activity of phenolics is primarily attributed to their redox properties that enable them to act as singlet oxygen quenchers, reducing agents and hydrogen donors (Rice-Evans et al. 1996, Galleano et al. 2010, Gharekhani et al. 2012). The hydroxyl (–OH) groups of phenolics are good H-donating antioxidants that disrupt the cycle of new radical generation by scavenging reactive oxygen species (ROS) (Castellano et al. 2012).

Types of Phenolic Compounds

The basic structure of a phenolic compound comprises of an aromatic ring with one or more –OH groups. However, phenolic compounds found in Nature are structurally diverse from simple phenolic molecules to complex polymerized compounds (Balasundram et al. 2006). Phenolics found in food material can be divided into three major groups: simple phenols and phenolic acids, hydroxycinnamic acid derivatives and flavonoids (Ho 1992). In addition, based on the number of carbons, the phenolic compounds commonly found in plants can be classified into several groups (Harborne 1989, Baxter et al. 1998, Robards et al. 1999, Balasundram et al. 2006) as shown in Table 1. Phenolic acids, flavonoids and tannins are considered as the main dietary phenolics (King and Young 1999, Balasundram et al. 2006). Flavonoids constitute the largest group of low-molecular-weight plant phenolics and have been studied most extensively (King and Young 1999). They are also the most important plant pigments. Over 4,000 different types of flavonoids are found in Nature (Harborne 1989, Craig 1999). Flavonoids usually occur bound to sugar molecules and consist mainly of catechins, proanthocyanins, anthocyanidins, flavons and flavonols and their glycosides (Ho 1992, King and Young 1999). According to the degree of hydroxylation and the presence of a C2–C3 double-bond in the heterocyclic pyrone ring, flavonoids can be divided into 13 classes (González 2002) and the most important ones are flavonols, flavones, isoflavones, anthocyanidins or anthocyanins and flavanones (Scalbert and Williamson 2000). Flavon-3-ols are most dominant in different kinds of tea (*Camellia sinensis*), berries, cherries, grapes, plums, apricots, red wine and chocolate (Lotito and Frei 2006, D'Archivio et al. 2007, Ratnasooriya et al. 2010). Anthocyanins are natural pigments in plants and exhibit blue, purple or red color (D'Archivio et al. 2007, Wojdyło et al. 2008). These compounds are abundant in purple berries, apples, cherries, red and purple grapes and pomegranates, red wine and certain vegetables,

Table 1. Classification of phenolic compounds.

Class	Number of carbon atoms	Basic structure	Examples
Simple phenolics Benzoquinones	6	C_6	Catechol, hydroquinone 2,6-Dimethoxybenzoquinone
Phenolic acids	7	C_6-C_1	Gallic, salicylic acids
Acethophenones Phenylacetic acids	8	C_6-C_2	3-Acetyl-6-methoxybenzaldehyde p-Hydroxyphenylacetic acid
Hydroxycinnamic acids Phenylpropanoids Coumarins Isocoumarins Chromones	9	C_6-C_3	Caffeic, ferulic acids Myristicin, eugenol Umbelliferone, aesculetin Bergenin Eugenin
Napthoquinones	10	C_6-C_4	Juglone, plumbagin
Xanthones	13	$C_6-C_1-C_6$	Mangiferin
Stilbenes Anthraquinones	14	$C_6-C_2-C_6$	Lunularic acid, resveratrol Emodin
Flavonoids Isoflavonoids	15	$C_6-C_3-C_6$	Quercetin, cyaniding Genistein
Lignans Neolignans	18	$(C_6-C_3)_2$	Pinoresinol Eusiderin
Biflavonoids	30	$(C_6-C_3-C_6)_2$	Amentoflavone, agathisflavone
Lignins	many	$(C_6-C_3)n$	Pinoresinol
Condensed tannins (proanthocyanidins or flavolans)	many	$(C_6-C_3-C_6)n$	Selligueain A, prodelphinidin

such as cabbage, onions and radish (Manach et al. 2004, D'Archivio et al. 2007, de Pascual-Teresa et al. 2010). Some examples of flavanones are: eriodictyol, hesperetin and naringenin. These compounds are commonly found in citrus fruits and to a lesser extent in tomatoes and mint (Manach et al. 2004, D'Archivio et al. 2007). They can also be seen in green leafy herbs, like parsley and chamomile. Flavanones comprise of the smallest group of compounds, starting with glycosides of naringenin which is present in grapefruit followed by hesperetin that is present in oranges (D'Archivio et al. 2007).

Phenolic acids are divided into two subgroups: hydroxybenzoic and hydroxycinnamic acids (Balasundram et al. 2006). Phenolic acids are significant components of fruit and vegetables. These compounds play an important role in color stability, aroma profile and antioxidant activity. They act as acids because of their carboxylic group (Fleuriet and Macheix 2003). Ellagic and gallic acids are two major dietary hydroxybenzoic acids in berries and nuts (Maas et al. 1991, King and Young 1999). Caffeic, ferulic, *p*-coumaric and sinapic acids are the most common hydroxycinnamic acids and aromatic in Nature (Bravo 1998, Balasundram et al. 2006). Chlorogenic acid is an important member of this group with regard to food material and is the key substrate for enzymatic browning of fruits, such as apples and pears (Eskin 1990, Ho 1992). Chlorogenic acid is commonly found in higher quantities in seeds, such as coffee beans, sunflower seeds and grains and is formed when caffeic acid is combined with quinic acid (Sondheimer 1958, King and Young 1999).

Tannins are the third important group of polyphenolics which can further be divided into two subcategories: condensed and hydrolysable tannins (Porter 1989). These are high-molecular-weight polymers. Fruits, grains and legumes consist of condensed tannins which are mainly polymers of catechins or epicatechins, whereas hydrolysable tannins are polymers of gallic or ellagic acid and found in berries and nuts (King and Young 1999). Condensed tannins are also known as proanthocyanidins and polyflavonoid, consisting of chains of flavan-3-ol units. They usually accumulate in the outer layers of plants (Ho 1992, King and Young 1999).

Potential health benefits

For centuries, plants and plant-derived products have played a key role in maintenance of human health by improving the quality of life (Craig 1999). Numerous studies have reported the potential health benefits of plant polyphenolics in particular. Due to their potent antioxidant properties, plant phenolics have scientifically proven to prevent various oxidative stress-related as well as chronic diseases, such as cancer, cardiovascular and neurodegenerative diseases. In spite of their wide distribution in the plant kingdom, researchers have directed their attention to the health benefits of phenolics only now (Dai and Mumper 2010).

Cancer

Oxidative stress plays an important role in carcinogenesis. Several mechanisms contribute to the overall formation of tumours from oxidative damage. Free radicals induce oxidative stress, which leads to DNA damage in the cell, which, in turn, can lead to base mutation, single and double strand breaks, DNA cross-linking and chromosomal abnormality, if left unrepaired (Liu 2003). Therefore, phenolics with antioxidant properties have been found to be beneficial in preventing or treating the oxidative damage that can induce cancer. In addition to antioxidant properties, polyphenols also modulate the activity of a number of enzymes and cell receptors, indicating other specific biological actions in prevention and treatment of several diseases, including cancer (Dai and Mumper 2010). For instance, different phenolic compounds associate with the regulation of gene expression in cell proliferation and apoptosis, both *in vitro* and *in vivo*. Several *in vitro* and *in vivo* studies show that flavonoids may interrupt different stages of carcinogenesis not only with antioxidant activity but also with other anticancer mechanisms (Hollman et al. 1996, Rice-Evans and Miller 1996, Tham et al. 1998, Yang et al. 2001, Kris-Etherton et al. 2002). Polyphenols may affect the molecular events in the initiation, promotion and progression stages of carcinogenesis and isoflavones and lignans may affect the estrogen-related activities related to tumour formation (Yang et al. 2001). Flavonoids have been reported to modulate key enzymes and receptors involved in signal transduction pathways of cellular proliferation, differentiation, apoptosis, inflammation, angiogenesis, metastasis and reversal of multidrug resistance (Ravishankar et al. 2013). Programmed cell death, commonly known as apoptosis, is required to maintain a balance between cell proliferation and cell loss (Zuzana 2011). Misregulation of this balance can lead to malignant transformation, whereas induction of apoptosis suppresses the development of cancer (Tang and Porter 1996, Bhat and Pezzuto 2002). Compounds, such as resveratrol, have shown to induce apoptosis in malignant cells and provide a promising natural strategy to prevent cancer (Katdare et al. 1999, Surh et al. 1999).

Loo (2003) suggested that highly invasive or metastatic cancer cells may require a specific amount of oxidative stress for maintaining their proliferation or apoptosis. Therefore, they generate high yet tolerable amounts of H_2O_2 (hydrogen peroxide) which act as signaling molecules (in the mitogen-activated protein kinase pathway) to activate redox-sensitive transcription factors and responsive genes. These transcription factors and genes are involved in the survival and proliferation of cancer cells. Loo (2003) also proposed that polyphenols with antioxidant capacity can either scavenge the H_2O_2 produced by the cancer cells or certain polyphenols (such as EGCG, quercetin and gallic acid) can induce the formation of H_2O_2 to achieve an intolerable level of high oxidative stress in cancer cells to inhibit their proliferation. Gopalakrishnan and Tony Kong (2008) suggested that phytochemicals, such as polyphenols, protect normal cells by Nrf2 which plays a key role in antioxidant response elements (ARE)-driven gene expression and on the other hand, modulate the transcription factors nuclear factor κB (NFκB) and AP-1 in abnormal cancer cells, which lead to cytotoxicity. Phenolics can modulate various components of the epigenetic machinery in humans (Link et al. 2010). These modulations include the changes in DNA methylation pattern, histone modifications and the expression of some non-coding miRNAs which lead to activation of tumor-suppressor genes and inactivation of oncogenes. Modulation of cytochrome p450 expression is another proposed anticancer mechanism of action of polyphenols. Cytochrome p450 mixed-function oxidases play an important role in the metabolic activation of chemical carcinogens. Phenolic compounds are hypothesized to stimulate cytochrome p450-conjugating enzymes with the ability to metabolically inactivate the chemical carcinogens produced by the mixed function oxidases (Vuong et al. 2014).

Many plant-derived phenolic compounds, for instance, tea polyphenols (green tea), gingerol (gingers), resveratrol (grapes), curcumin (turmeric), genistein (soybean), rosmarinic acid (rosemary), apigenin (parsley) and silymarin (milk thistle) are used in conjunction with chemotherapy and radiation therapy (Wang et al. 2012). Gingerol, a major phenolic compound derived from ginger (*Zingiber officinale*) and its derivative 6-shogaol have been found to possess anticancer activity against oral, kidney, lung, brain and breast cancer cells (Chen et al. 2008, Chen et al. 2010, Han et al. 2015, Lee et al. 2014, Hsu et al. 2015). The latter can induce stress in cancer cells by increasing cytosolic Ca^{2+} levels and cause apoptotic cell death of both human oral cancer cells and renal tubular cells (Chen et al. 2008, 2010).

The role of the flavonoid, quercetin, in anticancer research has been emphasized in a number of reports (Kris-Etherton et al. 2002). Quercetin is the most abundant flavonoid found in fruits after keamferol and myricetin (Vuong et al. 2014). The antioxidant activity of quercetin is well established as the most potent scavenger of ROS (superoxide) and reactive nitrogen species (nitric oxide and peroxynitrite) (Boots et al. 2008, Vuong et al. 2014). An *in vivo* study by Jin et al. (2006) revealed that quercetin had a significant preventive effect on benzo[a]pyrene-induced DNA damage along with a potential chemopreventive effect on the benzo[a]pyrene-induced carcinogenesis of lung cancer. The inhibition of cytochrome p4501A1 activity might be the possible mechanism of these effects of quercetin. Quercetin was also found to inhibit the matrix metalloproteinase-3 (MMP-3) activity and invasion of the MDA-MB-231 human breast carcinoma cell line (Phromnoi et al. 2009) and HGF/Met signaling in medulloblastoma cell line (Labbe et al. 2009).

Another dietary flavonoid—keamferol (also known as kaempferol), a type of phytoestrogen found in a number of fruits and vegetables—is shown to promote human health by reducing the risk of chronic diseases, especially cancer (Chen 2013, Kim and Choi 2013, Lee and Kim 2016). It is reported to regulate major elements of cellular signal transduction pathways associated with apoptosis, angiogenesis, inflammation and metastasis (Zhang et al. 2008, Chen and Chen 2013, Lee and Kim 2016). The anticancer effect of kaempferol in MIA PaCa-2, PANC-1, and SNU-213 human pancreatic cancer cells is mediated by inhibition of EGFR related Src, ERK1/2, and AKT pathways (Lee and Kim 2016). Furthermore, kaemferol is reported to inhibit the migration and invasion ability of medulloblastoma (Labbe et al. 2009) and breast cancer cells (Phromnoi et al. 2009).

Myricetin has been broadly studied to investigate its anticancer properties and mechanisms of action against different types of cancer (Maggiolini et al. 2005, Lu et al. 2006, Kumamoto et al. 2009, Sun et al. 2012, Devi et al. 2015). For instance, in human colon cancer cells, myricetin induces cell death via BAX/BCL2-dependent pathway (Kim et al. 2014), whereas in the colorectal carcinoma cells, it inhibits MMP-2 protein expression and enzyme activity (Ko et al. 2005). It also acts as an agonist for estrogen receptor alpha which leads to inhibition of hormone-dependent MCF7 breast-cancer cell proliferation (Maggiolini et al. 2005). Additionally, myricetin interacts with a number of oncoproteins, such as protein kinase B (PKB) (AKT), Fyn, MEK1, and JAK1–STAT3 (Janus kinase–signal transducer and activator of transcription 3), and reduces the neoplastic transformation of cancer cells (Kumamoto et al. 2009, Sun et al. 2012, Devi et al. 2015). Lu et al. (2006) showed that the plant flavonoids—quercetin and myricetin, inhibit thioredoxin reductase (TrxR) (which is overexpressed in many aggressive tumors) that induces cell death.

Apigenin, a flavone abundantly found in fruits and vegetables, is shown to possess anticancer properties against cancer of breast, cervix, colon, leukemia, lung, ovarian, prostate, skin, thyroid, gastric, liver and neuroblastoma, to name a few (Shukla and Gupta 2010). A study by Ruela-de-Sousa et al. (2010) suggested that apigenin can block proliferation in two types of leukemia cells—myeloid and erythroid subtypes through cell-cycle arrest in G_2/M phase (myeloid HL60) and G_0/G_1 phase (erythroid TF1 cells). Choudhury et al. (2013) also showed that apigenin and curcumin can synergistically induce cell death and apoptosis and block cell cycle progression at G_2/M phase of A549 lung epithelium cancer cells. They also established that both apigenin and curcumin can simultaneously bind at different sites of tubulin. Similarly, apigenin is also linked with inhibition of pancreatic cancer cell proliferation by G_2/M cell cycle arrest, down regulation of the overexpressed protein geminin, increase in growth inhibitory effects of gemcitabine and abrogation of gemcitabine resistance in multiple reports (Ujiki et al. 2006, Salabat et al. 2008, Strouch et al. 2009). Another study by Gomez-Garcia et al. (2013) demonstrated that both potassium apigenin and carnosic acid have chemoprotective effects against 7,12-dimethyl benzanthracene

(DMBA)-induced carcinogenesis in hamster. Moreover, the anti-proliferative and anti-angiogenic effects of the flavonoid, apigenin, were illustrated by Melstrom et al. (2011). They established that apigenin inhibits HIF-1α, GLUT-1, and VEGF mRNA and protein expression in pancreatic cancer cells in both normoxic and hypoxic conditions, proving its potential as a therapeutic agent for pancreatic cancer. Johnson et al. (2011) showed that citrus flavonoids, such as luteolin, apigenin and quercetin can inhibit glycogen synthase kinase-3β (GSK-3β), which leads to decreased cancer cell proliferation and survival by reducing NFκB activity. They made similar observations *in vivo*. Apigenin and luteolin were also shown to improve the efficacy of certain chemotherapeutic drugs—gemcitabine, cisplatin, 5-fluorouracil and oxaliplatin—in terms of their anti-proliferative activity against BxPC-3 human pancreatic cancer cells by Johnson and Gonzalez de Mejia in 2013. Moreover, Lee et al. (2008) made similar observations and proposed that gemcitabine in combination with apigenin resulted in enhanced apoptosis and growth inhibition by down-regulation of NFκB activity through suppression of AKT activation in pancreatic cancer cell lines *in vitro.*

Other phenolic compounds, such as catechin, epicatechin, epigallocatechin-3-gallate (EGCG), nariganin, chalcones, daidzein, gallic acid, protocatechuic acid (PCA), caffeic acid, genistein, stilbenes and anthocyanins were also investigated for their anticancer mechanisms against different types of cancer cell lines (Chahar et al. 2011, Li et al. 2013, Vuong et al. 2014).

An inverse relation between the consumption of flavonoid and risks for certain types of cancer was demonstrated in many epidemiologic studies (Kris-Etherton et al. 2002). A study of 9,959 men and women aged 15–99 years, in Finland, was conducted to demonstrate the relation between the intake of flavonoids and subsequent risk of cancer (Knekt et al. 1997). This study observed an inverse association between flavonoid intake and incidence of all the sites of cancer combined. The risk of lung cancer was reduced with flavonoid intake as found in the 24-year follow-up study. Another study conducted in Finland with a cohort of 27,110 male smokers aged 50–69 years without history of cancer, revealed that the intake of flavonols and flavones can be inversely associated with the risk of lung cancer, but not with that of other cancers (Hirvonen et al. 2001). Likewise, statistically significant inverse associations between the food sources rich in flavonoids quercetin (onions and apples) and naringin (white grapefruit) and lung cancer risk were observed by Le Marchand et al. (2000).

Cardiovascular diseases

Cardiovascular disease (CVD) is one the major killers in all developed countries with a rise in prevalence (Rangel-Huerta et al. 2015, Tome-Carneiro and Visioli 2016). Polyphenols from foods, such as tea, coffee, cocoa, olive oil, red wine and many fruits and vegetables were studied extensively to evaluate their effect on CVD risk (Tangney and Rasmussen 2013). Several epidemiologic studies and intervention trials suggest that polyphenols present in fruits and vegetables are associated with decreased risk of cardiovascular diseases (Morton et al. 2000, Kris-Etherton et al. 2002, Vita 2005, Tangney and Rasmussen 2013, Rangel-Huerta et al. 2015, Tome-Carneiro and Visioli 2016). Oxidative stress may play a role in the pathogenesis of CVD, like atherosclerosis (Morton et al. 2000). As polyphenols are known for their antioxidant activities, increased intake of dietary antioxidants may protect against the development of CVD. In addition, recent evidence suggests that polyphenols have immunomodulatory and vasodilatory properties which may also contribute in reducing the risk of CVD (Tangney and Rasmussen 2013).

Among all the classes of polyphenols, flavonoids have gained more attention in terms of their impact on CVD (Tangney and Rasmussen 2013). Recently Tome-Carneiro and Visioli (2016) reviewed the association between the consumption of polyphenol-rich preparation from food, such as grapes, apple, cocoa, olive, tea, coffee, berries and soy and the reduction of various CVD hallmarks like hypertension, endothelial dysfunction, arterial stiffness, dyslipidemia, inflammation, oxidative status and altered glycaemia. Arts et al. (2001) demonstrated that catechins (a major group of flavonoids) in apples and wine were inversely associated with coronary heart disease (CHD) death. Similarly, another study that had included 5,133 Finnish men and women aged 30–69 years in Finland, observed a significant inverse gradient between dietary intake of flavonoids and coronary mortality (Knekt et al. 1996). Mukamal et al. (2002) suggested that tea consumption in the year before acute myocardial infarction is linked with lower mortality after infarction. However, another study by Sesso et al. (2003) revealed that flavonoid intake was not significantly associated with a reduced risk of CVD. Hertog et al. (1993) found that intakes of tea, onions and apples (major sources of flavonoids: quercetin, kaempferol, myricetin, apigenin and luteolin) were inversely related to CHD mortality, but these associations were not significant. They evaluated 805 men, aged 65–84 years, and concluded that flavonoids found in regularly consumed foods may reduce the risk of death from CHD in elderly men.

It has been interpreted that phenolic compounds can inhibit oxidation of human low-density lipoproteins (LDL) *in vitro* (Visioli and Galli 1994, Visioli et al. 1995, Tuck and Hayball 2002). Oxidation of LDL is linked to the formation of atherosclerotic plaques *in vivo*, which can lead to the development of CHD (Tuck and Hayball 2002). Red wine lowers the risk of CHD in comparison to other alcoholic beverages, indicating a probable role of the polyphenols present in red wine in reducing thrombotic and atherogenic processes (Lopez et al. 2003a). Another study by the same researchers suggests (Lopez et al. 2003b) that red wine phenolics may inhibit platelet aggregation and prevent oxidation of the human LDL. The critical role of platelet aggregation in the pathogenesis of acute coronary syndromes has been emphasized (Vita 2005). Additionally, it is shown that alterations in endothelial function contribute to pathogenesis and clinical expression of CVD as the vascular endothelium plays a primary role in regulating the vascular homeostasis (Vita and Keaney 2002, Vita 2005). Phenolic compound resveratrol (3,5,4′-trihydroxystilbene) (a type stilbenes), found principally in grape skin and red wine, is shown to confer cardioprotective effects by acutely improving endothelial function in patients with CHD (Lekakis et al. 2005, Rangel-Huerta et al. 2015). Various studies suggest the effect of polyphenols, such as ascorbic acid and flavonoids, on endothelial and platelet functions (Vita 2005).

Anthocyanins are responsible for different colors of many flowers and fruits and have been found to be effective in inhibiting LDL-cholesterol oxidation and platelet aggregation, indicating their cardioprotective effects (Craig 1999). The impact of tea polyphenols on reducing the blood cholesterol concentrations and blood pressure in hypercholesterolemic and hypertensive rats was established (Dreosti 1996, Craig 1999). Catechins and gallate esters may confer some of these effects (Dreosti 1996). Catechins present in green tea leaves and theaflavins (catechin dimers) in black tea leaves have inhibitory effects against LDL-cholesterol oxidation (Ishikawa et al. 1997, Craig 1999). Hodgson et al. (2001) also illustrated that consumption of tea can lower the plasma concentrations of marker of platelet aggregation P-selectin.

The potential health benefits of eating a diet rich in olive oil to reduce the risks of CHD and atherosclerosis have also been documented widely (Visioli and Galli 1994, Visioli et al. 1995, Tuck and Hayball 2002, Tome-Carneiro and Visioli 2016). These health benefits are often linked with the presence of the phenolic compound, hydroxytyrosol, in olive oil which exhibits anti-platelet and anti-thrombotic properties *in vivo* (Tuck and Hayball 2002, Tome-Carneiro and Visioli 2016). Weinbrenner et al. (2004) showed that olive oil phenolics can modulate the oxidative/antioxidative status of healthy men who consume a very low-antioxidant diet. Visioli and Galli (1994) validated that the olive phenolic, oleuropein, can protect LDL from oxidation. Tripoli et al. (2005) also explained that the olive phenolics: oleuropein, hydroxytyrosol, caffeic acid, PCA and 3,4-dihydroxyphenylethanol-elenolic acid can hinder LDL oxidation, 3-hydroxy3-methylglutaryl reductase, thromboxane B2 and platelet aggregation, thus lowering the risk of CVD.

Diabetes

Defects in insulin secretion, insulin action, or both are primary causes of *diabetes mellitus*, which is a group of metabolic diseases. According to the American Diabetes Association (2004), this metabolic disease is characterized by chronic hyperglycemia (high levels of blood glucose), leading to long-term damage, dysfunction, and failure of various organs including eyes, kidneys, nerves, heart and blood vessels. Enzymes, such as pancreatic α-amylase and α-glucosidase play an important role in hydrolyzing dietary carbohydrates (such as starch) to glucose and its absorption in the intestine, respectively (Apostolidis et al. 2006). Therefore, inhibiting these enzymes offers an attractive strategy in managing postprandial hyperglycemia in patients with type-2 diabetes by regulating starch breakdown and intestinal glucose absorption (Kwon et al. 2008a). Phenolic compounds derived from different food sources like raspberry, apple, cranberry, maple syrup extracts, wheat, buckwheat, corn, oats, wine, tea, clove, mung beans, strawberries, red currants, black currants, red and green gooseberries, sage, rosemary, soybean and finger millet have been proven to be potent inhibitors of these enzymes in a number of *in vitro* studies (Apostolidis et al. 2006, Kwon et al. 2006, Apostolidis et al. 2007, Cheplick et al. 2007, Randhir Shetty 2007, Kwon et al. 2008a, Kwon et al. 2008b, Randhir et al. 2008, Shobana et al. 2009, Adyanthaya et al. 2010, da Silva Pinto et al. 2010a, da Silva Pinto et al. 2010b, Apostolidis et al. 2011, Vadivel and Biesalski 2011, Adefegha and Oboh 2012, Ademiluyi and Oboh 2013). Kwon et al. (2006) suggested that catechin had the highest α-glucosidase inhibitory activity of 99.6 per cent followed by caffeic acid, rosmarinic acid, resveratrol, catechol, PCA and quercetin. Likewise, naringenin, kaempferol, luteolin, apigenin, (+)-catechin/(–)-epicatechin, diadzein and EGCG are effective α-glucosidase and pancreatic amylase inhibitors (Tadera et al. 2006). Many *in vivo* studies imply that phenolic compounds, such as moracin M, steppogenin-4′-*O*-β-D-glucosiade, mullberroside A from the root bark of *Morus alba* and marsupsin and pterostilbene isolated from the heartwood of *Pterocarpus marsupium* can lower the blood glucose level of hyperglycemic rats (Manickam et al. 1997, Zhang et al. 2009). Similarly, olive leaf and green tea polyphenols improve insulin sensitivity in humans (Venables et al. 2008, de Bock et al. 2013).

Peroxisome proliferator-activated receptor gamma (PPAR-γ) has a vital role in glucose and fat metabolism. Thus, PPAR-γ agonists are widely used in the treatment of hyperglycemia, dyslipidaemia and their complications (Li et al. 2008). Phenolics with PPAR-γ ligand-binding activity have been obtained from licorice (*Glycyrrhiza uralensis* roots) which may help in the treatment of diabetes (Kuroda et al. 2003). Chronic sub-acute inflammation has also been accepted as an important factor in the development of insulin resistance and diabetes in animals and humans. Various non-flavonoid polyphenols have been shown to reduce the production of inflammatory mediators, such as IL-1β, IL-8, MCP-1, COX-2 or iNOS in these animal models of diabetes (Miranda et al. 2015).

A study by Montonen et al. (2004) involving 2,285 men and 2,019 women, 40–69 years of age and free of diabetes, concluded that vitamin E (α-tocopherol, γ-tocopherol, δ-tocopherol, and β-tocotrienol) intake was significantly associated with a reduced risk of type 2 diabetes. However, they did not find any correlation between intake of vitamin C and type 2 diabetes risk. Another study by Knekt et al. (2002) established that a reduction in risk of type 2 diabetes was associated with higher intakes of two flavonoids—quercetin and myricetin in 10,054 Finish men and women.

Obesity

Obesity is one of the biggest concerns in the world of public health and preventive medicine and has become a global epidemic in both developed and developing nations (Hsu and Yen 2008). Obesity is a major risk factor in developing many diseases, particularly diabetes, hypertension, osteoarthritis, heart disease and even cancer (Hsu and Yen 2008, Yun 2010). The potential *in vitro* and *in vivo* anti-obesity properties of phenolics present in natural sources have been documented widely in scientific literature (Moreno et al. 2003, Sharma et al. 2005, Ono et al. 2006, Birari and Bhutani 2007, Hsu and Yen 2008, Hsu et al. 2009, Yun 2010, Gonzalez-Castejon and Rodriguez-Casado 2011, Park et al. 2012, Sergent et al. 2012, Williams et al. 2013). Inhibition of dietary triglyceride absorption via inhibition of pancreatic lipase (PL) using natural products is one of the most recent approaches implemented in the treatment of obesity (Birari and Bhutani 2007). For example, phenolics present in the leaf extract of *Nelumbo nucifera* can inhibit the activities of α-amylase and lipase causing inhibition of digestion, impairment of lipid and carbohydrate absorption, acceleration of lipid metabolism and upregulation of energy expenditure in high-fat diet-induced obese mice (Ono et al. 2006). Similarly, bioactive phytochemicals present in grape seeds, *Eriochloa villosa*, *Orixa japonica* and *Setaria italic* exhibited strong anti-lipase activity *in vitro* (Moreno et al. 2003, Sharma et al. 2005). Phenolic-rich extracts of *Salacia reticulata*, peanut shell, *Mangifera indica* leaf and stem bark and proanthocyanidin rich fraction of aqueous ethanol extract of *Cassia nomame* fruits have also been reviewed for their anti-lipase activity (Yamamoto et al. 2000, Yoshikawa et al. 2002, Moreno et al. 2006a, b, Birari and Bhutani 2007). Sergent et al. (2012) demonstrated that EGCG, kaempferol and quercetin are potent PL inhibitors.

An increase in the number and size of adipocytes (the primary site of energy storage) differentiated from fibroblastic pre-adipocytes in adipose tissues defines obesity at cellular level (Furuyashiki et al. 2004, Hsu and Yen 2008). Decreased energy intake and increased energy expenditure, decreased pre-adipocyte differentiation and proliferation, decreased lipogenesis and increased lipolysis and fat oxidation are some of the recently studied anti-obesity mechanisms (Wang and Jones 2004, Hsu and Yen 2008). Resveratrol commonly found in red grapes, apples, peanuts, blueberries, and cranberries is proven to decrease adipogenesis and viability in maturing preadipocytes by regulating adipocyte-specific transcription factors and the expression of adipocyte specific genes like *PPARγ, C/EBPα, SREBP-1c, FAS, LPL*, and *HSL* (Rayalam et al. 2008, Baile et al. 2011, Gonzalez-Castejon and Rodriguez-Casado 2011). Park et al. (2012) also evaluated the anti-obesity effect of *Schisandra chinensis* in 3T3-L1 cells and high fat diet-induced obese rats and suggested that *S. chinensis* and its polyphenol and lignan extracts significantly exert anti-adipogenic and obesity effects in both cell and animal studies. Similar effects of phenolics, such as chlorogenic acid, coumaric acids, gallic acid, quercetin, luteolin, apigenin, naringenin, genistein, daidzein, catechin, EGCG, esculetin, procyanidin, rutin have been reviewed widely in scientific literature (Hsu and Yen 2008, Yun 2010, Gonzalez-Castejon and Rodriguez-Casado 2011, Williams et al. 2013).

Immune system

Our immune system is comprised of innate and adaptive immunity. Macrophage activation plays an important role in both innate and adaptive immunity. It is induced by interferon (IFN)-γ, an immune regulatory cytokine secreted mainly by appropriately stimulated natural killer or T cells (Kolodziej and Kiderlen 2005, Provenza and Villalba 2010). The immunomodulatory effects of phenolic compounds have been highlighted in numerous studies (Birt et al. 2001, John et al. 2011, Karasawa et al. 2011, Cuevas et al. 2013). The stimulation or suppression of the immune system by phenolic compounds can be attributed to the presence of –OH groups in the structure (Manosroi et al. 2003). The immunomodulatory properties can be partially linked with the antioxidant activity of polyphenols. A mediator's role of epigenetic mechanisms in the modulation of the immune response by polyphenols has also been postulated (Cuevas et al. 2013). Cinnamic acid, a phenolic acid commonly found in propolis, has been shown to activate macrophages, hence, affects the initial stages of the immune response (Ivanovska et al. 1995, Sforcin 2007).

Inflammation is a significant part of the body's immune response. However, an inappropriate immune response may lead to a prolonged inflammation which can be harmful to the body. A recent review by Ambriz-Pérez et al. (2016) cites phenolic compounds as natural alternatives in treating inflammatory diseases. Tumor necrosis factor α (TNF-α), produced primarily by macrophages, is one of the most important regulatory cytokine that facilitates a number of cell functions, such as stimulation of nitric oxide (NO) production which plays a key role in chronic inflammation (Wang and Mazza 2002). Carlos et al. (2009) showed immunostimulatory effects of phenolic compounds from lichens on NO and H_2O_2 production. A study by Sergent et al. (2010) established that genistein and EGCG can downregulate the inflammatory response in inflamed human intestinal epithelial cells by a post-transcriptional regulatory mechanism. Wang and Mazza (2002) reported that anthocyanidins/anthocyanins and anthocyanin-rich extracts induced TNF-α production in LPS/IFN-γ-activated RAW

264.7 macrophages. They also showed that gallic acid, (+)-catechin, kaempferol, myricetin and daidzein induced TNF-α production whereas, quercetin and genistein inhibited TNF-α production. Wadsworth and Koop (1999) demonstrated that resveratrol can stimulate TNF-α secretion in LPS-activated macrophages. Karasawa et al. (2011) described that chlorogenic acid, caffeic acid, pelargonin and ferulic acid isolated from the hot water extract from matured fruit of date palm tree (*Phoenix dactylifera* L.) can stimulate IFN-γ mRNA expression significantly in mouse Peyer's patch cell cultures. Moreover, they also showed that chlorogenic acid and caffeic acid stimulated significant increase in the number of IFN-γ(+)CD4(+) cells in mice.

Quercetin specifically affects the function of different enzymes involved in generating inflammatory and immune responses, such as serine protein kinases, phospholipases, lipoxygenase, cyclooxygenase and nitric oxide synthase (Baumann et al. 1980, Ferriola et al. 1989, Chiesi and Schwaller 1995, Gerritsen et al. 1995, Bugianesi et al. 2000). Sato et al. (1997) suggested that quercetin inhibits the induction of IL-8 and monocyte chemoattractant protein-1 expression by TNF-α in cultured human synovial cells. Yokoo and Kitamura (1997) showed that pretreatment of mesangial cells with quercetin can protect them from H_2O_2-induced apoptosis via suppression of the tyrosine kinase-c-Jun/AP-1 pathway triggered by oxidant stress.

Dietary flavonoid-mediated modulation of inflammatory response by inhibition of T lymphocytes has also been described (Middleton et al. 2000, Provenza and Villalba 2010). A study by John et al. (2011) stated that *Cassia auriculata*-derived polyphenols can enhance T cell immunity by increasing the number of T cells and its sensitivity towards stimulants and decreasing ROS production by neutrophils. Wong et al. (2011) showed that EGCG can significantly improve the regulatory T cell (important in maintaining immune tolerance and suppressing autoimmunity) frequencies and numbers in spleen and lymph nodes and inhibit T cell response in mice. Similarly, Yang et al. (2012) reported that baicalin, a phenolic compound isolated from the Chinese herb, Huangqin, can promote regulatory T cells differentiation, suggesting its role as a natural immunosuppressive agent for treatment of autoimmune inflammatory diseases. Chiang et al. (2003) found that chlorogenic acid, ferulic acid, *p*-coumaric acid, vanillic acid, baicalein and baicalin can enhance the activity of human lymphocyte proliferation and secretion of IFN-γ. However, vitexicarpin, a flavonoid from the fruits of *Vitex rotundifolia* inhibited B and T lymphocyte proliferation (You et al. 1998). T helper (T$_h$) cells are also associated with the regulation of autoimmunity and help in induction of the autoimmune process (Cuevas et al. 2013). Phenolic compounds, such as EGCG and baicalin, exert effects on the differentiation of T$_h$ cells (Yang et al. 2011, Cuevas et al. 2013, Wang et al. 2013).

Liu et al. (2012) established the stimulating effect of the phenolic compounds like geraniin, isocorilagin, quercetin 3-β-D-glucopyranoside, kaempferol 3-β-D-glucopyranoside, quercetin, kaempferol and rutin derived from emblica fruit (*Phyllanthus emblica* L.) on splenocyte proliferation. Gerritsen et al. (1995) showed that apigenin can inhibit cytokine-induced endothelial cell adhesion protein gene expression highlighting its therapeutic potential in the treatment of various inflammatory diseases that involve an increase in leukocyte adhesion and trafficking. This study also demonstrated that apigenin has potent anti-inflammatory activity in carrageenan-induced rat paw edema and delayed type of hypersensitivity in the mouse.

Infectious diseases

Infectious diseases, also known as communicable diseases or transmissible diseases, are ailments caused by microorganisms, such as fungi, bacteria, viruses, viroids, prions and parasites. The spread of these microorganisms in human can occur from animal, other human or even from the environment (food- and waterborne diseases). According to the World Health Organization (2016), infectious diseases, such as HIV/AIDS, tuberculosis (TB), malaria, neglected tropical diseases and hepatitis, are currently some of the biggest killers in the world. According to their report, in 2014 alone, 9.6 million people were diagnosed with TB, whereas, 2 million people were infected with HIV globally. Moreover, the development of drug resistance in a microorganism is one of the largest issues of medical science today. Development of multi drug resistant (MDR) bacteria has been described as a substantial threat to the US public health and national security (Ventola 2015). For instance, Methicillin Resistant *Staphylococcus aureus* (MRSA) kills more people annually in the US than HIV/AIDS, Parkinson's disease, emphysema, and homicide combined (Gross 2013, Golkar et al. 2014, Ventola 2015). Antibiotic resistance is observed increasingly in other bacterial pathogens, such as *Streptococcus pneumoniae*, *Mycobacterium tuberculosis*, *Klebsiella pneumoniae*, *Pseudomonas aeruginosa*, *Escherichia coli* and *Neisseria gonorrhoeae* due to several factors like overuse, inappropriate prescription, extensive agricultural use of antibiotics and availability of very few new antibiotics (Ventola 2015).

Therefore, the search for novel antimicrobial agents from natural sources is a current trend in medical microbiology. Plant phenolics have been studied extensively over the last few decades to evaluate their antimicrobial properties (Davidson and Branden 1981, Tomás-Barberán et al. 1990, Nicholson and Hammerschmidt 1992, Cowan 1999, Zhu et al. 2004, Proestos et al. 2005, Pereira et al. 2007, Estevinho et al. 2008, Süzgeç-Selçuk and Birteksöz 2011, Stojkovic et al. 2013, Gyawali and Ibrahim 2014, Guil-Guerrero et al. 2016). The ability of –OH group to bind to the active sites of key enzymes to modify microbial cell metabolism and the antioxidant properties of phenolics conferred by the –OH groups are some of the aspects that have been linked with the antimicrobial properties of phenolics (Cowan 1999, Gyawali and Ibrahim 2014). The –OH group of phenolic compounds can interact with the bacterial cell membrane and disrupt it, causing leakage of cellular components and cell death (Gyawali and Ibrahim 2014). Therefore, the antimicrobial properties of phenolics can be attributed to the presence of –OH groups in their structure. The number and position of double bonds in the structure have also been reported to play a major role in determining the relative toxicity of phenolics against microorganisms (Friedman et al. 2002, Gochev et al. 2010, Gyawali and Ibrahim 2014).

Finding compounds with the ability to inhibit quorum sensing (QS) of bacteria is another new approach for developing new antimicrobials. QS helps bacteria to carry out cell to cell communication by secreting chemical signal molecules called autoinducers or pheromones. Skogman et al. (2016) recently showed that flavones inhibit QS in *Escherichia coli* and *Pseudomonas aeruginosa* strains and restrict them to form mature biofilms from micro colonies. Phenolic compounds, ranging from simple phenols to condensed tannins, have been reported to be effective against microorganisms with various mechanisms of actions, such as substrate deprivation, membrane disruption, binding to adhesins, formation of complex with cell wall, inactivation of enzymes and metal ion complexation, to name a few (Cowan 1999). A review by Gyawali and Ibrahim (2014) highlighted the studies that have evaluated the antibacterial properties of phenolics present in plant byproducts, such as pomegranate fruit peels, green tea waste, apple peels, coconut husk, potato peels, grape and olive pomace, etc. With regard to the effectiveness of phenolics against MDR bacterial strains, Hatano et al. (2000) observed that 8-(gamma,gamma-dimethylallyl)-wighteone, 3'-(gamma,gamma-dimethylallyl)-kievitone, licochalcone A, gancaonin G, isoangustone A, glyasperins C and D, glabridin, licoricidin, glycycoumarin and licocoumarone isolated from commercial licorice showed significant antibacterial effects on the MRSA strains. They also reported that the phenolic compound, licoricidin, remarkably increased the oxacillin susceptibility of MRSA strains by affecting the enzymatic function of penicillin-binding protein 2'. Eerdunbayaer et al. (2014) also demonstrated that licorice phenolics: licoarylcoumarin, glycycoumarin, gancaonin I exhibit moderate to potent antibacterial effects on the vancomycin-resistant *Enterococcus faecium* and *E. faecalis* strains. Another study by Alves et al. (2013) revealed that 2,4-dihydroxybenzoic, vanillic, syringic acid and *p*-coumaric acid isolated from wild mushrooms inhibited the growth of MRSA. However, these compounds had no effects against methicillin-susceptible *S. aureus*. They also concluded that the presence of carboxylic acid, two – OH groups in para and ortho positions of the benzene ring and also a methoxyl group in the meta position are crucial for inhibition of MRSA. Tsuchiya et al. (1996) established that tetrahydroxyflavanones isolated from *Sophora exigua* and *Echinosophora koreensis* have a potent inhibitory activity against MRSA strains. They also suggested that anti-MRSA activity can be credited to the 2',4'- or 2',6'-dihydroxylation of the B ring and 5,7-dihydroxylation of the A ring in the flavanone structure. Bisignano et al. (1999) reported that hydroxytyrosol displayed broad antimicrobial activity against *Haemophilus influenzae*, *Moraxella catarrhalis*, *Salmonella typhi*, *Vibrio parahaemolyticus*, *Moraxella catarrhalis*, *Salmonella* species, *Vibrio cholerae*, *Vibrio alginolyticus*, *Vibrio parahaemolyticus*, penicillin-susceptible and penicillin-resistant *S. aureus* strains, emphasizing its potential to treat intestinal or respiratory tract infections. An interesting study by Stojkovic et al. (2013) evaluated the *in situ* antimicrobial activities of naturally-occurring phenolics: caffeic acid, *p*-coumaric acid and rutin, using food systems, revealed that they inhibited the development of *S. aureus* in chicken soup, indicating their probable use as preservatives in the food industry. Estevinho et al. (2008) elucidated antibacterial properties of phenolic compounds present in honey and found that *p*-hydroxibenzoic acid, cinnamic acid, naringenin, pinocembrin and chrysin inhibited the growth of *S. aureus* more significantly than that of *Bacillus subtilis*, *S. lentus*, *K. pneumoniae* and *E. coli*. Honda et al. (2010) investigated the anti-mycobacterial properties of phenolic compounds present in the lichens, *Parmotrema dilatatum*, *Parmotrema tinctorum*, *Pseudoparmelia sphaerospora* and *Usnea subcavata* and found that diffractaic acid (MIC = 15.6 µg/mL, 41.6 µM) had the greatest anti-tubercular activity against *Mycobacterium tuberculosis* followed by norstictic acid (MIC = 62.5 µg/mL, 168 µM) and usnic acid (MIC = 62.5 µg/mL, 182 µM). Similarly, Askun et al. (2013) presented the prospect of phenolics present in *Thymus sipthorpii*, *Satureja aintabensis* as promising anti-TB drugs. The same team earlier suggested that major phenolic compounds: carvacrol, rosmarinic acid, hesperidin and naringenin identified from the methanol extracts *Thymbra spicata* are responsible for high level of anti-TB activity (Askun et al. 2009). The anti-mycobacterial role of flavonoids, such as pinocembrin, cryptocaryone, isobachalcone, genistein, prunetin, (*2S*)-naringenin, amentoflavone, 4′ monomethoxyamentoflavone, EGCG, 4′-dihydroxy-3,7,8,3′-tetramethoxyflavone, 5,4′-dihydroxy-3,7,8-trimethoxyflavone, nevadensin, isothymusin, myricetin and quercitin-3-*O*-*β*-D-glucoside, isonivanone has been recently reviewed by Santhosh and Suriyanarayanan (2014).

Current literature also portrays numerous reports on the efficacy of phenolic compounds against human pathogenic fungi. Daglia (2012) reviewed the studies that evaluated the antifungal activity of flavan-3-ol, flavonol, condensed tannins and hydrolysable tannins. Magnolol (5,5′-diallyl-2,2′-dihydroxybiphenyl) and honokiol (5,5′-diallyl-2,4′-dihydroxybiphenyl) isolated from the stem bark of *Magnolia obovate* showed significant inhibitory activities against *Trichophyton mentagrophytes*, *Microsporium gypseum*, *Epidermophyton floccosum*, *Aspergillus niger*, *Cryptococcus neoformans*, and *Candida albicans* (Bang et al. 2000). Yamaguchi et al. (2011) reported that the ellagitannin, called tellimagrandin II isolated from leaves of *Ocotea odorifera*, has potent inhibitory activity against *C. parapsilosis*. They also observed a synergy between tellimagrandin II and other well-known antifungal agents, such as nystatin, amphotericin and fluconazole. Hirasawa and Takada (2004) also found that EGCG enhances the antifungal effect of amphotericin B or fluconazole against *C. albicans*, which is either resistant or susceptible to antimycotics. Eugenol, the major phenolic component of clove essential oil prevented filamentous growth of *C. albicans* and can be used as a therapeutic agent for biofilm-associated candidal infections (He et al. 2007). The antifungal activity of eugenol and vanillin against *C. albicans* and *C. neoformans* has also been validated before (Boonchird and Flegel 1982).

Flavonoids have also been reported to have antiviral properties against HIV, respiratory syncytial virus (RSV), herpes simplex virus type 1 (HSV-1), poliovirus type 1, parainfluenza virus type 3, coxsackie B virus type 1 (Cox B1) (Cowan 1999). Meyer et al. (1997) described that galangin (3,5,7-trihydroxyflavone) isolated from the shoots of *Helichrysum aureonitens* showed significant antiviral activity against HSV-1 and Cox B1. Flavonoids, catechins, tannins, condensed tannins, coumarins derived from plants have been described to interfere with different enzymes, such as reverse transcriptase, integrase and protease of HIV (Cowan 1999, Hassan Khan and Ather 2007). The tea catechin EGCG was found to be effective against both influenza A and B viruses in Madin-Darby canine kidney (MDCK) cells *in vitro* (Nakayama et al. 1993). Parallel observations were made by Song et al. (2005) suggesting the antiviral effect of EGCG and (–)-epicatechingallate (ECG) derived from green tea on influenza virus not only because of their specific interaction with hemagglutinin, but also their ability to alter the physical properties of the viral membrane. EGCG is reported to possess anti-adenoviral and anti-enteroviral properties (Weber et al. 2003, Ho et al. 2009). Another interesting study conducted by Hauber et al. (2009) proposed that EGCG counteracts semen-mediated enhancement of HIV infection and might be beneficial as a supplement to antiretroviral microbicides to reduce sexual transmission of HIV-1. Fassina et al. (2002) also suggested that EGCG can hinder HIV infection and replication in peripheral blood lymphocytes. Additionally, Williamson et al. (2006) showed that anti-HIV activity of EGCG is due to the high-affinity binding of

EGCG to the CD4 molecule which prevents the binding of HIV-1 glycoprotein 120 to the latter on T cells. The role of other phenolics, such as oxyresveratrol (a type of stilbene) isolated from *Artocarpus lakoocha*, has also been highlighted as anti-HSV and anti-HIV agents (Likhitwitayawuid et al. 2005). Gallic acid showed potent antiviral effect against HSV-1 and parainfluenza type-3 (Ozcelik et al. 2011). Flavan-3-ols and oligomeric proanthocyanidins from *Rumex acetose* were demonstrated as potent antivirals against HSV-1 (Gescher et al. 2011). Polyphenol-rich extract from the Mediterranean plant *Cistus incanus* also exhibited significant anti-influenza virus activity in A549 or MDCK cell cultures infected with prototype avian and human influenza strains of different subtypes (Ehrhardt et al. 2007). Non-tannin polyphenol fraction of *Melissa officinalis* produced an antiviral effect similar to caffeic acid against HSV and vaccinia viruses in egg and cell-culture systems (Kucera and Herrmann 1967).

Malaria is an infectious disease caused by four species of *Plasmodium*: *P. falciparum*, *P. vivax*, *P. ovale* and *P. malariae* and transmitted by the infected female *Anopheles* mosquito. About 90 per cent of cases and deaths related to malaria occur in sub-Saharan Africa (Kaur et al. 2009). The first antimalarial drug, an alkaloid named quinine, was derived from the bark of *Cinchona succiruba.* Quinine is still used for treating malaria but few cases of resistance have been reported. Flavonoids, such as EGCG, (–)-cis-3-Acetoxy-4′,5,7-trihydroxyflavanone, 6-hydroxyluteolin 7-O-rhamnoside, calycosin, acacetingenistein, abruquinone B and sikokianin B and C have been reviewed for their potent anti-plasmodial activities against malaria (Kaur et al. 2009). Laphookhieo et al. (2009) found that vismione B isolated from *Cratoxylum maingayi* and *Cratoxylum cochinchinense* had the highest activity against *P. falciparum* ($IC_{50} = 0.66$ µg/mL) among the isolated phenolic compounds. They also concluded that the chromene ring present in vismione B is particularly responsible for anti-malarial activity.

Bioavailability of phenolic compounds

According to the US Food and Drug Administration, the definition of bioavailability is "the rate and extent to which the active ingredient or active moiety is absorbed from a drug product and becomes available at the site of action" (2012). As per this principle, the rate of absorption and the availability at the site of action is important to be bioactive and henceforth be bioavailable (Surangi and Vasantha 2013). There are many factors which affect the bioavailability of ingested dietary flavonoids. When it comes to flavonoid absorption, the chemical structure in terms of molecular weight, glycosylation and esterification plays a pivotal role (Scalbert et al. 2002).

The three dominant methods for determination of bioavailability are: the *in vitro* digestion gastrointestinal (GIT) model, the *in vitro* CaCo-2 (Colon adeno carcinoma cell) model and the *in vivo* models (Oksuz 2013). *In vitro* digestion GIT model has two parts: digestion, which is performed by using commercial digestive enzymes, such as pepsin, pancreatin and absorption which is commonly based on CaCo-2 cells (Parada and Aguilera 2007). The *in vivo* model is time consuming and expensive because of ethical restrictions though it generates accurate results. *In vivo* studies can be affected by reducing properties of metabolites which are formed after absorption and excretion (Bermudez-Soto et al. 2004).

Phenolic compounds bind to other molecules, mostly sugars (glycosyl residues) and proteins (Morton et al. 2000). Though there are very few free-occurring molecules in plant tissues, they are toxic in their free state and detoxified when they are bound (Giada 2013). According to the nature of their carbon skeleton, the most dominant classes of phenolic compounds are phenolic acids and flavonoids and the less common ones are stilbenes and lignans.

According to Van het Hof et al. (1999), tea catechins increased in the protein-rich fraction of plasma (60 per cent) and also in high-density lipoprotein (HDL) (23 per cent) after consumption of eight cups of tea per day in an interval of two hours. However, this did not significantly enhance the resistance of LDL to oxidation *ex vivo* (Van het Hof et al. 1999). Blood and urine levels of flavanols in humans increased after consumption of 1.5, 3.0, and 4.5 g of green tea solids (Yang et al. 1998). There was a significant increase in plasma concentrations of EGCG with an increased dose from 1.5 to 3.0 g. A linear increase effect was not observed but it reached a plateau. It was also seen that pharmacokinetic parameters of excretion revealed no dose-response relationship.

Isoflavones are primarily found in leguminous plants and specifically in soy and its products (Beecher 2003, Manach et al. 2004). A higher bioavailability, assessed as the percentage of dose excreted through urine, was observed after ingestion of isoflavones from the fermented soy product, tempeh (containing mainly aglycones) than after ingestion of nonfermented soybean pieces (containing the naturally-occurring isoflavone glucosides) (Hutchins et al. 1995). Izumi et al. (2000) supported the observations to a great extent followed by Richelle et al. (2002), who found similar AUCs for isoflavone glucosides and aglycones provided from the same source with or without the previous hydrolysis. This finding was also supported, using purified compounds (Zubik and Meydani 2003). These experiments concluded that deglycosylation is not a rate-limiting step though it is considered to be quite an important step in absorption.

The percentage of intact anthocyanins excreted in urine was estimated to be less than 0.1 per cent in humans. Under these observations, it was concluded that anthocyanins undergo extensive metabolism in the body before being excreted in the urine (Jim 2014) but systemic bioavailability of the intact anthocyanins is probably not the best way to estimate the degree of absorption of anthocyanins (Fang 2014). High plasma concentrations of phenolic acid metabolites have been found following administration of fruits containing anthocyanins (Jim 2014). Between 30 per cent and 44 per cent of consumed Cy-3-glc was found as PCA in the human plasma following consumption of blood orange juice and black raspberries (Vitaglione et al. 2007, Chen et al. 2012).

The bioavailability of glycosides of eriodictyol (a type of flavanone), present in lemons, has never been studied in humans (D'Archivio et al. 2007). According to the studies of Bugianesi et al. (2002), C_{max} values for flavanone metabolites are usually measured after five hours of ingestion for any kind of citrus fruit as this is the time required for hydrolysis of the rhamnoglycosides hesperidin, naringin and narirutin by the human microflora, before the aglycones are absorbed in the colon. But, when tomato paste was ingested, which contains naringeninaglycon as aglycones, the C_{max} was reached at a more rapid rate of only two hours.

Absorption, bioavailability and metabolism of monomeric phenols have been extensively studied in both animals and humans (Kuhnle et al. 2000, Baba et al.

2001, Donovan et al. 2001, Abd El Mohsen et al. 2002), but little is known about the bioavailability of polymeric tannins. The results are controversial. As these compounds have high molecular weight, it is very unlikely that they would be absorbed intact. Almost 90 per cent of the consumed procyanidins from apple juice were recovered in the ileostomy effluent and therefore, would reach the colon under physiologic circumstances (Kahle et al. 2007). The molecules which are in blood or are excreted in urine are different from the ones which are ingested and hence, the tannin polymerization has quite a huge impact on the body. For example, more highly-polymerized proanthocyanidins typically present poor absorption through the gut barrier and limited metabolism by the intestinal microflora as compared to catechin (Gonthier et al. 2003).

Phloretin and phloridzin are the major chalcones present in apples followed by chalconaringen which is dominant in tomatoes and lastly, arbutin in pears. Arbutin is also commonly found in strawberry and bearberry, wheat and also in trace amounts in tea, coffee, red wine, and broccoli (Robards et al. 1999, Clifford 2000). Dietary phenolics were consumed in minute quantities (in mg dose) and when food is administered at pharmacological dose, they are found in the free form in blood. The predominant factor determining the site of metabolism or bioavailability would be the dose, while large doses are metabolized in the liver, but small in the intestinal mucosa (Hijova 2006).

Gallic acid is usually consumed via red wine or tea but after it is absorbed in the body, the major components found in the body are in the 4-omethylated and O-glucuronidated forms (Shahrzad and Bitsch 1998, Caccetta et al. 2000, Shahrzad et al. 2001, Cartron et al. 2003). Aglycone and its metabolites reach a consumption of around 4.71 M after consumption of 200 mL of Assam black tea in two hours which contained around 50 mg of gallic acid; hence, the urinary excretion amounted to 39.6 per cent of the ingested dose (Shahrzad and Bitsch 1998, Shahrzad et al. 2001).

When volunteers consumed tomatoes (Bourne and Rice-Evans 1998) or beer (Bourne et al. 2000), ferulic acid was rapidly detected in the plasma. Total urinary excretion (aglycone and metabolites) increased progressively and reached its maximum seven-to-eight hours after ingestion of beer or tomatoes, respectively. However, the main source of ferulic acid is whole-grain food in which it is esterified with araboxylans and hemicelluloses (Sophie and Angel 2008).

Side effects and toxicity

The simple phenolics, which are ubiquitous among plants, used as food additives and ingested daily in mg quantities, can induce double-strand DNA breaks, DNA adducts, mutations and chromosome aberrations in a great variety of test systems (Stich 1991). However, they can also suppress the genotoxic activity of numerous carcinogenic compounds in both *in vitro* and *in vivo* assays. This dual function of dietary phenolics also becomes evident when their carcinogenic or anti-carcinogenic potential is examined (Stich 1991). Moreover, Brune et al. (1989) suggested that galloyl groups present in phenolic compounds, such as tannic acid and gallic acid, can inhibit the absorption of iron. They also established that oregano, tea and coffee can inhibit iron absorption in proportion to their respective content of galloyl groups.

Carcinogenic activity

Even though flavonoids are often considered to be safe as they are of 'plant origin', consumption of flavonoids should be done cautiously as some flavonoids show the ability of direct interaction with DNA and/or enhance carcinogenic activation into DNA modifying agents (Hodek et al. 2006). Some flavonoids, such as quercetin, have mutagenic pro-oxidant effects which may interfere with essential biochemical pathways (Rietjens et al. 2005). Enhanced expression of cytochrome p450 by flavonoids in colon tissue might be responsible for increasing incidence of colorectal carcinoma in humans (Hodek et al. 2006).

Estrogenic activity

Though flavonoids are weakly estrogenic according to the studies of Miksicek (1995) and Ross and Kasum (2002), it has been seen that some specific flavones and flavonols like apigenin, kaempferol and narigenin act through estrogen-receptor mediated mechanisms and have shown to exhibit anti-estrogenic effects which are similar to those of isoflavones in breast cancer cells. The most active flavonoids, such as apigenin and kaempferol, are known to inhibit estrone reduction at 0.12 µM/L concentration (Makela et al. 1998) and some flavones also inhibit the oxidation of testosterone and estradiol to less active steroids (Zuzana 2011).

Cytotoxicity

Mostly the citrus flavanones, like hesperetin and naringenin, were reported to have inhibitory activity on a number of protein kinases (So et al. 1996, Huang et al. 1999, Fischer et al. 2000). Usually the cellular effects of flavonoids depend on the interaction with the membrane or uptake by the cytosol (Zuzana 2011). Information regarding uptake of flavonoids and their metabolites from circulation into various cell types and their further assumed modification by cell interactions have become increasingly important as attention focuses on the new concept of flavonoids as potential modulators of intracellular signaling cascades vital to cellular function (Williams et al. 2004). A most recent concept lays emphasis on flavonoids being considered as potential modulators of intracellular signaling which cascade vital cellular function and hence information regarding uptake, circulation and the metabolites is very important. Anthocyanidins were shown to have some kind of dual role according to different cell types. Flavonoids may also interact with mitochondria, interfere with pathways of intermediary metabolism and/or downregulate the expression of adhesion molecules (Panes et al. 1996, Soriani et al. 1998).

Genotoxic activity

Simple phenolics which are ingested daily as food at the molecular level can induce double strand DNA breaks *in vitro* (Yamada et al. 1985). The green tea catechin EGCG was recently shown to induce hydrogen peroxide (H_2O_2) generation and cause subsequent oxidative damage to isolated and cellular DNA in the presence of transition metal ions (Furukawa et al. 2003).

Conclusion

Current scientific literature provides plenty of information about the possible health benefits of consuming polyphenols and polyphenol-rich food. These reports lay emphasis on the role of phenolic compounds in the prevention and treatment of several chronic pathological conditions, such as cancer, CVD, diabetes, obesity and even infectious diseases. However, some of these studies are controversial and inconclusive as the mechanisms of action of phenolic compounds on various aspects of human health are not entirely understood yet. Many reports highlight the side effects and toxicity associated with phenolics. Several studies evaluated different potential health beneficial properties of phenolic compounds only *in vitro* and do not provide any information on how these compounds will act in 'real life' situations. Hence, more *in vivo* animal studies are warranted in order to obtain useful information for subsequent therapeutic and nutritional interventions. Nonetheless, the findings till date are quite encouraging and there is tremendous scope for more *in vivo* studies. The current knowledge about phenolic compounds will provide the foundation for further systematic investigations.

References

Abd El Mohsen, M. M., G. Kuhnle, A. R. Rechner and H. Schroeter. 2002. Uptake and metabolism of epicatechin and its access to the brain after oral ingestion. Free Rad. Biol. Med. 33: 1693–1702.

Adefegha, S. A. and G. Oboh. 2012. *In vitro* inhibition activity of polyphenol-rich extracts from *Syzygium aromaticum* (L.) Merr. & Perry (Clove) buds against carbohydrate hydrolyzing enzymes linked to type 2 diabetes and Fe^{2+}-induced lipid peroxidation in rat pancreas. Asian Pac. J. Trop. Biomed. 2: 774–781.

Ademiluyi, A. O. and G. Oboh. 2013. Soybean phenolic-rich extracts inhibit key-enzymes linked to type 2 diabetes (α-amylase and α-glucosidase) and hypertension (angiotensin I converting enzyme) *in vitro*. Exp. Toxicol. Pathol. 65: 305–309.

Adyanthaya, I., Y. I. Kwon, E. Apostolidis and K. Shetty. 2010. Health benefits of apple phenolics from postharvest stages for potential type 2 diabetes management using *in vitro* models. J. Food Biochem. 34: 31–49.

Alves, M. J., I. C. Ferreira, H. J. Froufe, R. M. Abreu, A. Martins and M. Pintado. 2013. Antimicrobial activity of phenolic compounds identified in wild mushrooms, SAR analysis and docking studies. J. Appl. Microbiol. 115: 346–357.

Ambriz-Pérez, D. L., N. Leyva-López, E. P. Gutierrez-Grijalva, J. B. Heredia and F. Yildiz. 2016. Phenolic compounds: Natural alternative in inflammation treatment—A review. Cogent Food Agric. 2: 1131412.

American Diabetes Association. 2004. Diagnosis and Classification of Diabetes Mellitus. Diabetes Care. doi: 10.2337/dc10-S062.

Apostolidis, E., Y. I. Kwon and K. Shetty. 2006. Potential of cranberry-based herbal synergies for diabetes and hypertension management. Asia. Pac. J. Clin. Nutr. 15: 433–441.

Apostolidis, E., Y. I. Kwon and K. Shetty. 2007. Inhibitory potential of herb, fruit, and fungal-enriched cheese against key enzymes linked to type 2 diabetes and hypertension. Innov. Food Sci. & Emerg. Technol. 8: 46–54.

Apostolidis, E., L. Li, C. Lee and N. P. Seeram. 2011. *In vitro* evaluation of phenolic-enriched maple syrup extracts for inhibition of carbohydrate hydrolyzing enzymes relevant to type 2 diabetes management. J. Funct. Foods 3: 100–106.

Arts, I. C., D. R. Jacobs, Jr., L. J. Harnack, M. Gross and A. R. Folsom. 2001. Dietary catechins in relation to coronary heart disease death among postmenopausal women. Epidemiology 12: 668–675.

Askun, T., G. Tumen, F. Satil and M. Ates. 2009. *In vitro* activity of methanol extracts of plants used as spices against *Mycobacterium tuberculosis* and other bacteria. Food Chem. 116: 289–294.

Askun, T., E. M. Tekwu, F. Satil, S. Modanlioglu and H. Aydeniz. 2013. Preliminary antimycobacterial study on selected Turkish plants (Lamiaceae) against *Mycobacterium tuberculosis* and search for some phenolic constituents. BMC Complement. Altern. Med. 13: 1–11.

Baba, S., N. Osakabe, M. Natsume, Y. Muto, T. Takizawa and J. Terao. 2001. Absorption and urinary excretion of (–)-epicatechin after administration of different levels of cocoa powder or (–)-epicatechin in rats. J. Agric. Food Chem. 49: 6050–6056.

Baile, C. A., J. Y. Yang, S. Rayalam, D. L. Hartzell, C. Y. Lai, C. Andersen and M. A. Della-Fera. 2011. Effect of resveratrol on fat mobilization. Ann. N.Y. Acad. Sci. 1215: 40–47.

Balasundram, N., K. Sundram and S. Samman. 2006. Phenolic compounds in plants and agri-industrial by-products: Antioxidant activity, occurrence and potential uses. Food Chem. 99: 191–203.

Bang, K. H., Y. K. Kim, B. S. Min, M. K. Na, Y. H. Rhee, J. P. Lee and K. H. Bae. 2000. Antifungal activity of magnolol and honokiol. Arch. Pharm. Res. 23: 46–49.

Baumann, J., F. von Bruchhausen and G. Wurm. 1980. Flavonoids and related compounds as inhibition of arachidonic acid peroxidation. Prostaglandins 20: 627–639.

Baxter, H., J. B. Harborne and G. P. Moss. 1998. Phytochemical Dictionary: A Handbook of Bioactive Compounds from Plants. CRC press.

Beecher, G. R. 2003. Overview of dietary flavonoids: Nomenclature, occurrence and intake. J. Nutr. 3: 3248S–3254S.

Bermúdez-Soto, M. and F. A. Thomás-Barberán. 2004. Evaluation of commercial red fruit juice concentrates as ingredients for antioxidant functional juices. Eur. Food Res. Technol. 219: 133–141.

Bhat, K. P. and J. M. Pezzuto. 2002. Cancer chemopreventive activity of resveratrol. Ann. N.Y. Acad. Sci. 957: 210–229.

Bhuyan, D. J., Q. V. Vuong, A. C. Chalmers, I. A. van Altena, M. C. Bowyer and C. J. Scarlett. 2015. Microwave-assisted extraction of *Eucalyptus robusta* leaf for the optimal yield of total phenolic compounds. Ind. Crops Prod. 69: 290–299.

Birari, R. B. and K. K. Bhutani. 2007. Pancreatic lipase inhibitors from natural sources: Unexplored potential. Drug Discov. Today 12: 879–889.

Birt, D. F., S. Hendrich and W. Wang. 2001. Dietary agents in cancer prevention: Flavonoids and isoflavonoids. Pharmacol. Ther. 90: 157–177.

Bisignano, G., A. Tomaino, R. L. Cascio, G. Crisafi, N. Uccella and A. Saija. 1999. On the *in-vitro* antimicrobial activity of oleuropein and hydroxytyrosol. J. Pharm. Pharmacol. 51: 971–974.

Boonchird, C. and T. W. Flegel. 1982. *In vitro* antifungal activity of eugenol and vanillin against *Candida albicans* and *Cryptococcus neoformans*. Can. J. Microbiol. 28: 1235–1241.

Boots, A. W., G. R. Haenen and A. Bast. 2008. Health effects of quercetin: From antioxidant to nutraceutical. Eur. J. Pharmacol. 585: 325–337.

Bourne, L., G. Paganga, D. Baxter, P. Hughes and C. Rice-Evans. 2000. Absorption of ferulic acid from low-alcohol beer. Free Radic. Res. 32: 273–280.

Bourne, L. C. and C. Rice-Evans. 1998. Bioavailability of ferulic acid. Biochem. Biophys. Res. Commun. 253: 222–227.

Bravo, L. 1998. Polyphenols: Chemistry, dietary sources, metabolism, and nutritional significance. Nutr. Rev. 56: 317–333.

Brune, M., L. Rossander and L. Hallberg. 1989. Iron absorption and phenolic compounds: Importance of different phenolic structures. Eur. J. Clin. Nutr. 43: 547–557.

Bugianesi, R., M. Serafini, F. Simone, D. Wu, S. Meydani, A. Ferro-Luzzi, E. Azzini and G. Maiani. 2000. High-performance liquid chromatography with coulometric electrode array detector for the determination of quercetin levels in cells of the immune system. Anal. Biochem. 284: 296–300.

Bugianesi, R., G. Catasta, P. Spigno, A. D'Uva and G. Maiani. 2002. Naringenin from cooked tomato paste is bioavailable in men. J. Nutr. 132: 3349–52.

Caccetta, R. A., K. D. Croft, L. J. Beilin and I. B. Puddey. 2000. Ingestion of red wine significantly increases plasma phenolic acid concentrations but does not acutely affect *ex vivo* lipoprotein oxidizability. Am. J. Clin. Nutr. 71: 67–74.

Carlos, I. Z., C. B. A. Carli, D. C. G. Maia, F. P. Benzatti, F. C. M. Lopes, F. M. Roese, M. Watanabe, A. C. Micheletti, L. C. D. Santos, W. Vilegas and N. K. Honda. 2009. Immunostimulatory effects of the phenolic compounds from lichens on nitric oxide and hydrogen peroxide production. Rev. Bras. Farmacogn. 19: 847–852.

Cartea, M. E., M. Francisco, P. Soengas and P. Velasco. 2011. Phenolic compounds in *Brassica* vegetables. Molecules 16: 251–280.

Cartron, E., G. Fouret, M. A. Carbonneau, C. Lauret, F. Michel, L. Monnier, B. Descomps and C. L. Leger. 2003. Red-wine beneficial long-term effect on lipids but not on antioxidant characteristics in plasma in a study comparing three types of wine—description of two O-methylated derivatives of gallic acid in humans. Free Radic. Res. 37: 1021–1035.

Castellano, G., J. Tena and F. Torrens. 2012. Classification of phenolic compounds by chemical structural indicators and its relation to antioxidant properties of *Posidonia oceanica* (L.) Delile. Match Commun. Math. Co. 67: 231–250.

Chahar, M. K., N. Sharma, M. P. Dobhal and Y. C. Joshi. 2011. Flavonoids: A versatile source of anticancer drugs. Pharmacogn. Rev. 5: 1–12.

Chen, A. Y. and Y. C. Chen. 2013. A review of the dietary flavonoid, kaempferol on human health and cancer chemoprevention. Food chem. 138: 2099–2107.

Chen, C. Y., C. H. Chen, C. H. Kung, S. H. Kuo and S. Y. Kuo. 2008. [6]-gingerol induces Ca2+ mobilization in Madin-Darby canine kidney cells. J. Nat. Prod. 71: 137–140.

Chen, C. Y., Y. H. Yang and S. Y. Kuo. 2010. Effect of [6]-shogaol on cytosolic Ca2+ levels and proliferation in human oral cancer cells (OC2). J. Nat. Prod. 73: 1370–1374.

Chen, W., D. Wang, L. S. Wang, D. Bei, J. Wang, W. A. See, S. R. Mallery, G. D. Stoner and Z. Liu. 2012. Pharmacokinetics of protocatechuic acid in mouse and its quantification in human plasma using LC-tandem mass spectrometry. J. Chromatogr. B 908: 39–44.

Cheplick, S., Y. -I. Kwon, P. Bhowmik and K. Shetty. 2007. Clonal variation in raspberry fruit phenolics and relevance for diabetes and hypertension management. J. Food Biochem. 31: 656–679.

Chiang, L. C., L. T. Ng, W. Chiang, M. Y. Chang and C. C. Lin. 2003. Immunomodulatory activities of flavonoids, monoterpenoids, triterpenoids, iridoid glycosides and phenolic compounds of *Plantago* species. Planta Medica 69: 600–604.

Chiesi, M. and R. Schwaller. 1995. Inhibition of constitutive endothelial NO-synthase activity by tannin and quercetin. Biochem. Pharmacol. 49: 495–501.

Choudhury, D., A. Ganguli, D. G. Dastidar, B. R. Acharya, A. Das and G. Chakrabarti. 2013. Apigenin shows synergistic anticancer activity with curcumin by binding at different sites of tubulin. Biochimie 95: 1297–1309.

Clifford, M. N. 2000. Miscellaneous phenols in foods and beverages—Nature, occurrence, and dietary burden. J. Sci. Food Agric. 80: 1126–1137.

Cowan, M. M. 1999. Plant products as antimicrobial agents. Clin. Microbiol. Rev. 12: 564–582.

Craig, W. J. 1999. Health-promoting properties of common herbs. Am. J. Clin. Nutr. 70: 491s–499s.

Cuevas, A., N. Saavedra, L. A. Salazar and D. S. P. Abdalla. 2013. Modulation of immune function by polyphenols: Possible contribution of epigenetic factors. Nutrients 5: 2314–2332.

D'Archivio, M., C. Filesi, R. Di Benedetto, R. Gargiulo, C. Giovannini and R. Masella. 2007. Polyphenols, dietary sources and bioavailability. Ann. Ist. Super. Sanita. 43: 348–361.

da Silva Pinto, M., J. E. de Carvalho, F. M. Lajolo, M. I. Genovese and K. Shetty. 2010a. Evaluation of antiproliferative, anti-type 2 diabetes, and antihypertension potentials of ellagitannins from strawberries (fragaria × ananassa duch.) using *in vitro* models. J. Med. Food 13: 1027–1035.

da Silva Pinto, M., Y. -I. Kwon, E. Apostolidis, F. M. Lajolo, M. I. Genovese and K. Shetty. 2010b. Evaluation of red currants (*Ribes rubrum* L.), black currants (*Ribes nigrum* L.), red and green gooseberries (*Ribes uva-crispa*) for potential management of type 2 diabetes and hypertension using *in vitro* models. J. Food Biochem. 34: 639–660.

Daglia, M. 2012. Polyphenols as antimicrobial agents. Curr. Opin. Biotechnol. 23: 174–181.

Dai, J. and R. J. Mumper. 2010. Plant phenolics: Extraction, analysis and their antioxidant and anticancer properties. Molecules 15: 7313–7352.

Davidson, P. M. and A. L. Branden. 1981. Antimicrobial activity of non-halogenated phenolic compounds. J. Food Prot. 44: 623–632.

de Bock, M., J. G. B. Derraik, C. M. Brennan, J. B. Biggs, P. E. Morgan, S. C. Hodgkinson, P. L. Hofman and W. S. Cutfield. 2013. Olive (*Olea europaea* L.) leaf polyphenols improve insulin sensitivity in middle-aged overweight men: A randomized, placebo-controlled, crossover trial. PLoS ONE 8: e57622.

de Pascual-Teresa, S., D. A. Moreno and C. Garcia-Viguera. 2010. Flavanols and anthocyanins in cardiovascular health: A review of current evidence. Int. J. Mol. Sci. 11: 1679–1703.

Del Rio, D., L. G. Costa, M. E. Lean and A. Crozier. 2010. Polyphenols and health: What compounds are involved? Nutr. Metab. Cardiovasc. Dis. 20: 1–6.

Devi, K. P., T. Rajavel, S. Habtemariam, S. F. Nabavi and S. M. Nabavi. 2015. Molecular mechanisms underlying anticancer effects of myricetin. Life Sciences 142: 19–25.

Dimitrios, B. 2006. Sources of natural phenolic antioxidants. Trends Food Sci. Tech. 17: 505–512.

Donovan, J. L., V. Crespy, C. Manach, C. Morand, C. Besson, A. Scalbert and C. Remesy. 2001. Catechin is metabolized by both the small intestine and liver of rats. J. Nutr. 131: 1753–1757.

Dreosti, I. E. 1996. Bioactive ingredients: antioxidants and polyphenols in tea. Nutr. Rev. 54: S51–58.

Eerdunbayaer, M., A. Orabi, H. Aoyama, T. Kuroda and T. Hatano. 2014. Structures of two new flavonoids and effects of licorice phenolics on vancomycin-resistant *Enterococcus* species. Molecules 19: 3883–3897.

Ehrhardt, C., E. R. Hrincius, V. Korte, I. Mazur, K. Droebner, A. Poetter, S. Dreschers, M. Schmolke, O. Planz and S. Ludwig. 2007. A polyphenol rich plant extract, CYSTUS052, exerts anti influenza virus activity in cell culture without toxic side effects or the tendency to induce viral resistance. Antiviral Res. 76: 38–47.

Eskin, N. A. M. 1990. Biochemistry of Food Spoilage: Enzymatic Browning. pp. 401–432. Biochemistry of Foods (2nd Ed.). Academic Press, San Diego.

Estevinho, L., A. P. Pereira, L. Moreira, L. G. Dias and E. Pereira. 2008. Antioxidant and antimicrobial effects of phenolic compound extracts of Northeast Portugal honey. Food Chem. Toxicol. 46: 3774–3779.

Fang, J. 2014. Some anthocyanins could be efficiently absorbed across the gastrointestinal mucosa: Extensive presystemic metabolism reduces apparent bioavailability. J. Agric. Food Chem. 62: 3904–3911.

Fassina, G., A. Buffa, R. Benelli, O. E. Varnier, D. M. Noonan and A. Albini. 2002. Polyphenolic antioxidant (–)-epigallocatechin-3-gallate from green tea as a candidate anti-HIV agent. AIDS 16: 939–941.

Ferriola, P. C., V. Cody and E. Middleton, Jr. 1989. Protein kinase C inhibition by plant flavonoids. Kinetic mechanisms and structure-activity relationships. Biochem. Pharmacol. 38: 1617–1624.

Fischer, P. M. and D. P. Lane. 2000. Inhibitors of cyclin-dependent kinases as anti-cancer therapeutics. Curr. Med. Chem. 7: 1213–1245.

Fleuriet, A. and J. J. Macheix. 2003. Phenolic acids in fruits and vegetables. pp. 1–101. *In*: C. A. Rice-Evans and L. Packer (eds.). Flavonoids in Health and Disease. Marcel Dekker Inc., New York.

Food and Drug Administration. 2012. Code of federal regulations, Title 21 Food and Drugs, Volume 5, Chapter 1 Food and Drug Administration, Department of Health and Human Services, Subchapter D—Drugs for Human Use, Part 320—Bioavailability and Bioequivalence Requirements.

Friedman, M., P. R. Henika and R. E. Mandrell. 2002. Bactericidal activities of plant essential oils and some of their isolated constituents against *Campylobacter jejuni, Escherichia coli, Listeria monocytogenes*, and *Salmonella enterica*. J. Food. Prot. 65: 1545–1560.

Fu, L., B. T. Xu, X. R. Xu, X. S. Qin, R. Y. Gan and H. B. Li. 2010. Antioxidant capacities and total phenolic contents of 56 wild fruits from South China. Molecules 15: 8602–8617.

Furukawa, A., S. Oikawa, M. Murata, Y. Hiraku and S. Kawanishi. 2003. Epigallocatechin gallate causes oxidative damage to isolated and cellular DNA. Biochem. Pharmacol. 66: 1769–1778.

Furuyashiki, T., H. Nagayasu, Y. Aoki, H. Bessho, T. Hashimoto, K. Kanazawa and H. Ashida. 2004. Tea catechin suppresses adipocyte differentiation accompanied by down-regulation of PPARgamma2 and C/EBPalpha in 3T3-L1 cells. Biosci. Biotechnol. Biochem. 68: 2353–2359.

Galleano, M., S. V. Verstraeten, P. I. Oteiza and C. G. Fraga. 2010. Antioxidant actions of flavonoids: Thermodynamic and kinetic analysis. Arch. Biochem. Biophys. 501: 23–30.

Gerritsen, M. E., W. W. Carley, G. E. Ranges, C. P. Shen, S. A. Phan, G. F. Ligon and C. A. Perry. 1995. Flavonoids inhibit cytokine-induced endothelial cell adhesion protein gene expression. Am. J. Pathol. 147: 278–292.

Gescher, K., A. Hensel, W. Hafezi, A. Derksen and J. Kuhn. 2011. Oligomeric proanthocyanidins from *Rumex acetosa* L. inhibit the attachment of herpes simplex virus type-1. Antiviral Res. 89: 9–18.

Gharekhani, M., M. Ghorbani and N. Rasoulnejad. 2012. Microwave-assisted extraction of phenolic and flavonoid compounds from *Eucalyptus camaldulensis* Dehn leaves as compared with ultrasound-assisted extraction. Latin Am. Appl. Res. 42: 305–310.

Giada, M. D. 2013. Food phenolic compounds: Main classes, sources and their antioxidant power. *In*: J. A. Morales-Gonzalez (ed.). Oxidative Stress and Chronic Degenerative Diseases— A Role for Antioxidants. InTech. doi: 10.5772/51687.

Gochev, V., A. Dobreva, T. Girova and A. Stoyanova. 2010. Antimicrobial activity of essential oil from *Rosa alba*. Biotechnol. Biotechnol. Equip. 24: 512–515.

Golkar, Z., O. Bagasra and D. G. Pace. 2014. Bacteriophage therapy: A potential solution for the antibiotic resistance crisis. J. Infect. Dev. Ctries. 8: 129–136.

Gomez-Garcia, F. J., M. P. Lopez-Jornet, N. Alvarez-Sanchez, J. Castillo-Sanchez, O. Benavente-Garcia and V. Vicente Ortega. 2013. Effect of the phenolic compounds—apigenin and carnosic acid on oral carcinogenesis in hamster induced by DMBA. Oral Dis. 19: 279–286.

Gonthier, M. -P., L. J. Donovan, O. Texier, C. Felgines, C. Remesey and A. Scalbert. 2003. Metabolism of dietary procyanidins in rats. Free Rad. Biol. Med. 35: 837–844.

González, C. S. M. 2002. Compuestos polifenólicos: estructura y classificación: Presencia en alimentos y consumo: biodisponibilidad y metabolismo. Alimentaria 329: 19–28.

Gonzalez-Castejon, M. and A. Rodriguez-Casado. 2011. Dietary phytochemicals and their potential effects on obesity: A review. Pharmacol. Res. 64: 438–455.

Gopalakrishnan, A. and A. N. Tony Kong. 2008. Anticarcinogenesis by dietary phytochemicals: Cytoprotection by Nrf2 in normal cells and cytotoxicity by modulation of transcription factors NF-kappa B and AP-1 in abnormal cancer cells. Food Chem. Toxicol. 46: 1257–1270.

Gross, M. 2013. Antibiotics in crisis. Curr. Biol. 23: R1063–1065.

Guil-Guerrero, J. L., L. Ramos, C. Moreno, J. C. Zúñiga-Paredes, M. Carlosama-Yepez and P. Ruales. 2016. Antimicrobial activity of plant-food by-products: A review focusing on the tropics. Livest. Sci. 189: 32–49.

Gyawali, R. and S. A. Ibrahim. 2014. Natural products as antimicrobial agents. Food Control 46: 412–429.

Han, M. A., S. M. Woo, K. J. Min, S. Kim, J. W. Park, D. E. Kim, S. H. Kim, Y. H. Choi and T. K. Kwon. 2015. 6-Shogaol enhances renal carcinoma Caki cells to TRAIL-induced apoptosis through reactive oxygen species-mediated cytochrome c release and down-regulation of c-FLIP(L) expression. Chem. Biol. Interact. 228: 69–78.

Harborne, J. 1989. General procedures and measurement of total phenolics. pp. 1–28. *In*: J. B. Harborne (ed.). Methods in Plant Biochemistry. Academic Press, London.

Hassan Khan, M. T. and A. Ather. 2007. Potentials of phenolic molecules of natural origin and their derivatives as anti-HIV agents. pp. 223–264. *In*: M. R. El-Gewely (ed.). Biotechnology Annual Review. Elsevier.

Hatano, T., Y. Shintani, Y. Aga, S. Shiota, T. Tsuchiya and T. Yoshida. 2000. Phenolic constituents of licorice. VIII. Structures of glicophenone and glicoisoflavanone, and effects of licorice phenolics on methicillin-resistant *Staphylococcus aureus*. Chem. Pharm. Bull. 48: 1286–1292.

Hauber, I., H. Hohenberg, B. Holstermann, W. Hunstein and J. Hauber. 2009. The main green tea polyphenol epigallocatechin-3-gallate counteracts semen-mediated enhancement of HIV infection. Proc. Natl. Acad. Sci. USA 106: 9033–9038.

He, M., M. Du, M. Fan and Z. Bian. 2007. *In vitro* activity of eugenol against *Candida albicans* biofilms. Mycopathologia 163: 137–143.

Hertog, M. G., E. J. Feskens, P. C. Hollman, M. B. Katan and D. Kromhout. 1993. Dietary antioxidant flavonoids and risk of coronary heart disease: The Zutphen elderly study. Lancet 342: 1007–1011.

Hijova, E. 2006. Bioavailability of Chalcones. Bratisl. Lek. Listy. 107: 80–84.

Hirasawa, M. and K. Takada. 2004. Multiple effects of green tea catechin on the antifungal activity of antimycotics against *Candida albicans*. J. Antimicrob. Chemother. 53: 225–229.

Hirvonen, T., J. Virtamo, P. Korhonen, D. Albanes and P. Pietinen. 2001. Flavonol and flavone intake and the risk of cancer in male smokers (Finland). Cancer Causes Control 12: 789–796.

Ho, C. -T. 1992. Phenolic compounds in food. pp. 2–7. Phenolic Compounds in Food and their Effects on Health I. ACS Symposium Series. American Chemical Society. Washington, DC.

Ho, H. Y., M. L. Cheng, S. F. Weng, Y. L. Leu and D. T. Chiu. 2009. Antiviral effect of epigallocatechin gallate on enterovirus 71. J. Agric. Food Chem. 57: 6140–6147.

Hodek, P., P. Hanustiak, J. Krizkova, R. Mikelova, S. Krizkova, M. Stiborova, L. Trnkova, A. Horna, M. Beklova and R. Kizek. 2006. Toxicological aspects of flavonoid interaction with biomacromolecules. Neuro. Endocrinol. Lett. 27(S2): 14–17.

Hodgson, J. M., I. B. Puddey, T. A. Mori, V. Burke, R. I. Baker and L. J. Beilin. 2001. Effects of regular ingestion of black tea on haemostasis and cell adhesion molecules in humans. Eur. J. Clin. Nutr. 55: 881–886.

Hollman, P. C., M. G. Hertog and M. B. Katan. 1996. Role of dietary flavonoids in protection against cancer and coronary heart disease. Biochem. Soc. Trans. 24: 785–789.

Honda, N. K., F. R. Pavan, R. G. Coelho, S. R. de Andrade Leite, A. C. Micheletti, T. I. Lopes, M. Y. Misutsu, A. Beatriz, R. L. Brum and C. Q. Leite. 2010. Antimycobacterial activity of lichen substances. Phytomedicine 17: 328–332.

Hsu, C. L. and G. C. Yen. 2008. Phenolic compounds: Evidence for inhibitory effects against obesity and their underlying molecular signaling mechanisms. Mol. Nutr. Food Res. 52: 53–61.

Hsu, C. L., C. H. Wu, S. L. Huang and G. C. Yen. 2009. Phenolic compounds rutin and o-coumaric acid ameliorate obesity induced by high-fat diet in rats. J. Agric. Food Chem. 57: 425–431.

Hsu, Y. L., J. Y. Hung, Y. M. Tsai, E. M. Tsai, M. S. Huang, M. F. Hou and P. L. Kuo. 2015. 6-shogaol, an active constituent of dietary ginger, impairs cancer development and lung metastasis by inhibiting the secretion of CC-chemokine ligand 2 (CCL2) in tumor-associated dendritic cells. J. Agric. Food Chem. 63: 1730–1738.

Huang, Y. T., J. J. Hwang, P. P. Lee, F. C. Ke, J. H. Huang, C. J. Huang, C. Kandaswami, E. Jr. Middleton and M. T. Lee. 1999. Effects of luteolin and quercetin, inhibitors of tyrosine kinase, on cell growth and metastasis-associated properties in A431 cells overexpressing epidermal growth factor receptor. Br. J. Pharmacol. 128: 999–1010.

Hutchins, A. M., J. L. Slavin and J. W. Lampe. 1995. Urinary isoflavonoid phytoestrogen and lignan excretion after consumption of fermented and unfermented soy products. J. Am. Diet. Assoc. 95: 545–551.

Ishikawa, T., M. Suzukawa, T. Ito, H. Yoshida, M. Ayaori, M. Nishiwaki, A. Yonemura, Y. Hara and H. Nakamura. 1997. Effect of tea flavonoid supplementation on the susceptibility of low-density lipoprotein to oxidative modification. Am. J. Clin. Nutr. 66: 261–266.

Ivanovska, N. D., V. B. Dimov, S. Pavlova, V. S. Bankova and S. S. Popov. 1995. Immunomodulatory action of propolis. V. Anticomplementary activity of a water-soluble derivative. J. Ethnopharmacol. 47: 135–143.

Izumi, T., M. K. Piskula, S. Osawa, A. Obata and K. Tobe. 2000. Soy isoflavone aglycones are absorbed faster and in higher amounts than their glucosides in humans. J. Nutr. 130: 1695–1699.

Jim, F. 2014. Bioavailability of anthocyanins. Drug. Metab. Rev. 46: 508–520.

Jin, N. Z., Y. P. Zhu, J. W. Zhou, L. Mao, R. C. Zhao, T. H. Fang and X. R. Wang. 2006. Preventive effects of quercetin against benzo[a]pyrene-induced DNA damages and pulmonary precancerous pathologic changes in mice. Basic Clin. Pharmacol. Toxicol. 98: 593–598.

John, C. M., P. Sandrasaigaran, C. K. Tong, A. Adam and R. Ramasamy. 2011. Immunomodulatory activity of polyphenols derived from *Cassia auriculata* flowers in aged rats. Cell. Immunol. 271: 474–479.

Johnson, J. L., S. G. Rupasinghe, F. Stefani, M. A. Schuler and E. Gonzalez de Mejia. 2011. Citrus flavonoids luteolin, apigenin, and quercetin inhibit glycogen synthase kinase-3beta enzymatic activity by lowering the interaction energy within the binding cavity. J. Med. Food. 14: 325–333.

Johnson, J. L. and E. Gonzalez de Mejia. 2013. Interactions between dietary flavonoids apigenin or luteolin and chemotherapeutic drugs to potentiate anti-proliferative effect on human pancreatic cancer cells, *in vitro*. Food Chem. Toxicol. 60: 83–91.

Kahle, K., W. Huemmer, M. Kempf, W. Scheppach, T. Erk and E. Richling. 2007. Polyphenols are intensively metabolized in the human gastrointestinal tract after apple juice consumption. J. Agric. Food Chem. 55: 10605–1061.

Karasawa, K., Y. Uzuhashi, M. Hirota and H. Otani. 2011. A matured fruit extract of date palm tree (*Phoenix dactylifera* L.) stimulates the cellular immune system in mice. J. Agric. Food Chem. 59: 11287–11293.

Katdare, M., H. Jinno, M. P. Osborne and N. T. Telang. 1999. Negative growth regulation of oncogene-transformed human breast epithelial cells by phytochemicals: Role of apoptosis. Ann. N.Y. Acad. Sci. 889: 247–252.

Kaur, K., M. Jain, T. Kaur and R. Jain. 2009. Antimalarials from Nature. Bioorg. Med. Chem. 17: 3229–3256.

Kim, M. E., T. K. Ha, J. H. Yoon and J. S. Lee. 2014. Myricetin induces cell death of human colon cancer cells via BAX/BCL2-dependent pathway. Anticancer Res. 34: 701–706.

Kim, S. -H. and K. -C. Choi. 2013. Anti-cancer effect and underlying mechanism(s) of kaempferol, a phytoestrogen, on the regulation of apoptosis in diverse cancer cell models. Toxicol. Res. 29: 229–234.

King, A. M. Y. and G. Young. 1999. Characteristics and occurrence of phenolic phytochemicals. J. Am. Diet. Assoc. 99: 213–218.

Knekt, P., R. Jarvinen, A. Reunanen and J. Maatela. 1996. Flavonoid intake and coronary mortality in Finland: A cohort study. BMJ 312: 478–481.

Knekt, P., R. Jarvinen, R. Seppanen, M. Hellovaara, L. Teppo, E. Pukkala and A. Aromaa. 1997. Dietary flavonoids and the risk of lung cancer and other malignant neoplasms. Am. J. Epidemiol. 146: 223–230.

Knekt, P., J. Kumpulainen, R. Järvinen, H. Rissanen, M. Heliövaara, A. Reunanen, T. Hakulinen and A. Aromaa. 2002. Flavonoid intake and risk of chronic diseases. Am. J. Clin. Nutr. 76: 560–568.

Ko, C. H., S. C. Shen, T. J. Lee and Y. C. Chen. 2005. Myricetin inhibits matrix metalloproteinase 2 protein expression and enzyme activity in colorectal carcinoma cells. Mol. Cancer Ther. 4: 281–290.

Kolodziej, H. and A. F. Kiderlen. 2005. Antileishmanial activity and immune modulatory effects of tannin and related compounds on *Leishmania* parasitised RAW 264.7 cells. Phytochemistry 66: 2056–2071.

Kris-Etherton, P. M., K. D. Hecker, A. Bonanome, S. M. Coval, A. E. Binkoski, K. F. Hilpert, A. E. Griel and T. D. Etherton. 2002. Bioactive compounds in foods: Their role in the prevention of cardiovascular disease and cancer. Am. J. Med. 113 Suppl 9B: 71s–88s.

Kucera, L. S. and E. C. Herrmann, Jr. 1967. Antiviral substances in plants of the mint family (labiatae). I. Tannin of *Melissa officinalis*. Proc. Soc. Exp. Biol. Med. 124: 865–869.

Kuhnle, G., J. P. Spencer, H. Schroeter, B. Shenoy, E. S. Debnam, S. K. Srai, C. Rice-Evans and U. Hahn. 2000. Epicatechin and catechin are O-methylated and glucuronidated in the small intestine. Biochem. Biophys. Res. Commun. 277: 507–512.

Kumamoto, T., M. Fujii and D. X. Hou. 2009. Myricetin directly targets JAK1 to inhibit cell transformation. Cancer Lett. 275: 17–26.

Kuroda, M., Y. Mimaki, Y. Sashida, T. Mae, H. Kishida, T. Nishiyama, M. Tsukagawa, E. Konishi, K. Takahashi, T. Kawada, K. Nakagawa and M. Kitahara. 2003. Phenolics with PPAR-γ ligand-Binding activity obtained from licorice (*Glycyrrhiza uralensis* Roots) and ameliorative effects of glycyrin on genetically diabetic KK-Ay mice. Bioorg. Med. Chem. Lett. 13: 4267–4272.

Kwon, Y. I., D. A. Vattem and K. Shetty. 2006. Evaluation of clonal herbs of *Lamiaceae* species for management of diabetes and hypertension. Asia. Pac. J. Clin. Nutr. 15: 107–118.

Kwon, Y. -I., E. Apostolidis and K. Shetty. 2008a. Inhibitory potential of wine and tea against α-amylase and α-glucosidase for management of hyperglycemia linked to type 2 diabetes. J. Food Biochem. 32: 15–31.

Kwon, Y. I., E. Apostolidis and K. Shetty. 2008b. *In vitro* studies of eggplant (*Solanum melongena*) phenolics as inhibitors of key enzymes relevant for type 2 diabetes and hypertension. Bioresour. Technol. 99: 2981–2988.

Labbe, D., M. Provencal, S. Lamy, D. Boivin, D. Gingras and R. Beliveau. 2009. The flavonols quercetin, kaempferol, and myricetin inhibit hepatocyte growth factor-induced medulloblastoma cell migration. J. Nutr. 139: 646–652.

Laphookhieo, S., W. Maneerat and S. Koysomboon. 2009. Antimalarial and cytotoxic phenolic compounds from *Cratoxylum maingayi* and *Cratoxylum cochinchinense*. Molecules 14: 1389–1395.

Le Marchand, L., S. P. Murphy, J. H. Hankin, L. R. Wilkens and L. N. Kolonel. 2000. Intake of flavonoids and lung cancer. J. Natl. Cancer Inst. 92: 154–160.

Lee, D. H., D. W. Kim, C. H. Jung, Y. J. Lee and D. Park. 2014. Gingerol sensitizes TRAIL-induced apoptotic cell death of glioblastoma cells. Toxicol. Appl. Pharmacol. 279: 253–265.

Lee, J. and J. H. Kim. 2016. Kaempferol inhibits pancreatic cancer cell growth and migration through the blockade of EGFR-related pathway *in vitro*. PLoS ONE 11: e0155264.

Lee, S. H., J. K. Ryu, K. Y. Lee, S. M. Woo, J. K. Park, J. W. Yoo, Y. T. Kim and Y. B. Yoon. 2008. Enhanced anti-tumor effect of combination therapy with gemcitabine and apigenin in pancreatic cancer. Cancer Lett. 259: 39–49.

Lekakis, J., L. S. Rallidis, I. Andreadou, G. Vamvakou, G. Kazantzoglou, P. Magiatis, A. L. Skaltsounis and D. T. Kremastinos. 2005. Polyphenolic compounds from red grapes acutely improve endothelial function in patients with coronary heart disease. Eur. J. Cardiovasc. Prev. Rehabil. 12: 596–600.

Li, H., B. Yang, J. Huang, T. Xiang, X. Yin, J. Wan, F. Luo, L. Zhang, H. Li and G. Ren. 2013. Naringin inhibits growth potential of human triple-negative breast cancer cells by targeting beta-catenin signaling pathway. Toxicol. Lett. 220: 219–228.

Li, Y., Y. Qi, T. H. Huang, J. Yamahara and B. D. Roufogalis. 2008. Pomegranate flower: A unique traditional antidiabetic medicine with dual PPAR-alpha/-gamma activator properties. Diabetes Obes. Metab. 10: 10–17.

Likhitwitayawuid, K., B. Sritularak, K. Benchanak, V. Lipipun, J. Mathew and R. F. Schinazi. 2005. Phenolics with antiviral activity from *Millettia erythrocalyx* and *Artocarpus lakoocha*. Nat. Prod. Res. 19: 177–182.

Link, A., F. Balaguer and A. Goel. 2010. Cancer chemoprevention by dietary polyphenols: Promising role for epigenetics. Biochem. Pharmacol. 80: 1771–1792.

Liu, L., L. Zubik, F. W. Collins, M. Marko and M. Meydani. 2004. The antiatherogenic potential of oat phenolic compounds. Atherosclerosis 175: 39–49.

Liu, R. H. 2003. Health benefits of fruit and vegetables are from additive and synergistic combinations of phytochemicals. Am. J. Clin. Nutr. 78: 517S–520S.

Liu, X., M. Zhao, K. Wu, X. Chai, H. Yu, Z. Tao and J. Wang. 2012. Immunomodulatory and anticancer activities of phenolics from emblica fruit (*Phyllanthus emblica* L.). Food Chem. 131: 685–690.

Loo, G. 2003. Redox-sensitive mechanisms of phytochemical-mediated inhibition of cancer cell proliferation (review). J. Nutr. Biochem. 14: 64–73.

Lopez, M., F. Martinez and C. Del Valle. 2003a. The study of phenolic compounds as natural antioxidants in wine. Crit. Rev. Food Sci. Nutr. 43: 233–244.

Lopez, M., F. Martinez, C. Del Valle, M. Ferrit and R. Luque. 2003b. Study of phenolic compounds as natural antioxidants by a fluorescence method. Talanta 60: 609–616.

Lotito, S. and B. Frei. 2006. Consumption of flavonoid-rich foods and increased plasma antioxidant capacity in humans: Cause, consequence, or epiphenomenon? Free Radi. Biol. Med. 41: 727–1746.

Lu, J., L. V. Papp, J. Fang, S. Rodriguez-Nieto, B. Zhivotovsky and A. Holmgren. 2006. Inhibition of mammalian thioredoxin reductase by some flavonoids: Implications for myricetin and quercetin anticancer activity. Cancer Res. 66: 4410–4418.

Maas, J. L., G. J. Galletta and G. D. Stoner. 1991. Ellagic acid, an anticarcinogen in fruits, especially in strawberries: A review. HortScience 26: 10–14.

Maggiolini, M., A. G. Recchia, D. Bonofiglio, S. Catalano, A. Vivacqua, A. Carpino, V. Rago, R. Rossi and S. Ando. 2005. The red wine phenolics piceatannol and myricetin act as agonists for estrogen receptor alpha in human breast cancer cells. J. Mol. Endocrinol. 35: 269–281.

Makela, S., M. Poutanen, M. L. Kostian, N. Lehtimaki, L. Strauss, R. Santti and R. Vihko. 1998. Inhibition of 17beta-hydroxysteroid oxidoreductase by flavonoids in breast and prostate cancer cells. Proc. Soc. Exp. Biol. Med. 217: 310–316.

Manach, C., A. Scalbert, C. Morand, C. Rémésy and L. Jiménez. 2004. Polyphenols: Food sources and bioavailability. Am. J. Clin. Nutr. 79: 727–747.

Manickam, M., M. Ramanathan, M. A. Farboodniay Jahromi, J. P. N. Chansouria and A. B. Ray. 1997. Antihyperglycemic activity of phenolics from *Pterocarpus marsupium*. J. Nat. Prod. 60: 609–610.

Manosroi, A., A. Saraphanchotiwitthaya and J. Manosroi. 2003. Immunomodulatory activities of *Clausena excavata* Burm. f. wood extracts. J. Ethnopharmacol. 89: 155–160.

Melstrom, L. G., M. R. Salabat, X. Z. Ding, M. J. Strouch, P. J. Grippo, S. Mirzoeva, J. C. Pelling and D. J. Bentrem. 2011. Apigenin down-regulates the hypoxia response genes: HIF-1alpha, GLUT-1, and VEGF in human pancreatic cancer cells. J. Surg. Res. 167: 173–181.

Meyer, J. J., A. J. Afolayan, M. B. Taylor and D. Erasmus. 1997. Antiviral activity of galangin isolated from the aerial parts of *Helichrysum aureonitens*. J. Ethnopharmacol. 56: 165–169.

Middleton, E., Jr., C. Kandaswami and T. C. Theoharides. 2000. The effects of plant flavonoids on mammalian cells: Implications for inflammation, heart disease and cancer. Pharmacol. Rev. 52: 673–751.

Miksicek, R. J. 1995. Estrogenic flavonoids: Structural requirements for biological activity. Proc. Soc. Exp. Biol. Med. 208: 44–50.

Miranda, J., A. Lasa, L. Aguirre, A. Fernandez-Quintela, I. Milton and M. P. Portillo. 2015. Potential application of non-flavonoid phenolics in diabetes: Anti-inflammatory effects. Curr. Med. Chem. 22: 112–131.

Montonen, J., P. Knekt, R. Jarvinen and A. Reunanen. 2004. Dietary antioxidant intake and risk of type 2 diabetes. Diabetes Care 27: 362–366.

Moreno, D. A., N. Ilic, A. Poulev, D. L. Brasaemle, S. K. Fried and I. Raskin. 2003. Inhibitory effects of grape seed extract on lipases. Nutrition 19: 876–879.

Moreno, D. A., N. Ilic, A. Poulev and I. Raskin. 2006a. Effects of *Arachis hypogaea* nutshell extract on lipid metabolic enzymes and obesity parameters. Life Sci. 78: 2797–2803.

Moreno, D. A., C. Ripoll, N. Ilic, A. Poulev, C. Aubin and I. Raskin. 2006b. Inhibition of lipid metabolic enzymes using *Mangifera indica* extracts. J. Food Agric. Environ. 4: 21–26.

Morton, L. W., R. A. A. Cacceta, I. B. Puddey and K. D. Croft. 2000. Chemistry and biological effects of dietary phenolic compounds: Relevance to cardiovascular disease. Clin. Exp. Pharmacol. Physiol. 27(3): 152–159.

Mukamal, K. J., M. Maclure, J. E. Muller, J. B. Sherwood and M. A. Mittleman. 2002. Tea consumption and mortality after acute myocardial infarction. Circulation 105: 2476–2481.

Nakayama, M., K. Suzuki, M. Toda, S. Okubo, Y. Hara and T. Shimamura. 1993. Inhibition of the infectivity of influenza virus by tea polyphenols. Antiviral Res. 21: 289–299.

Naumovski, N. 2015. Bioactive composition of plants and plant foods. pp. 81–115. *In*: C. J. Scarlett and Q. V. Vuong (eds.). Plant Bioactive Compounds for Pancreatic Cancer Prevention and Treatment. NOVA Publishers, New York.

Nicholson, R. L. and R. Hammerschmidt. 1992. Phenolic compounds and their role in disease resistance. Annu. Rev. Phytopathol. 30: 369–389.

Oksuz, T. 2013. Food Matrix Effect on Bioavailability of Phenolics and Anthocyanins in Sour Cherry. M.Sc. Thesis, Istanbul Technical University, Graduate School of Science Engineering and Technology.

Ono, Y., E. Hattori, Y. Fukaya, S. Imai and Y. Ohizumi. 2006. Anti-obesity effect of *Nelumbo nucifera* leaves extract in mice and rats. J. Ethnopharmacol. 106: 238–244.

Ozcelik, B., M. Kartal and I. Orhan. 2011. Cytotoxicity, antiviral and antimicrobial activities of alkaloids, flavonoids and phenolic acids. Pharm. Biol. 49: 396–402.

Panes, J., M. E. Gerritsen, D. C. Anderson, M. Miyasaka and D. N. Granger. 1996. Apigenin inhibits tumor necrosis factor-induced intercellular adhesion molecule-1 upregulation *in vivo*. Microcirculation 3: 279–286.

Parada, J. and J. M. Aguilera. 2007. Food microstructure affects the bioavailability of several nutrients. J. Food Sci. 72(2): R21–R32.

Park, H. J., J. -Y. Cho, M. K. Kim, P. -O. Koh, K. -W. Cho, C. H. Kim, K. -S. Lee, B. Y. Chung, G. -S. Kim and J. -H. Cho. 2012. Anti-obesity effect of *Schisandra chinensis* in 3T3-L1 cells and high fat diet-induced obese rats. Food Chem. 134: 227–234.

Penna, C., S. Marino, E. Vivot, M. C. Cruanes, D. M. J. de, J. Cruanes, G. Ferraro, G. Gutkind and V. Martino. 2001. Antimicrobial activity of Argentine plants used in the treatment of infectious diseases. Isolation of active compounds from *Sebastiania brasiliensis*. J. Ethnopharmacol. 77: 37–40.

Pereira, A., I. Ferreira, F. Marcelino, P. Valentão, P. Andrade, R. Seabra, L. Estevinho, A. Bento and J. Pereira. 2007. Phenolic compounds and antimicrobial activity of olive (*Olea europaea* L. Cv. Cobrançosa) leaves. Molecules 12: 1153–1162.

Phromnoi, K., S. Yodkeeree, S. Anuchapreeda and P. Limtrakul. 2009. Inhibition of MMP-3 activity and invasion of the MDA-MB-231 human invasive breast carcinoma cell line by bioflavonoids. Acta Pharmacol. Sin. 30: 1169–1176.

Porter, L. J. 1989. Tannins. pp. 389–419. *In*: J. B. Harborne (ed.). Methods in Plant Biochemistry. Academic Press.

Proestos, C., N. Chorianopoulos, G. J. E. Nychas and M. Komaitis. 2005. RP-HPLC analysis of the phenolic compounds of plant extracts. Investigation of their antioxidant capacity and antimicrobial activity. J. Agric. Food Chem. 53: 1190–1195.

Provenza, F. D. and J. J. Villalba. 2010. The role of natural plant products in modulating the immune system: An adaptable approach for combating disease in grazing animals. Small Ruminant Res. 89: 131–139.

Puupponen-Pimia, R., L. Nohynek, C. Meier, M. Kahkonen, M. Heinonen, A. Hopia and K. M. Oksman-Caldentey. 2001. Antimicrobial properties of phenolic compounds from berries. J. Appl. Microbiol. 90: 494–507.

Randhir, R. and K. Shetty. 2007. Mung beans processed by solid-state bioconversion improves phenolic content and functionality relevant for diabetes and ulcer management. Innov. Food Sci. Emerg. Technol. 8: 197–204.

Randhir, R., Y. -I. Kwon and K. Shetty. 2008. Effect of thermal processing on phenolics, antioxidant activity and health-relevant functionality of select grain sprouts and seedlings. Innov. Food Sci. Emerg. Technol. 9: 355–364.

Rangel-Huerta, O. D., B. Pastor-Villaescusa, C. M. Aguilera and A. Gil. 2015. A systematic review of the efficacy of bioactive compounds in cardiovascular disease: Phenolic compounds. Nutrients 7: 5177–5216.

Ratnasooriya, C., H. P. V. Rupasinghe and A. Jamieson. 2010. Juice quality and polyphenol concentration of fresh fruits and pomace of selected Nova Scotia-grown grape cultivars. Can. J. Plant Sci. 90: 193–205.

Rauha, J. P., S. Remes, M. Heinonen, A. Hopia, M. Kahkonen, T. Kujala, K. Pihlaja, H. Vuorela and P. Vuorela. 2000. Antimicrobial effects of Finnish plant extracts containing flavonoids and other phenolic compounds. Int. J. Food Microbiol. 56: 3–12.

Ravishankar, D., A. K. Rajora, F. Greco and H. M. I. Osborn. 2013. Flavonoids as prospective compounds for anti-cancer therapy. Int. J. Biochem. Cell Biol. 45: 2821–2831.

Rayalam, S., M. A. Della-Fera and C. A. Baile. 2008. Phytochemicals and regulation of the adipocyte life cycle. J. Nutr. Biochem. 19: 717–726.

Rice-Evans, C. A. and N. J. Miller. 1996. Antioxidant activities of flavonoids as bioactive components of food. Biochem. Soc. Trans. 24: 790–795.

Rice-Evans, C. A., N. J. Miller and G. Paganga. 1996. Structure-antioxidant activity relationships of flavonoids and phenolic acids. Free Radic. Biol. Med. 20: 933–956.

Richelle, M., S. Pridmore-Merten, S. Bodenstab, M. Enslen and E. A. Offord. 2002. Hydrolysis of isoflavone glycosides to aglycones by beta-glycosidase does not alter plasma and urine isoflavone pharmacokinetics in postmenopausal women. J. Nutr. 132: 2587–2592.

Rietjens, I. M., M. G. Boersma, H. Van der Woude, S. M. Jeurissen, M. E. Schutte and G. M. Alink. 2005. Flavonoids and alkenylbenzenes: Mechanisms of mutagenic action and carcinogenic risk. Mutat. Res. 574: 124–138.

Robards, K., P. D. Prenzler, G. Tucker, P. Swatsitang and W. Glover. 1999. Phenolic compounds and their role in oxidative processes in fruits. Food Chem. 66: 401–436.

Ross, J. A. and C. M. Kasum. 2002. Dietary flavonoids: Bioavailability, metabolic effects and safety. Annu. Rev. Nutr. 22: 19–34.

Ruela-de-Sousa, R. R., G. M. Fuhler, N. Blom, C. V. Ferreira, H. Aoyama and M. P. Peppelenbosch. 2010. Cytotoxicity of apigenin on leukemia cell lines: Implications for prevention and therapy. Cell Death Dis. 1: e19.

Salabat, M. R., L. G. Melstrom, M. J. Strouch, X. Z. Ding, B. M. Milam, M. B. Ujiki, C. Chen, J. C. Pelling, S. Rao, P. J. Grippo, T. J. McGarry and D. J. Bentrem. 2008. Geminin is overexpressed in human pancreatic cancer and downregulated by the bioflavanoid apigenin in pancreatic cancer cell lines. Mol. Carcinog. 47: 835–844.

Santhosh, R. S. and B. Suriyanarayanan. 2014. Plants: A source for new antimycobacterial drugs. Planta Medica 80: 9–21.

Sato, M., T. Miyazaki, F. Kambe, K. Maeda and H. Seo. 1997. Quercetin, a bioflavonoid, inhibits the induction of interleukin 8 and monocyte chemoattractant protein-1 expression by tumor necrosis factor-alpha in cultured human synovial cells. J. Rheumatol. 24: 1680–1684.

Scalbert, A. and G. Williamson. 2000. Dietary intake and bioavailability of polyphenols. J. Nutr. 130: 2073S–2085S.

Scalbert, A., C. Morand, C. Manach and C. Remesy. 2002. Absorption and metabolism of polyphenols in the gut and impact on health. Biomed. Pharmacother. 56: 276–282.

Sergent, T., N. Piront, J. Meurice, O. Toussaint and Y. J. Schneider. 2010. Anti-inflammatory effects of dietary phenolic compounds in an *in vitro* model of inflamed human intestinal epithelium. Chem. Biol. Interact. 188: 659–667.

Sergent, T., J. Vanderstraeten, J. Winand, P. Beguin and Y. -J. Schneider. 2012. Phenolic compounds and plant extracts as potential natural anti-obesity substances. Food Chem. 135: 68–73.

Sesso, H. D., J. M. Gaziano, S. Liu and J. E. Buring. 2003. Flavonoid intake and the risk of cardiovascular disease in women. Am. J. Clin. Nutr. 77: 1400–1408.

Sforcin, J. M. 2007. Propolis and the immune system: A review. J. Ethnopharmacol. 113: 1–14.

Shahrzad, S. and I. Bitsch. 1998. Determination of gallic acid and its metabolites in human plasma and urine by high-performance liquid chromatography. J. Chromatogr. B 705: 87–95.

Shahrzad, S., K. Aoyagi, A. Winter, A. Koyama and I. Bitsch. 2001. Pharmacokinetics of gallic acid and its relative bioavailability from tea in healthy humans. J. Nutr. 131: 1207–1210.

Sharma, N., V. K. Sharma and S. -Y. Seo. 2005. Screening of some medicinal plants for anti-lipase activity. J. Ethnopharmacol. 97: 453–456.

Shobana, S., Y. N. Sreerama and N. G. Malleshi. 2009. Composition and enzyme inhibitory properties of finger millet (*Eleusine coracana* L.) seed coat phenolics: Mode of inhibition of α-glucosidase and pancreatic amylase. Food Chem. 115: 1268–1273.

Shukla, S. and S. Gupta. 2010. Apigenin: A promising molecule for cancer prevention. Pharm. Res. 27: 962–978.

Skogman, M., S. Kanerva, S. Manner, P. Vuorela and A. Fallarero. 2016. Flavones as quorum sensing inhibitors identified by a newly optimized screening platform using *Chromobacterium violaceum* as reporter bacteria. Molecules 21: 1211–1221.

So, F. V., N. Guthrie, A. F. Chambers, M. Moussa and K. K. Carroll. 1996. Inhibition of human breast cancer cell proliferation and delay of mammary tumorigenesis by flavonoids and citrus juices. Nutr. Cancer 26: 167–181.

Sondheimer, E. 1958. On the distribution of caffeic acid and the chlorogenic acid isomers in plants. Arch. Biochem. Biophys. 74: 131–138.

Song, J. M., K. H. Lee and B. L. Seong. 2005. Antiviral effect of catechins in green tea on influenza virus. Antiviral Res. 68: 66–74.

Sophie, L. and Gil-I. Angel. 2008. Bioavailability of phenolic acids. Phytochem. Rev. 7: 301–311.

Soriani, M., C. Rice-Evans and R. M. Tyrrell. 1998. Modulation of the UVA activation of haem oxygenase, collagenase and cyclooxygenase gene expression by epigallocatechin in human skin cells. FEBS Lett. 439: 253–257.

Stanner, S. A., J. Hughes, C. N. Kelly and J. Buttriss. 2004. A review of the epidemiological evidence for the 'antioxidant hypothesis'. Public Health Nutr. 7: 407–422.

Stich, H. F. 1991. The beneficial and hazardous effects of simple phenolic compounds. Mutat. Res. 259: 307–324.

Stojkovic, D., J. Petrovic, M. Sokovic, J. Glamoclija, J. Kukic-Markovic and S. Petrovic. 2013. *In situ* antioxidant and antimicrobial activities of naturally occurring caffeic acid, *p*-coumaric acid and rutin, using food systems. J. Sci. Food Agric. 93: 3205–3208.

Strouch, M. J., B. M. Milam, L. G. Melstrom, J. J. McGill, M. R. Salabat, M. B. Ujiki, X. Z. Ding and D. J. Bentrem. 2009. The flavonoid apigenin potentiates the growth inhibitory effects of gemcitabine and abrogates gemcitabine resistance in human pancreatic cancer cells. Pancreas 38: 409–415.

Sun, F., X. Y. Zheng, J. Ye, T. T. Wu, J. Wang and W. Chen. 2012. Potential anticancer activity of myricetin in human T24 bladder cancer cells both *in vitro* and *in vivo*. Nutr. Cancer 64: 599–606.

Surangi, H. T. and H. P. Vasantha Rupasinghe. 2013. Flavonoid bioavailability and attempts of bioavailability enhancement. Nutrients 5(9): 3367–3387.

Surh, Y. J., Y. J. Hurh, J. Y. Kang, E. Lee, G. Kong and S. J. Lee. 1999. Resveratrol, an antioxidant present in red wine, induces apoptosis in human promyelocytic leukemia (HL-60) cells. Cancer Lett. 140: 1–10.

Süzgeç-Selçuk, S. and A. S. Birteksöz. 2011. Flavonoids of *Helichrysum chasmolycicum* and its antioxidant and antimicrobial activities. S. Afr. J. Bot. 77: 170–174.

Tadera, K., Y. Minami, K. Takamatsu and T. Matsuoka. 2006. Inhibition of alpha-glucosidase and alpha-amylase by flavonoids. J. Nutr. Sci. Vitaminol. 52: 149–153.

Talapatra, S. K. and B. Talapatra. 2015. Shikimic acid pathway. pp. 625–678. Chemistry of Plant Natural Products: Stereochemistry, Conformation, Synthesis, Biology and Medicine. Springer Berlin Heidelberg, Berlin, Heidelberg.

Tang, D. G. and A. T. Porter. 1996. Apoptosis: A current molecular analysis. Pathol. Oncol. Res. 2: 117–131.

Tangney, C. and H. E. Rasmussen. 2013. Polyphenols, inflammation, and cardiovascular disease. Curr. Atheroscler. Rep. 15: 324–324.

Tham, D. M., C. D. Gardner and W. L. Haskell. 1998. Potential health benefits of dietary phytoestrogens: A review of the clinical, epidemiological, and mechanistic evidence. J. Clin. Endocrinol. Metab. 83: 2223–2235.

Tomás-Barberán, F., E. Iniesta-Sanmartín, F. Tomás-Lorente and A. Rumbero. 1990. Antimicrobial phenolic compounds from three Spanish *Helichrysum* species. Phytochemistry 29: 1093–1095.

Tome-Carneiro, J. and F. Visioli. 2016. Polyphenol-based nutraceuticals for the prevention and treatment of cardiovascular disease: Review of human evidence. Phytomedicine 23: 1145–1174.

Tripoli, E., M. Giammanco, G. Tabacchi, D. Di Majo, S. Giammanco and M. La Guardia. 2005. The phenolic compounds of olive oil: Structure, biological activity and beneficial effects on human health. Nutr. Res. Rev. 18: 98–112.

Tsuchiya, H., M. Sato, T. Miyazaki, S. Fujiwara, S. Tanigaki, M. Ohyama, T. Tanaka and M. Iinuma. 1996. Comparative study on the antibacterial activity of phytochemical flavanones against methicillin-resistant *Staphylococcus aureus*. J. Ethnopharmacol. 50: 27–34.

Tuck, K. L. and P. J. Hayball. 2002. Major phenolic compounds in olive oil: Metabolism and health effects. J. Nutr. Biochem. 13: 636–644.

Ujiki, M. B., X. -Z. Ding, M. R. Salabat, D. J. Bentrem, L. Golkar, B. Milam, M. S. Talamonti, R. H. Bell, T. Iwamura and T. E. Adrian. 2006. Apigenin inhibits pancreatic cancer cell proliferation through G_2/M cell cycle arrest. Mol. Cancer 5: 76–76.

Vadivel, V. and H. K. Biesalski. 2011. Contribution of phenolic compounds to the antioxidant potential and type II diabetes related enzyme inhibition properties of *Pongamia pinnata* L. Pierre seeds. Process Biochem. 46: 1973–1980.

Van het Hof, K. H., S. A. Wiseman, C. S. Yang and L. B. Tijburg. 1999. Plasma and lipoprotein levels of tea catechins following repeated tea consumption. Proc. Soc. Exp. Biol. Med. 220: 203–209.

Vázquez, G., J. Santos, M. S. Freire, G. Antorrena and J. González-Álvarez. 2012. Extraction of antioxidants from eucalyptus (*Eucalyptus globulus*) bark. Wood Sci. Technol. 46: 443–457.

Venables, M. C., C. J. Hulston, H. R. Cox and A. E. Jeukendrup. 2008. Green tea extract ingestion, fat oxidation, and glucose tolerance in healthy humans. Am. J. Clin. Nutr. 87: 778–784.

Ventola, C. L. 2015. The antibiotic resistance crisis: Part 1: Causes and threats. PT 40: 277–283.

Visioli, F. and C. Galli. 1994. Oleuropein protects low density lipoprotein from oxidation. Life Sci. 55: 1965–1971.

Visioli, F., G. Bellomo, G. Montedoro and C. Galli. 1995. Low density lipoprotein oxidation is inhibited *in vitro* by olive oil constituents. Atherosclerosis 117: 25–32.

Vita, J. A. and J. F. Keaney, Jr. 2002. Endothelial function: A barometer for cardiovascular risk? Circulation 106: 640–642.

Vita, J. A. 2005. Polyphenols and cardiovascular disease: effects on endothelial and platelet function. Am. J. Clin. Nutr. 81: 292S–297S.

Vitaglione, P., G. Donnarumma, A. Napolitano, F. Galvano, A. Gallo, L. Scalfi and V. Fogliano. 2007. Protocatechuic acid is the major human metabolite of cyanidinglucosides. J. Nutr. 137: 2043–2048.

Vuong, Q. V., S. Hirun, P. A. Phillips, T. L. K. Chuen, M. C. Bowyer, C. D. Goldsmith and C. J. Scarlett. 2014. Fruit-derived phenolic compounds and pancreatic cancer: Perspectives from Australian native fruits. J. Ethnopharmacol. 152: 227–242.

Wadsworth, T. L. and D. R. Koop. 1999. Effects of the wine polyphenolics quercetin and resveratrol on pro-inflammatory cytokine expression in RAW 264.7 macrophages. Biochem. Pharmacol. 57: 941–949.

Wang, H., T. O. Khor, L. Shu, Z. Su, F. Fuentes, J. -H. Lee and A. -N. T. Kong. 2012. Plants against cancer: A review on natural phytochemicals in preventing and treating cancers and their druggability. Anticancer Agents Med. Chem. 12: 1281–1305.

Wang, J. and G. Mazza. 2002. Effects of anthocyanins and other phenolic compounds on the production of tumor necrosis factor alpha in LPS/IFN-gamma-activated RAW 264.7 macrophages. J. Agric. Food Chem. 50: 4183–4189.

Wang, J., M. Pae, S. N. Meydani and D. Wu. 2013. Green tea epigallocatechin-3-gallate modulates differentiation of naive CD4(+) T cells into specific lineage effector cells. J. Mol. Med. 91: 485–495.

Wang, Y. W. and P. J. Jones. 2004. Conjugated linoleic acid and obesity control: Efficacy and mechanisms. Int. J. Obes. Relat. Metab. Disord. 28: 941–955.

Weber, J. M., A. Ruzindana-Umunyana, L. Imbeault and S. Sircar. 2003. Inhibition of adenovirus infection and adenain by green tea catechins. Antiviral Res. 58: 167–173.

Weinbrenner, T., M. Fito, R. de la Torre, G. T. Saez, P. Rijken, C. Tormos, S. Coolen, M. F. Albaladejo, S. Abanades, H. Schroder, J. Marrugat and M. I. Covas. 2004. Olive oils high in phenolic compounds modulate oxidative/antioxidative status in men. J. Nutr. 134: 2314–2321.

Williams, D. J., D. Edwards, I. Hamernig, L. Jian, A. P. James, S. K. Johnson and L. C. Tapsell. 2013. Vegetables containing phytochemicals with potential anti-obesity properties: A review. Food Res. Int. 52: 323–333.

Williams, R. J., J. P. Spencer and C. Rice-Evans. 2004. Flavonoids: antioxidants or signaling molecules? Free Radic. Biol. Med. 36(7): 838–849.

Williamson, M. P., T. G. McCormick, C. L. Nance and W. T. Shearer. 2006. Epigallocatechin gallate, the main polyphenol in green tea, binds to the T-cell receptor, CD4: Potential for HIV-1 therapy. J. Allergy Clin. Immunol. 118: 1369–1374.

Wojdylo, A., J. Oszmianski and P. Laskowski. 2008. Polyphenolic compounds and antioxidant activity of new and old apple varieties. J. Agric. Food Chem. 56: 6520–6530.

Wong, C. P., L. P. Nguyen, S. K. Noh, T. M. Bray, R. S. Bruno and E. Ho. 2011. Induction of regulatory T cells by green tea polyphenol EGCG. Immunol. Lett. 139: 7–13.

World Health Organization. 2016. World Health Statistics 2016: Monitoring health for the SDGs. Global Health Observatory (GHO) data.

Yamada, K., S. Shirahata, H. Murakami, K. Nishiyama, K. Shinohara and H. Omura. 1985. DNA breakage by phenyl compounds. Agric. Biol. Chem. 49: 1423–1428.

Yamaguchi, M. U., F. P. Garcia, D. A. Cortez, T. Ueda-Nakamura, B. P. Filho and C. V. Nakamura. 2011. Antifungal effects of Ellagitannin isolated from leaves of *Ocotea odorifera* (Lauraceae). Antonie van Leeuwenhoek 99: 507–514.

Yamamoto, M., S. Shimura, Y. Itoh, T. Ohsaka, M. Egawa and S. Inoue. 2000. Anti-obesity effects of lipase inhibitor CT-II, an extract from edible herbs, *Nomame herba*, on rats fed a high-fat diet. Int. J. Obes. Relat. Metab. Disord. 24: 758–764.

Yang, C. S., L. Chen, M. J. Lee, D. Balentine, M. C. Kuo and S. P. Schantz. 1998. Blood and urine levels of tea catechins after ingestion of different amounts of green tea by human volunteers. Cancer Epidemiol. Biomarkers Prev. 7: 351–354.

Yang, C. S., J. M. Landau, M. T. Huang and H. L. Newmark. 2001. Inhibition of carcinogenesis by dietary polyphenolic compounds. Annu. Rev. Nutr. 21: 381–406.

Yang, J., X. Yang, Y. Chu and M. Li. 2011. Identification of baicalin as an immunoregulatory compound by controlling T_H17 cell differentiation. PLoS ONE 6: e17164.

Yang, J., X. Yang and M. Li. 2012. Baicalin, a natural compound, promotes regulatory T cell differentiation. BMC Complement. Altern. Med. 12: 64.

Yokoo, T. and M. Kitamura. 1997. Unexpected protection of glomerular mesangial cells from oxidant-triggered apoptosis by bioflavonoid quercetin. Am. J. Physiol. Renal. Physiol. 273: F206–F212.

Yoshikawa, M., H. Shimoda, N. Nishida, M. Takada and H. Matsuda. 2002. *Salacia reticulata* and its polyphenolic constituents with lipase inhibitory and lipolytic activities have mild antiobesity effects in rats. J. Nutr. 132: 1819–1824.

You, K. M., K. H. Son, H. W. Chang, S. S. Kang and H. P. Kim. 1998. Vitexicarpin, a flavonoid from the fruits of Vitex rotundifolia, inhibits mouse lymphocyte proliferation and growth of cell lines *in vitro*. Planta Medica 64: 546–550.

Yun, J. W. 2010. Possible anti-obesity therapeutics from nature—A review. Phytochemistry 71: 1625–1641.

Zhang, M., M. Chen, H. -Q. Zhang, S. Sun, B. Xia and F. -H. Wu. 2009. *In vivo* hypoglycemic effects of phenolics from the root bark of *Morus alba*. Fitoterapia 80: 475–477.

Zhang, Y., A. Y. Chen, M. Li, C. Chen and Q. Yao. 2008. Ginkgo biloba extract kaempferol inhibits cell proliferation and induces apoptosis in pancreatic cancer cells. J. Surg. Res. 148: 17–23.

Zhu, X., H. Zhang and R. Lo. 2004. Phenolic compounds from the leaf extract of artichoke (*Cynara scolymus* L.) and their antimicrobial activities. J. Agric. Food Chem. 52: 7272–7278.

Zubik, L. and M. Meydani. 2003. Bioavailability of soybean isoflavones from aglycone and glucoside forms in American women. Am. J. Clin. Nutr. 77: 1459–1465.

Zuzana, K. 2011. Toxicological aspects of the use of phenolic compounds in disease prevention. Interdiscip. Toxicol. 4: 173–183.

CHAPTER 3

Alkaloids
Potential Health Benefits and Toxicity

Renée A. Street,[1,2,*] *Gerhard Prinsloo*[3] and *Lyndy J. McGaw*[4]

Introduction

Alkaloids are most abundant in higher plants and are present in at least a quarter of all higher plant species (Aniszewski 2015). Since time immemorial, alkaloids have been used as medicines, poisons and protection agents. Plants have the ability to synthesize phytochemicals of medicinal value and these yields can in turn be utilized for therapeutic purposes (Waisundara et al. 2015). Today, alkaloids remain important compounds that serve as a 'rich reservoir' for drug discovery (Qiu et al. 2014, Jiang et al. 2016). Alkaloids may be used as natural or modified compounds or be completely synthesized based on the model of the natural molecule (Aniszewski 2015). Alkaloids, such as morphine, codeine, quinine, nicotine, vinblastine and cocaine are well known for their various pharmacological properties, such as analgesic, central nervous depressant, antipyretic, antitumour and antimalarial activity. Newly discovered alkaloids, such as berberine and galanthamine, with uses as AChE inhibitors, anti-diabetics and antioxidants are becoming better known (Cushnie et al. 2014). Berberine has been used as an oral drug to treat gastroenteritis and diarrhoea for more than 1400 years (Zhang and Chen 2012) and regulates glucose and lipid metabolism via multiple pathways to restore insulin sensitivity (Yang et al. 2014). It inhibits liver gluconeogenesis and promotes the differentiation of adipocytes. It also possesses anti-inflammation properties and has protective effects against damage to pancreatic islets (Chueh and Lin 2012a). It can be used in the treatment of convulsion and epilepsy as

---

[1] South African Medical Research Council, Environment and Health Research Unit, 491 Ridge Road, Durban, South Africa.
[2] University of Johannesburg, Department of Environmental Health, John Orr Building, Corner of Siemert and Beit Street, Doornfontein, Johannesburg, South Africa.
[3] University of South Africa (UNISA), Private Bag x 6, Florida 1710 Johannesburg, South Africa. E-mail: prinsg@unisa.ac.za
[4] University of Pretoria, Department of Paraclinical Sciences, Private Bag X04, Onderstepoort 0110, Pretoria, South Africa. E-mail: lyndy.mcgaw@up.ac.za
* Corresponding author: renee.street@mrc.ac.za

it exhibits anticonvulsant activity by modulating neurotransmitter systems (Bhutada et al. 2010). Administration of berberine is a potential therapeutic approach for the treatment of various disorders, given the level of evidence available from various biological assays (Pirillo and Catapano 2015). It is also known that many alkaloids are highly toxic and the presence of pyrrolizidine alkaloids in foods and food supplements is highly undesirable. This chapter describes the sources and types of alkaloids and highlights their potential health benefits besides their toxicity.

Sources and Types of Alkaloids

The current definition of alkaloids is that they are nitrogen-containing natural products which are not otherwise classified as peptides, antibiotics, non-protein amino acids, amines, cyanogenic glycosides, glucosinolates, cofactors, phytohormones or primary metabolites, such as purine or pyrimidine bases (Van Wyk and Wink 2004). Approximately 25 per cent of plants scientifically investigated thus far contain alkaloids, but alkaloids may also occur in bacteria, fungi and animals (Leal et al. 2012, Aniszewski 2015). Some plant families contain more alkaloids than others, with alkaloid-rich families including Amaryllidaceae, Berberidaceae, Colchicaceae, Fabaceae, Papaveraceae, Asteraceae, Ranunculaceae, Rubiaceae, Solanaceae and others.

Alkaloids are multi-purpose compounds active in various environmental interactions, such as defense against herbivores, bacteria, fungi, viruses or competing plants. They represent the active constituents of many medicinal plants but are also well-known animal toxins. The toxicity of many alkaloids and their pharmacological activity in humans and animals appears to be correlated with their interactions with particular molecular targets (Van Wyk and Wink 2004). Alkaloids are frequently derived from the same amino acid precursor as the mammalian neurotransmitters serotonin, dopamine, noradrenaline, GABA, glutamic acid and histamine, so their structures are often similar to those of such neurotransmitters. Other alkaloids are planar and lipophilic and are thus able to intercalate DNA (for example, berberine and emetine) while pyrrolizidine alkaloids may alter metabolic activation in the liver. Inhibition of the assembly or disassembly of microtubules occurs after exposure to alkaloids, such as colchicine, vinblastine and taxol.

Plant tissues important for survival and reproduction are the main storage areas for alkaloids, including actively-growing young tissues, roots and stem bark, flowers, seeds, seedlings and photosynthetically active tissues, with senescing tissues containing much reduced alkaloid levels. In some herbaceous plant species, alkaloids are stored in epidermal or subepidermal tissues that are the first line of defense against insects and microorganisms. Examples of such alkaloids are aconitine, nicotine, cocaine, colchicine and coniine. Some plants produce latex containing defensive chemicals including alkaloids, such as morphine and related alkaloids in Papaveraceae. Concentrations and patterns of alkaloids often change during the development of the plant, and alkaloids can be degraded to serve as a nitrogen source.

Ergot alkaloids are well-known mycotoxins produced by fungi of the *Claviceps* genus, especially *Claviceps purpurea* which parasitizes the seed heads of living plants at flowering time (Krska and Crews 2008). These alkaloids are also produced by some plant species, particularly those of the morning glory family (Wilkinson et al. 1988).

Intoxication resulting from accidental ingestion of *Claviceps purpurea* has been documented in Europe for many centuries, causing intense pain due to vasoconstriction (hence the name St. Anthony's Fire or Holy Fire) and other symptoms (Krska and Crews 2008). In a review of marine invertebrates as a source of new natural products, Leal et al. (2012) noted that alkaloids comprised 22.1 per cent of a total of 5,286 natural products discovered from marine sources in 2000 to 2009. Alkaloids were reportedly dominant in Phyla Chordata (tunicates) and Porifera (sponges) (Leal et al. 2012).

In plants, alkaloids are synthesized from amino acids and display an extraordinary array of structural types (Fig. 1). They are generally classified according to the structure of the nitrogen-containing ring in the molecule; for example, tropane alkaloids have a tropane ring system and pyridine alkaloids have a pyridine ring system (Van Wyk et al. 2002). Many unusual alkaloids occur which are not placed in specific groups, such as the tripeptide alkaloid sanjoinine A, or frangufoline isolated from *Ziziphus* species (Van Wyk et al. 1997).

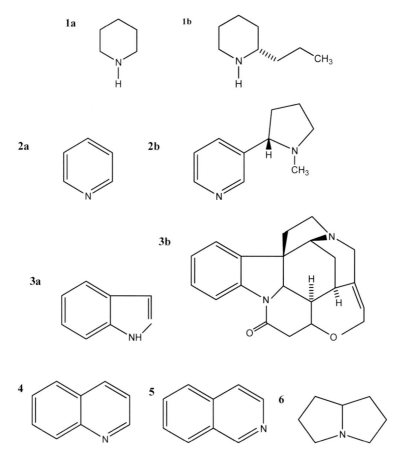

Figure 1. Representative structures of some alkaloid types (1a = piperidine, 1b = coniine, 2a = pyridine, 2b = nicotine, 3a = indole, 3b = strychnine, 4 = quinoline, 5 = isoquinoline, 6 = pyrrolizidine).

Tropane alkaloids (atropine type alkaloids)

Tropane alkaloids occur predominantly in the Solanaceae family and contain a tropane nucleus derived from the amino acid ornithine. Cocaine, from the coca plant (*Erythroxylum coca*) is one such example. *Atropa belladonna*, or deadly nightshade, has long been known to possess poisonous substances. The major alkaloids in this species, hyoscyamine and atropine, affect the autonomous nervous system by, among other effects, increasing heart rate and decreasing intestinal tone and peristalsis (Van Wyk et al. 2002). Hallucinations and delirium occur following acute poisoning.

Piperidine and pyridine type alkaloids (tobacco alkaloids)

These alkaloids are synthesized from the amino acid, lysine. Coniine, a piperidine alkaloid, which causes paralysis of the motor nerve endings, is an extremely toxic compound found in *Conium maculatum*, the hemlock plant used centuries ago in Greece to execute criminals. Nicotine, a pyridine alkaloid, is absorbed through mucous membranes of the nose, mouth and lungs and acts on the central nervous system, smooth muscle of the intestine and the cardiovascular system (Van Wyk et al. 2002).

Indole alkaloids

Indole alkaloids, derived from tryptamine, are found mainly in four plant families—Apocynaceae, Gelsemiaceae, Rubiaceae and Strychnaceae. The seeds of *Strychnos nux-vomica* contain strychnine, a bitter and extremely poisonous alkaloid. Strychnine poisoning causes anxiety, increased sensitivity to light and noise, convulsive attacks and finally death following asphyxia caused by contraction of the diaphragm (Van Wyk et al. 2002). Other examples of indole alkaloids include reserpine from *Rauvolfia* and vincristine from *Catharanthus roseus*, the Madagascar periwinkle.

Quinoline alkaloids

These alkaloids are synthesized from anthranilic acid. An example of a quinoline alkaloid is quinine, the anti-malarial substance isolated from cinchona species, mainly *Cinchona pubescens*.

Isoquinoline alkaloids

Isoquinoline alkaloids are derived from the amino acids—phenylalanine and tyrosine. The opium poppy, *Papaver somniferum*, is the source of morphine which is one of the most well-known isoquinoline alkaloids. Morphine affects the central nervous system, leading to a pain-relieving analgesic effect, but uncontrolled use leads to addiction. Tolerance is rapidly gained and the dose must be increased to yield the same effect. Heroin is a synthetic derivative of morphine. Another famous example is curare, or (+)-tubocurarine, which is a bisbenzyltetrahydroisoquinoline alkaloid used as a forest hunting poison. It has a strong muscle-relaxing effect so that prey animals, such as monkeys and birds fall to the ground. The meat of the prey is safe to eat as curare is only poisonous when injected. Another example of an isoquinoline alkaloid (Amaryllidaceae type) is lycorine (Van Wyk et al. 1997).

Quinolizidine (lupin) alkaloids

These are derived from the amino acid lysine and are common in family Fabaceae, comprising several structural types. These include lupinine (bicyclic), cytosine (tricyclic) and sparteine (tetracyclic). Sparteine affects the heart and also causes an increase in the tone and strength of uterine contractions (an oxytoxic compound). Cytisine and anagyrine are quite toxic and cause teratogenic effects in cattle with contaminated milk, potentially resulting in human poisoning.

Pyrrolizidine (Senecio) alkaloids

Pyrrolizidine alkaloids (PAs) occur in several plant families as amino alcohols (necines, such as retronecine) or as macrocyclic ring structures, such as retrorsine. The base is unsaturated in all hepatotoxic compounds (for example, retrorsine) but saturated in all less harmful macrocyclic pyrrolizidines, such as platyphylline. Chronic poisoning of livestock in South Africa may be caused by PAs of *Senecio* spp. (Asteraceae) and *Crotolaria* spp. (Fabaceae). Medicinal use of *Crotolaria* has caused chronic intoxication, symptoms of which include abdominal pain, loss of appetite and oedema of the extremities (Van Wyk et al. 2002).

Steroid alkaloids

Steroidal alkaloids comprise two main types—firstly C_{21} alkaloids derived from pregnane (occurring in the Apocynaceae) and C_{24} alkaloids, which are closely related to steroidal saponins and found in Solanaceae and Melianthaceae. Solanine is present in small quantities in species of Solanaceae, including potatoes and tomatoes. Poisoning with this substance causes throat irritation, headache, fatigue, diarrhoea, coma and convulsions, potentially caused by inhibition of cholinesterase.

Terpenoid alkaloids

These alkaloids are sometimes classed as pseudo alkaloids as they are not derived from amino acids but from terpenoids, which incorporate a nitrogen atom during biosynthesis. Aconitine, the main alkaloid of *Aconitum napellus* (aconite or monkshood), is extremely poisonous, with a lethal dose to humans of 3–6 mg (Van Wyk et al. 2002). Terpenoid alkaloids also occur in *Delphinium* species (larkspur) and *Erythrophleum* species (African war and ordeal poisons).

Potential Health Benefits

Antioxidant activity

Antioxidant activity is based on the availability of electrons to neutralize free radicals (Gülçin 2012). Increasing evidence shows that plants containing antioxidants may be of major importance in disease prevention (Gülçin 2012, Alam et al. 2013). Isoline isolated from *Ligularia duciformis* was administered at 100 mg/kg for 36 h to mice, resulting in a decrease in the total antioxidant capacity of the liver, brain and lung (Liu et al. 2010). The antinociceptive and free radical scavenging properties of six aporphine and two benzyltetrahydroisoquinoline alkaloids isolated from the root of *Lindera angustifolia* have been investigated. All alkaloids except magnocurarine exhibited remarkable radical scavenging effects (36–90 per cent scavenging at 25–100 μg/ml) in the DPPH radical scavenging test. Boldine and norboldine showed significant antinociceptive activity with 76.3 per cent and 74.6 per cent writhing inhibition, respectively. The alkaloids therefore possess free radical scavenging and antinociceptive activities. The antinociceptive activity seems to be related to the free radical scavenging effect (Zhao et al. 2006).

Cepharanthine and fangchinoline from *Stephania rotunda* showed potent radical scavenging activity in the DPPH, DMPD and $O_2^{\cdot-}$ assays. Inhibition of 94.6 and 93.3 per cent on lipid peroxidation of linoleic acid emulsion at 30 μg/ml was found for the two compounds respectively (Gülçin et al. 2010). Linearilobin, lycoctonine, 14-acetyltalatizamine, browniine, cammaconine, talatizamine, cochlearenine and linearilin from the roots of *Delphinium linearilobum* were tested for their antioxidant activity in DPPH and metal chelating activity assays with positive results (Kolak et al. 2006). From *Mahonia aquifolium*, the alkaloids berberine, jatrorrhizine and magnoflorine were tested for their DPPH radical scavenging activity. Those compounds, bearing free phenolic groups (jatrorrhizine and magnoflorine), proved to be better antioxidants than magnoflorine (Račková et al. 2004). From Coptidis Rhizoma (dried rhizome of *Coptis chinensis*), compounds such as jateorrhizine and groenlandicine exhibited significant ONOO⁻ scavenging activities with IC_{50} values of 0.78 and 0.84 μM, respectively (Jung et al. 2009).

Anti-inflammatory activity

Rising levels of reactive oxygen species (ROS) and systemic inflammation are often serious concerns in many disease conditions, like obesity. Therefore, compounds with both antioxidant and anti-inflammatory activities are considered beneficial in treating several chronic human diseases. Berberine, with various biological activities as discussed in the other sections, also has an effect on dopamine D1- and D2-like receptors and the ensuing anti-inflammatory response (Kawano et al. 2015). Supplementation with this compound also significantly decreased the expression ratios of pro-anti-inflammatory and/or Th1/Th2 cytokines in the spleen, liver and kidney (Chueh and Lin 2012b). In addition to the effect observed with berberine in the anti-inflammatory response, two derivatives, namely 13-methylberberine and 13-ethylberberine, decreased nitrite production concentration-dependently. The IC_{50} values of the two compounds were 11.64 and 9.32 μM, respectively (Lee et al. 2003).

Annotine has been shown to increase differentiation of allogeneic CD4+ T cells towards a Th2/Treg phenotype, which is noteworthy in the development of new treatments for Th1- and /or Th17-mediated inflammatory diseases (Hardardottir et al. 2015). Oxymatrine, isolated from *Sophora flavescens*, possesses anti-inflammatory, anti-oxidative and anti-apoptotic properties. It has been used for the treatment of chronic viral hepatitis and many other diseases. The effect of oxymatrine on inflammation is mediated by toll-like receptor4 (TLR4) and nuclear factor kappa-B (NF-κB) oxidative injury. Lycorine inhibits LPS-induced production of pro-inflammatory mediators. Acetylcorynoline, a major alkaloid component derived from *Corydalis bungeana*, has the ability to regulate lipopolysaccharide-stimulated activation of mouse bone marrow, a major modulator in the immune system (Qiu et al. 2014). Norbelladine a precursor for Amaryllidaceae alkaloids (e.g., belladine, crinamine, lycorine and galanthamine) reduced DPPH-radical by 31 per cent and superoxide radicals from xanthine oxidase by 33 per cent at a concentration of 10 μM. At a concentration of 0.25 μM, it also inhibits both COX-1 and COX-2 enzymes by 51 per cent and 25 per cent respectively. Norbelladine may be a compound with the ability to quench radicals, inhibit COX enzymes as well as to suppress NF-κB activation at relatively low concentrations (Park 2014).

Tripterygium wilfordii is commonly used to treat rheumatoid arthritis with alkaloids identified as the main constituents. The alkaloid fraction significantly reduces paw swelling and suppresses articular cartilage degeneration. It also could inhibit the production of IL-6, IL-8 and TNF-α in serum and the expression of IL-6, IL-8, NF-κB and TNF-α in synovial tissue targeting inflammation (Zhang et al. 2013). Cepharanthine is a biscoclaurine alkaloid isolated from *Stephania cepharantha*, which displays anti-inflammatory, anti-allergic and immunomodulatory activities *in vivo*. Cepharanthine suppresses the production of inflammatory cytokines and a chemokine, TNF-a, IL-1b, IL-6, and IL-8, in human monocytic cell cultures, including primary monocyte/macrophage cultures. This effect of cepharanthine was concentration-dependent, and significant suppression was observed at 0.1 mg/ml (Okamoto et al. 2001). Treatment with 4,8-dimethoxy-1-vinyl-b-carboline and 4-methoxy-1-vinyl-b-carboline from *Melia azedarach* decreased the levels of iNOS protein and mRNA in a concentration-dependent manner. Suppression of iNOS and cyclooxygenase-2 induction by lipopolysaccharide through selective inhibition of the expression of genes, was reported as the inflammatory mechanism (Lee et al. 2000). Aloperine is an alkaloid isolated from the plant *Sophora alopecuroides*. It is documented that aloperine can suppress pain-like behaviors in numerous inflammatory pain models in mice. Aloperine (40 mg/kg or 80 mg/kg) had an anti-nociceptive effect and reduced writhing behavior. Aloperine can suppress the pro-inflammatory cytokines tumor necrosis factor (TNF), IL-1 beta and prostaglandin E2 (Gao and Lagerström 2015).

Cancer

Accounting for 8.8 million deaths in 2015, cancer is one of the leading causes of death globally (WHO 2017). The most common causes of cancer deaths are as a result of lung, liver and colorectal cancers (1.69 million, 788,000 and 774,000 deaths respectively). A large proportion of cancer deaths could be prevented by avoiding or modifying key risk factors with one of the most important risk factors being tobacco use (WHO 2017).

Microtubule-binding natural products, such as vinca alkaloids and taxanes, contribute to the improvement of cancer therapy. With advances in new formulations, targets and sources, microtubule-binding products will be considered important for cancer therapy (Yue et al. 2010). The oldest and most well-known group of plant alkaloids used to treat cancer are the vinca alkaloids. The mechanism of action of this group is to inhibit cell proliferation by affecting the microtubular dynamics during mitosis, thereby causing a distinctive block, leading to apoptosis (Nirmala et al. 2011). The first naturally occurring vinca alkaloids to advance into clinical use were vinblastine (VLB) and vincristine (VCR), isolated from the Madagascar periwinkle (*Catharanthus roseus*) (Cragg and Newman 2005). In combination chemotherapy VLB is used to treat bladder and breast cancer and VCR is used in combination chemotherapy of acute lymphoblastic leukaemias and lymphomas (Kingston 2009). Vinorelbine and vindesine are semi-synthetic analogs acquired from the active compounds (Nirmala et al. 2011). Vinflunine is the newest synthetic vinca alkaloid available for clinical practice (Moudi et al. 2013). Vinflunine is used in platinum–resistant transitional cell cancer of the urothelial tract patients (Castellano et al. 2014). Taxanes are a class of structurally complex diterpene alkaloids found in the genus *Taxus* (Cuendet and Pezzuto 2008). Taxol, discovered in the bark of the relatively rare *Taxus brevifolia* (Pacific Yew), is one of the most noteworthy compounds in cancer treatment. The success of this has brought about comprehensive studies on the synthesis of analogues. Paclitaxel, and an analogue, docetaxel, are active against a range of human cancers (de Weger et al. 2014). Taxanes are unique as they stabilize microtubules rather than destabilizing them as vinca alkaloids do (Nobili et al. 2009, Cragg and Newman 2013, Mukhtar et al. 2014). Nonetheless, with increasing clinical use of taxanes, the development of resistance is an important issue (Wang et al. 2015). There is currently a significant amount of interest in nano-based formulations for the safe and effective delivery of taxanes (Wang et al. 2012, Feng and Mumper 2013).

Berberine is an isoquinoline alkaloid belonging to the structural class, protoberberines. Berberine is present in plants of *Berberis* and *Coptis* species (Ding et al. 2014) among others. Studies show that berberine exerts both *in vitro* and *in vivo* anticancer activity through different mechanisms (Sun et al. 2009, Liu et al. 2016). Palmatine is a close analog of berberine and both compounds are habitually found in the same plant species. A growing body of work shows that berberine and palmatine make up strong intercalation complexes with both DNA and RNA (Kumar 2015). Over the past few years, there has been a spike in studies on berberine for human cancer treatment with particular emphasis on antineoplastic action of berberine (Wang et al. 2015).

Steroidal alkaloids on investigations revealed significant and potent anticancer effects on different types of cancer (Iovine et al. 2016, Jiang et al. 2016). Therapeutic potential of both *in vitro* and *in vivo* preclinical studies look promising and there are also encouraging clinical trial results (Iovine et al. 2016, Jiang et al. 2016). Nonetheless, teratogenicity remains an important issue, and with steroidal alkaloids further pharmacokinetic and clinical studies have become necessary to define the efficacy and safety of these alkaloids in cancer treatment (Jiang et al. 2016). Cyclopamine is an example of a steroidal alkaloid that has been identified as the first inhibitor of the Hedgehog signaling pathways, which are implicated in embryonic development and tumorigenesis (Beutler 2013). The key synthetic strategies to produce coclopamine and its derivatives as well as their biological activity have been comprehensively reviewed by Iovine et al. (2016).

Noscapine is a phthalideisoquinoline alkaloid from *Papaver somniforum* (opium poppy). Recent progress shows that a number of semi-synthetic derivatives are emerging as possible candidates for novel anticancer therapies (DeBono et al. 2015, Schläger and Dräger 2016).

Cardiovascular diseases

Cardiovascular diseases (CVD) are a group of disorders of the heart and blood vessels. Heart attacks and strokes are typically acute events and are primarily caused by a blockage that prevents the flow of blood to the heart or brain (WHO 2016). In 2012, an estimated 17.5 million individuals died from CVDs, accounting for 31 per cent of all global deaths. The cardioprotective effects of numerous natural plant products may be due to their antioxidative, antiangiogenic, anti-ischemic, antihypercholesterolemic inhibition of platelet aggregation and/or anti-inflammatory properties that decrease the risk of cardiovascular disorders (Vasanthi et al. 2012).

Berberine protects against heart failure, hypertension, hyperlipidemia, insulin resistance, arrhythmias and platelet aggregation (Caliceti et al. 2016, Xia and Luo 2016). Endoplasmic reticulum (ER) stress plays a fundamental role in myocardial ischemia/reperfusion (MI/R)-induced apoptosis (Zhao et al. 2016). A recent study investigated whether the protective effects of berberine resulted from modulating ER stress levels during MI/R injury and defined the signaling mechanisms. The study concluded that berberine ameliorates MI/R injury in rats by activating the AK2/STAT3 signaling pathway and reducing ER stress-induced apoptosis (Zhao et al. 2016). A further study revealed that berberine treatment protects against I/R ischemia-reperfusion-induced myocardial infarction by selectively inhibiting excessive autophagy (Huang et al. 2015). A further study revealed that berberine has been shown to reduce blood cholesterol levels by inhibiting cholesterol absorption and promoting its excretion. Treatment with berberine (50, 100 and 150 mg/kg) in rats fed an atherogenic diet for eight weeks reduced plasma total cholesterol and non-high-density lipoprotein cholesterol levels by 29–33 per cent and 31–41 per cent, respectively (Wang et al. 2014). Based on clinical trial evidence, *in vivo* animal experiments, and *in vitro* cellular studies, berberine was scientifically proven to be a safe and effective cholesterol-lowering drug (Liu et al. 2016). Strong evidence suggests that berberine may be a promising tool in response to cardiovascular disorders and will have an important role in the treatment of cardiovascular disease in future (Xia and Luo 2016).

The first major study investigating the cardio-protective characteristics of an indole alkaloid, vincosamide, isolated from *Moringa oleifera* leaves revealed that the mechanism may involve the prevention of cardiac myofibril disruption. This may possibly be through a reduction of oxidative stress and by this means enhancing the cardiac contractile function (Panda et al. 2013). Three new alkaloids and nine known analogues were recently isolated from marine-derived fungus *Penicillium expansum* Y32 and each compound assessed for cardiovascular effects with a live zebrafish model. All compounds exhibited significant mitigative effects on bradycardia, caused by astemizole, in the heart rate experiments. The results suggested that these compounds may be likely lead candidates for the treatment of CVD (Fan et al. 2015). Studies show that rutaecarpine is a promising drug in the treatment of CVD (Jia and Hu 2010, Jayakumar and Sheu 2011).

Diabetes

Diabetes, one of the earliest diseases to be recognized, can be traced back to 400 BC where it was described by Indian physicians. The term 'diabetes' or 'to pass through' was first used in 230 by the Greek Appollonius of Memphis 1, but effective treatments were not available until the early part of the 20th century (Zhang and Chen 2012). With the application and development of insulin in 1921 and 1922 by two Canadians, Frederick Banting and Charles Best, the first treatments were realised and today it is estimated that 347 million people worldwide are affected, resulting in 3.4 million deaths in 2010 (Zhang and Chen 2012). Diabetes mellitus is a metabolic disorder characterized by elevated blood glucose levels (hyperglycemia) due to the failure of the pancreatic beta cells to produce enough insulin, insulin action or loss of an effective target tissue response to insulin. It arises from complex interactions between genetic and lifestyle factors (Subramanian and Prasath 2014). The World Health Organization (WHO) predicts that diabetes will be the 7th leading cause of death in 2030 (Yang et al. 2014).

Since this disease places an enormous burden on health care systems all over the world (Subramanian and Prasath 2014), numerous studies have been conducted to determine the effect of plant remedies. Therapeutic options for type 2 diabetes include a strict regulation of diet, oral hypoglycemics and insulin, but all have their advantages and disadvantages (Zhang and Chen 2012). From many investigations on plant medicines, it was revealed that alkaloid-containing plants as well as isolated alkaloids are very effective in treating this disease (Zhang and Chen 2012). Berberine, from the root extract of *Coptis chinensis* and the crude extract, was found to promote glucose uptake in differentiated C2C12 cells at 6.25 µg/ml and 50 µg/ml respectively. Coptisonine and octadecyl caffeate from this plant also stimulated glucose uptake at 25 and 50 µg/ml (Yang et al. 2014). Derivatives of berberine, such as chloroberberine and bromoberberine, showed better results than berberine with 90 per cent increased activity at levels of 1 µg/ml (Ding et al. 2014). The compound, 16,17-dihydro-17b-hydroxy isomitraphylline alkaloid, was isolated from *Mitragyna parvifolia* with inhibitory activity on DPP IV at 100 mg/kg. It was found to induce proliferation of pancreatic cells and increase the formation of beta-cells (Shukla and Srinivasan 2012). The DPP-IV inhibitory potential of the seed extract of *Castanospermum australe* extract showed an IC_{50} value of 13.96 µg/ml with the three isolated alkaloids showing comparable DPP-IV inhibition with berberine. The seed extract (150 mg/kg body weight) normalizes hyperglycemia in T2DM rats with strong DPP-IV inhibitory potential and among the three alkaloids isolated, the compound 7-deoxy-6-epi-castanospermine showed the highest activity (Bharti et al. 2012).

Various alkaloids (alstiphyllanines E–G, alstiphyllanine H and 16 related alkaloids) were isolated from the leaves of *Alstonia macrophylla*. Alstiphyllanines E and F showed moderate Na+-glucose cotransporter (SGLT1 and SGLT2) inhibitory activity. This cotransporter is a membrane protein that plays an important role in the re-absorption of glucose in the kidneys (Arai et al. 2010). From fenugreek (*Trigonella foenum-graecum*) trigonelline was isolated with known diuretic, cardiotonic, hypotensive, hypoglycemic and hypolipidemic effects. Trigonelline at 150 mg/kg body weight resulted in an improvement in hepatic and muscle glycogen content of insulin-resistant diabetic rats and effectively normalized the status of the lipid profile (Subramanian and Prasath 2014). An alkaloid-rich fraction of *Capparis decidua* also significantly inhibited acute elevation of blood glucose levels and reduced total cholesterol and triglyceride (Sharma et al. 2010).

Several plants have shown activity against α-glucosidase in the search for anti-diabetic drugs. From *Hericium erinaceus*, eight alkaloids were isolated with activity against α-glucosidase (Wang et al. 2015). *Ziziphus oxyphylla* Edgw (Rhamnaceae) contains the alkaloids, nummularine-R, nummularin-C, and hemsine-A, which showed potent α-glucosidase inhibition and moderate anti-glycation activities with IC_{50} values of 212.1 mM, 215.1 mM, and 394.0 mM respectively with no toxicity to a PC-3 cell line (Choudhary et al. 2011). The well-known plant *Catharanthus roseus* contains numerous other alkaloids, such as vindogentianine, vindoline, vindolidine, vindolicine, vindolinine, perivine and serpentine. Vindogentianine exhibited potential hypoglycemic activity in β-TC6 and C2C12 cells by inducing higher glucose uptake and significant *in vitro* PTP-1B inhibition (Tiong et al. 2015). Another well-known alkaloid, piperine, isolated from *Piper nigrum* (black pepper), showed statistically significant antihyperglycemic activity (Qiu et al. 2014). Luciferine, extracted from *Nelumbo nucifera*, can stimulate both phases of insulin secretion in isolated islets. It is found to stimulate insulin secretion by closing potassium-adenosine triphosphate channels (Qiu et al. 2014). From *Lupinus* species, 2-thionosparteine and lupanine showed enhanced insulin secretion at 8.3 and 16.7 mM, and 13-a-OH lupanine or 17-oxo-lupanine at 16.7 mM (López et al. 2004). In another study, 2-thionosparteine was confirmed to be a possible plasma-glucose-lowering agent with insulin secreting activity (Bobkiewicz-Kozłowska et al. 2007).

Obesity

Obesity is a complex chronic condition in which excess body fat accumulates into adipocytes, leading to adverse metabolic effects on blood pressure, cholesterol, an increased risk for type 2 diabetes T2DM, coronary heart disease and hypertension (Choi et al. 2014, WHO 2016). In obesity, there is an increase in the number of differentiated mature cells, which are regulated by genetic and environmental factors. The prevalence of obesity has been increasing in both the developed and developing world (Choi et al. 2014). In 2014, nearly 2 billion adults worldwide were overweight (39 per cent of adults 18+) and, of these, more than half a billion were obese. In Americas, Europe and Eastern Mediterranean regions, 50 per cent of women are overweight and 25 per cent of these are obese. Obesity is, however, more prevalent in low and lower middle-income countries with more women obese than men (WHO 2016). Screening natural products for anti-obesity potential is important in the search for treatment of this worldwide disease (Choi et al. 2014). The fruits of *Piper retrofractum* have been used for their anti-flatulent, expectorant, antitussive, antifungal and appetizing properties in traditional medicine, and they are reported to possess gastroprotective and cholesterol-lowering properties. Piperidine alkaloids from *Piper retrofractum,* namely piperine, pipernonaline and dehydropipernonaline were isolated as anti-obesity constituents of this plant (Kim et al. 2011).

Five alkaloids, berberine, epiberberine, coptisine, palmatine and magnoflorine were isolated from an extract of the rhizome of *Coptis chinensis* (Coptidis Rhizoma) and it was found to significantly inhibit adipocyte differentiation and lipid contents by downregulation of protein levels (Choi et al. 2014). Repeated oral treatment of the alkaloid fraction of *Hunteria umbellata* seed extract produced weight loss, antihyperlipidemic and cardioprotective effects significantly. Additionally, dose-dependent weight loss and decreases in the serum triglyceride, total cholesterol and low density lipoprotein cholesterol were observed. Furthermore, increased serum levels of high density lipoprotein cholesterol fraction were also found (Adeneye and Crooks 2015). Bouchardatine significantly reduced lipid accumulation and mainly inhibited early differentiation of adipocytes through proliferation inhibition and cell cycle arrest in a dose-dependent manner, without any observed cytotoxicity (Rao et al. 2015). Two evodia alkaloids, rutaecarpine and evodiamine, exhibited anti-lipogenic and anti-gluconeogenic effects which may have therapeutic potential for treatment of hyperglycemia and T2DM (Yu et al. 2015).

Immune system response

Modulation of the immune system can be addressed through a variety of specific and non-specific approaches for a stimulatory, suppressive or regulatory effect (Patwardhan and Gautam 2005). Botanicals are chemically complex and produce a diverse range of natural products with antimicrobial and immunomodulating potential, including isoflavonoids, indoles, phytosterols, polysaccharides, sesquiterpenes, alkaloids, glucans and tannins (Chattopadhyay 2006). Modulation can be targeted to affect the immune system through cytokine secretion, histamine release, immunoglobulin secretion, class switching, cellular co-receptor expression, lymphocyte expression and phagocytosis. Many plant-based preparations alter the immune function and display an array of immunomodulatory effects (Patwardhan and Gautam 2005).

The well-known alkaloid morphine used in pain treatment, also showed expression of cannabinoid receptor 1 and CB2-R in culture cells, indicating that it may be involved in abnormal immune function (Zhang et al. 2012). Similarly cocaine, its metabolites and related alkaloids from *Erythroxylon coca* were found to be responsible for suppression of the immune response at doses of 15 to 60 mg/kg (Watson et al. 1983). The alkaloids (20 S)-(bennzamido)-3b-(N,N-dimethyamino)-pregnane and (20 S)-(bennzamido)-pregnane-3-one, pachysanaximine A and 20a-diacetamido-5a-pregnane were isolated from *Sarcococca saligna* and the immunomodulatory potential was found to be between IC_{50} = 10 mg/ml (95 per cent) and 1.6 mg/ml (Iqbal et al. 2015). Tomatine from *Lycopersicon pimpinellifolium* showed antigen-specific IFN-γ secretion and cytotoxic T lymphocyte activity *in vitro* which were both significantly enhanced compared to responses detected from similarly stimulated splenocytes from naive and tomatine-saline-immunised control mice (Heal et al. 2001). A hot water infusion of the root bark of *Cissampelos sympodialis* is used traditionally to treat several inflammatory disorders, including asthma, a chronic inflammatory allergic disease. The anti-allergic and immunoregulatory properties of this plant inhibit the airway hyper-reactivity and lung remodeling. The most active compound was found to be warifteine (Bezerra-Santos et al. 2012, Costa et al. 2013). The fruits of *Morus alba* have been traditionally used as a tonic to boost the immune responses. Five alkaloids isolated from the plant namely 5-(hydroxymethyl)-1H-pyrrole-2-carboxaldehyde, 2-formyl-1H-pyrrole-1-butanoicacid, 2-formyl-5-(hydroxymethyl)-1H-pyrrole-1-butanoicacid, 2-formyl-5-(methoxymethyl)-1H-pyrrole-1-butanoic acid and Morrole A and two compounds 2-formyl-5-(hydroxymethyl)-1H-pyrrole-1-butanoicacid and 2-formyl-5-(methoxymethyl)-1H-pyrrole-1-butanoic acid showed macrophage activity through enhancement of nitric oxide, TNF-α and IL-12 production and the stimulation of phagocytic activity (Kim et al. 2013). Sinomenine, is an alkaloid extracted from *Sinomenium acutum*, suppressed the production of cytokines and the

Th1 and Th2 immune responses (Feng et al. 2006). The immunomodulatory effects of punarnavine (40 mg/kg body weight) were found to enhance the total white blood cell count and plague-forming cells. Punarnavine also showed enhanced proliferation of splenocytes, thymocytes and bone marrow cells and induced elevated levels of pro-inflammatory cytokines, such as TNF-α, IL-1β, and IL-6 (Manu and Kuttan 2009).

Infectious diseases

Infectious diseases are caused by pathogenic microorganisms including bacteria, viruses, parasites or fungi and are among the most significant causes of morbidity and mortality worldwide (WHO 2012). Annually infectious diseases kill nearly 9 million individuals, many of them being children under five years of age (WHO 2012). Infections such as pneumonia, diarrheal diseases and malaria account for nearly half of the deaths in these children, while in the adult population of developing countries, approximately a quarter of the disease burden is caused by infectious diseases associated with HIV/AIDS (Mahady 2005). Infectious disease mortality rates are also increasing in industrial nations, such as the United States, a trend exacerbated by emerging antibiotic resistance (Mahady 2005). The development of new treatments, together with prevention and improved monitoring, are vital in the fight against infectious diseases.

A recent review (Savoia 2012) highlighted the antimicrobial activity of natural products sourced from plants, focusing on the mechanisms of action of these compounds (including several alkaloids) that may in future serve as a basis for the development of substances to be used against bacterial and fungal infections. In reviewing the potential of Cameroonian medicinal plants and natural products against microbial infections, Kuete (2010) found that most of the bioactive compounds isolated were phenolics, but alkaloids also featured strongly. In Cameroonian medicinal plant studies, alkaloids were mostly isolated from three families, including Rutaceae, Caesalpiniaceae and Apocynaceae (Kuete 2010). Norcassaide and norerythrosuaveolide isolated from *Erythrophleum suaveolens* had significant (MIC < 20 µg/mL) activities against various microbial species, including *Klebsiella pneumoniae*, *Neisseria gonorrhoea*, *Candida albicans* and *C. krusei* (Ngounou et al. 2005). Kokusaginine, masculine and nkolbisine isolated from the stem bark of *Tecla afzelii* were also active against various bacterial species (Kuete et al. 2008).

Some examples are given below of other useful alkaloids which possess interesting anti-infective activity. *Chelidonium majus* is used to treat various infectious diseases, and constituents from this species has *in vitro* antifungal activity against drug-resistant yeast isolates (Meng et al. 2009). Bioassay-guided fractionation led to the isolation of 8-hydroxylated alkaloids with potent activity against the drug-resistant fungi (Meng et al. 2009). Diterpenoid alkaloids, which are commonly found in plants belonging to Ranunculaceae or buttercup family, often have antimicrobial properties (Omulokoli et al. 1997, Atta-ur and Choudhary 1999).

Many natural products isolated from plants are shown to have antiviral activity and alkaloids feature strongly in this area (Cos et al. 2008). Mechanisms of action of these antiviral alkaloids include inhibition of reverse transcriptase and inhibition of HIV-induced cellular fusion (Cos et al. 2008). As an example, michellamine B is an anti-HIV naphthylisoquinoline alkaloid dimer from the tropical liana *Ancistrocladus korupensis* (White et al. 1999). This compound has undergone extensive preclinical evaluation but was considered too toxic to be advanced to clinical trials as a potential anti-HIV drug, so efforts have been made to prepare synthetic derivatives with less toxicity and enhanced potency (Cos et al. 2008). A glycoalkaloid from *Solanum khasianum* berries, solamargine and other alkaloids may be useful against HIV infections (McMahon et al. 1995).

Chikungunya virus (CHIKV), a mosquito-transmitted virus, has re-emerged as a significant public health threat in recent years and there is currently no treatment available for this disease (Kaur et al. 2013). Following screening of a natural products library for CHIKV antiviral activity, harringtonine, a cephalotaxine alkaloid, displayed potent inhibition of CHIKV infection with minimal cytotoxicity. The alkaloid inhibited an early stage of the CHIKV replication cycle after viral entry into cells, and it was shown that it affects the CHIKV RNA production as well as the viral protein expression (Kaur et al. 2013).

Berberine is an important alkaloid because of its diverse properties. It is potentially effective against trypanosomes (Freiburghaus et al. 1996) and plasmodia (Omulokoli et al. 1997). The mechanism of action of highly aromatic planar quaternary alkaloids, such as berberine and harmane, has been attributed to their ability to intercalate with DNA (Phillipson and O'Neil 1989). Berberine is well-known for its antibacterial activity and this isoquinoline alkaloid is found in many plant species, such as *Coptis chinensis*, *Berberis vulgaris* and *Hydrastis canadensis* (Mahady 2005). Berberine was shown to inhibit the growth of *Staphylococcus aureus* with an MIC of 25 µg/mL (Chi et al. 1991). Structural modification of this alkaloid have also been prepared to analyze structure-activity relationships; antibacterial activity was higher with 8-alkyl-berberine and 8-alkyl-12-bromo-berberine, and it also increased with lengthening of the aliphatic chain of these two derivatives (Iwasa et al. 1998). Of the compounds tested, 12-bromo-8-n-hexyl-berberine (12-BHB) was 64 times more active than clinically used berberine against *Staphylococcus aureus* (Iwasa et al. 1998). Berberine is an excellent DNA intercalator and is active against several microorganisms, targeting RNA polymerase, gyrase and topoisomerase IV and nucleic acid (Yi et al. 2007).

A review focusing on the potential of various types of alkaloids, together with structure-activity relationships, in the treatment of tuberculosis was recently published (Kishore et al. 2009). A large number of different types of alkaloids, mainly isolated from higher plants or marine sponges, have been reported to be active with very low MIC values against *Mycobacterium tuberculosis* and other mycobacterial species (Kishore et al. 2009).

Artemisinin is a well-known success story of an antimalarial drug developed from *Artemisia annua* (Mueller et al. 2004, Newman and Cragg 2007). This follows from the long-term use of quinine (from *Cinchona* species) against malaria. A sponge of the Spongosorites, which contains the nortopsentin and topsentin class of bisindole imidazole alkaloids, was shown to strongly inhibit the growth of *Plasmodium falciparum* (Alvarado et al. 2013). Alkaloids have been found to have microbicidal effects against *Giardia* and *Entamoeba* protozoal species (Ghoshal et al. 1996), but the major antidiarrhoeal effect is most likely due to their effects on transit time in the small intestine (Cowan 1999).

Alkaloids have also been shown to have schistosomocidal activity, with solasodine and solamargine isolated from *Solanum lycocarpum* fruit causing separation of coupled worms, extensive disruption of worm teguments and death of *Schistosoma mansoni* adult worms (Keiser and Utzinger 2007). An imidazole alkaloid, epiisopiloturine, isolated from *Pilocarpus microphyllus* (Rutaceae) leaves, affected the survival time of *S. mansoni* adult worms and schistosomula but was not cytotoxic to mammalian cells (Veras et al. 2012).

Infections of macrophages by protozoan parasites of the genus *Leishmania* result in the disease leishmaniasis—a neglected tropical disease prevalent in 88 (mostly developing) countries worldwide (Mishra et al. 2009). Alkaloids are important to plants for defence against microorganisms and herbivory, and have diverse effects on parasites with several showing excellent antileishmanial activity (Mishra et al. 2009). These include quinolone, indole, steroidal, acridone and diterpene alkaloids, with some having very low effective concentrations against *Leishmania* parasites and minimal toxicity to mammalian cells (Mishra et al. 2009). Wink (2012) further reviewed the potential of medicinal plants against human infections caused by endoparasites, including protozoa, nematodes, trematodes and cestodes. As with bacteria and fungi, resistance development is a serious problem in controlling such parasite infections; hence, alternative sources of effective drugs are being sought. Many plant-derived secondary metabolites, including alkaloids, have been shown to interfere with central targets in parasites, such as membrane integrity, DNA (alkylation or intercalation), microtubules and neuronal signal transduction (Wink 2012).

Bioavailability of Alkaloids

To be medicinally useful, a compound or drug must be bioavailable and should not be toxic to the host. With prokaryotic organisms, it is potentially easier to find targets for antimicrobial action that are not shared with the host. For eukaryotic organisms, such as parasites, which share many molecular and biochemical properties with their hosts, it is more difficult to find effective, non-toxic anti-parasitic drugs (Wink 2012). There are numerous reports of *in vitro* activity of plant compounds, such as alkaloids, against various disease-causing organisms, but factors such as their bioavailability must necessarily be kept in mind when evaluating their potential efficacy and safety. An initial step is to determine the selectivity index, which compares the activity of a drug against a target organism to cytotoxicity against a panel of human cells. This may help to identify which compounds are useful for further studies and those which may be discarded at an early stage. Selectivity index calculations are intrinsic to antiviral and other anti-infective testing, and refer to the ratio of the maximum drug concentration causing either 50 per cent or 90 per cent inhibition of growth of normal cells (CC_{50}, CC_{90}) and the minimum drug concentration at which 50 per cent or 90 per cent of the virus (or other infectious organisms) is inhibited (IC_{50}, IC_{90}) (Cos et al. 2008). However, it is important to keep in mind that *in vitro* cytotoxicity does not always translate to *in vivo* toxicity as the metabolism effects need to be taken into account.

Following ingestion, a drug or active compound needs to penetrate the intestinal mucosa and be able to withstand the effect of enzymes in the gut wall and liver that threaten to inactivate it. Bioavailability (*F*) is generally referred to as the fraction of an orally administered dose that reaches the systemic circulation intact, taking into account absorption as well as local metabolic degradation (Rang et al. 2007).

It is determined by measuring the plasma drug concentration versus time curves in a group of test subjects following oral and (separately) intravenous administration. The fraction absorbed following an intravenous dose is 1 by definition. The areas under the plasma concentration curve (AUC) are used to estimate bioavailability as AUC_{oral}/ $AUC_{intravenous}$ (Rang et al. 2007). Absolute bioavailability compares the bioavailability of the active compound in the systemic circulation after non-intravenous administration (for example, oral, sublingual or subcutaneous administration), with bioavailability of the same drug after intravenous administration. A compound given by the intravenous route will have an absolute bioavailability of 100 per cent ($F = 1$).

As well as being affected by the drug preparation, variations in enzyme activity in the gut wall or liver, in gastric pH or intestinal motility will also influence bioavailability. The concept of bioavailability does not take into account the rate of absorption, but only relates to the total proportion of the drug that reaches the systemic circulation (Rang et al. 2007). There are many ways of administering drugs apart from ingestion, including intravenous injection, which is the fastest and most certain route of drug administration, as well as subcutaneous or intramuscular injection, sublingually, rectally or by application to epithelial surfaces (Rang et al. 2007).

Relatively little information has been published on bioavailability of alkaloids compared to their biological activity. Rat specimens are commonly employed in detection of bioavailability and pharmacokinetic parameters of plant compounds in medicinal plants. Chen (2016) studied the bioavailability in rats of Kumu injection (KMI)—a treatment made from the branches and stems of *Picrasma quassiodes* which has been used clinically against upper respiratory tract infection, acute tonsillitis, enteritis and bacillary dysentery. The active ingredients of this preparation are three— canthinone alkaloids, and intramuscular injection showed over 90 per cent absolute bioavailability while for oral administration, the values were lower than 50 per cent, suggesting that intramuscular injection of KMI was suitable in clinical usage (Chen et al. 2016). Wu et al. (2013) compared the pharmacokinetics and bioavailability in rats of protopine, tetrahydropalmatine, bicuculline and egenine in different formulations prepared from the rhizomes of *Corydalis decumbens*, a traditional Chinese medicine. The hydrochloride freeze-dried powder was the best among the three formulations for the alkaloid extract of *Corydalis decumbens*, where protopine and tetrahydropalmatine (the major bioactive components) showed highest absorption and bioavailability, while bicuculline and egenine (the toxic compounds) were detected at lower concentrations in rat plasma. This provided useful information on a potential formulation for further development of the alkaloid extract of *Corydalis decumbens* as a new drug.

Capsaicin, a naturally-occurring alkaloid from red pepper, has wide application as a food additive and medicine. A pharmacokinetic study in rats showed that the absorption of capsaicin, which is weakly water soluble, in micelle form was facilitated *in vivo* with enhanced oral bioavailability (Zhu et al. 2014).

In vitro models have also been used to study bioavailability of alkaloids. A recent publication made use of the Caco-2 intestinal cell monolayer model—a recognised tool in analyzing drug absorption, to study compatibility of the transport of *Veratrum nigrum* alkaloids in different proportions with *Panax ginseng* (Ma et al. 2016). It was concluded that the effects of *Veratrum* alkaloids could be improved when combined with *Panax ginseng* in suitable proportions (Ma et al. 2016). As most multi-herb preparations involve more than two herbs, the herbs may interact with each other to initiate therapeutic action or to modulate toxicological effects of the constituent herbs (Chan 1995). Zheng et al. (2015) used the Caco-2 cell monolayer model to explore the absorption of Ephedra alkaloids alone and in combination with other components of the Mahuang decoction used in traditional Chinese medicine. It was found that Cassia twig, bitter apricot kernel and prepared licorice in the Mahuang decoction decreased the absorption of Ephedra alkaloids, potentially alleviating the drastic diaphoretic effect and toxicity of Ephedra (Zheng et al. 2015).

Alkaloids may also be used to improve the bioavailability of other compounds. For example, to increase the bioavailability of curcumin, co-administration of the alkaloid piperine, a constituent of black pepper (*Piper nigrum*) and long pepper (*Piper longum*), may be performed. Piperine (20 mg) given with 2 g of curcumin increased serum curcumin bioavailability 20 times in humans and 1.56 times in rats (Shoba et al. 1998).

The well-known alkaloid, berberine, has been the subject of many studies which indicate that it may have several potential uses, such as regulation of lipid and glucose metabolism, suppression of tumor cell proliferation, induction of apoptosis and antimicrobial activity (Liu et al. 2016). The oral bioavailability of berberine is very poor, however (below 1 per cent) (Chen et al. 2011, Liu et al. 2016), so its medical benefits are limited when taken as a medical treatment orally. Bioavailability enhancement using various techniques may perhaps be an effective solution. Berberine is shown to be safe in human subjects studied in the short-term and chronically, but is mainly used as an antidiarrhoeal agent and thus is only needed to act topically inside the gastrointestinal lumen (Liu et al. 2016), so the long-term safety of this alkaloid is dependent on little or no absorption. Some adverse effects have been observed after high–dose administrations of berberine, which is concerning and further research is needed (Liu et al. 2016). To improve the oral bioavailability of berberine, suitable excipients which enable permeability improvement and P-glycoprotein efflux inhibition could allow increased absorption and reduction of the intestinal first-pass effect (Liu et al. 2016). It is clear that alkaloids hold much promise with regard to therapeutic applications, but various problems, such as toxicity and bioavailability, may need to be overcome before their potential is fully realized.

Side effects and toxicity

Plants may contain substances capable of producing varying degrees of discomfort and adverse physical or chemical effects or even death (Fuller and McClintock 1986). One of the key roles of alkaloids in plants is toxicity against predators and pathogens (Matsuura and Fett-Neto 2015). Alkaloids can influence the nervous system of an animal, thereby changing the functionality of the organism (Aniszewski 2015). The specific actions of these toxins may be exploited for the development of new drugs (Philippe and Angenot 2005, Montaser and Luesch 2011) as common biological property of alkaloids is their cytotoxicity (Aniszewski 2015).

Owing to the meticulous efforts of veterinarians and other researchers over the last few decades, most of the plant poisonings affecting livestock have been identified and described (Botha and Penrith 2008). However, human poisonings are less well documented especially when traditional plant-based medicines are concerned (Kuete 2014). The toxic alkaloids which are frequently associated with fatal and non-fatal poisonings of humans include aconitine, atropine, coniine, colchicine, cytisine, dimethyltryptamine, harmine, harmaline, ibogaine, kawain, mescaline, scopolamine and taxine (reviewed by Beyer et al. 2009).

Pyrrolizidine alkaloid containing plants is among the most significant sources of human and animal exposure to plant toxins and carcinogens (Cheeke 1989, Fu et al. 2004). Pyrrolizidine alkaloids (PAs) are found mainly in plants of three families: Boraginaceae, Compositae and Leguminosae. The genera which exhibit the greatest toxicity to humans and livestock include *Senecio*, *Crotalaria* and *Heliotropium* (Rizk 1990). The PAs of significance to human adverse effects are the hepatotoxic PAs which are the esters of 1-hydroxymethyl dehydropyrrolizidine. These compounds are metabolized in the liver to electrophilic derivatives known as pyrroles (ANZFA 2001). The pyrroles give rise to hepatocellular damage, cirrhosis and veno-occlusive disease (ANZFA 2001). The deliberate use of PA-containing plants as traditional plant-based remedies or herbal teas and subsequent toxicity is well documented (Steenkamp et al. 2002). This has resulted in regulatory restrictions on the use of PA-containing plants and plant-derived products over the years. An example is the internal use of comfrey (*Symphytum officinale*) dietary supplements which have been banned from the United States market since 2002 (Fu et al. 2004). With the exception of the deliberate use of traditional plant-based remedies and nutritional supplements, humans may also be unintentionally exposed through the ingestion of PA contaminated food (ANZFA 2001, Allgaier and Franz 2015). Contaminated food may include grain-based products, vegetables and honey. An analysis of retail honeys on the German/European market detected PAs (calculated as retronecine equivalents) in 19 (9%) of the 216 samples analyzed (range of 0.019–0.120 µg/g) (Kempf et al. 2008).

Tropane alkaloids (TAs) are among the plant toxins with potential health risks to humans. The term tropane alkaloids refers to a group of over 200 compounds renowned for their occurrence in the family Solanaceae. Of special interest are the *Datura* species which readily synthesize high amounts of tropane alkaloids. One of the most notorious poisonous plants is *Datura stramonium* L., a wild-growing herb commonly known as Jimson weed (Adegoke and Alo 2013, Sanlidag et al. 2014). Plants of the genus *Datura* are often abused for their hallucinogenic properties, resulting in acute poisoning and even death (Boumba et al. 2004, Diker et al. 2007). *Datura* species produce numerous tiny seeds encapsulated in apple-shaped fruit capsules. Upon release, the seeds have been found as noteworthy impurities in soybean and linseed products, introducing a variable amount of tropane alkaloids into the feed material (Alexander et al. 2008). Another notorious TA-containing plant is the perennial *Atropa belladonna*, commonly known as deadly nightshade. It contains toxic TAs in all parts, the more dangerous of which are the berries owing to their attractive appearance and sweet taste (Beyer et al. 2009).

Recent incidents with intoxications by TAs in humans have resulted in the need for more information on the relevance for humans (Adamse et al. 2014). Human foods that potentially contain TAs are herbal teas, plant-based preparations, blue or black berries and edible flowers. Contamination has been found in beans, buckwheat, linseed and soybean (Adamse et al. 2014). A recent European study detected TAs in

22 per cent of the 113 cereal-based foods for infants and young children. Mean TA levels were 3.9, 2.4 and 0.4 µg kg^{-1}, in 2011, 2012 and 2014 respectively. The acute TA reference dose, derived by European Food Safety Authority in 2013, would have been surpassed by young children when consuming certain products sampled in 2011 and 2012 (Mulder et al. 2015).

Aconitum alkaloids are widely renowned to contain a series of diester diterpene alkaloids, such as aconitine, mesaconitine, and hypaconitine. The alkaloids include cardiotoxins and neurotoxins; therefore, most incidents of aconite poisoning result in neurological and cardiovascular manifestations (Li et al. 2016). All species of the genus *Aconitum* (*Ranunculaceae* family) are highly toxic (Beyer et al. 2009) and owing to this have been used in cases of suicide and homicide (Niitsu et al. 2013).

Detection of toxic plant alkaloids in human and plant samples plays an imperative role in clinical and forensic toxicology investigation of suspected herbal poisoning cases (Beyer et al. 2009, Ng et al. 2013). Analytical methods of some toxic alkaloids in biological samples have been developed and described; however, most of these methods detect only one particular group of alkaloids. Commonly used methods include high performance liquid chromatography (HPLC), gas chromatography–mass spectrometry (GC–MS), capillary electrophoresis, liquid chromatography–mass spectrometry (LC–MS), and liquid chromatography–tandem mass spectrometry (LC–MS/MS) (Beyer et al. 2009, Ng et al. 2013). Ng et al. (2013) recently developed and validated a LC–MS/MS method for simultaneous detection of 22 toxic plant alkaloids in herbal and urine samples. The alkaloids included aconitum alkaloids and their hydrolyzed products (aconitine, hypaconitine, mesaconitine, yunaconitine, crassicauline A, benzoylaconine, benzoylmesaconine, benzoylhypaconine, deacetylyunaconitine, deacetylcrassicauline A), solanaceous tropane alkaloids (atropine, anisodamine, scopolamine, anisodine), sophora alkaloids (matrine, sophoridine, oxymatrine, cytisine, N-methylcytisine), strychnos alkaloids (brucine, strychnine) and colchicine.

Conclusion

Alkaloids derived from plants are an important source of potentially useful chemical lead compounds for drug discovery. Their distinctive and diverse structures have previously been used as a major source of novel, effective therapeutic agents (Qiu et al. 2014). Since the discovery of alkaloids, such as morphine and codeine, more alkaloids, such as vinblastine, vincristine and nicotine have become important in treatment of diseases. Newer alkaloids, such as lycorine, piperine and berberine are quickly being recognized for their healing properties and show the importance of investigating natural products for new drug leads. The majority of natural therapeutics are derived from higher plants; however, the marine environment is an understudied area for new molecules (Montaser and Luesch 2011).

Through alkaloid modification and synthesis, contemporary drug discovery tends to develop compounds that have higher bioactivity than those in nature. Nonetheless, natural compounds are important sources of novel skeletons for drug leads (Aniszewski 2015). Scientific literature as well as evidence-based practices involving alkaloids is a rapidly growing field (Aniszewski 2015). Apart from the healing ability of alkaloids, their toxicity and mutagenicity necessitates quality assessment and control of food and medicine products.

References

Adamse, P., H. Van Egmond, M. Noordam, P. Mulder and M. De Nijs. 2014. Tropane alkaloids in food: Poisoning incidents. Quality Assurance and Safety of Crops & Foods 6: 15–24.

Adegoke, S. and L. Alo. 2013. Datura stramonium poisoning in children. Nigerian Journal of Clinical Practice 16: 116–118.

Adeneye, A. A. and P. A. Crooks. 2015. Weight losing, antihyperlipidemic and cardioprotective effects of the alkaloid fraction of Hunteria umbellata seed extract on normal and triton-induced hyperlipidemic rats. Asian Pacific Journal of Tropical Biomedicine 5: 387–394.

Alam, M. N., N. J. Bristi and M. Rafiquzzaman. 2013. Review on *in vivo* and *in vitro* methods evaluation of antioxidant activity. Saudi Pharmaceutical Journal 21: 143–152.

Alexander, J., D. Benford, A. Cockburn, J. Cravedi, E. Dogliotti, A. Di Domenico, M. Fernandez-Cruz, F. Fürst, J. Fink-Gremmels and C. Galli. 2008. Tropane alkaloids (from *Datura* sp.) as undesirable substances in animal feed. The EFSA Journal 691: 1–55.

Allgaier, C. and S. Franz. 2015. Risk assessment on the use of herbal medicinal products containing pyrrolizidine alkaloids. Regulatory Toxicology and Pharmacology 73: 494–500.

Alvarado, S., B. F. Roberts, A. E. Wright and D. Chakrabarti. 2013. The bis(indolyl)imidazole alkaloid nortopsentin a exhibits antiplasmodial activity. Antimicrob Agents Chemother 57: 2362–2364.

Aniszewski, T. 2015. Alkaloids: Chemistry, Biology, Ecology and Applications. Elsevier.

ANZFA. 2001. Pyrrolizidine Alkaloids in Food—A Toxicological Review and Risk Assessment. Technical report series.

Arai, H., Y. Hirasawa, A. Rahman, I. Kusumawati, N. C. Zaini, S. Sato, C. Aoyama, J. Takeo and H. Morita. 2010. Alstiphyllanines E–H, picraline and ajmaline-type alkaloids from Alstonia macrophylla inhibiting sodium glucose cotransporter. Bioorganic & Medicinal Chemistry 18: 2152–2158.

Atta-ur, R. and M. I. Choudhary. 1999. Diterpenoid and steroidal alkaloids. Nat. Prod. Rep. 16: 619–635.

Beutler, J. A. 2013. Natural products as tools for discovering new cancer targets. pp. 213–237. Natural Products and Cancer Drug Discovery. Springer.

Beyer, J., O. H. Drummer and H. H. Maurer. 2009. Analysis of toxic alkaloids in body samples. Forensic Science International 185: 1–9.

Bezerra-Santos, C. R., A. Vieira-de-Abreu, G. C. Vieira, R. Jaime Filho, J. M. Barbosa-Filho, A. L. Pires, M. A. Martins, H. S. Souza, C. Bandeira-Melo and P. T. Bozza. 2012. Effectiveness of Cissampelos sympodialis and its isolated alkaloid warifteine in airway hyperreactivity and lung remodeling in a mouse model of asthma. International Immunopharmacology 13: 148–155.

Bharti, S. K., S. Krishnan, A. Kumar, K. K. Rajak, K. Murari, B. K. Bharti and A. K. Gupta. 2012. Antihyperglycemic activity with DPP-IV inhibition of alkaloids from seed extract of Castanospermum australe: Investigation by experimental validation and molecular docking. Phytomedicine 20: 24–31.

Bhutada, P., Y. Mundhada, K. Bansod, P. Dixit, S. Umathe and D. Mundhada. 2010. Anticonvulsant activity of berberine, an isoquinoline alkaloid in mice. Epilepsy & Behavior 18: 207–210.

Bobkiewicz-Kozłowska, T., M. Dworacka, S. Kuczyński, M. Abramczyk, R. Kolanoś, W. Wysocka, P. M. G. Lopez and H. Winiarska. 2007. Hypoglycaemic effect of quinolizidine alkaloids—lupanine and 2-thionosparteine on non-diabetic and streptozotocin-induced diabetic rats. European Journal of Pharmacology 565: 240–244.

Botha, C. and M. -L. Penrith. 2008. Poisonous plants of veterinary and human importance in southern Africa. Journal of Ethnopharmacology 119: 549–558.

Boumba, V. A., A. Mitselou and T. Vougiouklakis. 2004. Fatal poisoning from ingestion of Datura stramonium seeds. Veterinary and Human Toxicology 46: 81–82.

Caliceti, C., P. Franco, S. Spinozzi, A. Roda and A. FG Cicero. 2016. Berberine: New insights from pharmacological aspects to clinical evidences in the management of metabolic disorders. Current Medicinal Chemistry 23: 1460–1476.

Castellano, D., J. Puente, G. de Velasco, I. Chirivella, P. López-Criado, N. Mohedano, O. Fernández, I. García-Carbonero, M. B. González and E. Grande. 2014. Safety and effectiveness of vinflunine in patients with metastatic transitional cell carcinoma of the urothelial tract after failure of one platinum-based systemic therapy in clinical practice. BMC Cancer 14: 1.

Chan, K. 1995. Progress in traditional Chinese medicine. Trends Pharmacol Sci. 16: 182–187.

Chattopadhyay, D. 2006. Ethnomedicinal antivirals: Scope and opportunity. Modern Phytomedicine: Turning Medicinal Plants into Drugs, 313–339.

Cheeke, P. R. 1989. Toxicants of Plant Origin: Alkaloids. CRC Press.

Chen, L., X. Miao, Z. Peng, J. Wang and Y. Chen. 2016. The pharmacokinetics and bioavailability of three canthinone alkaloids after administration of Kumu injection to rats. J. Ethnopharmacol. 182: 235–241.

Chen, W., Y. Q. Miao, D. J. Fan, S. S. Yang, X. Lin, L. K. Meng and X. Tang. 2011. Bioavailability study of berberine and the enhancing effects of TPGS on intestinal absorption in rats. AAPS Pharm. Sci. Tech. 12: 705–711.

Chi, H. J., Y. S. Woo and Y. J. Lee. 1991. Effect of berberine and some antibiotics on the growth of microorganisms. Korean Journal of Pharmacognosy 22: 45–50.

Choi, J. S., J. -H. Kim, M. Y. Ali, B. -S. Min, G. -D. Kim and H. A. Jung. 2014. Coptis chinensis alkaloids exert anti-adipogenic activity on 3T3-L1 adipocytes by downregulating C/EBP-α and PPAR-γ. Fitoterapia 98: 199–208.

Choudhary, M. I., A. Adhikari, S. Rasheed, B. P. Marasini, N. Hussain and W. A. Kaleem. 2011. Cyclopeptide alkaloids of Ziziphus oxyphylla Edgw as novel inhibitors of α-glucosidase enzyme and protein glycation. Phytochemistry Letters 4: 404–406.

Chueh, W. -H. and J. -Y. Lin. 2012a. Berberine, an isoquinoline alkaloid, inhibits streptozotocin-induced apoptosis in mouse pancreatic islets through down-regulating Bax/Bcl-2 gene expression ratio. Food Chemistry 132: 252–260.

Chueh, W. -H. and J. -Y. Lin. 2012b. Protective effect of isoquinoline alkaloid berberine on spontaneous inflammation in the spleen, liver and kidney of non-obese diabetic mice through downregulating gene expression ratios of pro-/anti-inflammatory and Th1/Th2 cytokines. Food Chemistry 131: 1263–1271.

Cos, P., L. Maes, A. Vlietinck and L. Pieters. 2008. Plant-derived leading compounds for chemotherapy of human immunodeficiency virus (HIV) infection—An update (1998–2007). Planta Med. 74: 1323–1337.

Costa, H. F., F. C. Leite, A. F. Alves, J. M. Barbosa-Filho, C. R. B. dos Santos and M. R. Piuvezam. 2013. Managing murine food allergy with Cissampelos sympodialis Eichl (Menispermaceae) and its alkaloids. International Immunopharmacology 17: 300–308.

Cowan, M. M. 1999. Plant products as antimicrobial agents. Clin. Microbiol. Rev. 12: 564–582.

Cragg, G. M. and D. J. Newman. 2005. Plants as a source of anti-cancer agents. Journal of Ethnopharmacology 100: 72–79.

Cragg, G. M. and D. J. Newman. 2013. Natural products: A continuing source of novel drug leads. Biochimica et Biophysica Acta (BBA)-General Subjects 1830: 3670–3695.

Cuendet, M. and J. M. Pezzuto. 2008. Antitumor alkaloids in clinical use or in clinical trials. Modern Alkaloids: Structure, Isolation, Synthesis and Biology 25–52.

Cushnie, T. T., B. Cushnie and A. J. Lamb. 2014. Alkaloids: An overview of their antibacterial, antibiotic-enhancing and antivirulence activities. International Journal of Antimicrobial Agents 44: 377–386.

de Weger, V. A., J. H. Beijnen and J. H. Schellens. 2014. Cellular and clinical pharmacology of the taxanes docetaxel and paclitaxel—A review. Anti-cancer Drugs 25: 488–494.

DeBono, A., B. Capuano and P. J. Scammells. 2015. Progress toward the development of noscapine and derivatives as anticancer agents. Journal of Medicinal Chemistry 58: 5699–5727.

Diker, D., D. Markovitz, M. Rothman and U. Sendovski. 2007. Coma as a presenting sign of Datura stramonium seed tea poisoning. European Journal of Internal Medicine 18: 336–338.

Ding, Y., X. Ye, J. Zhu, X. Zhu, X. Li and B. Chen. 2014. Structural modification of berberine alkaloid and their hypoglycemic activity. Journal of Functional Foods 7: 229–237.

Fan, Y. -Q., P. -H. Li, Y. -X. Chao, H. Chen, N. Du, Q. -X. He and K. -C. Liu. 2015. Alkaloids with cardiovascular effects from the marine-derived fungus Penicillium expansum Y32. Marine Drugs 13: 6489–6504.

Feng, H., K. Yamaki, H. Takano, K. -i. Inoue, R. Yanagisawa and S. Yoshino. 2006. Suppression of Th1 and Th2 immune responses in mice by Sinomenine, an alkaloid extracted from the chinese medicinal plant Sinomenium acutum. Planta Medica 72: 1383–1388.

Feng, L. and R. J. Mumper. 2013. A critical review of lipid-based nanoparticles for taxane delivery. Cancer Letters 334: 157–175.

Freiburghaus, F., R. Kaminsky, M. H. Nkunya and R. Brun. 1996. Evaluation of African medicinal plants for their *in vitro* trypanocidal activity. J. Ethnopharmacol. 55: 1–11.

Fu, P. P., Q. Xia, G. Lin and M. W. Chou. 2004. Pyrrolizidine alkaloids—genotoxicity, metabolism enzymes, metabolic activation, and mechanisms. Drug Metabolism Reviews 36: 1–55.

Fuller, T. C. and E. M. McClintock. 1986. Poisonous Plants of California. Univ of California Press.

Gao, T. and M. C. Lagerström. 2015. The anti-inflammatory alkaloid aloperine in Chinese herbal medicine is potentially useful for management of pain and itch. Scandinavian Journal of Pain 8: 25–26.

Ghoshal, S., B. N. Prasad and V. Lakshmi. 1996. Antiamoebic activity of Piper longum fruits against Entamoeba histolytica *in vitro* and *in vivo*. J. Ethnopharmacol. 50: 167–170.

Gülçin, I. 2012. Antioxidant activity of food constituents: An overview. Archives of Toxicology 86: 345–391.

Gülçin, İ., R. Elias, A. Gepdiremen, A. Chea and F. Topal. 2010. Antioxidant activity of bisbenzylisoquinoline alkaloids from Stephania rotunda: Cepharanthine and fangchinoline. Journal of Enzyme Inhibition and Medicinal Chemistry 25: 44–53.

Hardardottir, I., E. S. Olafsdottir and J. Freysdottir. 2015. Dendritic cells matured in the presence of the lycopodium alkaloid annotine direct T cell responses toward a Th2/Treg phenotype. Phytomedicine 22: 277–282.

Heal, K. G., N. A. Sheikh, M. R. Hollingdale, W. J. W. Morrow and A. W. Taylor-Robinson. 2001. Potentiation by a novel alkaloid glycoside adjuvant of a protective cytotoxic T cell immune response specific for a preerythrocytic malaria vaccine candidate antigen. Vaccine 19: 4153–4161.

Huang, Z., Z. Han, B. Ye, Z. Daj, P. Shan, Z. Lu, K. Daj, C. Wang and W. Huanga. 2015. Berberine alleviates cardiac ischemia/reperfusion injury by inhibiting excessive autophagy in cardiomyocytes. European Journal of Pharmacology 762: 1–10.

Iovine, V., M. Mori, A. Calcaterra, S. Berardozzi and B. Botta. 2016. One hundred faces of cyclopamine. Current Pharmaceutical Design.

Iqbal, N., A. Adhikari, N. Kanwal, O. M. Abdalla, M. A. Mesaik and S. G. Musharraf. 2015. New immunomodulatory steroidal alkaloids from Sarcococa saligna. Phytochemistry Letters 14: 203–208.

Iwasa, K., D. U. Lee, S. I. Kang and W. Wiegrebe. 1998. Antimicrobial activity of 8-alkyl- and 8-phenyl-substituted berberines and their 12-bromo derivatives. J. Nat. Prod. 61: 1150–1153.

Jayakumar, T. and J. -R. Sheu. 2011. Cardiovascular pharmacological actions of rutaecarpine, a quinazolinocarboline alkaloid isolated from Evodia rutaecarpa. Journal of Experimental & Clinical Medicine 3: 63–69.

Jia, S. and C. Hu. 2010. Pharmacological effects of rutaecarpine as a cardiovascular protective agent. Molecules 15: 1873–1881.

Jiang, Q. -W., M. -W. Chen, K. -J. Cheng, P. -Z. Yu, X. Wei and Z. Shi. 2016. Therapeutic potential of steroidal alkaloids in cancer and other diseases. Medicinal Research Reviews 36: 119–143.

Jung, H. A., B. -S. Min, T. Yokozawa, J. -H. Lee, Y. S. Kim and J. S. Choi. 2009. Anti-alzheimer and antioxidant activities of Coptidis Rhizoma alkaloids. Biological and Pharmaceutical Bulletin 32: 1433–1438.

Kaur, P., M. Thiruchelvan, R. C. Lee, H. Chen, K. C. Chen, M. L. Ng and J. J. Chu. 2013. Inhibition of chikungunya virus replication by harringtonine, a novel antiviral that suppresses viral protein expression. Antimicrob. Agents Chemother. 57: 155–167.

Kawano, M., R. Takagi, A. Kaneko and S. Matsushita. 2015. Berberine is a dopamine D1- and D2-like receptor antagonist and ameliorates experimentally induced colitis by suppressing innate and adaptive immune responses. Journal of Neuroimmunology 289: 43–55.

Keiser, J. and J. Utzinger. 2007. Advances in the discovery and development of trematocidal drugs. Expert Opin. Drug Discov. 2: S9–S23.

Kempf, M., T. Beuerle, M. Bühringer, M. Denner, D. Trost, K. von der Ohe, V. B. Bhavanam and P. Schreier. 2008. Pyrrolizidine alkaloids in honey: Risk analysis by gas chromatography-mass spectrometry. Molecular Nutrition & Food Research 52: 1193–1200.

Kim, K. J., M. -S. Lee, K. Jo and J. -K. Hwang. 2011. Piperidine alkaloids from Piperretrofractum Vahl. protect against high-fat diet-induced obesity by regulating lipid metabolism and activating AMP-activated protein kinase. Biochemical and Biophysical Research Communications 411: 219–225.

Kim, S. B., B. Y. Chang, Y. H. Jo, S. H. Lee, S. -B. Han, B. Y. Hwang, S. Y. Kim and M. K. Lee. 2013. Macrophage activating activity of pyrrole alkaloids from Morus alba fruits. Journal of Ethnopharmacology 145: 393–396.

Kingston, D. G. 2009. Tubulin-interactive natural products as anticancer agents (1). Journal of Natural Products 72: 507–515.

Kishore, N., B. B. Mishra, V. Tripathi and V. K. Tiwari. 2009. Alkaloids as potential anti-tubercular agents. Fitoterapia 80: 149–163.

Kolak, U., M. Öztürk, F. Özgökçe and A. Ulubelen. 2006. Norditerpene alkaloids from Delphinium linearilobum and antioxidant activity. Phytochemistry 67: 2170–2175.

Krska, R. and C. Crews. 2008. Significance, chemistry and determination of ergot alkaloids: A review. Food Addit. Contam. Part A Chem. Anal. Control Expo. Risk Assess. 25: 722–731.

Kuete, V. 2010. Potential of cameroonian plants and derived products against microbial infections: A review. Planta Med. 76: 1479–1491.

Kuete, V. 2014. Toxicological Survey of African Medicinal Plants. Elsevier.

Kuete, V., J. D. Wansi, A. T. Mbaveng, M. M. Kana Sop, A. T. Tadjong, V. P. Beng, F. X. Etoa, J. Wandji, J. J. M. Meyer and N. Lall. 2008. Antimicrobial activity of the methanolic extract and compounds from Teclea afzelii (Rutaceae). South African Journal of Botany 74: 572–576.

Kumar, G. S. 2015. Isoquinoline alkaloids and their analogs: Nucleic acid and protein binding aspects, and therapeutic potential for drug design. Bioactive Natural Products: Chemistry and Biology.

Leal, M. C., C. Madeira, C. A. Brandao, J. Puga and R. Calado. 2012. Bioprospecting of marine invertebrates for new natural products—A chemical and zoogeographical perspective. Molecules 17: 9842–9854.

Lee, B. G., S. H. Kim, O. P. Zee, K. R. Lee, H. Y. Lee, J. W. Han and H. W. Lee. 2000. Suppression of inducible nitric oxide synthase expression in RAW 264.7 macrophages by two β-carboline alkaloids extracted from Melia azedarach. European Journal of Pharmacology 406: 301–309.

Lee, D. -U., Y. J. Kang, M. K. Park, Y. S. Lee, H. G. Seo, T. S. Kim, C. -H. Kim and K. C. Chang. 2003. Effects of 13-alkyl-substituted berberine alkaloids on the expression of COX-II, TNF-α, iNOS, and IL-12 production in LPS-stimulated macrophages. Life Sciences 73: 1401–1412.

Li, H., L. Liu, S. Zhu and Q. Liu. 2016. Case reports of aconite poisoning in mainland China from 2004 to 2015: A retrospective analysis. Journal of Forensic and Legal Medicine 42: 68–73.

Liu, C. -S., Y. -R. Zheng, Y. -F. Zhang and X. -Y. Long. 2016. Research progress on berberine with a special focus on its oral bioavailability. Fitoterapia 109: 274–282.

Liu, T. -Y., Y. Chen, Z. -Y. Wang, L. -L. Ji and Z. -T. Wang. 2010. Pyrrolizidine alkaloid isoline-induced oxidative injury in various mouse tissues. Experimental and Toxicologic Pathology 62: 251–257.

López, P. M. G., P. G. de la Mora, W. Wysocka, B. Maiztegui, M. E. Alzugaray, H. Del Zotto and M. I. Borelli. 2004. Quinolizidine alkaloids isolated from Lupinus species enhance insulin secretion. European Journal of Pharmacology 504: 139–142.

Ma, Y. -H., M. -Y. Wei, Y. -Y. Liu, F. -R. Song, Z. -Y. Liu and Z. -F. Pi. 2016. Study on intestinal transport of Veratrum alkaloids compatible with Panax ginseng across the Caco-2 cell monolayer model by UPLC-ESI-MS method. Chinese Chemical Letters 27: 215–220.

Mahady, G. B. 2005. Medicinal plants for the prevention and treatment of bacterial infections. Curr. Pharm. Des. 11: 2405–2427.

Manu, K. A. and G. Kuttan. 2009. Immunomodulatory activities of Punarnavine, an alkaloid from Boerhaavia diffusa. Immunopharmacology and Immunotoxicology 31: 377–387.

Matsuura, H. N. and A. G. Fett-Neto. 2015. Plant Alkaloids: Main Features, Toxicity and Mechanisms of Action.

McMahon, J. B., M. J. Currens, R. J. Gulakowski, R. W. Jr., Buckheit, C. Lackman-Smith, Y. F. Hallock and M. R. Boyd. 1995. Michellamine B, a novel plant alkaloid, inhibits human immunodeficiency virus-induced cell killing by at least two distinct mechanisms. Antimicrob. Agents Chemother. 39: 484–488.

Meng, F., G. Zuo, X. Hao, G. Wang, H. Xiao, J. Zhang and G. Xu. 2009. Antifungal activity of the benzo[c]phenanthridine alkaloids from Chelidonium majus Linn against resistant clinical yeast isolates. J. Ethnopharmacol. 125: 494–496.

Mishra, B. B., R. R. Kale, R. K. Singh and V. K. Tiwari. 2009. Alkaloids: Future prospective to combat leishmaniasis. Fitoterapia 80: 81–90.

Montaser, R. and H. Luesch. 2011. Marine natural products: A new wave of drugs? Future Medicinal Chemistry 3: 1475–1489.

Moudi, M., R. Go, C. Y. S. Yien and M. Nazre. 2013. Vinca alkaloids. International Journal of Preventive Medicine 4.

Mueller, M. S., N. Runyambo, I. Wagner, S. Borrmann, K. Dietz and L. Heide. 2004. Randomized controlled trial of a traditional preparation of Artemisia annua L. (Annual Wormwood) in the treatment of malaria. Trans. R Soc. Trop. Med. Hyg. 98: 318–321.

Mukhtar, E., V. M. Adhami and H. Mukhtar. 2014. Targeting microtubules by natural agents for cancer therapy. Molecular Cancer Therapeutics 13: 275–284.

Mulder, P. P., D. P. Pereboom-de Fauw, R. L. Hoogenboom, J. de Stoppelaar and M. de Nijs. 2015. Tropane and ergot alkaloids in grain-based products for infants and young children in the Netherlands in 2011–2014. Food Additives & Contaminants: Part B 8: 284–290.

Newman, D. J. and G. M. Cragg. 2007. Natural products as sources of new drugs over the last 25 years. J. Nat. Prod. 70: 461–477.

Ng, S. W., C. K. Ching, A. Y. W. Chan and T. W. L. Mak. 2013. Simultaneous detection of 22 toxic plant alkaloids (aconitum alkaloids, solanaceous tropane alkaloids, sophora alkaloids, strychnos alkaloids and colchicine) in human urine and herbal samples using liquid chromatography—Tandem mass spectrometry. Journal of Chromatography B 942-943: 63–69.

Ngounou, F. N., R. N. Manfouo, L. A. Tapondjou and D. Lontsi. 2005. Antimicrobial diterpenoid alkaloids from Erythrophleum suaveolens (Guill. & perr.) Brenan. Bulletin of the Chemical Society of Ethiopia.

Niitsu, H., Y. Fujita, S. Fujita, R. Kumagai, M. Takamiya, Y. Aoki and K. Dewa. 2013. Distribution of Aconitum alkaloids in autopsy cases of aconite poisoning. Forensic Science International 227: 111–117.

Nirmala, M. J., A. Samundeeswari and P. D. Sankar. 2011. Natural plant resources in anti-cancer therapy—A review. Research in Plant Biology 1.

Nobili, S., D. Lippi, E. Witort, M. Donnini, L. Bausi, E. Mini and S. Capaccioli. 2009. Natural compounds for cancer treatment and prevention. Pharmacological Research 59: 365–378.

Okamoto, M., M. Ono and M. Baba. 2001. Suppression of cytokine production and neural cell death by the anti-inflammatory alkaloid cepharanthine: A potential agent against HIV-1 encephalopathy. Biochemical Pharmacology 62: 747–753.

Omulokoli, E., B. Khan and S. C. Chhabra. 1997. Antiplasmodial activity of four Kenyan medicinal plants. J. Ethnopharmacol. 56: 133–137.

Panda, S., A. Kar, P. Sharma and A. Sharma. 2013. Cardioprotective potential of N, α-l-rhamnopyranosyl vincosamide, an indole alkaloid, isolated from the leaves of Moringa oleifera in isoproterenol induced cardiotoxic rats: *In vivo* and *in vitro* studies. Bioorganic & Medicinal Chemistry Letters 23: 959–962.

Park, J. B. 2014. Synthesis and characterization of norbelladine, a precursor of Amaryllidaceae alkaloid, as an anti-inflammatory/anti-COX compound. Bioorganic & Medicinal Chemistry Letters 24: 5381–5384.

Patwardhan, B. and M. Gautam. 2005. Botanical immunodrugs: scope and opportunities. Drug Discovery Today 10: 495–502.

Philippe, G. and L. Angenot. 2005. Recent developments in the field of arrow and dart poisons. Journal of Ethnopharmacology 100: 85–91.

Phillipson, J. D. and M. J. O'Neil. 1989. New leads to the treatment of protozoal infections based on natural product molecules. Acta Pharmaceutica. Nordica 1: 131–144.

Pirillo, A. and A. L. Catapano. 2015. Berberine, a plant alkaloid with lipid-and glucose-lowering properties: From *in vitro* evidence to clinical studies. Atherosclerosis 243: 449–461.

Qiu, S., H. Sun, A. H. Zhang, H. Y. Xu, G. L. Yan, Y. Han and X. J. Wang. 2014. Natural alkaloids: Basic aspects, biological roles, and future perspectives. Chin. J. Nat. Med. 12: 401–406.

Račková, L., M. Májeková, D. Košť'álová and M. Štefek. 2004. Antiradical and antioxidant activities of alkaloids isolated from Mahonia aquifolium. Structural aspects. Bioorganic & Medicinal Chemistry 12: 4709–4715.

Rang, H. P., M. M. Dale, J. M. Ritter and R. J. Flower. 2007. Rang and Dale's Pharmacology. Elsevier, Churchill Livingstone.

Rao, Y., H. Liu, L. Gao, H. Yu, J. -H. Tan, T. -M. Ou, S. -L. Huang, L. -Q. Gu, J. -M. Ye and Z. -S. Huang. 2015. Discovery of natural alkaloid bouchardatine as a novel inhibitor of adipogenesis/ lipogenesis in 3T3-L1 adipocytes. Bioorganic & Medicinal Chemistry 23: 4719–4727.

Rizk, A. -F. 1990. Naturally Occurring Pyrrolizidine Alkaloids. CRC press.

Sanlidag, B., O. Derinöz and N. Yildiz. 2014. A case of pediatric age anticholinergic intoxication due to accidental Datura stramonium ingestion admitting with visual hallucination. The Turkish Journal of Pediatrics 56: 313.

Savoia, D. 2012. Plant-derived antimicrobial compounds: Alternatives to antibiotics. Future Microbiol. 7: 979–990.

Schläger, S. and B. Dräger. 2016. Exploiting plant alkaloids. Current Opinion in Biotechnology 37: 155–164.

Sharma, B., R. Salunke, C. Balomajumder, S. Daniel and P. Roy. 2010. Anti-diabetic potential of alkaloid rich fraction from Capparis decidua on diabetic mice. Journal of Ethnopharmacology 127: 457–462.

Shoba, G., D. Joy, T. Joseph, M. Majeed, R. Rajendran and P. S. Srinivas. 1998. Influence of piperine on the pharmacokinetics of curcumin in animals and human volunteers. Planta Med. 64: 353–356.

Shukla, A. and B. Srinivasan. 2012. 16, 17-Dihydro-17b-hydroxy isomitraphylline alkaloid as an inhibitor of DPP-IV, and its effect on incretin hormone and β-cell proliferation in diabetic rat. European Journal of Pharmaceutical Sciences 47: 512–519.

Steenkamp, P., F. Van Heerden and B. -E. Van Wyk. 2002. Accidental fatal poisoning by Nicotiana glauca: Identification of anabasine by high performance liquid chromatography/photodiode array/mass spectrometry. Forensic Science International 127: 208–217.

Subramanian, S. P. and G. S. Prasath. 2014. Antidiabetic and antidyslipidemic nature of trigonelline, a major alkaloid of fenugreek seeds studied in high-fat-fed and low-dose streptozotocin-induced experimental diabetic rats. Biomedicine & Preventive Nutrition 4: 475–480.

Sun, Y., K. Xun, Y. Wang and X. Chen. 2009. A systematic review of the anticancer properties of berberine, a natural product from Chinese herbs. Anti-cancer Drugs 20: 757–769.

Tiong, S. H., C. Y. Looi, A. Arya, W. F. Wong, H. Hazni, M. R. Mustafa and K. Awang. 2015. Vindogentianine, a hypoglycemic alkaloid from Catharanthus roseus (L.) G. Don (Apocynaceae). Fitoterapia 102: 182–188.

Van Wyk, B. -E. and M. Wink. 2004. Medicinal Plants of the World: An Illustrated Scientific Guide to Important Medicinal Plants and their Uses. Timber Press.

Van Wyk, B., B. Van Oudtshoorn and N. Gericke. 1997. Medicinal Plants of South Africa.

Van Wyk, B., F. Van Heerden and B. Van Oudtshoorn. 2002. Poisonous Plants of South Africa.

Vasanthi, R. H., N. ShriShriMal and D. K. Das. 2012. Phytochemicals from plants to combat cardiovascular disease. Current Medicinal Chemistry 19: 2242–2251.

Veras, L. M., M. A. Guimaraes, Y. D. Campelo, M. M. Vieira, C. Nascimento, D. F. Lima, L. Vasconcelos, E. Nakano, S. S. Kuckelhaus, M. C. Batista, J. R. Leite and J. Moraes. 2012. Activity of epiisopiloturine against Schistosoma mansoni. Curr. Med. Chem. 19: 2051–2058.

Waisundara, V., M. Watawana and N. Jayawardena. 2015. Costus speciosus and Coccinia grandis: Traditional medicinal remedies for diabetes. South African Journal of Botany 98: 1–5.

Wang, A. Z., R. Langer and O. C. Farokhzad. 2012. Nanoparticle delivery of cancer drugs. Annual Review of Medicine 63: 185–198.

Wang, K., L. Bao, K. Ma, N. Liu, Y. Huang, J. Ren, W. Wang and H. Liu. 2015. Eight new alkaloids with PTP1B and α-glucosidase inhibitory activities from the medicinal mushroom Hericium erinaceus. Tetrahedron 71: 9557–9563.

Wang, Y., X. Yi, K. Ghanam, S. Zhang, T. Zhao and X. Zhu. 2014. Berberine decreases cholesterol levels in rats through multiple mechanisms, including inhibition of cholesterol absorption. Metabolism 63: 1167–1177.

Watson, E., J. C. Murphy, H. N. ElSohly, M. A. ElSohly and C. Turner. 1983. Effects of the administration of coca alkaloids on the primary immune responses of mice: interaction with Δ 9-tetrahydrocannabinol and ethanol. Toxicology and Applied Pharmacology 71: 1–13.

White, E. L., W. R. Chao, L. J. Ross, D. W. Borhani, P. D. Hobbs, V. Upender and M. I. Dawson. 1999. Michellamine alkaloids inhibit protein kinase C. Arch. Biochem. Biophys. 365: 25–30.

WHO. 2012. Global Report for Research on Infectious Diseases of Poverty.

WHO. 2016. Cardiovascular Diseases (CVD).

WHO. 2017. Cancer Fact Sheet.

Wilkinson, R. E., W. S. Hardcastle and C. S. McCormick. 1988. Psychotomimetic ergot alkaloid contents of seed from Calonyction muricatum, Jacquemontia tamnifolia, Quamoclit lobata and Q. sloteri. Botanical Gazette 149: 107–109.

Wink, M. 2012. Medicinal plants: A source of anti-parasitic secondary metabolites. Molecules 17: 12771–12791.

Wu, C., R. Yan, R. Zhang, F. Bai, Y. Yang, Z. Wu and A. Wu. 2013. Comparative pharmacokinetics and bioavailability of four alkaloids in different formulations from Corydalis decumbens. Journal of Ethnopharmacology 149: 55–61.

Xia, L. -M. and M. -H. Luo. 2016. Study progress of berberine for treating cardiovascular disease. Chronic Diseases and Translational Medicine.

Yang, T. -C., H. -F. Chao, L. -S. Shi, T. -C. Chang, H. -C. Lin and W. -L. Chang. 2014. Alkaloids from Coptis chinensis root promote glucose uptake in C2C12 myotubes. Fitoterapia 93: 239–244.

Yi, Z. B., Y. Yan, Y. Z. Liang and Z. Bao. 2007. Evaluation of the antimicrobial mode of berberine by LC/ESI-MS combined with principal component analysis. J. Pharm. Biomed. Anal. 44: 301–304.

Yu, L., Z. Wang, M. Huang, Y. Li, K. Zeng, J. Lei, H. Hu, B. Chen, J. Lu and W. Xie. 2015. Evodia alkaloids suppress gluconeogenesis and lipogenesis by activating the constitutive androstane receptor. Biochimica et Biophysica Acta (BBA)—Gene Regulatory Mechanisms.

Yue, Q. -X., X. Liu and D. -A. Guo. 2010. Microtubule-binding natural products for cancer therapy. Planta Medica 76: 1037.

Zhang, M. and L. Chen. 2012. Berberine in type 2 diabetes therapy: A new perspective for an old antidiarrheal drug? Acta Pharmaceutica Sinica B 2: 379–386.

Zhang, Q. -Y., M. Zhang and Y. Cao. 2012. Exposure to morphine affects the expression of endocannabinoid receptors and immune functions. Journal of Neuroimmunology 247: 52–58.

Zhang, Y., W. Xu, H. Li, X. Zhang, Y. Xia, K. Chu and L. Chen. 2013. Therapeutic effects of total alkaloids of Tripterygium wilfordii Hook f. on collagen-induced arthritis in rats. Journal of Ethnopharmacology 145: 699–705.

Zhao, G. -l., L. -M. Yu, W. -l. Gao, W. -X. Duan, B. Jiang, X. -D. Liu, B. Zhang, Z. -H. Liu, M. -E. Zhai and Z. -X. Jin. 2016. Berberine protects rat heart from ischemia/reperfusion injury via activating JAK2/STAT3 signaling and attenuating endoplasmic reticulum stress. Acta Pharmacologica. Sinica 37: 354–367.

Zhao, Q., Y. Zhao and K. Wang. 2006. Antinociceptive and free radical scavenging activities of alkaloids isolated from Lindera angustifolia Chen. Journal of Ethnopharmacology 106: 408–413.

Zheng, M., H. Zhou, H. Wan, Y. L. Chen and Y. He. 2015. Effects of herbal drugs in Mahuang decoction and their main components on intestinal transport characteristics of Ephedra alkaloids evaluated by a Caco-2 cell monolayer model. J. Ethnopharmacol. 164: 22–29.

Zhu, Y., W. Peng, J. Zhang, M. Wang, C. K. Firempong, C. Feng, H. Liu, X. Xu and J. Yu. 2014. Enhanced oral bioavailability of capsaicin in mixed polymeric micelles: Preparation, *in vitro* and *in vivo* evaluation. Journal of Functional Foods 8: 358–366.

CHAPTER 4

Analytic Methods for the Bioactive Compounds in Waste

Mark Tarleton

Thin-layer Chromatography

All methods of chromatography utilize the same general principal. A mixture, containing two or more compounds, is treated in same manner, resulting in each compound being separated and purified. Compounds are separated based on their degree of interaction between a mobile (carrier) phase and a stationary phase. When the correct sets of parameters are selected, each individual compound has a unique level of interaction between the two phases, resulting in separation from the remaining reaction mixture.

The simplest form of chromatography is the thin layer chromatography (TLC). This analytical technique is relatively easy, quick, cheap, requires a very small amount of sample and can give very useful information when performed correctly. It is primarily a qualitative analysis. A TLC plate consists of a glass or aluminium-backed strip that is coated with an adsorbent material, most commonly silica or alumina, which acts as the stationary phase in the separation. Depending on application, the stationary phase material (polarity, particle size) and plate size can be altered to achieve the desired outcome. The compound mixture is dissolved in a solvent, preferably volatile and spotted via micro capillary onto a baseline of the TLC plate approximately 10 mm from the base. The solution of the mixture can be concentrated or diluted to apply a useful amount of material onto the plate; too much will cause the material to streak and poorly separate, while too little material may not be visible after separation. The carrier solvent is then evaporated, leaving the dried mixture on the 'baseline' of the TLC plate (Fig. 1A) (Komsta 2013, Lewis and Evans 2011, Naumoska and Vovk 2015).

Chemistry, School of Environmental & Life Sciences, Faculty of Science and Information Technology, The University of Newcastle, University Drive, Callaghan, 2308, NSW, Australia. E-mail: mark.tarleton@newcastle.edu.au

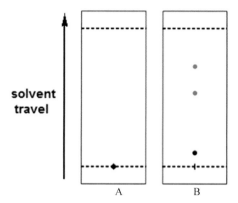

Figure 1. An undeveloped TLC plate containing one spot of a mixture of compounds on the baseline (A) and the same plate after being developed using a mobile phase to carry and separate the individual compounds of the mixture (B).

The spotted TLC plate is placed in a chamber containing a solvent or a mixture of solvents (~ 5 mm depth) that will be used as the mobile phase of separation. Again, depending on application, the solvent composition can be altered to achieve the desired outcome. Once the TLC plate is placed inside the chamber, it is sealed and the solvent (mobile phase) moves up the TLC plate (stationary phase) via capillary action. It is removed from the chamber once the solvent front is approximately 10 mm from the top of the plate. The mixture of compounds is separated based on the level of interaction each compound has with the mobile and stationary phases. Compounds that have a greater interaction with the stationary phase, will stay low on the plate, whereas compounds that have a strong interaction with the mobile phase, will travel much further up the plate (Fig. 1B) (Komsta 2013, Lewis and Evans 2011, Oros and Cserhati 2013).

More complex mixtures of materials may be subject to 2D TLC in order to achieve complete separation. In comparison to traditional 1D TLC, where the mobile phase and mixture travel in one direction, 2D TLC utilizes two different mobile phases perpendicular to each other (Fig. 2). This method is useful when partial separation is produced from the first TLC. After drying off the initial mobile phase mixture, the TLC plate is turned 90 degrees and placed into another mobile phase and a second separation is performed (Milz and Spangenberg 2013, Sherma 2000, Sherma 2012).

Using high-performance TLC (HPTLC), a quantitative analysis of a reaction mixture can also be obtained, though the advantages of being quick and easy are lost. This technique is carried out using a thinner stationary phase with a smaller particle size (5 μm) compared to traditional TLC (20 μm). Uniform quantities of each mixture must be applied to the TLC plate for the quantitative analysis to be performed. This is achieved by using an instrumental automated sampler to apply the mixture to the TLC plate. A densitometric scanner must then be used to quantitatively measure the amount of material present in each separated spot after the TLC has been run. This instrument measures a sample using a slit-scanning densitometer. However, an alternative quantification method based on video or camera based image analysis is also a

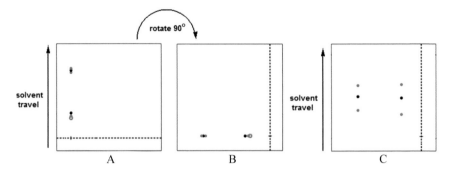

Figure 2. A TLC mixture is incompletely separated in one direction (A), then turned 90° (B), and run again in a different direction (C).

common practice (Bochenska and Pyka 2013, Dhumal 2015, Milz and Spangenberg 2013, Shah et al. 2014, Staszek et al. 2013, Vlajkovic et al. 2013).

A slight variation to HPTLC uses radiolabelled derivatives of analytes. This technique is known as thin layer radiochromatography (TLRC). As in HPTLC, sample application is completed by using an instrumental automated sampler in order to perform a quantitative analysis. The most common detection methods for TLRC are either autoradiography or liquid scintillation counting (LSC). When using autoradiography, the TLC plate that has separated the radiolabelled compounds, is placed directly onto a photographic film with the more radiolabelled compound present resulting in a more developed part or greater optical density on the film. This can then be quantified by completing a density measurement of the spots on the developed film by using a densitometer that has been calibrated with the films exposed to a set of standard solutions with known radioactivity (Milz and Spangenberg 2013, Sherma 2000, Sherma 2012).

When using LSC detection methodology, the spots on the developed TLC plate emitting radiation (alpha, beta, or gamma), are scraped off and mixed with a solvent, an emulsifying agent and a scintillator. The radioactive emissions are converted by the scintillator into light energy that can be quantified using an instrument containing a photomultiplier tube (PMT) and converted into a plotted spectrum between 0–2000 keV. The produced spectrum can provide information about the energy and the amount of radioactive material present (Sherma 2000, Sherma 2012).

Insight into the identity of separated materials can also be obtained by using the TLC analysis. A retention factor (R_f) can be assigned to any spot or zone on a developed plate. Simply, the R_f value is a ratio of how far the compound migrates up the TLC plate as compared to how far the mobile phase migrates, as shown in Equation 1 below (Dhumal 2015, Lewis and Evans 2011):

$$R_f = \frac{\textit{distance travelled by compound} \text{ (cm)}}{\textit{distance travelled by mobile phase} \text{ (cm)}}$$

Equation 1: R_f equation used to calculate a R_f value for individual spots on a TLC by dividing the distance travelled from the baseline by the compound by the distance travelled by the mobile phase solvent front.

Compounds that travel as far as the solvents in the mobile phase are assigned an R_f of 1.0, whereas compounds that stay on the baseline are assigned an R_f of 0. When performed under the same conditions (stationary phase/mobile phase), the R_f value for a particular compound will remain the same, regardless of the length of the TLC plate. If a pure sample of a known compound can be obtained, it can be run parallel to an unknown mixture, and if a spot within the mixture has the same R_f of the known compound, it is a good indication that the known compound is present within the mixture (Fig. 3) (Komsta 2013, Lewis and Evans 2011, Naumoska and Vovk 2015).

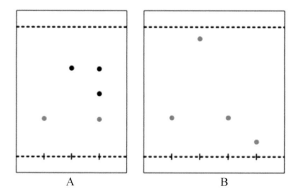

Figure 3. TLC plates can show if a reaction has formed a new compound by comparing a separated reaction mixture (A right) with the two starting materials (A left and centre). TLC A indicates that some of the both starting materials are still present and an unknown compound spot with a new R_f. Alternatively, TLC analysis can indicate the identity a separated compound (B left) when run against pure samples of known compounds (B centres and right) by comparing R_f values and colour/staining of each spot.

Selecting a stationary phase

The degree of separation in TLC analysis is greatly affected by selection of the stationary phase. Varying the particle size, layer thickness and chemical composition of the stationary phase will alter the speed and degree of separation to be performed. There is a wide range of stationary phases available for use in TLC analysis, including ion exchange, polar chemically bonded, non-polar chemically bonded, adsorption, and partition phases used to separate inorganic and organic mixtures (Komsta 2013, Sherma 2000, Sherma 2012).

Most often, TLC analysis is completed by using normal phase conditions, which consist of a polar stationary phase and a non-polar mobile phase. There are numerous commercially available TLC plates for use under normal phase conditions. Commercial TLC plates most commonly will consist of a pre-coated silica gel (60 nm pore diameter) containing a fluorescent indicator on an aluminium backed sheet. The indicator is added to allow many compounds to be observed on the plates at 254 nm without affecting separation. Apart from silica, aluminia and cellulose are also used in food-based-analysis depending on application (Komsta 2013, Naumoska and Vovk 2015, Oros and Cserhati 2013, Sherma 2012).

Alternatively, reverse phase TLC consists of a non-polar stationary phase and a polar mobile phase. Reverse phase TLC incorporates the use of C-18, C-2, C-8, or diphenyl stationary phases with an aluminium-backing plate. The difference in chemical composition of the stationary phase surface determines polarity and type of phase utilised (see Fig. 4 below) (Komsta 2013, Oros and Cserhati 2013, Sherma 2012).

Selecting a mobile phase

A wide range of solvents and solvent mixtures are used in TLC analysis, depending on the compound mixture being assessed. When carrying out normal-phase chromatography, the mobile phase is the non-polar component, with the stationary phase being the polar component. Reverse phase chromatography is the opposite; mobile phase polar, stationary phase is non-polar. In both cases, once the stationary phase is selected, it is fixed and not able to be adjusted. However, the mobile phase can be adjusted infinitely. The mobile phase can consist of either a single solvent or a combination of miscible solvents, and can be adjusted to produce the desired polarity. The solvents used are selected to keep the R_f values of all compounds between approximately 0.2–0.8. Solvents are generally chosen based on trial and error but can be influenced by previous experience or past applications found within corresponding literature. For example, simply changing ratio between two solvents, such as hexane and ethyl acetate, can range from quite a non-polar mobile phase (100 per cent hexane) right through to significantly more polar mobile phase (100 per cent ethyl acetate). If a more polar mobile phase is needed, mixtures of methanol and chlorinated solvents are commonly used. However, care must be taken not to exceed ~ 10 per cent methanol v/v as the silica present in the stationary phase may start to dissolve under these conditions (Komsta 2013, Naumoska and Vovk 2015, Sherma 2000).

Depending on application, some compounds being separated, may interact with the stationary phase and become either protonated (amines) or deprotonated (carboxylic acids). In normal-phase chromatography, this results in a 'streaking' of the compound up the plate instead of remaining circular, as they were when applied to the plate (see Fig. 5). This can result in incomplete separation of materials. However,

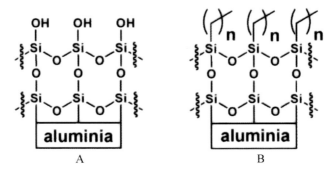

Figure 4. Structure of the surface of a typical normal phase silica (A) and a typical reverse phase silica (B). For example, in C-18 silica n = 18.

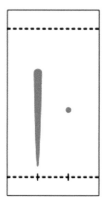

Figure 5. Comparison of a compound that has 'streaked' during TLC analysis resulting in an elongated spot (left) against the size and shape of a typical example of a desired TLC spot (right).

this may be overcome by adding a small amount of a non-nucleophillic base, such as triethylamine (TEA), to the mobile phase to act as a proton sink (Oros and Cserhati 2013, Sherma 2000, Sherma 2012, Staszek et al. 2013).

Viewing compounds

After a successful TLC separation has been achieved, the individual compounds need to be able to be seen. If TLC is being used as a qualitative analytical tool, then this involves performing a simple procedure to make the compounds visible to the human eye. Often with plant-based natural product chemistry, there will be some compounds that already reflect light in the visible spectrum, such as chlorophylls and carotenoids. Other compounds can be viewed if they react with UV radiation at 254 nm (short wave) or 366 nm (long wave) by simply holding the commercial TLC plates with fluorescent indicator under a UV lamp with the corresponding bulb illuminated. Generally, this will result in the compounds forming a dark red spot on a light green fluorescent background, as shown in Fig. 6 (Sherma 2000, Sherma 2012, Vlajkovic et al. 2013).

Non-UV active compounds need to be treated further to produce coloured spots on separated TLC plates. There are an extensive number of methods throughout scientific literature that have been used for this purpose, with this number always increasing. The determining factor when selecting a staining or viewing method will primarily depend on what functional groups are present within the structures being investigated. For example, a separated TLC plate containing various amino acids can be viewed by dipping the plate in a ninhydrin solution in a volatile organic solvent, such as acetone and heating. If a free amino group is present, this will result in a purple spot being generated (Sherma 2000, Sherma 2012, Vlajkovic et al. 2013).

An aqueous solution of potassium permanganate ($KMnO_4$), potassium carbonate (K_2CO_3) and sodium hydroxide (NaOH) is useful when investigating unsaturated organic compounds and alcohols, producing yellow to orange spots on TLC plates when stained (Sherma 2012).

When investigating unknown chemical mixtures, such as those present in various natural products or food products, it may be necessary to preform numerous separate staining or viewing techniques in order to make an accurate conclusion on how many compounds are present in these extractions. Alternatively, selection of a less selective staining technique may be used, depending on application, which reacts with the majority of functional groups being analyzed. An example of this is an acidic vanillin in ethanol solution. This is a good overall staining reagent and gives a range of coloured spots, depending on the functional groups present. A phosphomolybic acid ($H_3PO_3Mo_4$) and cerium sulfate ($Ce(SO_4)_2$) dip followed by heating produces dark blue spots for a wide range of functional groups on a yellow background, making visualisation of each compound easy. Examples of colours obtained using common staining techniques are shown in Fig. 7 (Sherma 2000, Sherma 2012).

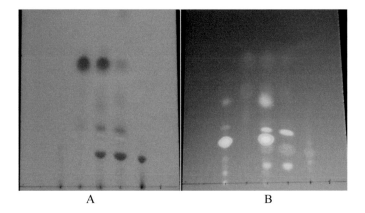

A   B

Figure 6. Developed TLC plate viewed under short wave UV light (A) and long wave UV light (B) on plates containing a fluorescent indicator within the silica backing plate.

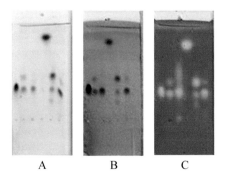

A   B   C

Figure 7. TLC plates containing various compounds with numerous functional groups stained with phosphomolybdic acid/cerium sulfate (A), vanillin (B), and permanganate (C) solutions.

For quantitative analysis of TLC, present in HPTLC and TRLC, compounds are viewed using similar staining/dipping techniques as required before being quantified using instrumentation discussed previously within this chapter.

Documentation of results

Results from TLC analysis are often recorded by taking a photograph of the separated and stained TLC plate. The R_f values of each identified spot is also needed for future comparisons with the documented TLC plate. The TLC conditions, such as the mobile phase and stationary phase composition, are also needed if reproduction of the TLC separation is to be completed. Using photography to document completed TLC plates is currently being phased out by other techniques, such as video camera recording and computer scanning in order to decrease processing time (Sherma 2000, Sherma 2012).

Upscale of TLC

Determination of experimental conditions (mobile phase, stationary phase) that successfully separate a reaction mixture can be rapidly determined by using TLC. However, if purification of large volumes of material is needed, the process needs to be unscaled. This can be achieved by using preparative layer chromatography (PLC). Preparative layer chromatography uses the same principles as traditional TLC, but uses larger silica plates with thicker layers of stationary phase. The larger and thicker plates allow a much larger amount of material (up to 1 g dried weight) to be loaded onto the plate for separation. Instead of placing a single spot on the plate baseline via micro capillary, the entire baseline of the plate is covered with the mixture. Using the same mobile phase and conditions discovered using TLC, the PLC plate separates the mixture into individual components. The silica bands containing the desired compound are isolated, the compound extracted and separated from the stationary phase, yielding a purified compound (Sherma 2000, Sherma 2012).

Separation and purification of large volumes of materials can also be achieved by using column liquid chromatography. Again, by using conditions highlighted during preliminary TLC analysis, large volumes can be purified. The amount of material that can be loaded onto these conventional columns is determined by the stationary phase used (particle size, type) and the diameter of the column of stationary phase once packed (Sherma 2000, Sherma 2012).

Gas chromatography

Gas chromatography (GC) is an example of a destructive instrumental analytical technique. When using a destructive analytical method, the analyte mixture being assessed, is destroyed and not returned to the user. However, this analytical technique requires a very small amount of material in order to perform a successful analysis of an unknown mixture (Cortes et al. 2009, Lewis and Evans 2011, Paio 2014, Pierce 2012).

As with all other methods of chromatography, separation of a mixture into its individual components is achieved by exposing the reaction mixture to a specific stationary phase and a specific mobile phase. The degree of interaction or noninteraction with each phase corresponds to what functional groups are present within the scaffold of each analyte and can be tailored by fine tuning the instrumental method used, so that each compound has a unique interaction with each phase, leading to a complete separation of each individual compound within the original mixture (Cortes et al. 2009, Lewis and Evans 2001, Piao 2014).

Initially, the compound mixture is dissolved in a volatile organic solvent. The sample to be separated and analyzed is injected (~ 0.1–2 μL) into a thin column of mobile phase—in this case, an inert gas. The sample is then passed through a heated column of varying lengths and diameter, depending on application. At this point, any carrier solvents and compounds are transferred into the gas phase. While passing though the column, coated with a stationary phase material, the individual compounds interact in a unique manner, causing separation of the mixture. After passing though the column, compounds are eluted at different times, depending on stationary phase interaction, on to a suitable detector. The instrument program then converts information gathered from the detector and displays it on the screen as a series of peaks (individual compounds) against an axis that determines the retention time of each compound (Agilent technologies 2012, Harris 2003, Pierce 2012, Shimadzu 2014).

As with all analysis methods, the results that are obtained are greatly affected by each variable in the process. Sample preparation, the process used to inject the sample, type of column, mobile phase, stationary phase, and detectors used will all influence the outcome of any GC analysis. Each of these factors will be discussed.

Sample preparation

Sample preparation for GC analysis is fairly simple. For a solid or liquid sample to be analysed using GC, it needs to be soluble and form a dilute solution in a volatile solvent. It also needs to be able to turn into vapour at the temperatures possible within the instrument. If this is not the case, some degree of chemical manipulation or pre-treatment may be necessary to form derivatives of analytes that are volatile under the possible temperature range available. As the solvent is injected into the system, the heated column instantly converts it to gas, and it is passed through the column. This leaves the mixture of compounds, which have also been heated and converted to the gas phase. These vapours are carried though the column and stationary phase by the mobile phase/carrier gas. Gas samples may be injected directly by using a gas tight syringe (Agilent technologies 2012, Cortes et al. 2009, Harris 2003).

Sample injection

Samples are introduced into a CG instrument via glass syringe injection through a rubber septum. Septum's degrade with use and must be periodically replaced. Depending on the injection method, they can be used at least 20 times (Hamilton 2012).

GC glass syringes generally inject between 0.1–2 μL of liquid; however, larger syringes are available for application outside this range. Gas syringes are typically larger as the concentrations of analyte are generally lower and can inject as much as 2 litres of gas analyte. Needle size and shape can also be tailored to the size of the sample being injected and what materials the needle needs to penetrate when preparing samples. Generally, the length of the needle is around 50 mm, with the internal and external diameters dependent on injecting volume (Agilent technologies 2012, Hamilton 2012, Harris 2003, Shimadzu 2014).

Samples can either be injected manually or by using an autosampler. When injecting manually, all needle rinsing, filling and the injection volume needs to be reproducible in order to achieve accurate results. The reproducibility of time between injections and starting the analysis is also highly importance, and is best left to a single operator for each analysis to reduce this error as much as possible (Agilent technologies 2012, Hamilton 2012, Shimadzu 2014, ThermoFischer Scientific 2011).

Automated sample injectors are able to run numerous samples through a GC sequentially. The number of samples that can be logged depends on the size of the sample holding carosel. Automated sample injectors rinse the needle between each sample injection and when calibrated correctly, can greatly minimize errors associated with injection volumes and starting the analysis after the sample has been injected. When using automated sample injectors, samples are loaded in small vials and enclosed with a sealing lid. The lid has a septum top that seals the sample vial to avoid evaporation. These lids can be penetrated by the injector needle and are commercially available, commonly comprising a soft silica material with a thin polytetrafluoroethylene sealing layer (Agilent technologies 2012, Hamilton 2012, Shimadzu 2014, ThermoFischer Scientific 2011).

Depending on the sample concentration, injected samples can be introduced to the GC column in one of the three main ways. When the analyte has a concentration of greater than 0.1 per cent of the sample solution and does not decompose at temperatures slightly higher than their boiling point, a split injection is generally used so as not to overload the column stationary phase (Fig. 8A). When using a split injection, the analyte is injected into a heated evaporation zone with a glass liner. High temperatures

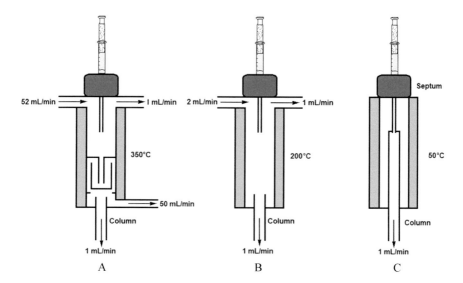

Figure 8. Diagram comparing split (A), splitless (B), and on column (C) injection methods used in GC.

(300–400°C) are used to ensure rapid evaporation but must not cause decomposition of the analyte. The mobile phase then pushes the gaseous sample through a mixing chamber first to ensure homogenous mixing with the carrier gas, then through a split vent. At the split vent, a small amount of the gas (containing 0.2–2 per cent of the original sample) is passed into the column, with the remainder of the gas discarded through a small valve to a waste vent. The amount of sample that is discarded and what is known as the split ratio, can range from 40:1 to 600:1, depending on the instrument and application (Agilent technologies 2012, Harris 2003, Shimadzu 2014).

Samples containing a lower concentration of analyte (less than 0.05 per cent) are introduced into the instrument column via the splitless injection method (Fig. 8B). Using this method, the sample is injected to a heated glass liner at lower temperatures of around 200–220°C when compared to those used during the split injection. Splitless injection does not use a mixing chamber before the start of the column. Instead, this method holds the analyte solution for a longer time in the glass-lined chamber at lower temperatures in order to avoid decomposition. Approximately 80 per cent of the original sample reaches the column when using splitless injection. To avoid broad peaks due to the slow introduction of samples, solvent trapping is generally used. The temperature of the start of the column is set well below the boiling point of the solvent selected, causing it to condense at the beginning of the column. As gaseous analyte reaches the start of the column, it becomes trapped in a narrow band. Once this process has occurred (~ 1–2 minutes), the separation commences by raising the column temperature, vaporising the solvent and band of analyte that can then pass through the column and be separated into individual components (Agilent technologies 2012, Harris 2003, Shimadzu 2014).

If analytes are known to decompose above their boiling point, the on-column injection method is used (Fig. 8C). This involves the analyte solution being directly injected into the column containing the stationary phase, without going through a heated injector port. When compounds are known to be susceptible to decomposition, the column temperature is initially set to allow solutes to condense in a narrow band during introduction into the column. The separation commences by raising the temperature to levels as low as possible to produce vaporisation. Depending on the column diameter, narrow syringes made from silica are often used to inject into the narrow columns (~ 0.2–0.3 mm) in order to increase resolution (Agilent technologies 2012, Harris 2003, Shimadzu 2014).

Selecting a column and reaction conditions

Gas chromatography is a highly practical analytical method. There are numerous variables that can be altered in order to give the optimum separation conditions for each individual application. When performing the initial analysis on small amounts of the material, open tubular columns are used; however, when larger amounts of material need to be separated, packed columns are utilized.

Open tubular columns are most commonly used in GC analysis. The columns can vary in length (10–105 m), but are long and narrow. These columns are sealed with a high temperature-resistant polyimide coating which supports both the brittle fused silica columns within and protects the column and separation from water vapor present in the surrounding atmosphere. The stationary phase, generally a non-volatile liquid, is covalently bound to the fused silica surface. There are numerous materials available for use as a stationary phase in GC. However, specific stationary phases will be discussed further below. As well as altering the stationary phase, judicious selection of the correct GC open tubular column, depending on application, can be achieved by altering the column length, internal column diameter and thickness of stationary phase present within the column (Agilent technologies 2012, Harris 2003, Shimadzu 2014).

The three main open tubular columns are the wall-coated open tubular column, the support-coated open tubular column, and the porous-layer open tubular column. Each type has its own advantages and disadvantages with the structural differences illustrated below in Fig. 9.

Wall-coated open tubular columns have a thin film of liquid stationary phase (0.1–5 μm) coating the inner wall of the column. Depending on application, the thickness of the stationary phase is altered. Reducing the thickness of the film results in an increase in the resolution and reduction in retention time as the mass transfer resistance is minimized. However, the sample capacity is also significantly reduced. These columns are useful when small amounts of high-boiling-point compounds need to be separated that get retained for too long on thicker columns. Highly sensitive detectors need to be coupled with these columns in order to derive benefit from their true potential (Agilent technologies 2012, Harris 2003, Shimadzu 2014).

Support-coated open tubular columns consist of solid particles coated with the stationary phase. These particles are attached to the inner wall which results in an increase of the stationary-phase surface area compared to the wall-coated variety and resulting in an increase in sample capacity. The reduction in retention time and resolution with these columns can be overcome or reduced by increasing the overall column length. This column is a good overall column with performance levels in between the two other open tubular columns (Agilent technologies 2012, Harris 2003, Shimadzu 2014).

Porous-coated tubular columns also have solid particles attached to the inner wall of the column. However, with this column, the particles are not coated but instead double as the stationary phase. This architecture increases the stationary phase surface area over the support-coated columns, increasing the sample capacity. However, a reduction in resolution is a consequence (Agilent technologies 2012, Shimadzu 2014).

Packed columns are the final and far less common type of GC column. They contain fine particles of solid support packed uniformly. The solid support may be

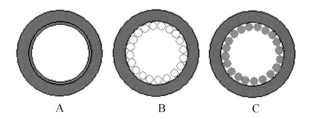

A   B   C

Figure 9. Cross sectional comparisons between wall-coated (A), support-coated (B) and porous-layer open tubular columns. The blue component represents the column wall, red represents the liquid stationary phase and orange represents the solid support particles.

coated in a non-volatile liquid stationary phase, similar to the support-coated open tubular columns, or the solid support may be the stationary phase, similar to the porous-coated tubular columns. Generally the columns are built from stainless steel or occasionally glass, and have a much thicker internal diameter (2–6 mm) and shorter length (1–5 m) than open tubular alternatives. The particle size of the solid support is measured in either mesh size, the size of mesh screens the particles will pass through, or micrometers. For example a column containing 60/80 mesh solid support results in the particles passing through a 60-mesh screen, but not an 80-mesh screen. Use of smaller sized solid support particles results in an increase in the surface area which results in a better resolution of peaks and a higher sample capacity. Unfortunately, it also results in less space between the particles which require higher operating pressures to pass the carrier gas through the column and longer analysis time. Regardless of the solid support used, packed columns are significantly less sensitive and have lower resolution of peaks due to peak broadening (Fig. 10), than the open tubular alternative but can be useful in some applications, such as separating vapors that are not retained in open tubular columns (Agilent technologies 2012, Harris 2003, Shimadzu 2014).

Selecting a stationary phase

Liquid stationary phases are selected, based on the polarity of the analyte mixture to be separated. When separating a mixture of non-polar compounds, such as saturated or aromatic hydrocarbons, a non-polar stationary phase is selected, containing high percentages of non-polar functional groups on the silica backbone (Fig. 11A). When separating compounds with intermediate polarity, such as ketones or alcohols, a stationary phase is selected with intermediate polarity. Intermediate stationary phases will vary in ratios of polar: non-polar functional groups, which can be tailored to suit the needs of the separation (Fig. 11B). Separating a mixture of polar compounds, such as polyphenols or polyhydroxyalcohols, a stationary phase with high percentages of polar functional groups is selected (Fig. 11C) (Agilent technologies 2012, Harris 2003, Shimadzu 2014).

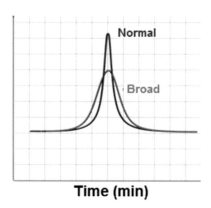

Figure 10. Comparison between a normal chromatography peak and a undesired broadened chromatography peak.

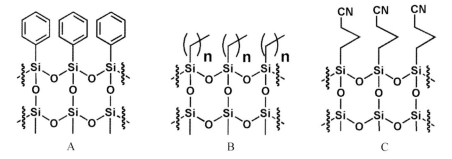

Figure 11. An example of the structure of non-polar (A), intermediate (B), and polar (C) silica stationary phase used in GC.

The stationary phases of GC columns have a finite life expectancy. Continuous and repetitive heating or exposure to oxygen will eventually result in stationary phase baking, in the exposure of increasing amounts of highly polar and reactive silanol (Si-OH) functional groups on the stationary phase backbone. As more of these groups form, the column becomes more acidic and has greater ability to form hydrogen bonds which result in undesirable tailing of analyte peaks in the chromatogram (Agilent technologies 2012, Shimadzu 2014).

Selecting a mobile phase

In gas chromatography three main gases are used as the mobile phase or carrier gas. They are hydrogen (H_2), helium (He), and nitrogen (N_2). The mobile phase is selected due to a number of parameters. This includes what detector is present within the instrument, what flow rate is required and the diameter of the column or stationary phase (Agilent technologies 2012).

Using hydrogen as the mobile phase results in the fastest separation of all the gases with little effect on resolution. However, there are also negative aspects of selecting hydrogen as the carrier gas. If unsaturated compounds are present within the analyte, the hydrogen gas, along with metal surfaces within the instrument, will result in catalytic hydrogenation of these compounds. This results in incorrect product identification. Hydrogen gas is also not applicable when a mass spectrum is employed as the detection method, as seen in GC-MS, when the oil of the vacuum pump is degraded and broken down by the presence of hydrogen. Most significantly, hydrogen forms explosive mixtures when mixed with air at ratios above 4 vol percentage, which is undesirable in a laboratory environment (Harris 2003).

By far the most common carrier gas utilized in GC analysis is helium. Helium can be run at comparable flow rates to hydrogen without any of the associated negative side-effects. Most detectors are compatable with helium. However, some specific detectors have a lower detection limit with other gases when compared to helium. For example, flame ionisation detectors have the lowest detection limit when nitrogen is used as the carrier gas (Agilent technologies 2012, Harris 2003).

Nitrogen gas can also be used as the mobile phase in GC. However, in order to give comparable resolution to using hydrogen and helium gases, the flow rate must be reduced in order to maintain the same plate height. This is due to the ability of solutes to diffuse much easily through hydrogen and helium carrier gases and resulting in a smaller mass transfer. As stated previously, the use of nitrogen as the carrier gas is advantageous when a flame ionisation detector is used (Harris 2003).

Care must be taken when selecting the grade of gas used in GC analysis. Any impurities present within the carrier gas have the ability to significantly degrade the stationary phase of the column. Most commercial gas manufacturers provide a highly pure instrument grade of gas that should be used in this method of analysis. It is also a common practice to employ the use of in-line purifiers or scrubbers between the gas sources and instrument to remove residual contaminants, such as water, hydrocarbons/ organic compounds and oxygen gas (Agilent technologies 2012, Harris 2003).

Selecting a detector

Detectors are selected on the basis of the analyte mixture and the information required from the analysis. If identifying an unknown compound is the primary objective, a qualitative technique is all that is required. This can be achieved by using one of the two main qualitative detectors. Passing separated unknown peaks into either a Fourier transform infrared spectrometer (FT-IR) or a mass spectrometer (MS) will allow the detector to compare the corresponding FT-IR or MS spectrum to libraries of spectra of known compounds. This method of analysis is effective only if the analyte has already been discovered elsewhere. However, it may provide some insight into some fragments or functional groups that combine to give the overall compound. If an unknown peak is identified during comparison with the spectral libraries, it can be confirmed by 'spiking' the sample with a pure sample of the identified compound. If the corresponding chromatogram issues a single peak when combined, then the structure identification is correct. Mass spectrometry can also be used to produce quantitative information, which is discussed in gas chromatography/mass spectrometry (GC-MS) and liquid chromatography/mass spectrometry (LC/MS) sections below (Blackman et al. 2008, Harris 2003).

A quantitative determination of an analyte requires the measurement of the area under the peak in question against an added internal standard of known concentration. The concentration of the analyte can be calculated using Equation 2, once the response factor has been calculated.

$$\frac{A_x}{[X]} = F \frac{A_s}{[S]}$$

Equation 2: Calculation of analyte in chromatogram where A_x = area of analyte, [X] = concentration of analyte, F = response factor, A_s = area of internal standard, [S] concentration of internal standard.

The only GC detector that is able to detect all types of analytes is the Thermal Conductivity Detector (TCD). Because of this, they are the most common detectors found within these instruments. TCD detectors measure the change in thermal conductivity between the carrier gas (He) and when the carrier gas is diluted with analyte to result in reduction of thermal conductivity. It consists of a block and very hot tungsten-rhenium filament. The carrier gas containing analyte passes through the filament, reducing conductivity which results in the filament increasing in temperature. As a result of this increase in temperature, the resistance in the filament decreases, thus changing the voltage across the filament which is what the detector measures. The sensitivity of this detector can be increased by decreasing the flow rate, or by increasing the current applied to the filament, or by increasing the temperature difference between the filament and the surrounding block, which is kept as cold as possible but warm enough so that all the analytes remain in the gas phase. TCD detectors are not sensitive enough to be used with narrow open tubular columns less than 0.53 nm. However, they can be used when larger open tubular columns or packed columns are installed (Harris 2003).

Electron capture detectors (ECD) are useful when analyzing compounds that contain functional groups with high electron affinity, such as conjugated carbonyls, halogens, nitriles, nitro compounds and organometallic compounds. Depending on the source and composition of the food waste being assessed, this detector will have varied levels of use. The carrier gas for this detector needs to be free of moisture to avoid reduction in sensitivity and will be either nitrogen or methane (5 per cent) in argon mixture. This detector contains a radioactive foil of ^{63}Ni that emits beta particles (electrons), which ionize the incoming carrier gas stream and result in more free electrons being present. The formed electrons are accelerated towards an anode (positively charged) which produces a constant current. When the carrier gas contains compounds with a high electron affinity (sample), the amount of free electrons is reduced, as they are captured by these compounds, resulting in less reaching the anode with a reduction in the current. The voltage of the detector between the anode and cathode is then varied to account for the change in current due to the analyte removing some of the free electrons (Harris 2003).

Flame ionisation detectors (FID) measure the level of organic material (hydrocarbons) that burn in a hydrogen gas and air flame. With the exception of carbonyl (C = O) and carboxyl (COOH) carbons, all other types of carbon atoms produce CHO+ ions when burnt in the flame. Although a very low percentage of carbon atoms produce this ion (1:100,000), the ion production is proportional to the amount of carbon atoms present in the flame. These CHO+ ions are collected at the cathode where electrons from the anode flow to neutralize these particles in the burning flame. The current between the anode and cathode is the signal for this detector. The detection limit of a FID is significantly smaller (100 X) than the conventional TCD. Because of this, these detectors can be used when narrow open tubular columns are installed. This can be reduced further when nitrogen is added into the hydrogen gas stream before it enters this detector. Both hydrogen and air are mixed into the carrier gas coming from the GC. However, care must be taken to keep the carrier gas temperature high enough for the analytes to remain in the gas phase (Harris 2003).

A photoionisation detector (PID) is also of use when analyzing food waste that is high in aromatic and unsaturated hydrocarbon content. This detector uses the same principles as the electron-capture detectors with a different radiation source. Photoionisation detectors use an ultraviolet source in the form of a lamp to ionise the gas molecules. These lamps are filled with a low pressure inert gas which is most commonly krypton. However, lamps containing argon and xenon can also be utilized. The final component unique to this detector is the window at the discharge end of the lamp that allows the spectral emissions (radiation) to pass through to ionise the passing gas stream. These windows are composed of materials that do not absorb UV radiation but their composition will vary, depending on the inert gas used within the lamps. When krypton is used, the windows are generally composed of magnesium fluoride, while higher powered argon lamps contain the lithium fluoride windows (Harris 2003, Verner 1984).

Gas Chromatography/Mass Spectrometry (GC-MS)

GC-MS couples two contrasting instruments together and is a very popular analysis method of organic compounds. In these instruments, the carrier gas and analyte mixture are heated and passed through an adequate column within the gas chromatogram component of the instrument, as described above. The individual compounds that have been separated, are then fed into the mass spectrophotometer component of the instrument, for either qualitative or quantitative analysis (Akande 2012, Chauhan et al. 2014, Lakshmi et al. 2013, Sahil et al. 2011).

There is a significant obstacle that must be overcome when using GC-MS instrumentation. The effluent exiting from the GC component is a high pressure mixture of carrier gas with a very low amount of analyte. However, the MS component operates under a high vacuum and requires a higher concentration of analyte than the CG component provides, and without the carrier gas present (Harris 2003, Lakshmi et al. 2013, McLafferty and Turecek 1993, Sahil et al. 2011).

An interface is needed to convert the output of the GC into what is required as an input for the MS. These interfaces all increase the concentration of analyte by removing carrier gas from the feed stock and can enrich the analyte concentration five fold to 20-fold depending on the application (Akande 2012, Griffiths et al. 2012, Sahil et al. 2011).

This can be achieved by using an effusion separator that passes the GC effluent though a microporous tube with a pore size of 1×10^{-4} cm, as shown in Fig. 12A. A vacuum is created on the outer side of the tube which results in the smaller and lighter carrier gas molecules being sucked out of the tube, with the heavier and larger organic

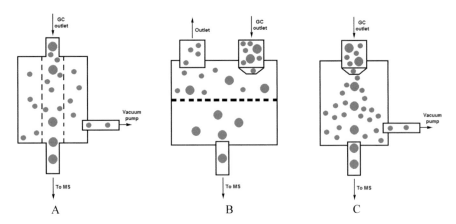

Figure 12. Comparison of effusion (A), selectively permeable membrane (B), and jet separator (C) interfaces that are utilised in GC-MS instruments.

analyte molecules remaining unaffected and continuing into the MS component (Griffiths et al. 2012, Lakshmi et al. 2013, Sahil et al. 2011).

Alternatively, a selectively permeable membrane interface is used to obtain the same desired outcome. The GC effluent enters a cavity containing a permeable membrane. The organic molecules present, with high permeability, pass through the membrane and continue into the vacuum of the mass spectrometer component. The carrier gas (for example, helium) has low permeability and does not pass through the membrane; instead it exits an outlet on the same side of the membrane as the inlet, as shown in Fig. 12B (Griffiths et al. 2012, Lakshmi et al. 2013, Sahil et al. 2011).

Interfaces containing a jet or an orifice separator pass the GC effluent through a fine jet, resulting in the stream of carrier gas and analyte molecules expanding rapidly into a vacuum chamber. The carrier gas, having a higher diffusion rate, expands further than the analyte, leaving a more concentrated beam of analyte in line with the jet opening. This beam is aligned at a second jet inlet that continues into the mass spectrometer instrumentation, whilst the carrier gas is diffused and exits the chamber into a vacuum pump outlet, as shown in Fig. 12C (Griffiths et al. 2012, Lakshmi et al. 2013, Sahil et al. 2011).

Mass Spectrometers

Numerous types of mass spectrometers are utilized within current GC-MS and LC-MS instrumentation. In Time-of-Flight spectrometers (T-o-F), separated compounds sourced from the relevant chromatography component, are ionised by using high voltage potentials of upto 5000 V. This high voltage is rapidly applied to a backing plate between the order of around 5000–20,000 times per second. This voltage potential is uniformly applied to all the ions varying in size, ideally resulting in all ions having the same kinetic energy. This causes all ionised compounds to accelerate away from the plate into the first drift region. Within this part of the spectrometer, lighter ions are accelerated much faster than heavier ions, as the velocity is directly proportional to mass when seen by Equation 3 below (Blackman et al. 2008, Griffiths et al. 2012, Harris 2003, McLafferty and Turecek 1993):

$$E - \frac{1}{2}mv^{\flat}$$

Equation 3: Calculation of the kinetic energy of an ion where: E = Kinetic energy, m = mass of ion, v = velocity.

Unfortunately, the main limitation of T-o-F spectrometers is that they operate at constant resolving power, which results in all the ions not passing into the first drift region with identical kinetic energy. Compounds ionised closer to the backing plate will accelerate faster than the compound of identical mass that is ionised further away from the backing plate since it will pass through a higher voltage difference. In order to negate this disadvantage and improve the resolving power, ions travel from the first drift region into a reflectron. This consists of a series of hollow rings increasing in potential, with ions containing more kinetic energy penetrating further into the reflectron. All ions are slowed down and reflected back in another direction by passing through a second drift region towards the detector grid. The reflected ions of the same mass, regardless of their initial kinetic energy, reach the detector grid at the same time. This type of mass spectrometer also has the advantage of having very high resolving power and mass to charge accuracy as well as having the capacity to measure ions with high masses. However, this spectrometer requires a significantly lower vacuum pressure (1×10^{-12} bar) to operate than the quadrupole instruments (1×10^{-9} bar) mentioned below (Akande 2012, Harris 2003, McLafferty and Turecek 1993).

The most common mass spectrometer currently in use is the Quadrupole mass spectrometer. Instruments containing this separator take in the separated compounds coming from a chromatography column and ionize them using an electron ionisation source under a vacuum pressure of 1×10^{-9} bar. The ions are accelerated through a potential between 5–15 V before entering the quadrupole component. This simply consists of four metal tubes that have both a constant voltage and a radio-frequency voltage applied to them. The formed electric field allows ions with only one mass to charge the ratio to pass through the quadrupole and reach the detector with other ions present being deflected. These other ions, without a correct trajectory, collide with the metal tubes and do not reach the detector. The applied voltages on the metal tubes can be changed rapidly, which creates conditions for other ions to be able to reach the detector. These spectrometers run at a constant resolution. So the ions with a lower mass to charge ratio are separated just as well as the ions with higher mass to charge ratios (Chauhan et al. 2014, Griffiths et al. 2012, Harris 2003, McLafferty and Turecek 1993).

Quadrupole Ion-Trap mass spectrometers work on a similar principle to the quadrupole spectrometer described above but do not pass ions through four charged metal tubes, but instead are ionised and separated in a void space between two end caps and a ring electrode. Separated compounds enter the spectrometer and are first ionised. This can occur using electron ionisation, where a gate electrode periodically emits electrons that pass through gaps in the encasing end cap and come in contact with the compounds present. Chemical ionisation can also be achieved provided, a reagent gas like methane is added to the system (Chauhan et al. 2014, Harris 2003, McLafferty and Turecek 1993).

Once the ions are formed, they are kept in circulating stable trajectories by applying a constant frequency radio frequency voltage to the central ring electrode. Specific ions are expelled from the void space by changing the radio frequency voltage. This results in expelled ions passing through holes within an end cap and colliding with the electron multiplier detector inside. The main advantage of this type of mass spectrometer is that it can be 100 times more sensitive than the other quadrupole mentioned previously since approximately half the ions present react with the detector in the Ion trap instrument as compared to a very small fraction reaching the detector in the Quadrupole instrument (Griffiths et al. 2012, Harris 2003, McLafferty and Turecek 1993, Niessen 2012).

Liquid Chromatography/Mass Spectrometry (LC-MS)

Combining Liquid Chromatography with mass spectrometry also requires a significant obstacle to be overcome. The liquid mobile phase and separated compounds coming from the CG component need to be heated and ionized at the interface in order to be used in the mass spectrometer component of the instrument. Vaporising the liquid mobile phase, containing the separated analytes, results in a significant increase in volume that is not desired in a contained system. This high difference in pressures between the LC component and the MS component, which is operated under a vacuum, has to be overcome. Non-volatile solutes, such as buffering salts, are often added to mobile phases used within liquid chromatography but must be omitted when these instruments are coupled with mass spectrometry so as not to cause a blockage in the system (Chandu et al. 2013, Couchman et al. 2011, Griffiths et al. 2012, Harris 2003, McLafferty and Turecek 1993).

The two most common interfaces used in LC-MS are Ion Spray and Atmospheric Pressure Chemical Ionisation. Ion Spray or Pneumatically Assisted Electrospray involves passing the liquid eluent from the LC through a charged (0 V) steel nebuliser capillary which projects a fine spray into a spray chamber. Depending on application, the chamber is charged at either –3500 V (positive ion MS) or +3500 V (negative ion MS). A coaxial stream of N_2 gas is passed over the outside of the capillary. When the liquid passes through the capillary, it forms a very fine spray of charged particles. The ions are then attracted to a charged glass capillary that leads to the mass spectrometer. The glass capillary has a higher potential (–4500 V for +MS, +4500 V for –MS) than the spray chamber in order to achieve this attraction. The ion stream then passes through numerous skinner cones and areas, decreasing in vacuum pressure before they pass into one of the mass spectrometers detailed above. When using the ion spray, negligible fragmentation of the parent ions occurs, resulting in relatively simple mass spectra. However, the level of fragmentation can be altered by altering the acceleration of the ions, which increases the rate of collisionally activated dissociation, when these ions collide with N_2 gas molecules between the glass capillary and the first skimmer cone. The pressure in this section of the instrument is kept around 3 mbar. Adjusting the potential applied to the skimmer cone results in a variance in potential, and acceleration between the glass capillary and skimmer cone result in a change in the kinetic energy of the ions (Antignac et al. 2004, Bojko et al. 2014, Gelpi 2003, Griffiths et al. 2012, Harris 2003, Knutsson et al. 2013, McLafferty and Turecek 1993, Niessen 2012, Sharma et al. 2014).

Atmospheric Pressure Chemical Ionisation passes the LC eluent through a fine needle and when exiting, mixes with a coaxial flow of N_2 gas, forming a fine mist. The formed mist is instantaneously passed through a small heater which forms a fine aerosol of compounds. The aerosol is sprayed onto the tip of a metal needle that has a very high (+6000 V) potential applied. This high voltage results in an electric corona forming around the needle, which injects electrons into the aerosol, forming both positive and negative ions through a sequence of numerous reactions. The formed ions are passed through a skimmer cone and into a high vacuum mass spectrometer. This system of ionisation is beneficial to LC-MS as it can take a relatively high eluent flow rate of 2 mL/minute. Compounds that can ionize to form a MH^+ ion, where M is the parent compound, have the greatest chance of detection within this technique. This type of ionisation, traditionally produces ions with a single charge and because of this, is not suitable in the analysis of macromolecules like proteins and peptide chains. Also limited fragmentation of ions is observed by using this technique under normal operation parameters. However, as in ion spray, the voltage can be adjusted (on the skimmer cone), to increase the kinetic energy of the ions and to increase the level of collisionally activated dissociation (Antignac et al. 2004, Griffiths et al. 2012, Harris 2003, McLafferty and Turecek 1993, Niessen 2012, Thomson 1998).

Quantitative analysis using mass spectrometry

Through the use of either selected ion monitoring or selected reaction monitoring, quantitative information can be obtained on a targeted analyte, be it a complex and poorly resolved mass spectrum. These techniques can increase the detection limit significantly (up to 100 times) as compared to monitoring the entire mass spectrum, because only the ions of interest are measured. They cannot be lost under peaks of the other ions present (Chandu et al. 2013, Griffiths et al. 2012, McLafferty and Turecek 1993, Niessen 2012).

When using ion selected monitoring, the detector only measures and records ions with a specific m/z (e.g., m/z 92) instead of measuring all ions eluted from the adjoining chromatogram with the range selected (e.g., mz/25–500). This results in a very well resolved chromatogram with a single peak attributed to the ions being measured at m/z 92. To produce quantitative information on this peak, a known quantity of an internal standard with a unique m/z (e.g., m/z 56) is added to the reaction. The same experiment is then repeated, but targets ions of m/z 56. Comparing the area of the targeted peak (m/z 92) in the first spectrum with the area of the internal standard peak (m/z 56) in the second spectrum results in production of quantitative information from the experimental data (Bojko et al. 2014, Chandu et al. 2013, Griffiths et al. 2012, Harris 2003, Li and Tse 2010, McLafferty and Turecek 1993, Moreno-Bondi et al. 2009).

Similarly, selected reaction monitoring can produce quantitative information on the ions present. Using this technique, an initial screen is set to allow only a specific ion (e.g., m/z 156) into a collision cell where fragmentation occurs. A fragment that is unique to the parent compound is then selected for analysis and comparison with the peak area of an added internal standard as described for ion selected monitoring. Using these techniques in the analysis of an ion chromatogram enables most interferences to be eliminated, which results in an increase in the signal to noise ratio (Griffiths et al. 2012, Harris 2003, Moreno-Bondi et al. 2009).

High Performance Liquid Chromatography (HPLC)

High Performance Liquid Chromatography (HPLC) relies on the same principles as the other methods of chromatography previously discussed. A mixture of analytes is introduced into the system, where each individual compound has a unique interaction between the mobile and stationary phases. Compounds with the least amount of interaction with the stationary phase are eluted first, while compounds that have a stronger interaction with the stationary phase require greater mobile phase and a longer time to be eluted. Once eluted, the analytes are passed through a detector in order to quantify how much of each individual compound is present. However, there are some components of the separation process that are unique to HPLC.

Sample introduction

HPLC is performed under high pressures and because of this, there is a common misconception that HPLC stands for High Pressure Liquid Chromatography. In order to achieve the application of uniform high pressure to this separation system, quality pumps must be employed. HPLC pumps consist of a twin piston design that draws in the mobile phase under atmospheric pressure, and by using a series of electronic valves, uniformly pushes the solvent through the column at high pressures of up to 400 bar (Harris 2003, Lee and Coates 1999, Mant and Hodges 1996, Meyer 2014).

The sample also needs to be introduced at atmospheric pressure into a highly pressurised liquid system. This is achieved through the use of interchangeable sample loops and a six-input valve system. The size of the sample loop can be selected on the basis of the volume of analyte being injected. When set in the input or loading position, the column is isolated from the atmosphere and the sample loop. A syringe is used to add the sample mixture into the sample loop. The valve is then rotated, causing the sample loop to be sealed from the atmosphere and exposed to the high pressure of the solvent inlet resulting in the solvent injecting the contents of the sample loop into the column to be separated (Harris 2003, Kalili and de Villiers 2011, Lee and Coates 1999, Meyer 2014).

HPLC columns

Columns used in HPLC are different to the general open columns seen in GC analysis. The closed columns employed in HPLC are either steel or plastic coated depending on the manufacturer. On the basis of the application, they can have an internal diameter of 1 to 5 mm and are generally between 5 to 30 cms long. The mobile phase material is not bonded to the wall of the HPLC columns, instead it is held within the column by the presence of an inert porous frit located at each end of the column. Manufacture of the HPLC columns is expensive and they get easily damaged or degraded. When unwanted material, such as dust or other contaminants are allowed to pass into the column, it causes irreversible blockages or damage to the stationary phase surface. Because of this issue, HPLC columns are fitted with much smaller guard columns that contain the same mobile phase as the main column. Guard columns are easily replaced and act as a cheap alternative to trap contaminants before they reach the main column where serious damage can occur (Harris 2003, Kalili and de Villiers 2011, Lynch and Young 2000, Mant and Hodges 1996, Meyer 2014, Watson 2014).

Stationary phases

The stationary phase utilised in HPLC columns is also unique when compared to other instrumental separation techniques. It consists of spherical microporous particles with a very high surface area. Although silica can be used directly as the stationary phase, it is far more common to use a bonded stationary phase that is covalently bound to the surface of the silica microporous particles. The structure of the bonded phase substituents can be altered to suit the type of separation required, forming a polar phase, a non-polar phase or an intermediate stationary phase somewhere in between. A common example of a polar and non-polar bonded stationary phase is shown below in Fig. 13 (Harris 2003, Mant and Hodges 1996, Meyer 2014, Watson 2014, Zhang and Qiu 2015).

Stationary phase degradation results in bond hydrolysis of the O-Si bonds, resulting in the formation of highly polar and reactive silanol (Si-OH) functional groups on the stationary phase surface. This can be limited or slowed by careful

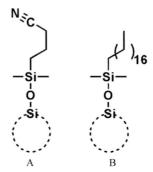

Figure 13. Example of the surface structure found on a polar bonded stationary phase particle (A) and a non-polar bonded stationary phase particle (B).

monitoring of environmental pH, keeping temperatures from rising for prolonged time periods and by adding difficult-to-cleave bulky alkyl groups to the bonded stationary phase. However, as with all stationary phases used in chromatography applications, each column has a finite lifespan (Angel de la Fuente et al. 2006, Harris 2003, Meyer 2014, Watson 2014, Zhang and Qiu 2015).

Operating parameters

Using a liquid mobile phase results in a significant decrease in the diffusion of analytes from within the liquid mobile phase to where can interact with the solid stationary phase present. To account for this significant disadvantage, HPLC uses a stationary phase with a much higher surface area comprising spherical microporous particles (3–10 µm). This results in an increase in the rate that analyte molecules equilibrate between mobile and stationary phases and with an increase in efficiency (Harris 2003, Lynch and Young 2000, Meyer 2014, Watson 2014).

As a consequence of this increase in surface area and increase of column efficiency, the stationary phase particles are smaller with less area between each stationary phase particle. The mobile phase requires a significantly higher pressure (upto 400 bar) to overcome this resistance in order to maintain a useful flow rate of between 0.5–5 mL/minute (Harris 2003, Lynch and Young 2000, Mant and Hodges 1996, Watson 2014).

The column of HPLC instruments is housed within an oven in order to maintain a uniform temperature, a uniform rate of mass transfer and a uniform separation level of analytes. Increasing the temperature will result in an increase in mass transfer rate. However, it increases the risk of irreversible column damage through the formation of highly polar and reactive silanol (Si-OH) functional groups and a reduction in column life expectancy (Harris 2003, Lynch and Young 2000, Meyer 2014).

In order to monitor column efficiency, the manufacturer will generally provide a set of parameters to run a standardized separation. It is important to perform this before any unknown separations are completed in order to have a useful baseline to calibrate the column with. Periodic replication of this process will enable users to monitor the column efficiency, to operate within the manufacturer's specifications and determine when the column needs replacing. Washing the columns and replacing the guard column periodically will prolong the column life by removing a build-up of any solutes or particles that have become strongly bound to the stationary phase, before they become permanently bound (Fernandez-Ramos et al. 2014, Harris 2003, Watson 2014).

Some solvent and stationary phase combinations may cause the column stationary phase to expand or contract in the column, leaving a void space. When this occurs, a doubling of each eluent peak can often be observed. Constant monitoring of this space at the top of the column will avoid these issues, as any void present can be filled in with more of the appropriate stationary phase as required (Fernandez-Ramos et al. 2014, Watson 2014).

Solvent elution

HPLC column efficiency and lifespan are heavily dependent on what is passed through them. The presence of any impurities can cause irreversible damage to the columns either by irreversibly reacting with the stationary phase, or by physically blocking the flow, resulting in a reduction of uniform flow and higher pressures. HPLC solvents of a higher purity than analytical grade solvents must be purchased for use in these instruments (Harris 2003, Meyer 2014, Watson 2014, Zacharis and Tzabavaras 2013).

There are also other safeguards that must be performed to avoid introducing detrimental material into HPLC columns. The solvent tubing has a microfilter or frit installed at the inlet that takes the solvent into the system from the solvent bottle, which must be monitored and replaced periodically. The solvents need to be devoid of any dissolved gases (O_2) that will react with the stationary phase material. Before use, all commercially sourced HPLC solvents need to be de-gassed. This is achieved either through purging the solvent with an inert gas, such as helium, or evacuating the solvent to remove any dissolved air. The presence of air bubbles in the solvent can cause serious issues within the HPLC pumps that deliver the uniform pressure necessary for analysis (Fernandez-Ramos et al. 2014, Harris 2003, Molnar-Perl 1999, Watson 2014).

HPLC can be performed by using a single solvent to separate and elute all analytes, known as isocratic elution, or more commonly through the use of multiple solvents, known as gradient elution. Isocratic elution may be possible, depending on the analyte composition. However, long run times and large volumes of solvent are generally required. To decrease run times and solvent use, some applications allow the use of gradient elution. In the simplest case, this involves starting the chromatogram run with 100 per cent of solvent A and slowly adding increasing the percentage of solvent B, which alters the solvent gradient until the end of the separation where 100 per cent of solvent B is passing through the column. A successful solvent gradient can be obtained through the use of previous experience with similar separations, or through the use of trials and experimentation (Harris 2003, Meyer 2014).

The ability to maintain an ideal symmetric sharp peak in HPLC chromatograms is also heavily solvent dependant. Often, some degree of peak tailing is observed when compounds containing exchangeable protons are separated. The incorporation of various additives into the mobile phase can often minimise or eliminate this issue. For streaking or tailing attributed to basic functional groups, such as amines, adding a low concentration of non-nucleophillic base, such as triethylamine to the mobile phase may alleviate the issue. When tailing is due to organic acids, adding a low concentration of ammonium acetate has been shown to reduce or eliminate this issue in some applications. However, the composition of mixtures to be separated is often unknown. In these situations, the addition of triethylammonium acetate can be a useful starting point (Harris 2003, Meyer 2014, Molnar-Perl 1999).

Finally, the process of line broadening after separation must be minimised or avoided as much as possible. The best way to achieve this is to select a suitable flow rate, minimise the tube dead volume between the column and the detector as much as possible and finally keep the detector volume below 20 per cent of the column volume.

HPLC detectors

A wide range of detectors can be used in HPLC. The composition of the analytes determines their selection in each application. Numerous examples, such as mass spectrometry, FT-IR, conductivity and electrochemical detectors have already been described in previous sections.

An ultraviolet (UV) detector is the most common type of detector found in HPLC instruments. These detectors utilise a flow cell that maintains a uniform sample path length. Modern HPLC UV detectors have a range of radiation sources and a monochromator. So a precise wavelength can be selected based on interaction with the separated analytes. The most common wavelength used is 254 nm (short wave) UV radiation. This radiation source is selected as most compounds being assessed from biological origins, including food products, absorb this wavelength. These detectors work on the same principle and contain the same components that are discussed in detail in the next section. More importantly, these detectors have a linear range of over 5 orders of magnitude (Franke and Custer 1999, Harris 2003, Lai and Franke 2013, Meyer 2014, Saka 2009).

However, using a single wavelength to measure the concentration of separated analytes may give a false indication of purity. A single peak in a chromatogram simply indicates that at least one compound absorbs radiation at that specified wavelength; it does not indicate how many other compounds may be present that have been eluted from the column at the same time and do not absorb this unique radiation band at all or absorb a negligible amount that is lost under the primary signal. It may also be difficult to quantify the amount of material attributed to each peak in the chromatogram when using this detector type. A small amount of analyte that absorbs radiation strongly may produce a peak with much more area underneath it than a large amount of analyte that poorly absorbs the selected radiation (Franke and Custer 1999, Harris 2003, Meyer 2014, Saka 2009).

Use of a refractive index detector in HPLC is advantageous as it is able to respond to most analytes. A parallel beam of visible light radiation is passed through a flow cell consisting of two triangular compartments—one containing the pure mobile phase and the other containing the elutant coming from the column. When only the mobile phase is eluted from the column, the beams of light passing between each compartment are unchanged and pass through a deflection plate to hit the photocell detector on the other side. When an analyte enters the cell, the refractive index between the two compartments changes, resulting in a different output that is received by the detector. However, it has numerous disadvantages as compared to other detectors.

When compared to the UV detector, it has a significantly lower detection limit (up to 1000 x). Because this detector measures the difference in light refraction between two liquid cells, it can only be used in separations that utilize isocratic solutions. Systems using an elution gradient will never have the same refractive index. So this detector becomes ineffective under these conditions. Such detectors are also linear over a short concentration range and inapplicable for low concentration analysis (Harris 2003, Meyer 2014).

The evaporative light-scattering detector is an example of a destructive HPLC detector. The eluent from the column is mixed with a source of nitrogen gas and forced through a nebuliser to facilitate an even dispersion of micro-droplets. These droplets are passed through a heated column where the mobile phase is evaporated, leaving particles of analyte to fall through a beam of radiation. The detector measures any change in the amount of radiation that reaches a photodetector on the other side of the sample stream, which is attributed to particles of analyte. One disadvantage of this method is that the detector measures only the mass of analyte passing through the beam; it does not give any information on its structure or molar mass. This detector can only be used on analytes that are significantly less volatile than the mobile phase being used. The observed signal response is also non-linear in this detector and because of this, requires complex polynomials when producing calibration curves (Harris 2003, Meyer 2014).

Evaporative light-scattering detectors do have the advantage of being effective in measuring separations that use mobile phase gradient elution and have the added benefit of no solvent peak forming at the beginning of the chromatogram that can hide some analytes with short retention times. Because the solvent is removed before detection, any additives or buffers present within the mobile phase would be much more volatile than the analytes, or this detector cannot be used (Harris 2003).

Collection of isolated compounds

Most modern HPLC instruments have some amount of collector that can be plumbed into the system. Primarily, these are automated collectors that are signalled to collect fractions containing analytes as shown by the detector, or alternatively, the eluent is simply pumped directly into a waste bottle. Ideally non-destructive detectors are used when analytes need to be collected, which is why the UV detector is the most common detector employed in HPLC instrumentation. However, when mass spectrometry is used, a splitter is employed to send a small amount of the eluent to the detector and the remainder to the fraction collector. The amount of material that can be passed through and collected can vary significantly, depending on application when using 3.5 mL to 50 mL vials. Once collected, the mobile phase must be removed to yield the purified and desired analyte.

There are systems available that can separate larger volumes of starting material in a single attempt. Initial separations are completed on a small analytical scale, which provides information on the concentration or the identity of an analyte. If a large volume of material needs to be purified, the processes of semi-preparative (less than 0.5 grams), preparative (several grams), or even industrial (several grams to kilograms) HPLC can be used. The larger scale HPLC systems use columns with increasingly larger internal diameters and lengths. This increase in column volume also requires an equal increase in the power of the solvent pump to maintain the correct flow rate and be able to withstand the higher associated pressure with an increase in the mobile phase volume. In order to alleviate the required pressures as much as possible, the particle size of the mobile phase is also increased slowly but must stay within the limits of maintaining an acceptable separation.

Spectrophotometric Detection (UV/Vis)

Spectrophotometry using the Ultraviolet (UV) and Visible regions (Vis) of the electromagnetic spectrum can also provide useful information on some organic compounds found within food waste material. It is possible to obtain both qualitative and quantitative information on using this technique. However, this analytical tool is best used in conjunction with other analytical methods and provides limited information when used alone (Hornback 2006, Lepot et al. 2016, Onate-jaen 2006).

The electromagnetic spectrum consists of numerous kinds of radiation, ranging from weak (radio waves) to high (gamma rays) energy. UV/Visible spectrophotometry utilises radiation in the UV (200–400 nm) and visible (400–700 nm) range (Blackman et al. 2008, Lewis and Evans 2011).

When UV/Visible radiation impinges on a substance, certain wavelengths of this radiation are absorbed by the molecules or ions that are present. The remaining radiation is transmitted from these substances, such as that seen by the human eye, and is deficient in the absorbed wavelengths. For example, a substance will appear purple to the eye as it absorbs the green-yellow wavelength of the visible spectrum and reflects red and blue wavelengths of light (Lewis and Evans 2011, Williams and Fleming 1995).

The absorption of a quantum of radiation (UV or Visible) excites an electron from its ground state to a higher-energy-excited state within the absorbing molecule or ion. The exact difference in energy between these two energy states (ΔE) is supplied to the electron by the light quantum, which has a definite frequency (v) and wavelength (λ) as shown in Equation 4 (Blackman et al. 2008, Williams and Fleming 1995):

$$\Delta E = hv = \frac{hc}{\lambda}$$

Equation 4: Calculating the change in energy ΔE between electron energy states where: h = Planck's constant (6.626×10^{-34} Js), c = velocity of light in a vacuum (2.998×10^8 ms^{-1}).

The absorption of radiation by a molecule or ion does not occur at a single wavelength in the UV/Visible spectrum. It occurs over a range of wavelengths which give rise to a unique absorption band, like a fingerprint for a molecule, which is centred around a single wavelength where maximum absorption occurs (λ_{max}) as shown below in Fig. 14 (Brown et al. 2010, Gurkan and Altunay 2016, Hornback 2006, Lepot et al. 2016, Lewis and Evans 2001, Onate-jaen 2006, Pham et al. 2011).

There are many sub-levels associated with the ground and excited states, which correspond to alternative vibrational energy levels within a specific molecule or ion. The origin of the absorption band therefore resides in the multiple electronic transitions of different energies which can occur when involving the vibrational states belonging to the ground and excited states. The most likely excitation that corresponds to λ_{max} is usually between the lowest vibrational sub-levels of the ground and excited electronic states as shown in Fig. 15 (Brown et al. 2010, Lewis and Evans 2001, Williams and Fleming 1995).

The degree of radiation absorbed by a solution is directly proportional to the concentration of analyte in solution and the path length the radiation has to travel through. Laws describing this relationship and its effect on the absorption of radiation were originally formulated by Lambert (1760) and Beer (1852). These laws are combined to quantitatively relate the level of radiation absorption to the concentration of absorbing analyte in solution and is known as the Beer-Lambert law (Equation 5). This law refers to using a single wavelength (usually λ_{max}) and not the entire absorption band (Ali et al. 2014, Blackman et al. 2008, Brown et al. 2010, Gurkan and Altunay 2016, Lepot et al. 2016, Williams and Fleming 1995).

$$A = \varepsilon c l$$

Equation 5: Beer-Lambert law where: A = absorbance, ε = molar absorptivity (mol/L/cm), l = pathlength (cm).

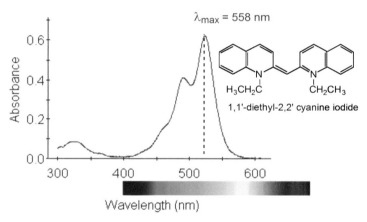

Figure 14. A wavelength scan of the organic compound 1,1'–diethyl–2,2'cyanine iodide between 300–600 nm.

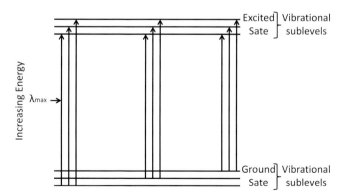

Figure 15. Comparison of vibrational ground and excited state energy levels. The green arrow represents energy difference associated with λ_{max} absorption.

The amount of UV/Visible radiation that is absorbed by solutions is carried out by using a spectrophotometer. In a spectrophotometer, radiation (UV or Visible) from a source bulb (deuterium lamp for UV, tungsten/halogen lamp for Visible) is passed through a monochromators, which selects a single wavelength to pass through the solution to be examined. The intensities of the incident light beam and transmitted light are then measured by a photoelectric device and processed by a computer. The output of the instrument is calibrated to read directly the absorbance of percentage transmission (Brown et al. 2010, Lepot et al. 2016, Lewis and Evans 2001, Pham et al. 2011).

For an organic compound to be measured by UV/Visible spectrophotometry, the energy gap between the excited and ground state for a group of electrons must range between 200–700 nm. The presence of electrons that are able to utilize this radiation are known as chromophores. In organic scaffolds, there are very few unconjugated chromophores that can be measured in UV/Visible spectrophotometry as they require higher energies (< 200 nm) to achieve transitions from a ground to an excited state. The most useful of these is the transition attributed to ketone (C = O) function groups which occurs around 300 nm (Ali et al. 2014, Hornback 2006, Williams and Fleming 1995).

The most important chromophores in organic compounds contain varying levels of conjugation. The more the conjugation within an organic scaffold, the smaller the energy gap between the excited and ground states. This results in electrons requiring less energetic radiation to make the transition, utilizing longer wavelengths. Organic compounds with significant levels of conjugation become coloured to the naked eye as various parts of the visible spectrum are absorbed, with the remainder of the unused wavelengths reflected as colour. There are also important colourless compounds that absorb radiation in the UV range (200–400 nm). These often contain a polyene, polyyne, or substituted aromatic ring systems (Williams and Fleming 1995).

As stated previously, limited information can be deduced from a UV/Visible spectrum when used to elucidate the structure of an unknown organic compound. It is best used to supplement other more detailed spectroscopic techniques when trying to achieve this outcome. However, this analytical method can provide some valuable qualitative, less specific information about an analyte. If a UV/Visible spectrum shows the presence of numerous bands in both the UV and the visible region, a long conjugated system or a polycyclic aromatic system is present within the organic scaffold. However, if there is limited bands or a single band present in the UV range (200–400 nm), only a low level chromophore consisting of two to three conjugated unsaturated bonds are present. Molar absorptivity (ε) values of spectral bands in the range of 1000–10000 mol/L/cm strongly support that some type of aromatic system is present within the scaffold of the organic compound (Hornback 2006, Williams and Fleming 1995).

Spectrophotometry can also be used to produce quantitative information on a solution, provided the analyte is purified and is the only compound present that absorbs radiation at the wavelength of maximum absorption (λ_{max}). In order to achieve this, the correct sample cuvette must be used. This is dependent on the region of the electromagnetic spectrum used as these materials also absorb electromagnetic radiation. Plastic cuvettes are used between 400–780 nm, glass cuvettes between 450–700 nm and quartz cuvettes between 200–400 nm. The selection of the correct solvent is also highly important. A solvent must be selected that dissolves the organic compound being assessed, but does not absorb strongly in the same region of the spectrum. It is also important to select a compatible organic solvent with the cuvette being some organic solvent to dissolve certain types of cuvette. The polarity of solvent in spectrophotometry also has an effect on the observed spectrum, as electrons transitioning between ground and excited states cause a slight change in the polarity of the organic analyte. When spectrophotometry of an analyte is carried out in solvents with differing polarity, a slight red or blue shift in wavelength maximum may be observed (Lepot et al. 2016, Williams and Fleming 1995).

The quantitative experiments are carried out in two parts. Initially a wavelength scan is carried out in order to find the wavelength of maximum absorption to be used in the calibration curve part of the analysis. This is achieved for a number of reasons. By running the experiment at λ_{max}, any errors in wavelength selection by the monochromators are minimized. At λ_{max}, a slight variance in wavelength results in a small change in absorbance when compared to other areas of the absorption band, as shown in Fig. 16. The analysis sensitivity is also maximised when running experiments at λ_{max}, as a lower amount of material is needed to achieve a higher absorbance at this wavelength (Brown et al. 2010, Lepot et al. 2016, Pham et al. 2011).

Once the λ_{max} has been found, the instrument is set at this wavelength for the final part of the analysis. After removing any absorbance attributed to the solvent, known as zeroing the instrument, a calibration curve is generated by measuring the absorbance of a range of standard solutions with a known concentration of analyte. This produces a directly proportional linear trendline between absorbance and concentration, the gradient of which is equal to the molar absorptivity (ε). Using the equation of this calibration curve, the concentration of an analyte in an unknown solution can be calculated from the measured absorbance. The amount of material present can then be calculated based on the level of dilution carried out between the analyzed solution and the stock solution of unknown organic compound (Pham et al. 2011).

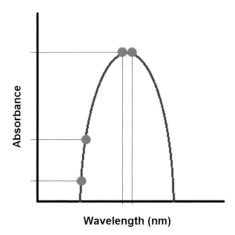

Figure 16. Comparison of the effect of monochromators error between readings taken at λ_{max} (negligible change in absorbance) and taken at other areas (significant change in absorbance) of the spectrum.

Nuclear Magnetic Resonance (NMR)

For any given nuclei to exhibit the nuclear magnetic resonance phenomenon they must have a nuclear spin (I) greater than zero. The spin quantum number is associated with the mass number and the atomic number of the species and the presence of a nuclear spin makes the nucleus behave like tiny spinning bar magnets. This occurs as the nucleus possesses both an electric charge and a mechanical spin which generates a magnetic field (Brown 2010, Hemminga 1992, Hornback 2006, Williams and Fleming 1995).

When an applied magnetic field is placed on the nucleus, these tiny magnets orient themselves in multiple ways depending on the associated nuclear spin number ($2I + 1$). Important nuclei, such as ^1H and ^{13}C are used extensively and have a spin of half which only allows them two orientations in a magnetic field. The nuclei can either be aligned with the external field being the low energy orientation or it can be opposed to the applied field creating a high energy orientation. The difference in energy between these two orientations, also called the transition energy, can be found by using Equation 6 (Brown 2010, Hemminga 1992, Williams and Fleming 1995):

$$\Delta E = \frac{\gamma h B}{2\pi}$$

Equation 6: Energy difference between the two possible orientations in a magnetic field where: B is the strength of the applied magnetic field, γ is the magnetogyric ratio of the nucleus and h is Planck's constant.

Under normal conditions, the difference between the number of nuclei in the higher and lower energy states is quite low and the ratio is given by the Boltzmann distribution shown in Equation 7 (Williams and Fleming 1995).

$$\frac{upper\ (N_\beta)}{lower\ (N_\alpha)} = \exp\left(-\frac{\Delta E}{kT}\right)$$

Equation 7: Calculation of the Boltzmann distribution where: N_β = the high energy state, N_α = the low energy state, k = Boltzmann's constant, and T is temperature (kelvin).

This ratio is changed when a radio signal is applied at the same frequency at which the nuclear magnets naturally gyrate in the magnetic field. When this occurs, some of the lower energy nuclei are promoted to the higher energy level which changes the Boltzmann distribution of the system. This resonance frequency is dependent on the applied field strength and the nature of the nucleus being analyzed (Brown 2010, Hornback 2006, Williams and Fleming 1995).

The resulting energy released when the higher energy nuclei return back to lower energy nuclei is then measured, but the difference between the number of nuclei in the upper and lower state is usually small. Because of this, NMR spectroscopy is quite an insensitive technique, but by changing the magnetic field strength there is a greater difference resulting in a more sensitive analysis being carried out.

Measuring a spectrum

There are two possible was to measure a NMR spectrum—the first is to treat the sample with a continuous frequency wave with the small amount of applied signal that is absorbed to promote some nuclei into the high energy state being detected by a receiving coil. The examined frequency range can be scanned from one end to the other by changing either the transmitter frequency or the applied magnetic field. The obtained spectrum is then directly plotted with absorbance vs. frequency but the resonances however, can only be excited sequentially (Blackman et al. 2008, Brown et al. 2010, Hemminga 1992, Hohmann et al. 2014, Hornback 2006).

The second way to measure the spectrum is to treat the sample with a single pulse of the frequency that covers the entire range being analyzed. This method can excite all resonances simultaneously with the pulses lasting a few microseconds. The pulse generates an oscillating magnetic field that is perpendicular to the applied magnetic field. The magnetization then changes its orientation and precesses 90° to the applied field. This is detected by placing a receiver coil that detects magnetization in the plane perpendicular to the applied field. The frequency of oscillation detected is the difference between the NMR resonance frequency and the excitation frequency. As the magnetization oscillates around the perpendicular axis, it slowly returns back to the original applied direction with each chemical environment producing a unique exponentially decaying cosine frequency signal. The decaying waves can then be separated and converted to frequency signals by Fourier transformation and plotted in the same way as the continual wave method, but being more sensitive because of a better signal to noise ratio (Blackman et al. 2008, Hemminga 1992, Williams and Fleming 1995).

There are many different machines available that are able to measure an NMR spectrum. All high resolution machines, however, must have a radio frequency source and a magnetic field which are stable and homogenous.

When measuring a sample, it is placed in a probe that is sucked down the bore by air pressure suction and positioned between the poles of the magnets inside the probe of the machine. The sample is spun around its vertical axis to improve the homogeneity of the sample and the radio frequency waves are transmitted by a coil connected to the probe. The radiation is then detected by either the same coil on the opposite side or a separate coil, depending on machine design. When the resonance conditions are met and the nuclei absorbs the radio frequency, the resulting signal is detected by the receiver coil, amplified and recorded. The spectrum is produced using knobs and switches on a continual wave instrument or a keyboard on a FT instrument and either printed or displayed digitally (FT) (Blackman et al. 2008, Marcone et al. 2013, Williams and Fleming 1995).

High resolution NMR spectrometers have superconducting magnets that operate at higher field strengths, but require liquid helium to cool them and liquid nitrogen to keep the helium in the liquid phase (Brown 2010, Hemminga 1992).

Sample preparation

The sample itself is prepared by firstly dissolving the compound (~ 10 mg) in a deuterated solvent that is chosen depending on the solubility and where the solvent gives a signal in the NMR spectrum. The most common solvents used in food product analysis are $CDCl_3$, MeOD, Acetone-d_6, D_2O and DMSO-d_6. To the solution of analyte, a small amount of a reference compound, usually tetramethylsilane (TMS), is added and all the resulting peaks in the spectrum are compared to the peak from the reference. However, most commercial deuterated solvents contain TMS and do not need this to be added. The formed solution is filtered through glass wool to remove any insoluble residues and placed inside the NMR sample tube and filled to a depth of 2–4 cm depending on the instrument. This tube is either placed directly into the bore or into an auto sampler, depending on the instrumentation, and lowered by air pressure into the probe through a bore running through the centre of the instrument (Brown 2010, Brown et al. 2010, Hohmann et al. 2014, Hornback 2006, Marcone et al. 2013).

Spectra

The most common nuclei assessed in NMR are ^1H, ^{13}C, ^{14}N, ^{19}F, ^{29}Si, and ^{31}P. This chapter will give a general introduction of ^1H and ^{13}C, one-dimensional and two-dimensional spectra as these spectra are primarily used in the majority of organic compound identification sourced from food products. As stated previously, each carbon and hydrogen nuclei produce a unique signal. These signals are referenced to the internal standard, TMS. This way all spectra can be compared to one another. The frequency that each nuclei resonates at is influenced by the applied field and the unique magnetic environment surrounding it. The surrounding magnetic environment is greatly influenced by the variance in surrounding atoms within the scaffold which alters the electron density surrounding the nucleus being measured. Modern instruments are able to distinguish these slight differences in magnetic environment, resulting in very few overlaps, if any, of signals from different nuclei (Blackman et al. 2008, Brown et al. 2010, Hemminga 1992, Williams and Fleming 1995).

Modern NMR instrumentation operates at frequencies between 300–900 MHz and are selected based on application. It becomes very inconvenient when comparing individual peaks in NMR spectrum when having to state very slight differences in frequency, especially when these numbers may vary between instruments and even between scans in the same instrument at different times. Instead, the difference in frequency is measured in relation to the internal standard (TMS) and divided by the operating frequency of the instrument as shown in Equation 8. This enables spectra from varying instruments to be accurately compared, as it produces a number in a useful range that is field independent (Blackman et al. 2008, Brown et al. 2010, Marcone et al. 2013, Williams and Fleming 1995).

$$\text{Chemical shift } (\delta) = \frac{v_s(Hz) - v_{TMS}(Hz)}{operating\ frequency\ (MHz)}$$

Equation 8: Calculation of the chemical shift δ.

This is known as the chemical shift (δ) and is expressed in fractions of the applied magnetic field in parts per million (ppm) relative to TMS, which is designated a value of 0 in all spectra. For ease of comparison, all NMR spectra are written right to left, with 0 values furthest right. Peaks or signals for each unique nucleus is said to be either downfield, towards the left of the spectrum at higher chemical shifts, or upfield, towards the right of the spectrum at lower chemical shifts (Blackman et al. 2008, Brown et al. 2010, Hohmann et al. 2014, Williams and Fleming 1995).

The degree of shielding of each nucleus from the magnetic field dictates the position of the signal for each nucleus. Nuclei that produce signals downfield in the spectrum are said to be de-shielded whilst nuclei with signals upfield in the spectrum are said to be shielded from the applied magnetic field. This concept is illustrated in Fig. 19 (Blackman et al. 2008, Brown et al. 2010, Hemminga 1992, Williams and Fleming 1995).

Each signal in an NMR spectrum results from a nucleus or equivalent nuclei being in a unique chemical environment based on varying proximities to other functional groups within the organic scaffold of the compound. Nuclei that are closer to electron-donating groups with low electron density (alkyl chains) are more shielded from the external magnetic field, resulting in the observed signal being shifted upfield in the spectrum. Alternatively, nuclei that are closer to electron withdrawing groups with high electron density (nitro, carbonyl) are less shielded from the external magnetic field, resulting in the observed signal being shifted downfield in the spectrum (Blackman et al. 2008, Brown et al. 2010, Hemminga 1992, Williams and Fleming 1995).

A far more narrow frequency range is required to resonate protons in all chemical environments and as such, most organic compounds produce a ^1H spectrum will all signals between 0–12 ppm (Fig. 17). The resonance of all carbon nuclei requires a wider range of frequencies, with ^{13}C spectra generally measured between 0–210 ppm (Fig. 17). However, there are unique cases where individual hydrogen or carbon nuclei resonate outside these parameters, in the case of nuclei exposed to an extremely de-shielded local environment. The parameters of the instrument can be varied by using the software package to scan outside the generic pre-set parameters in order to observe these signals (Brown et al. 2010, Hemminga 1992, Williams and Fleming 1995).

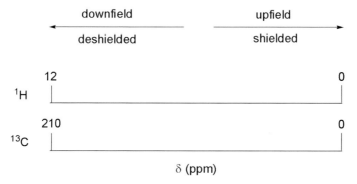

Figure 17. Comparison of the ranges commonly used in ^1H and ^{13}C NMR spectra. Signals are downfield and more de-shielded the further left they are positioned, and further up-field and more shielded the further right they are positioned on the NMR spectrum.

One-dimensional NMR

The simplest NMR spectrum is a one-dimensional spectrum (1D), consisting of a single spectrum of data on the X-axis and the relative peak intensity on the Y-axis. In NMR spectra, it has already been established that an individual nucleus is influenced by other nuclei with a spin. These interactions lead to the signal in the spectrum shown as some degree of a multiplet instead of a single (singlet) signal. The observed multiplet is directly related to the number and identity of the surrounding atoms with a nuclear spin, and the magnitude of the multiplet is directly related to the number of equivalent nuclei it represents (Brown et al. 2010, Hemminga 1992, Hornback 2006, Williams and Fleming 1995).

In 1D proton spectra, a wide range of coupling is possible and visible depending on the surrounding scaffold of the organic compound. The most important coupling is 3-bond coupling to another hydrogen atom (vicinal coupling) represented by $^3J_{CH}$, generally having values between 0–20 Hz, or two bond coupling to another hydrogen (geminal coupling) represented by $^2J_{HaHb}$, generally having values between 0–25 Hz. Geminal coupling is far less common, as hydrogen's connected to the same carbon atom need to be in different chemical environments. This is seen when the structure of a 3-D molecule is locked in a specific unsymmetrical conformation. Both coupling types and an example of structures containing each type are illustrated below in Fig. 18 (Blackman et al. 2008, Brown 2010, Monakhova et al. 2013, Williams and Fleming 1995).

The general rule for the degree of multiplet splitting observed is that a nucleus with a spin of $I = ½$ (H) equally coupled to n other H atoms, will give a signal with $n + 1$ lines. For example, a hydrogen atom coupled to 2 protons in the same chemical environment, will give rise to a triplet signal. The gap between the three peaks of the triplet is the coupling constant (J) between the hydrogen nuclei. However, it should be noted that coupling between protons with identical chemical shifts is not visible in NMR (Brown et al. 2010, Cagliani et al. 2013, Hemminga 1992, Hornback 2006, Williams and Fleming 1995).

Coupling to hydrogen atoms of alcohols (–OH), amines (–NH) and thiols (–SH) is often not visible in NMR spectra. These protons are known as 'exchangable' protons that exchange intermolecularly. Careful selection of deuterated solvent can stop this occurring and 'lock' the protons in place (DMSO), resulting in the coupling being seen; alternatively, performing a D_2O shake results in the signal being completely removed. Other solvents, such as $CDCl_3$, produce a non-coupled broad singlet within the spectrum (Brown 2010, Monakhova et al. 2013, Williams and Fleming 1995).

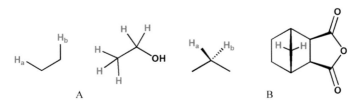

Figure 18. Generic and specific examples of $^3J_{CH}$ vicinal (A), and $^2J_{HaHb}$ geminal coupling (B).

Coupling is also possible in ^{13}C spectra between the carbon (^{13}C) and attached hydrogen atoms (^{1}H). However, this results in a spectrum with overlapping peaks as the coupling constants are large, which becomes difficult to interpret, especially in organic compounds with numerous carbon atoms having very similar chemical environments as close signals in the NMR spectrum. This problem is overcome by running a proton decoupled spectrum, which is the most common ^{13}C program used in NMR. To achieve this, the method is setup to irradiate the protons with frequencies that cause the high (N_β) and low (N_α) energy protons to exchange rapidly during measurement of the carbon signal. Because of this, the carbon nucleus only observes an average state of protons, which gives rise to a sharp single signal in the spectrum (Hornback 2006, Monakhova et al. 2013, Williams and Fleming 1995).

In some cases, it is necessary to see ^{13}C-^{1}H coupling to match the NMR signal to the correct carbon atom within the organic scaffold based on the number of attached hydrogens. These experiments alter the method of the scan by irradiating the sample with a frequency that is close, but does not coincide with the resonance of the protons. This narrows the coupling constants, resulting in less signal overlap and is known as off-resonance decoupling. This method is limited to compounds with a limited number of carbon atoms that are in chemical environments and varied enough to not cause signal overlap (Hemminga 1992, Hornback 2006, Monakhova et al. 2013).

Modern NMR instruments are able to provide the same information as off-resonance decoupling by running a Distortionless Enhancement through Polarisation Transfer (DEPT) experiment. By varying angle of the final frequency pulse, this experiment can differentiate between primary (CH), secondary (CH$_2$), tertiary (CH$_3$) and quaternary (C) carbons. By comparing each DEPT experiment to the original ^{13}C spectrum containing signals for all carbon atoms, the identity of each can be assigned. A 45° pulse results in peaks for all carbons with attached hydrogens. A 90° pulse results in peaks for primary (CH) carbons only and a 135° pulse gives signals in the normal direction for primary (CH) and tertiary (CH$_3$) and inverted signals for secondary (CH$_2$) carbons. Quaternary carbons are emitted from all DEPT spectra (Hornback 2006, Williams and Fleming 1995).

Two-dimensional NMR

Two-dimensional (2D) NMR experiments combine the coupling and splitting information from a single experiment with the corresponding spatial information from the analyzed compound. This is achieved by plotting two spectra on orthogonal axes and looking down on the plot from the top of the peaks. The intensity of each peak is visualized by contour lines, similar to those found on topographical maps (Hemminga 1992, Williams and Fleming 1995).

2-D NMR can produce either a homonuclear (comparing coupling between the same types of nucleus H-H) or a heteronuclear (comparing coupling between different types of nucleus C-H) spectrum. The most important homonuclear experiment is known as Correlation Spectroscopy (COSY). This experiment indicates which hydrogen atoms in an organic scaffold are coupling up to three bonds with each other. Plotting the 1H spectrum on both axes produces a grid with varying contours visible. These contours depict the degree of coupling between hydrogens attributed to each signal in the NMR spectrum. A contour is visible at the intersection between two peaks on the grid where coupling occurs. An example of a COSY spectrum is shown below in Fig. 19 (Williams and Fleming 1995).

A Nuclear Overhauser Effect Spectroscopy (NOESY) homonuclear spectrum shows which signals in a ¹H NMR are producing long range coupling through space, known as the Nuclear Overhauser Effect (NOE). The spectrum looks similar to a COSY spectrum, but the presence of a contour at the intersection of two signals symbolises coupling through space, instead of coupling through bonds as shown in the COSY spectrum (Williams and Fleming 1995).

Heteronuclear correlations can also be viewed in 2D NMR. This is a very important diagnostic tool in determining the structure of an unknown compound as it is able to show which hydrogen atoms are attached to which carbons atoms within the organic scaffold. Running a Heteronuclear Single Quantum Coherence (HSQC) experiment indicates the presence of all one bond coupling between ¹H and ¹³C nuclei within the scaffold. In HSQC spectra, the ¹H and ¹³C spectra are placed on orthogonal axes, with the presence of any contours at the intersection of a ¹H signal and a ¹³C signal indicating that those nuclei are joined through a chemical bond. An example of a HSQC spectrum is shown below in Fig. 20 (Williams and Fleming 1995).

Finally, there are also methods to observe the presence of long-range coupling (approximately 2–4 bonds) between 13C and 1H nuclei. Running a Heteronuclear Multiple-bond Correlation (HMBC) experiment surpresses all one-bond coupling (as shown in HSQC experiments), leaving only coupling visible that occurs between 2–4 bonds. This can often be useful in the complete elucidation of an unknown structure as there can be numerous carbon nuclei present that are not directly attached to a hydrogen

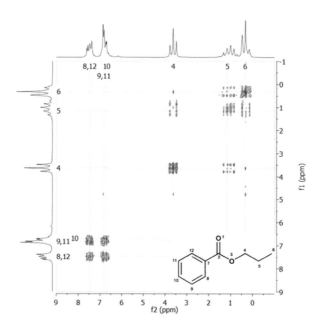

Figure 19. A sample COSY spectrum of propyl benzanoate. The structure is shown with numbered carbons to link the numbered signals that correspond to hydrogen atoms at each location (Magritek 2015).

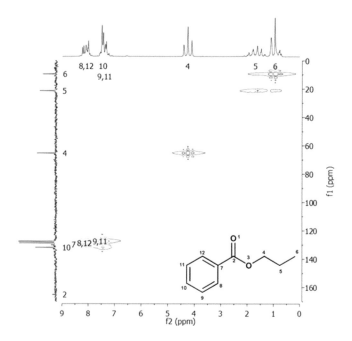

Figure 20. A sample HSQC spectrum of propyl benzanoate. The structure is shown with numbered carbons to link the numbered signals that correspond to hydrogen and carbon atoms at each location (Magritek 2015).

nucleus, but still couple to hydrogen nuclei numerous bonds away. Highlighting the presence of these carbon atoms and determining their place in an unknown organic scaffold can be the final piece of information required to determine the structure of an unknown organic compound sourced from food product waste (Williams and Fleming 1995).

Conclusion

As shown in this chapter, there are numerous methods and instrumentation that can be used for the analysis of bioactive compounds isolated from food waste products. There is, however, no single method of analysis that provides all the information required under every circumstance. The analytical technique needs to be carefully selected, depending on what information is required and what the chemical identity of the analyte mixture is. When embarking on a new research project, there are only two useful ways to develop successful experimental and instrumental techniques. Firstly, previous experience or access to individuals that have worked on similar projects is an invaluable resource to have. Otherwise the only other way to develop successful techniques is through logical trials and experimentation. Resources found within this chapter are aimed at providing at least a general starting point, though, each new project presents a new set of challenges.

References

Agilent technologies. 2012. Agilent J&W GC column selection guide: Speed your selection with this one-stop resource. Agilent Technologies Incoprorated, USA.

Akande, W. G. 2012. A review of experimental procedures of gas chromatography-mass spectrometry (gc-ms) and possible sources of analytical errors. Earth-Sci. Rev. 1(1): 1–9.

Ali, A. A., K. I. Hussain, A. H. Hamed, A. H. Moslah and R. R. Naema. 2014. Determination of polycyclic aromatic hydrocarbons, nitrate and nitrate in Iraqi vegetables by HPLC and UV/Vis spectrophotometer. Int. J. Pharm. 4(2): 165–171.

Angel de la Fuente, M., P. Luna and M. Juarez. 2006. Chromatographic techniques to determine conjugated linoleic acid isomers. Trends. Anal. Chem. 25(9): 917–926.

Antignac, J. P., F. Monteau, J. Negriolli, F. Andre and B. LeBizec. 2004. Application of hyphenated mass spectrometric techniques to the determination of corticosteroid residues in biological matrices. Chromatographia 59: S13–S22.

Blackman, A., S. E. Bottle, S. Schmid, M. Mocerino and U. Wille. 2008. Chemistry, 2nd Edition, John Wiley and Sons, Australia.

Bochenska, P. and A. Pyka. 2013. Use of TLC for the quantitative determination of acetylsalicylic acid, caffeine, and ethoxybenzamide in combined tablets. J. Liq. Chromatogr. Relat. Technol. 36(17): 2405–2421.

Bojko, B., M. Wasowicz and J. Pawliszyn. 2014. Metabolic profiling of plasma from cardiac surgical patients concurrently administered with tranexamic acidL DI-SPME-LC-MS analysis. J. Pharm. Anal. 4(1): 6–13.

Brown, T. L., H. E. Lemay, B. E. Bursten, C. J. Murphy, S. J. Langford and D. Sagatys. 2010. Chemistry the Central Science: A Broad Perspective, 2nd Edition, Pearson, Australia.

Brown, W. H. 2000. Introduction to Organic Chemistry, 2nd Edition, Harcourt Brace and Company, United States of America.

Cagliani, L. R., G. Pellegrino, G. Giugno and R. Consonni. 2013. Quantification of coffea arabica and coffes canephora var. robusta in roasted and ground coffee blends. Talanta 106: 169–173.

Chandu, B. R., K. Kanala, N. T. Hwise, P. Katakam and M. Khagga. 2013. Bioequivalence and pharmacokinetic study of febuxostat in human plasma by using LC-MS/MS with liquid liquid extraction method. Springerplus 2: 194.

Chauhan, A., M. K. Goyal and P. Chauhan. 2014. GC-MS technique and its analytical applications in science and technology. J. Anal. Bioanal. Tech. 5(6): 222.

Cortes, H. J., B. Winniford, J. Luong and M. Pursch. 2009. Comprehensive two-dimensional gas chromatography review. J. Sep. Sci. 32: 883–904.

Couchman, L., R. P. Vincent, L. Ghataore, C. F. Moniz and N. F. Taylor. 2011. Challenges and benefits of endogenous steroid analysis by LC-MS/MS. Bioanalysis 3(22): 2549–2572.

Dhumal, B. R., K. P. Bhusari, A. Patra, S. Thareja and N. S. Jain. 2015. Stability indicating high performance thin layer chromatographic method for the determination on tramadol hydrochloride in pharmaceutical formulation. J. Liq. Chromatogr. Relat. Technol. 38(10): 1088–1093.

Fernandez-Ramos, C., D. Satinsky, B. Smidova and P. Solich. 2014. Analysis of trace organic compounds in environmental, food and biological matrices using large-volume sample injection in column switching liquid chromatography. Trends. Anal. Chem. 62: 69–85.

Franke, A. A. and L. J. Custer. 1999. Application of HPLC coupled with ultraviolet-photodiode array detection for the analysis of phytoestrogens in biological samples. J. Med. Food. 2(3-4): 119–123.

Gelpi, E. 2003. Contributions of liquid chromatography-mass spectrometry to 'highlights' of biomedical research. J. Chromatogr. A 1000: 567–581.

Griffiths, W. J., M. Ogundare, A. Meljon and Y. Wang. 2012. Mass Spectrometry for Steroid Analysis in Mass Spectrometry Handbook, 297–337.

Gurkan, R. and N. Altunay. 2016. A simple efficient ultrasonic-assisted extraction procedure combined with UV-Vis spectrophotometry for the pre-concentration and determination of folic acid (vitamin B9) in various sample matrices. Food. Addit. Contam. Part A 33(7): 1127–1138.

Hamilton. 2012. Syringe Selection Guide. Hamilton Company. United States of America.

Harris, D. C. 2003. Quantitative Chemical Analysis, 6th Edition, W.H. Freeman and Company, United States of America.

Hemminga, M. A. 1992. Introduction to NMR. Trends. Food. Sci. Tech. 3: 179–186.

Hohmann, M., C. Felbinger, N. Christoph, H. Wachter, J. Wiest and U. Holzgrabe. 2014. Quantification of taurine in energy drinks using ^{1}H NMR. J. Pharm. Biomed. Anal. 93: 156–160.

Hornback, J. M. 2006. Organic Chemistry, 2nd Edition, Thompson Brooks/Cole, United States of America.

Kalili, K. M. and A. de Villiers. 2011. Recent developments in the HPLC separation of phenolic compounds. J. Sep. Sci. 34: 854–876.

Knutsson, M., R. Schmidt and P. Timmerman. 2013. LC-MS/MS of large molecules in a regulated bioanalytical environment—which acceptance criteria to apply? Bioanalysis 5(18): 2211–2214.

Komsta, L., R. Skibinski, E. Gowin and P. Maczka. 2013. Exploring hidden trends in classic and micellar thing layer chromatographic retention of model compounds by chemometric methods. J. Liq. Chromatogr. Relat. Technol. 36(17): 2348–2362.

Lai, J. F. and A. A. Franke. 2013. Analysis or circulating lipid-phase micronutrients in humans by HPLC: Review and overview of new developments. J. Chromatogr. B 931: 23–41.

Lakshmi Himabindu, M. R., S. Angala Parameswari and C. Gopinath. 2013. A review of GC-MS and methods development and validation. Int. J. Pharm. Qual. Ass. 4(3): 42–51.

Lee, H. S. and G. A. Coates. 1999. Measurement of total vitamin C activity in citrus products by HPLC: A review. J. Liq. Chrom. & Rel. Technol. 22(15) 2367–2387.

Lepot, M., A. Torres, T. Hofer, N. Caradot, G. Gruber, J. -B. Aubin and J. -L. Bertrand-Krajewski. 2016. Calibration of UV/Vis spectrophotometers: A review and comparison of different methods to estimate TSS and total and dissolved COD concentrations in sewers, WWTPs and rivers. Water. Res. 101: 519–534.

Lewis, R. and W. Evans. 2011. Chemistry, 4th Edition, Palgrave Foundation, United Kingdom.

Li, W. and F. L. S. Tse. 2010. Dried blood spot sampling in combination with LC-MS/MS for the quantitative analysis of small molecules. Biomed. Chrom. 24: 49–65.

Lynch, P. L. M. and I. S. Young. 2000. Determination of thiamine by high-performance liquid chromatography. J. Chromatogr. A 881: 267–284.

Magritek. 2015. Example NMR spectra. Viewed 21 September 2016: http://www.magritek.com/products/spinsolve/nmr-spectra-examples/.

Mant, C. T. and R. S. Hodges. 1996. Analysis of peptides by high-performance liquid chromatography. Methods Enzymol. 271: 3–50.

Marcone, M. F., S. Wang, W. Albabish, S. Nie, D. Somnarain and A. Hill. 2013. Diverse food-based application of nuclear magnetic resonance (NMR) technology. Food. Res. Int. 51: 729–747.

McLafferty, F. W. and F. Turecek. 1993. Interpretation of Mass Spectra, 4th Edition, University Science Books, Sausalito.

Meyer, V. R. 2014. HPLC and ultra HPLC: Basic Concepts from Handbook of Chemical and Biological Plant Analytical Methods. John Wiley and Sons, United States of America.

Milz, B. and B. Spangenberg. 2013. 2D-thin layer chromatography (2D-TLC) flash test of a 17α-ethinylestradiol and related steroids detected by fluorescence densitometry. J. Liq. Chromatogr. Relat. Technol. 36(17): 2378–2386.

Molnar-Perl, I. 1999. Simultaneous quantitation of acids and sugars by chromatography: Gas or high-performance liquid chromatography? J. Chromatogr. A 845: 181–195.

Monakhova, Y. B., T. Kuballa and D. W. Lachenmeier. 2013. Chemometric methods in NMR spectroscopic analysis of food products. Anal. Chem. 68(9): 755–766.

Moreno-Bondi, M. C., M. D. Marazuela, S. Herranz and E. Rodriguez. 2009. An overview of sample preparation procedures for LC-MS multiclass antibiotic determination in environmental and food samples. Anal. Bioanal. Chem. 395: 921–946.

Naumoska, K. and I. Vovk. 2015. Analysis of triterpenoids and phytosterols in vegetables by thin layer chromatography coupled to tandem mass spectrometry. J. Chromatogr. A 1381: 229–238.

Niessen, W. M. A. 2012. Fragmentation of toxicologically relevant drugs in negative-ion liquid chromatography-tandem mass spectrometry. Mass. Spectrom. Rev. 31: 626–665.

Onate-jaen, A., D. Bellido-milla and M. P. Hernandez-artiga. 2006. Spectrophotometric methods to differentiate beers and evaluate beer ageing. Food. Chem. 97: 361–369.

Oros, G. and T. Cserhati. 2013. Support related differential impact of substituents on performance of (alkoxyphenyl)benzamides in normal phase TLC. J. Liq. Chromatogr. Relat. Technol. 36(17): 2363–2377.

Pham, P. J., R. Hernandez, W. T. French, B. G. Estill and A. H. Mondala. 2011. A spectrophotometric method for quantitative determination of xylose in fermentation medium. Biomass Bioenergy 35: 2814–2821.

Piao, C., L. Chen and Y. Wang. 2014. A review of the extraction and chromatographic determination methods for the analysis of parabens. J. Chromatogr. B 969: 139–148.

Pierce, K. M., B. Kehimkar, L. C. Marney, J. C. Hoggard and R. E. Synovec. 2012. Review of chemometric analysis techniques for comprehensive two dimensional separations data. J. Chromatogr. A 1255: 3–11.

Sahil, K., B. Prashant, M. Akanksha, S. Premjeet and R. Devashish. 2011. Gas chromatography-mass spectrometry: applications. Int. J. Pharm. Biol. Sci. Arch. 2(6): 1544–1560.

Saka, C. 2009. High-performance liquid chromatography methods to simultaneous determination of anti-retroviral drugs in biological matrices. Crit. Rev. Anal. Chem. 39: 108–125.

Shah, D. A., D. J. Suthar, C. D. Nagada, U. K. Chhalotiya and K. K. Bhatt. 2014. Development and validation of HPTLC method for estimation of ibuprofen and famotidine in pharmaceutical dosage form. J. Liq. Chromatogr. Relat. Technol. 37(7): 941–950.

Sharma, P., P. Contractor, S. Guttikar, D. P. Patel and P. S. Shrivastav. 2014. Development of a sensitive and rapid method for quantitation of (S)-(–)- and (R)-(+)-metoprolol in human plasma by chiral LC-ESI-MS/MS. J. Pharm. Anal. 4(1): 63–79.

Sherma, J. 2000. Thin-layer chromatography in food and agricultural analysis. J. Chromatogr. A 880: 129–147.

Sherma, J. 2012. Biennial review of planar chromatography: 2009–2011. J. AOAC. Int. 95(4): 992–1009.

Shimadzu. 2014. GC Column Guide. Shimadzu Corporation, Japan.

Staszek, D., M. Orlowska, M. Waksmundzka-Hajnos, M. Sajewicz and T. Kowalska. 2013. Marker fingerprints originating from TLC and HPLC for selected plants from the lamiacae family. J. Liq. Chromatogr. Relat. Technol. 36(17): 2463–2475.

Thermo Fischer Scientific. 2011. Chromatography Vials and Closures. Thermo Fischer Scientific Incorporated, Australia.

Thomson, B. A. 1998. Atmospheric pressure ionization and liquid chromatography/mass spectrometry–together at last. J. Am. Soc. Mass. Spectrom. 9: 187–193.

Verner, P. 1984. Photoionization detection and its application in gas chromatography. J. Chromatogr. 300: 249–264.

Vlajkovic, J., F. Andric, P. Ristivojevic, A. Radoicic, Z. Tesic and D. Milojkovic-Opsenica. 2013. Development and validation of a TLC method for the analysis of synthetic foodstuff dyes. J. Liq. Chromatogr. Relat. Technol. 36(17): 2476–2488.

Watson, R. R. 2014. Polyphenols in Plants: Isolation, Purification and Extract Preparation. Academic Press, United States of America.

Williams, D. H. and I. Fleming. 1995. Spectroscopic Methods in Organic Chemistry, 5th Edition, McGraw-Hill Publishing, Berkshire, England.

Zacharis, C. K. and P. D. Tzabavaras. 2013. Trends and applications of fast liquid chromatography in bioanalysis. J. Chromatogr. B 927: 1–2.

Zhang, M. and H. Qiu. 2015. Progress in stationary phases modified with carbonaceous nanomaterials for high-performance liquid chromatography. Trends. Anal. Chem. 65: 107–121.

CHAPTER 5

Extraction and Utilisation of Bioactive Compounds from Agricultural Waste

Shamina Azeez, C.K. Narayana and *H.S. Oberoi**

Introduction

Production of food waste continues through the entire food life cycle: from agriculture to industrial manufacturing and processing, retail and household consumption. In developed countries, 42 per cent of food waste is produced by households, while 39 per cent of losses occur in the food-manufacturing industry, 14 per cent in food service sector and remaining 5 per cent in retail and distribution. Increasingly, industrial ecology concepts, such as cradle to cradle and circular economy, are considered leading principles for eco-innovation, aimed at 'zero waste economy' in which waste is used as raw material for new products and applications. Many of these residues, however, have the potential to be reused into other production systems, for e.g., biorefineries. The main applications of functional ingredients derived from this transformation are in the nutraceutical and pharmaceutical industry (Fig. 1) (Schieber et al. 2001).

Agro-industrial residues are the most abundant and renewable resource on earth that is poorly valorized or left to decay on the land. Accumulation of this biomass in large quantities every year results not only in deterioration of the environment but also in the loss of potentially valuable material which can be processed to yield a number of value-added products, such as food, fuel, feed and a variety of chemicals. Agro wastes include a wide variety of residues, like molasses, bagasse, oilseed cakes, milling by-products, such as straw, stem, stalk, leaves, husk, shell, peel, lint, seed/stones, pulp, whole pomace, stubble, etc. originating from cereals, pulses, legumes, fruits, vegetables, oil seeds, coffee, tea, etc. (Table 1). Disposal of residue in open spaces or in municipal bins contributes to environmental pollution (Babbar et al. 2011). The best alternative, therefore, is the recovery of phytochemicals/bioactive compounds from such agro-processing residue which can be used in food, cosmetics

Division of Post Harvest Technology & Agricultural Engineering, ICAR-Indian Institute of Horticultural Research, Hessaraghatta Lake PO, Bengaluru-560089, Karnataka, India.
* Corresponding author: harinder@iihr.res.in; hari_manu@yahoo.com

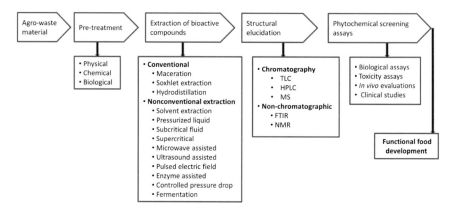

Figure 1. Schematic diagram showing the extraction, isolation and characterization of bioactive compounds from agro-biowaste.

Table 1. By-products/wastes obtained from different crops.

Crop	Waste
Coconut	Fronds, husk, shell
Coffee	Hull, husk, ground
Corn	Cob, stover, stalks, leaves
Cotton	Stalks
Nuts	Hulls
Peanuts	Shells
Rice	Bran, Hull/husk, straw, stalks
Sugarcane	Bagasse
Agricultural crops	Mixed agricultural crops, not limited to crop waste
Mixed type	Agricultural crops and waste including non-organic wastes
Fruits and vegetables	Peel, kernel, rind, stalk, seeds

and the pharmaceutical industry. The recovery of bioactive compounds from such wastes for development of functional/health foods is an efficient way to reuse waste. The agro-industrial residues have alternative uses or markets, though in developing countries they are directly used as fuel along with animal waste and forest litter. Of late, these materials are recycled as a cheap source of renewable feedstock for the production of value-added compounds. They are used as solid substrates in Solid State Fermentation (SSF) processes for the production of different bioactive phenolic compounds (Hernández et al. 2008, Robledo et al. 2008, Vattem and Shetty 2003).

Research on value addition and recycling of the agro- and food industries, localization of bioactive compounds and modification during their processing has received much attention and enormous literature exists on these areas. However, only a few by-products-derived antioxidants have been developed successfully from the vast quantities of plant residues produced by the food processing industry in Europe, primarily grape seed and olive waste extracts (Alonso et al. 2002, Amro et al. 2002). Potential crop candidates with a high annual production and already confirmed to be of high antioxidant potential include apple (Du Pont et al. 2002), tomato (Fuhrman et al. 2000) and artichoke (Jiménez-Escrig et al. 2003). Recycling of by-products has been justified by the fact that polyphenols are located extensively in the peels (Wolfe et al. 2003) and that processing conditions are known to influence the phenolic content (Wang et al. 2003). Although the antioxidant potential of less important crops, such as strawberry (Kähkönen et al. 2001), pear (Imeh and Khokhar 2002), red beet (Kujala et al. 2001), or broccoli (Kurilich et al. 2002) is known, only scanty literature is available on utilisation of their by-products for phenolic recovery. This is caused by three limiting factors: the effectiveness of recovery and extraction, the marketability of resulting extracts and the practical suitability as food, cosmetic or pharmaceutical products.

Bioactive compounds are extra nutritional constituents that occur naturally in small quantities in plant and food products. Most common bioactive compounds include secondary metabolites, such as antibiotics, mycotoxins, alkaloids, food grade pigments, plant growth factors and phenolic compounds (Kris-Etherton et al. 2002). Among the bioactive compounds, research on few pigments and phenolic compounds due to their therapeutic potential is gaining momentum. Polyphenols and flavonoids are ubiquitous bioactive compounds universally present in higher plants, which belong to a diverse group of secondary metabolites, with significant antioxidant capacities that can protect the human body from reactive free radicals (Robards et al. 1999). Reactive free radicals, such as superoxide anion, hydroxyl radical and peroxy radical may cause the disruption of membrane fluidity, protein denaturation, lipid peroxidation, oxidation of DNA and alteration of platelet functions in the human body (Fridovich 1978), resulting in many chronic health problems, such as cancer, inflammation and atherosclerosis. The search for plant-derived biomaterials has stimulated research interest in extracting polyphenolic compounds from underutilized bulk agro-waste. Some prominent ones are the olive oil industry in Australia, which generates large quantities of olive mill waste rich in biophenols having antioxidant, antimicrobial and molluscicidal activities (Obied et al. 2007); solid by-products from the wine industry are also a potential source of antioxidant phytochemicals (Makris et al. 2007); peels from banana, rambutan and mangosteen are rich in polyphenols. Agro-biowaste must go through pretreatment and extraction processes before valuable bioactive compounds present in them are derived. Thus, the main aim of this chapter is to provide comprehensive information about the nature of residues/by-products, pretreatment methods and extraction of bioactive compounds for futuristic applications in the food and pharmaceutical industry.

Pretreatment of Materials

Pretreatment is an important prerequisite for breakdown of the structure of agro-residues, which are mainly composed of cellulose, hemicellulose and lignin (Fig. 2). Lignocellulosic materials, such as agricultural wastes, forestry residues, grasses and woody materials have good potential for bio-fuel production. Typically, agricultural lignocellulosic biomass comprises about 10–25 per cent lignin, 20–30 per cent hemicellulose and 40–50 per cent cellulose (Iqbal et al. 2011). Cellulose is present in

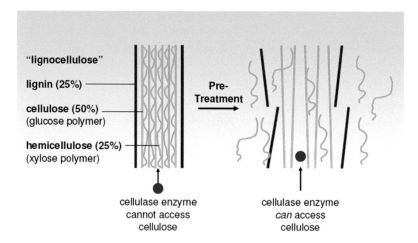

Figure 2. Structural disintegration in the lignocellulosic biomass after pretreatment. (Percentage of different polymers mentioned in the figure are just indicative and their actual composition differs with the lignocellulosic biomass).

large quantities in agro-industrial residues. As hemicellulose and cellulose are present in the cell wall, they undergo lignification. Hence, there is a need to have an effective and economic method to separate cellulose and hemicellulose from the cell wall. Various pretreatment methods, such as physical, chemical, biological (enzymatic) and their combinations are used. Physical and chemical treatments are used to break down the materials present in the agro-waste residues. Biological pretreatment with microorganisms is also recommended as the glucose present in cellulose is readily used by the microorganisms. Mosier et al. (2005) have provided a comprehensive overview of the promising pretreatment technologies for lignocellulosic biomass. Enzymes like phytase, laccase, lignin peroxidase, manganese peroxidase, produced by micro-organisms help in delignification, bleaching and manufacture of animal feed, etc. (Nigam et al. 2009). Effective pretreatment is characterized by the following parameters: preserving hemicellulose fractions to yield maximum fermentable sugars, limiting the loss of carbohydrate to minimize the formation of inhibitors due to degradation products, minimizing energy input and ensuring that the process is economically efficient and cost-effective (Asgher et al. 2013).

Physical Pretreatment Strategies

Physical pretreatment methods include comminution (mechanical size reduction), steam explosion, hydrothermolysis and microwave treatments where no chemical agents are used. Cellulose fibers absorb water readily and swell, the swelling being limited to the amorphous regions of the fiber, with the crystalline regions counteracting this action. For effective hydrolysis of cellulose, a pretreatment that causes swelling is desirable. Increased swelling can be obtained by physical treatments, such as steam treatment, milling and ultrasonic treatment. After this physical pretreatment, the number of glucosidic bonds available for subsequent chemical and/or biological pretreatments will substantially increase (Jostein and Jonny 1980). Pretreatment using extrusion processing has been reported for lignocellulosic biomass (Yoo et al. 2011, Zheng and Rehmann 2014). Hydrothermal treatments, like liquid hot water treatment or hot water compression treatments are ideal for low lignin substrates, especially orange peels, banana peels, etc. (Oberoi et al. 2011b, Oberoi et al. 2011c).

Microwaves are radio waves (1 m to 1 mm; 0.3 GHz to 300 GHz frequency) which on interacting with organic matter get absorbed by water, fats and sugars. Their energy gets transferred to organic molecules, generating enormous amount of heat. Microwaves used in pretreatment of lignocellulosic biomass cause localized heating, leading to disruption of lignocelluloses architecture and creating greater accessibility for cellulose and hemicellulose for enzymatic hydrolysis (Ooshima et al. 1984, Sarkar et al. 2012).

Chemical pretreatments

Chemical pretreatment includes alkali, acid, lime (calcium hydroxide) and ammonia treatments. To date chemical pretreatment methods have been most extensively used for delignification of cellulosic materials, for recovery of sugar monomers from cellulose and hemicellulose polymers from lignocellulosic biomass. Acid- and alkali-based hydrolysis are the most commonly used chemical pretreatments (Anwar et al. 2014).

Acid hydrolysis of cellulosic biomass with concentrated hydrochloric acid or sulphuric acid can be performed at very low temperatures as compared to dilute-acid pretreatment. The drawbacks of this process are that it requires acids in higher concentration (30–70 per cent), therefore causing highly corrosive reactions. This increases the infrastructure cost in terms of specialized non-metallic or non-corrosive material (ceramic or carbon-brick lining) and high operating and environmental costs in comparison to pretreatment with dilute-acid hydrolysis (Wyman 1999). Acid pretreatment has been applied on biomass feedstocks, like herbaceous material (grass), hardwoods and agricultural wastes, effectively solubilizing the hemicellulose (Liao et al. 2007). Other factors, including temperature and incubation time during acid pretreatment, also had important impact on alteration of the structure of biomass. Oberoi et al. (2010a) reported significant ethanol production from rice straw through fermentation of hydrolysate obtained through dilute acid hydrolysis. Studies on orange peels using successive acid pretreatments for ethanol production have been reported previously (Oberoi et al. 2010b). A major disadvantage of this process is the formation of secondary products which can lower the ethanol yield due to formation of furfural and hydroxyl-methyl furfural compounds as these compounds interfere in the fermentation process.

Alkali hydrolysis involves pretreatment with sodium, calcium and ammonium hydroxide, resulting in structural alterations inside the lignocellulosic material, such as the depletion of lignin barrier, cellulose swelling and partial decrystallization and solvation of cellulose and hemicellulose, respectively (Sills and Gossett 2011). Alkali pretreatment has been successfully used for corn stover, switch-grass, bagasse, wheat, and rice straw and cotton stalk (Hu et al. 2008, Zhao et al. 2008, Zhu et al. 2010, Oberoi et al. 2011a, Kaur et al. 2012, Rawat et al. 2014). Zhao et al. (2008) reported that sodium hydroxide pretreatment is effective for hardwood, wheat straw, switch-grass and soft-wood with less than 26 per cent lignin content.

Ammonia pretreatment involves aqueous ammonia treatment at high temperatures, which sufficiently reduces lignin content with loss of some hemicellulose, while cellulose is decrystallized. There are three types of ammonia pretreatment techniques, viz., the Ammonia Recycle Percolation (ARP) (Kim et al. 2003); soaking in Aqueous Ammonia treatments (SAA) (Kim et al. 2008) and Ammonia Fiber Explosion method (AFE) (Teymouri et al. 2005). In ARP treatment, the biomass is pretreated with aqueous ammonia in a flow-through column reactor. In SAA, the process at low temperature removes lignin efficiently by minimizing the interaction with hemicellulose and thereby increasing the surface area and pore size; the retained hemicellulose/cellulose can be hydrolyzed to fermentable sugars by commercial enzymes. Ammonia fiber explosion is a potential technique for pretreatment of lignocellulosic material, wherein the biomass is treated with liquid anhydrous ammonia at 60–100ºC and high pressure (250–300 psi) for 5 min, after which the pressure is released rapidly. The combined effects of ammonia and high pressure lead to swelling of lignocellulose biomass, disruption of lignocellulose architecture leading to hemicellulose hydrolysis and decrystallization of cellulose (Holtzapple et al. 1992).

Organosolv pretreatment process involves extraction of lignin by employing organic solvent or a mixture of solvents, such as ethanol, methanol, acetone and ethylene glycol, in combination with water (Ichwan and Son 2011). Temperature may vary from 100 to 250ºC; catalysts, such as inorganic or organic acids, may be used. This process causes hydrolysis of the internal bonds in lignin and also between lignin and hemicellulose, and hydrolysis of the glycosidic bonds in hemicellulose and to a lesser extent in cellulose. After removal of lignin, the cellulose-rich biomass is used for enzymatic hydrolysis (Zhao et al. 2009).

Delignification of lignocellulose can also be achieved by treatment with oxidizing agents, like hydrogen peroxide, ozone, oxygen or air, where lignin is converted to acids (Hammel et al. 2002, Nakamura et al. 2004). However, these acids can act as inhibitors during the fermentation process and therefore, acids need to be removed (Alvira et al. 2010); further, oxidative treatment also damages the hemicellulose fraction of the lignocellulose complex and a major portion of the hemicellulose gets degraded and becomes unavailable for fermentation (Lucas et al. 2012).

Biological pretreatment

Biological pretreatment employs wood-degrading microorganisms, such as white rot fungi, brown or soft-rot fungi and bacteria to modify the chemical composition and/ or structure of the lignocellulosic biomass. Biodelignification replaces or supplements chemical-based pretreatments detailed above; it is simple, economical, eco-friendly, does not require huge infrastructure and is less health hazardous, compared to physico-chemical or chemical-based pretreatment approaches. Therefore, research on biological pretreatment of lignocellulosic biomass is gaining momentum. Cellulose is commonly degraded by cellulases which are produced by several microorganisms. Many bacteria and fungi produce significant quantities of extracellular enzymes, capable of completely hydrolyzing crystalline cellulose *in vitro.* Fungi are the main cellulose-producing microorganisms. Cellulase is capable of breaking down a highly ordered cellulose polymer into sufficiently smaller sugars which are able to pass through the microbial cell wall and are expelled out in the medium. Enzymatic degradation of cellulose is a complex process that requires the participation of at least three types of enzymes: Endoglucanases (Cx) (E.C.3.2.1.4), Exoglucanase (C1) (E.C.3.2.1.91) and β-glucosidase (E.C.3.2.2.21) (Bhavsar et al. 2015). Bioconversion of lignocellulosic materials to useful products is normally a multi-step process which includes pretreatment, enzymatic hydrolysis and fermentation (Xiao et al. 2012), so that the modified or pretreated biomass is more amenable to enzyme digestion. The limitations of biological pretreatment are that it is a very slow process, requiring careful control of growth conditions and a large area for treatment. In addition, most lignolytic microorganisms solubilise/consume not only lignin but also hemicellulose and cellulose. Because of these techno-economic barriers, biological pretreatment is less attractive commercially (Eggeman and Elander 2005).

Extraction techniques

Bioactive compounds are extra-nutritional constituents that naturally occur in small quantities in plant and food products. Natural bioactive compounds have diverse structures and functionalities with molecules having enormous potential for the production of nutraceuticals, functional foods and food additives. Some of these compounds can be found in nature in high concentrations, such as polyphenols, but others are found at very low levels so that mass harvesting is required to obtain sufficient amounts, thus making chemical synthesis unprofitable (Joana Gil-Chavez et al. 2013). Phenolic compounds, include flavonoids, phenolic acids and tannins, among others. Flavonoids are the largest group of plant phenolics, comprising over half of the eight thousand naturally-occurring phenolic compounds, and include flavonols, flavones, flavanones, flavanols, isoflavones and anthocyanidins. Phenolic acids are another bioactive group of phenolic compounds found in plant and food products and comprise of the subgroups, i.e., hydroxybenzoic and the hydroxycinnamic acids (Harborne et al. 1999).

Bioactive compounds are generally recovered from natural sources by solid-liquid extraction employing organic solvents in heat-reflux systems. The classical techniques to obtain bioactive compounds from plants are: (1) Soxhlet extraction, (2) Maceration and (3) Hydrodistillation.

Soxhlet extractor was first proposed by the German chemist, Franz Ritter Von Soxhlet (1879), designed initially for extraction of lipids but now applied to other valuable bioactive compounds from various natural sources. In this method, a small amount of the dry sample is placed in a thimble in a distillation flask containing a specific solvent. On heating and condensing to an overflow level, the solution of the thimble-holder is aspirated by a siphon which unloads the solution back into the distillation flask. This solution carries extracted solutes into the bulk liquid. The solute remains in the distillation flask and the process runs repeatedly for exhaustive extraction.

Maceration has been used in homemade preparation of tonics since long. It became a popular and inexpensive way to obtain essential oils and bioactive compounds. For small-scale extraction, maceration generally consists of grinding plant materials to increase the surface area for proper mixing with an appropriate solvent in a closed vessel. The solvent is strained off but the marc (solid residue of this extraction process) is pressed to recover large amounts of occluded solutions. The strained and the pressed out liquid are filtered for further processing.

Hydrodistillation is a traditional method for extraction of bioactive compounds and essential oils from fresh plant materials using water as a solvent. Water distillation, water and steam distillation and direct steam distillation are the types of hydrodistillation (Vankar 2004). In hydrodistillation, the plant materials packed in a still compartment are boiled in sufficient water. Alternatively, direct steam is injected into the plant sample. Indirect cooling by water condenses the vapour mixture of water and oil. The condensed mixture flows to a separator, where oil and bioactive compounds separate automatically from water (Silva et al. 2005). Hydrodistillation involves three main physicochemical processes; hydrodiffusion, hydrolysis and decomposition by heat. A drawback limiting its use for thermo labile compounds is that at high extraction temperatures some volatile components are lost. Recently, more efficient clean and green techniques have been employed to obtain these compounds, such as the supercritical fluids, high pressure processes, microwave-assisted extraction ultrasound-assisted extraction and fermentation processes (Cortazar et al. 2005, Markom et al. 2007, Wang and Weller 2006, Martins et al. 2011).

Solvent extraction

In solvent extraction (SE) the pretreated raw material is exposed to different solvents which selectively extract compounds of interest and also other agents (flavors and colorings). Samples are usually centrifuged and filtered to remove solid residue and the extract is used as additive, food supplement or for preparation of functional foods (Starmans and Nijhuis 1996). Some of the most commonly used solvents in the extraction process are hexane, ether, chloroform, acetonitrile, benzene, methanol and ethanol in different ratios with water for extraction of both polar and nonpolar organic compounds, such as alkaloids, organochlorine pesticides, phenols, aromatic hydrocarbons, fatty acids and oils, among others (Plaza et al. 2010c) (Table 2). The disadvantages are that large amounts of solvent are required over extended periods, and since most of these solvents are toxic to humans and the environment (and the

Table 2. Bioactive compounds extracted using different solvents.

Water	Ethanol	Methanol	Chloroform	Dichloromethanol	Ether	Acetone
Anthocyanins	Tannins	Anthocyanin	Terpenoids	Terpenoids	Alkaloids	Flavonoids
Tannins	Polyphenols	Terpenoids	Flavonoids		Terpenoids	
Saponins	Flavonol	Saponins				
Terpenoids	Terpenoids	Tannins				
	Alkaloids	Flavones				
		Polyphenols				

Adapted from Cowan (1999)

extraction conditions are sometimes laborious), the solvent must be completely separated from the final extract by evaporation or concentration, before the product is used in food applications (Starmans and Nijhuis 1996). There is also the possibility of thermal degradation of bioactive compounds due to high temperatures and longer extraction time. The advantages are low processing cost and ease of operation. Solvent extraction has been improved by other methods, such as ultrasound or microwave extraction and super critical fluid extraction (SCFE), among others, to obtain better yields (Szentmihalyi et al. 2002).

Aludatt et al. (2016) found that the acetone-water solvent mixture (1:1; v/v) helped in efficient extraction of phenolics from the extracts of *Thymus vulgaris* L., with antioxidant and inhibitory activities on angiotensin-converting enzyme and α-glucosidase. The optimal extraction temperature for maximum phenolic content and antioxidant activity associated with methanol extraction was 60°C, whereas a lower temperature of 40°C was required to maximize activities for antihypertensive (inhibitory activities of ACE) and antidiabetic properties (inhibitory activities on α-glucosidase and α-amylase). Generally, major differences were noticed in phenolic profiles among the tested extraction conditions with thymol as the predominant phenolic seen in most extractions, while gallic acid, rosmarinic acid or diosmin were predominant in other extracts. Extracts with the same predominant phenolic compound and similar phenolic content showed major disparities in their ACE, glucosidase and α-amylase inhibitory activities, indicating that the major phenolic profiles of thyme extracts may not be necessarily related to the degree of inhibition of ACE, glucosidase and α-amylase enzymes. This study highlights the effect of varying extraction conditions, such as solvent type and combination of extraction time and temperature as related to free and bound phenolic content and profiles of the extracts with their therapeutic effects.

The *n*-hexane and dichloromethane fractions obtained from sequential extraction of *Lophostemon suaveolens*, a relatively unexplored endemic medicinal plant of Australia, exhibited antibacterial activity against *Streptococcus pyogenes* and methicillin sensitive and resistant strains of *Staphylococcus aureus*. GC-MS analysis of the *n*-hexane fraction by Naz et al. (2016) showed the presence of the antibacterial compounds, aromadendrene, spathulenol, β-caryophyllene, α-humulene and α-pinene and the anti-inflammatory compounds, β caryophyllene and spathulenol. Fractionation of the dichloromethane extract led to the isolation of eucalyptin and the known anti-inflammatory compound betulinic acid. The profiles of bioactive compounds (including phenolics and flavonoids in free and bound fractions, anthocyanins, proanthocyanidins, vitamin E, and γ-oryzanol) of outer and inner rice bran from six colored rice samples was obtained using 80 per cent ethanol (Huang and Ng 2012). The authors further reported that the free fraction of the extracts dominated the total phenolics (72–92 per cent) and the total flavonoids (72–96 per cent) of colored rice bran.

Pressurized liquid extraction

Pressurized liquid extraction (PLE) is referred to as accelerated SE and pressurized SE. It uses organic liquid solvents at high temperature (50 to 200°C) and pressure (1450 to 2175 psi) to ensure rapid extraction rate. As the temperature increases, the dielectric constant of the solvent decreases, lowering the polarity of the solvent. Thus, temperature can be used to match the polarity of a solvent to that of the compounds of interest to be recovered could be extracted (Dunford et al. 2010, Miron et al. 2010). High pressure helps the extraction cells to be filled faster and forces the liquid into the solid matrix. PLE allows a faster extraction, with lesser solvents and higher yields, compared to traditional SE. In addition, using PLE, food-grade extracts are obtained when water or other GRAS solvents, such as ethanol, are used (Plaza et al. 2010a). Despite these advantages, PLE is not suitable for thermolabile compounds, since high temperature damages their structure and functional activity (Ajila et al. 2011).

Subcritical water extraction

The use of water under high temperature and pressure below supercritical conditions in extraction processes is referred to as subcritical water extraction (SWE). Subcritical water extraction is carried out by using hot water (from 100 to 374°C, the latter being the critical temperature of water) under high pressure (usually from 10 to 60 bar) to maintain water in the liquid state (Herrero et al. 2006). Solvent parameters, such as dielectric constant, solubility and temperature are affected when the liquid state is maintained. The dielectric constant of water at room temperature, which is nearly 80°C, decreases to about 30 at 250°C, similar to some organic solvents like ethanol or methanol (Adil et al. 2007). Therefore, this technique can be used for the extraction of non-polar phytochemicals and replace organic solvents, with due consideration for the variability of dielectric constants in different types of compounds. This technique has a number of important advantages over the traditional extraction techniques—it is faster, yields higher, facilitates use of lower solvent volumes and is environment-friendly (Plaza et al. 2010b). Subcritical water extraction has been successfully applied to the extraction of different bioactive compounds (mainly antioxidants) from several vegetable and other matrices. Hassas-Roudsari et al. (2009) reported that SWE at 160°C yielded the highest total phenolic content and antioxidant capacity per gram of canola seed meal. Highest recovery of catechins and proanthocyanidins from wine-related products was observed in the material subjected to three sequential extractions at 50, 100, and 150°C. Selective extraction of compounds with different degrees of polymerization can be achieved by using one-step extraction at different temperatures (Garcia-Marino et al. 2006).

Supercritical fluid extraction (SCFE)

Supercritical fluid extraction (SCFE) is an environmentally-safe technology for extraction of bioactive compounds from plants, food by-products, algae and microalgae, etc. The advantages of this technique are high selectivity, faster extraction, less pollution and the use of nontoxic organic solvents. The technique is based on properties of fluids, such as density, diffusivity, dielectric constant and viscosity, and usually involves modifying conditions, such as pressure and temperature to reach a supercritical fluid state (SF). Under these conditions, a fluid is between the gas and liquid states because the density of SF is similar to that of the liquid and its viscosity is similar to that of a gas. Thus the supercritical state of a fluid is the state in which liquid and gas are identical to each other. In addition, SFs have better transport properties than liquids due to their density which, unlike liquid solvents, is adjustable by changing pressure and temperature (Herrero et al. 2006). In SCFE, the raw material is placed in a temperature-controlled and pressurized extractor vessel, the fluid and the dissolved compounds are transported to separators and the products are collected through a tap located in the lower part of the separator. The fluid is regenerated and cycled or released into the environment (Sihvonen et al. 1999). Though there are many compounds that can be used as SFs (ethylene, methane, nitrogen, xenon, or fluorocarbons), carbon dioxide is the most popular due to its safety and low cost and causes minimal alteration in the bioactive compounds, thus preserving their functional properties (Daintree et al. 2008, Cavero et al. 2006). Supercritical carbon dioxide is an attractive alternative to organic solvents, because it is GRAS, non-explosive, nontoxic, inexpensive and can solubilize lipophilic substances to be easily removed from the final products, since CO_2 is a gas at room temperature and pressure (Wang and Weller 2006). However, because of its low polarity, it is less effective in extracting highly polar compounds from their matrices (Herrero et al. 2006). Therefore, in order to enhance solubility and selectivity of the process, solubility enhancers, called co-solvents or modifiers such as hexane, methanol, ethanol, isopropanol, acetonitrile, or dichloromethane, are added in small quantities (Sihvonen et al. 1999). Of these, ethanol is a recommended co-solvent in SCFE because of its lower toxicity and miscibility in CO_2 (Liza et al. 2010). This method has been applied to extract lipid compounds, such as tocopherols, phytosterols, policosanols and free fatty acids from sorghum (Liza et al. 2010). Cavero et al. (2006) reported that oregano leaf extracts obtained by SC-CO_2 possessed high antioxidant capacity, especially when ethanol was used as a co-solvent. SF-CO_2 is important for the extraction of natural compounds in the food industry because it allows the extraction of thermally labile or easily oxidized compounds (Herrero et al. 2010). SFE is freely used in many industrial applications, including coffee decaffeination, fatty acid refining and the extraction of essential oils and flavors with the potential use in nutraceuticals and functional foods. The conditions for extraction, recovery and characterization of bioactive compounds using supercritical extraction are listed in Table 3. Thus, this method is an important alternative to conventional extraction methods for extracting biologically active compounds and active substances in microparticles as dry powders (Daintree et al. 2008). Subcritical carbon dioxide soxhlet extraction (SCDS) is an innovative

Table 3. Conditions for the extraction, recovery and characterization of natural compounds from plants using supercritical extraction.

Sample	Temperature (°C)	Pressure (Bar)	Co-solvent (%)	Target compounds
Red grape pomace	45	100–250	Methanol (5)	Pro-anthocyanidins
Spearmint leaves	40–50–60	100–300	Ethanol (10)	Flavonoids
Green tea leaves	60	310	Ethanol (10)	Catechins
Tomato paste waste	880	300	Ethanol (5)	Carotenoids
Tomato skin	75	350	Ethanol (10)	Carotenoids
Grape seed	30–40	130	Ethanol (30)	Anthocyanins

Adapted from Babbar et al. (2015)

green technology with the capacity to extract 10 times more bioactive compounds than hexane, from rice bran oil, viz., total tocopherol, total tocotrienol including α-, β-, γ- and δ-tocopherol; α-, γ- and δ-tocotrienol and γ-oryzanol (Chia et al. 2015).

Microwave assisted extraction

As detailed in the section on pretreatment, microwave assisted extraction (MAE) is more advanced than the traditional SE methods since it works on the principle of heating the moisture inside the cells and evaporating it to produce a high pressure on the cell wall. The pressure builds up inside the biomaterial, which modifies the physical properties of the biological tissues (cell wall and organelles disrupter) by improving the porosity of the biological matrix. It thereby allows better penetration of extracting solvent and improves the yield of the desired compounds (Routray and Orsat 2011). Microwave-assisted extraction is suitable for the recovery of a vast array of bioactive compounds, especially those with antioxidant capacity, such as phenolic compounds (Moreira et al. 2012), carotenoids (Pasquet et al. 2011), terpenoids, alkaloids and saponins (Zhang et al. 2011). Higher antioxidant activity is obtained in peel extracts of citrus mandarin (Hayat et al. 2009), peanut skins (Ballard et al. 2010), tomatoes (Li et al. 2011) and onions (Zill-e-Huma et al. 2011), compared to rotary extraction. Microwave-assisted extraction is affected by a large number of factors, such as power, frequency, exposure time, moisture content and particle size of the sample matrix, type and concentration of solvent, ratio of solid to liquid, extraction temperature, extraction pressure and the number of extraction cycles (Mandal et al. 2007), the solvent being the most critical in terms of its solubility, dielectric constant and dissipation factors. Solvents with high dielectric constant, such as water and polar solvents which can absorb high microwave energy, are usually better than nonpolar solvents (Wang and Weller 2006). In addition, the dissipation factor (the efficiency with which different solvents heat up under microwave) plays an important role. Recovery of phenolic compounds is greater when using solvents, such as ethanol or methanol as compared to water, which is associated with a higher dissipation factor (Ajila et al. 2011). Though water has a higher dielectric constant than ethanol or methanol, since its dissipation factor is low, water is slower in heating the moisture inside the sample matrix and generate the pressure to trigger the leaching out of the phytochemicals. Therefore, it is advisable to use solvents with a high dielectric constant as well as a high dissipation factor which can be achieved by using a mixture of water with other solvents (ethanol or methanol). Microwave-assisted extraction has many advantages over conventional extraction techniques, including lower environmental pollution, higher extraction efficiency and shorter extraction time. However, in order to be considered for industrial applications, some of the important limitations which need to be overcome include recovery of nonpolar compounds and modification of the chemical structure of target compounds which may alter their bioactivity. Carvalho et al. (2016) reported that the quantitative profile of phenolic compounds in goji (*Lycium barbarum*) extracts was strongly dependent on microwave-assisted extraction conditions, with significant correlations found between the presence of several flavonoids and solvent composition, as well as between phenolic acids with methoxy group and the response to DNA-based sensors. The authors suggest that targeted extractions for specific compounds would result in extracts that are richer in antioxidant capacity.

Ultrasonic assisted extraction

Though the ultrasound technology is not new, it has only recently been exploited for the recovery of phytochemicals from natural sources. Ultrasound-assisted extraction (UAE) is a viable alternative to conventional SE, providing higher recovery with better bioactivity and lower solvent consumption. Its extraction efficiency is due to the phenomenon called acoustic cavitation, wherein with sufficient ultrasound intensity the expansion cycle creates cavities or microbubbles in the liquid. Once formed, bubbles will absorb the energy from the sound waves and grow during the expansion cycles to recompress during the compression cycle. Further, bubbles may start another rarefaction cycle or collapse, leading to shock waves of extreme conditions of pressure and temperature (several hundred atmospheres and around 5000 K temperature) (Leighton 2007, Soria and Villamiel 2010, Esclapez et al. 2011). The implosion of cavitation bubbles can hit the surface of the solid matrix and disintegrate the cells, causing the release of the desired compounds. Ultrasound assisted extraction has been used for the extraction of proteins (Qu et al. 2012), sugars (Karki et al. 2010), polysaccharides-protein complex (Cheung et al. 2012) and oil (Adam et al. 2012). Recent studies revealed that with UAE, phenolic compounds were less degraded (Dobias et al. 2010) and sometimes no degradation was observed under optimized conditions (Pingret et al. 2012). However, UAE should be carefully used in the extraction of unstable compounds, such as carotenoids (Zhao et al. 2006).

Pulsed electric field extraction (PEF)

The pulsed electric field (PEF) treatment is useful in improving the pressing, drying, extraction and diffusion processes (Vorobiev and Lebovka 2005). When a living cell is suspended in an electric field, the electric potential passes through the cell membrane and based on the dipole nature of the membrane molecules, electric potential separates molecules according to their charge. After exceeding a critical value of approximately one volt of transmembrane potential, repulsion occurs between the charge carrying molecules that form pores in weak areas of the membrane and cause drastic increase in permeability (Bryant and Wolfe 1987). Usually a simple circuit with exponential decay pulses is used for PEF treatment of plant materials in a chamber consisting of two electrodes in either continuous or batch mode (Puértolas et al. 2010). The effectiveness of PEF treatment strictly depends on the parameters, such as field strength, specific energy input, pulse number, treatment temperature and properties of the materials to be treated (Heinz et al. 2003).

Pulsed electric field can increase mass transfer during extraction by destroying the membrane structure of the plant materials for enhancing extraction and decreasing extraction time. Pulsed electric field treatment at a moderate electric field (500 and 1000 V/cm; for $10^{-4}-10^{-2}$ s) can damage the cell membranes with little temperature increase and minimize the degradation of heat-sensitive compounds (Fincan and Dejmek 2002). Pulsed electric field can also be used as a pretreatment process prior to conventional extraction to lower the extraction effort (López et al. 2009). Pulsed Electric Field treatment (at 1 kV/cm with low energy consumption of 7 kJ/kg) in a solid liquid extraction process for extraction of betanin from beetroot showed maximum extraction as compared with freezing and mechanical pressing (Fincan et al. 2004).

Guderjan et al. (2005) showed that the recovery of phytosterols from maize increased by 32.4 per cent and isoflavonoids (genistein and daidzein) from soybeans increased by 20–21 per cent when PEF was used as pretreatment process. Corralesa et al. (2008) extracted bioactive compounds, such as anthocyanins from grape by-products by using various techniques and found better extraction of anthocyanin monoglucosides by PEF. The application of PEF treatment on grape skin before maceration step can reduce the duration of maceration and improve the stability of bioactives (anthocyanin and polyphenols) during vinification (López et al. 2008).

Enzyme-assisted extraction

Most phytochemicals, such as flavonoids, are present in different forms, interacting with the cell wall components (cellulose, hemicellulose and pectin) (Fu et al. 2008). Enzymes, such as cellulase, β-glucosidase, xylanase, β-gluconase and pectinase help to degrade the cell wall structure and depolymerize the plant cell wall polysaccharides, facilitating the release of these linked compounds, since they can hydrolyze the ester-linked phenolic acids (Chen et al. 2010). There are several reports on enzymatic treatment of plant tissues for the extraction of natural bioactive compounds. Chandini et al. (2011) found that the enzyme tannase was superior to pectinase for improving the quality of black tea extracts. Extraction of luteolin and apigenin from pigeonpea leaves was facilitated by pectinase, cellulase and β-glucosidase. Enzyme-assisted extraction was used to improve the antioxidant composition of black carrot juice and to obtain vegetable oils (Khandare et al. 2010, Szydlowska-Czerniak et al. 2010). Enzyme extraction is used to extract compounds from algae as well, where the structural complexity and rigidity of the algal cell wall prove an obstacle (Wang et al. 2010). Enzyme-assisted extraction is valuable in extracting precious by-products, such as gallic acid from agricultural waste (Curiel et al. 2010), which can be used for the preparation of food additives, such as pyrogallol and propyl gallate (Yu and Li 2008) and also to serve as an intermediate for the synthesis of antibacterial drug, trimethroprim, by pharmaceutical chemistry (Curiel et al. 2010).

Instant controlled pressure drop-assisted extraction

Instant controlled pressure-drop (DIC) was defined by Allaf and Vidal (1988) and since then, the technology has been refined for the extraction of volatile compounds and antioxidants (Berka-Zougalia et al. 2010). Instant controlled pressure-drop consists of thermo-mechanical effects induced by subjecting the raw material for a short period of time to saturated steam followed by an abrupt pressure drop towards a vacuum (Ben Amor and Allaf 2009). The pressure-drop applied provokes the auto-vaporization of volatile compounds, instantaneous cooling of the products which stop thermal degradation and expansion of the cell wall, thus enhancing the mass transfer and improving the recovery of the desired compounds. The auto-vaporization of volatile compounds makes DIC suitable for the recovery of essential oils (Allaf et al. 2012). Kristiawan et al. (2008) extracted essential oils of Indonesian Kananga in less than 6 min with a yield of 2.8 g/100 g dry matter compared with a similar yield (2.5 g/100 g dry matter) but after a prolonged 16 hours steam distillation. Allaf et al. (2012) found that DIC extracts from orange peels had better essential oil quality (major oxygenated compounds) and antioxidant capacity (about 13 per cent) when compared with hydro-distilled extracts—the common way to obtain essential oils. Using DIC, Ben Amor and Allaf (2009) optimized the parameters (vacuum, pressure and temperature) to improve the recovery of anthocyanins from Roselle by up to 135 per cent and by using water as a solvent. Allaf et al. (2012) used DIC and UAE sequentially for the extraction of phenolic compounds (naringin and hesperidin) from orange peels with a very high yield, best kinetics and antioxidant capacity compared to standard SE.

Fermentation methods

Extraction/production of bioactive compounds by fermentation is a promising alternative which can provide high quality and high activity extracts, while precluding any toxicity associated with the extractants, especially the organic solvents. During fermentation, bioactive compounds are obtained as secondary metabolites produced by microorganisms after the microbial growth is completed (Nigam 2009). Studies on liquid culture show that the production of these compounds starts when growth is limited by the exhaustion of one key nutrient—carbon, nitrogen or phosphate source (Barrios-González et al. 2005).

Fermentation processes are of two types: submerged fermentation (SmF), where microorganisms are cultivated in a liquid medium containing nutrients and solid state fermentation (SSF), where the microbial growth and product formation occurs on solid particles in the absence (or near absence) of directly available water. However, the substrate contains sufficient moisture to allow growth and metabolism of microorganisms (Pandey 2003). Solid state fermentation has gained more importance since this process may lead to a higher yield and productivity or better product characteristics than SmF. In addition, since low-cost agricultural and agro-industrial residues are utilized as substrates, capital and operating costs are lower as compared to SmF. Low water volume in SSF has also a large impact on the economy of the process due to smaller fermenter-size, reduced downstream processing, reduced stirring and lower sterilization costs (Nigam 2009, Pandey 2003). The main limitation of SSF, however, is the scaling-up of the process, largely due to heat transfer and culture homogeneity problems (Di Luccio et al. 2004, Mitchell et al. 2000). Although many bioactive compounds are still produced by SmF, in the last few decades, the SSF technique is being preferred for its efficiency.

Pomegranate wastes contain phenolic compounds, including anthocyanins (derived from delphinidin, cyanidin and pelargonidin), hydrolysable tannins (catechin, epicatechin, punicalin, pedunculagin, punicalagin, gallic and ellagic acid esters of glucose) (Cuccioloni et al. 2009, Gil et al. 2000), and several lignans (isolariciresinol, medioresinol, matairesinol, pinoresinol, syringaresinol and secoisolariciresinol) (Bonzanini et al. 2009), with antioxidant, anti-mutagenic, anti-inflammatory and anticancer activities (Naveena et al. 2008). Pomegranate husks are successfully used as a matrix to produce almost 8 kg of ellagic acid per ton of waste, by SSF with *Aspergillus niger* GH1. This process is economical and quite profitable from the industrial point of view, considering the commercial price of this acid and the low cost and abundance of the husks. Elagitannin acyl hydrolase is responsible for bioconversion of elagitannin into ellagic acid during SSF of pomegranate husks (Robledo et al. 2008). Cranberry pomace, the by-product of cranberry juice processing industry, is also a good source of ellagic acid and other phenolic compounds. Bioprocessing of this waste by SSF with *Lentinus edodes*, using its esterase enzyme, was useful to increase the ellagic acid content by being an alternative for the production of bioactive compounds (Vattem and Shetty 2003). In India, Teri pod (*Caesalpinia digyna*) cover, the solid residue obtained during processing of the pod for recovery of oil, contains tannin that can be used as substrate for microbial bioconversion to gallic acid by SSF with *Rhizopus oryzae* (Kar et al. 1999). Green coconut husk, an abundant agro-industrial residue, is a potential source of ferulic acid from which vanillin can be obtained *via* microbial conversion by the basidiomycete *Phanerochaete chrysosporium* under SSF, where the production of lignolytic enzymes released ferulic acid from the coconut husk cell wall and subsequently, vanillin was obtained with a high yield by the ferulic acid conversion (Barbosa et al. 2008). The action of enzymes, such as α-amylase, laccase and β-glucosidase, tannin acyl hydrolase, ellagitanin acyl hydrolase, among others, plays an important role in the mobilization of bioactive phenolic compounds during SSF (Cho et al. 2009, Zheng and Shetty 2000). Lignocellulosic enzymes are mainly produced by fungi, since these microorganisms have two extracellular enzymatic systems—a hydrolytic system that can degrade polysaccharides and an oxidative ligninolytic system, which degrades lignin and opens phenyl rings, increasing the free phenolic content (Sánchez 2009). During SSF of soybean with *Bacillus pumilus* HY1, Cho et al. (2009) reported a significant increase in flavanols and gallic acid content associated with bacterial β-glucosidase and esterase activities. Similarly, improvement in the antioxidant potential of fermented rice is associated with phenolic compound increases by β-glucosidase and α-amylase activities during SSF (Bhanja et al. 2008).

Agricultural or forestry biowastes, such as cereal and vegetable wastes, like straw, bagasse, stover, cobs, husks, among others, are lignocellulosic materials composed mainly of cellulose, hemicellulose and lignin. Lignin contains numerous phenolic components, such as ferulic, *p*-coumaric, syringic, vanillic and *p*-hydroxybenzoic acids (Mussatto et al. 2007), which can be recovered by SSF, using filamentous fungi like the white-rot fungi *Phanerochaete chrysosporium*, *Trametes versicolor*, *Trametes hirsuta* and *Bjerkandera adusta*, which can degrade lignin. These fungi utilize the polysaccharides after lignin degradation to grow and reproduce, thus increasing the nutritional value of the agro-industrial substrates that is generally low. After SSF, the materials can be used as animal feed or soil fertilizer (Nigam et al. 2009, Oberoi et al. 2011b). The main extracellular enzymes participating in lignin degradation are lignin peroxidase, manganese peroxidise and laccase (Philippoussis 2009).

In order to harness the potential of phytochemicals as bioactive compounds, development of better extraction methods using advanced technologies help to obtain the greatest yield in the shortest processing time and at a low cost, but an eye on the environmental cost is imperative. The main differences between these extraction technologies depend on the design of the reactors, solvents used, time and temperature of the processes and yields. The physico-chemical properties of desired compounds and availability of their sources decide the strategy to be adopted. Also, the cost-benefit analysis of the use of a single or combination of extraction technologies should be taken into consideration by the food and pharmaceutical industries.

Functional Food Development

Consumers are increasingly demanding safe and healthy natural food ingredients. There is increasing awareness about the health-promoting benefits of antioxidants in foods, and this in combination with the realization that a number of common synthetic preservatives may have hazardous effects, has led to a vast number of reports on natural antioxidants (Babbar and Oberoi 2013). Functional foods are those foods or ingredients which, when consumed regularly, produce a specific beneficial health effect beyond their basic nutritional properties. These include a variety of bioactive compounds included in food formulations with a specific purpose (Day et al. 2009, Roberfroid 2002, Nobili et al. 2009, Bernal et al. 2010).

From a practical point of view, a functional food can be:

- A natural food
- A food to which a component has been added
- A food from which a component has been removed
- A food in which the bioavailability of one or more components has been modified, or
- A combination of the above possibilities.

All claims for functional foods should be based on the scientific classification of markers (indicators and/or factors) for target functions. The effect must go beyond what is normally the established role of diet, such as a target function or a biological activity or for a disease risk reduction (Diplock et al. 1999). This evidence should rely on three categories of data:

- Biological observations,
- Epidemiological data, and
- Intervention studies, mostly based on markers.

Plaza et al. (2008) emphasized three important aspects of a functional food:

1) The functional effect is different from that of normal nutrition,
2) The functional effect must be demonstrated satisfactorily, and
3) The benefit can lead to improvement of physiological function or to reduction of risk of developing a pathological process.

Since all bioactive compounds do not cover these aspects, some are not considered functional food ingredients. Food and Drug Administration in the USA has grouped several compounds with potential as new bioactive food ingredients that are considered generally safe (GRAS), including vegetable oil, sterol esters, phytostanol esters, lactoferrin, fish oil concentrate, tuna oil, diacylglycerol and inulin, among others (Burdock et al. 2006). There remains a wide range of bioactive compounds with beneficial effects on health but they need government clearance. The most remarkable case among these is the use of polyphenols. Whole foods, such as fruits and vegetables represent the simplest form of functional foods because they are rich in several natural bioactive compounds, like polyphenols and carotenoids. The health benefits include decrease in cholesterol levels, alleviation of lactose intolerance, maintaining remission of Crohn's disease, faster relief from diarrhoea and inhibition of cancer cell proliferation *in vivo* and *in vitro*, antioxidant, antiviral, antihypertensive (Marette et al. 2010).

Until recently, natural antioxidants have found application mostly as food supplements, like green tea extract (Buetler et al. 2002) or as food preservatives obtained from aromatic plants, like rosemary and salvia extracts (Zupko et al. 2001). They also find application in health products in the cosmetic and pharmaceutical industry due to the customer's demand for 'non-chemical' ingredients. The phenolic antioxidants are a substitute for synthetic preservatives or as active ingredients; for example, a skin-protecting additive in dermatology. These three sectors are promoted together as functional foods: food supplements, nutraceuticals or cosmeceuticals. Food-derived extracts as ingredients in food and cosmetic products are perceived as non-toxic antioxidants and find commercial application as radical scavengers and flavonoids with anti-ageing and photoprotection benefits, in both hydrophilic and lipophilic systems (Katiyar and Elmets 2001). Examples of functional food products that are currently in market are drinks, cereals, bakery products, spreads, meat products and eggs, among others (Siro et al. 2008).

Processing of mango fruits generates a significant amount of by-products, such as peels and seeds which represent up to 60 per cent of the fresh fruit. Mango by-products are an important source of bioactive phenolic compounds. Phenolic characterization is an essential step for utilisation of mango by-products as food ingredients, and so provides an added value to mango production. Dorta et al. (2014) have characterized the phenolic compounds from peel and seed extracts of three mango varieties (Keitt, Sensation and Gomera 3) produced in Spain through different microwave-assisted extraction conditions. Thirty compounds were identified that belong to five phenolic families: gallates and gallotannins; flavonoids mainly quercetin derivatives; ellagic acid and derivatives; xanthones principally mangiferin; benzophenones and derivatives of maclurin, variation was attributed to the variety of mango, the part of fruit studied (peel or seed) and the extraction conditions, with the latter having a greater influence on the phenolic content.

Licorice (*Glycyrrhiza glabra*) is a popular herbal supplement used for the treatment of chronic inflammatory conditions and possesses anticancer and antiviral activities. It contains a number of phytochemicals including terpenoids, saponins, flavonoids, polyamines and polysaccharides. There are over 30 species in the Glycyrrhiza genus world-wide, most of which have been little characterized in terms of phytochemical or pharmacological properties to explain their medicinal use. UV, MS and NMR spectral analyses of extracted components from *Glycyrrhiza glabra*, *G. uralensis*, *G. inflata* and *G. echinata* have identified the major components as glycyrrhizin, 4-hydroxyphenyl acetic acid and glycosidic conjugates of liquiritigenin/isoliquiritigenin. Primary metabolites profiling use of GCMS revealed the presence of cadaverine, an amino acid, exclusively found in *G. inflata* roots (Farag et al. 2012).

Potential ingredients for use in the food industry, such as napin, cruciferon, oleosin, inulin, cynarin, and fiber have been extracted from canola seed meal, artichokes and other important sources, such as algae and microalgae (Plaza et al. 2008, Lattanzio et al. 2009, Aider and Barbana 2010). Functional foods include probiotic microorganisms, such as lactic acid bacteria *Lactobacillus acidophilus*, *L. johnsonni*, *L. casei shirota*, and also various species of bifidobacterium (Sanders 1998) and prebiotic products mainly presented through dairy products and milk which are good vehicles to deliver the probiotics (Day et al. 2009, Saulnier et al. 2009). Probiotics can modify the immune system, help in the improvement and maintenance of intestinal flora and prevention of gastrointestinal disorders (Hekmat et al. 2009). However, the bioavailability of bioactive ingredients after consumption, their mechanism and mode of action, biological effects *in vivo* and *in vitro* and structure-to-function relationship and physiological mechanism by which the health benefits manifest are not well understood. So research on these aspects need greater attention.

Functional foods from residues/by-products of food processing

The growing interest in replacement of synthetic food antioxidants by natural ones has fostered research on screening of plant-derived raw materials for identifying new antioxidants. Attention is now focused on inexpensive or residual sources from agricultural industries. Fruit peels, often the waste parts of various fruits from consumption and food industry, are rich sources of powerful natural antioxidants. Table 4 gives the composition and content of bioactive compounds observed in methanolic extracts of various plant wastes. Dehydrated waste grape skins from the juice industry were added to aged and young red wines as an innovative way of compensating for colour loss before bottling and increasing colour intensity of wines by 11 to 31 per cent. Total polyphenols increased by 10 to 20 per cent, which mostly comprised of gallic acid, catechin, epicatechin and resveratrol; its anthocyanins content also increased by 50 mg/l (Pedroza et al. 2013). Among the fruit peels, mangosteen peel is an important source of natural antioxidants, like phenolic acids and flavonoids, which possess biological and medicinal properties (Suttirak and Manurakchinakorn 2014). Schieber et al. (2001) have reviewed the scope for obtaining functional compounds from by-products of plant food-processing industry. Pectin production is an economical use for apple pomace with superior gelling properties compared to citrus pectins; the limitation being the slightly brown hue of apple pectins caused by enzymatic browning (Renard et al. 1997). Apple pomace is a good source of polyphenols which are predominantly localized in the peels, such as catechins, hydroxycinnamates, phloretin glycosides, quercetin glycosides and procyanidins. Apple pomace phenols are rich in antioxidant activity and cariogenicity of *Streptococci.* Therefore they have possible application in dentifrices (Yanagida et al. 2000). Release of phenolics can be enhanced by using pectinases and cellulases.

Apple, golden rod and artichoke by-products were extracted on a pilot plant scale by Peschel et al. (2006) and their antioxidant activity was confirmed to be similar to the established antioxidants (Oxynex® 0.1 per cent, Controx® KS 0.15 per cent and butylated hydroxytoluene (BHT) 0.01 per cent), demonstrating the possibility of recovering high amounts of phenolics with antioxidant properties from fruit and vegetable residuals not only for food but also cosmetic applications. Broccoli by-products—leaves and stalks—are rich in bioactive compounds, including nitrogen-sulphur compounds (glucosinolates and isothiocyanates) and phenolics (chlorogenic and sinapic acid derivatives, and flavonoids), as also essential nutrients (minerals and vitamins). They are of huge interest as a source of health-promoting compounds useful as ingredients for the development of functional foods. Green tea enriched with broccoli concentrates showed improved physical quality, phytochemical composition and antioxidant capacity, encouraging the production of novel products

Table 4. Composition and content of bioactive compounds in methanolic extracts of various plant materials.

Plant material	Compound	Concentration (mg/g)
Tomato peel	Cis-lycopene	22.02
	Beta-carotene	6.87
	Trans-lycopene	36.49
	Lutein	1.08
	Ascorbic acid	12.27
	Quercetin	2.89
	Kaempferol	7.20
Cucumber peel	Chlorophyll	3.46
	Pheophytin	1.95
	Phellandrene	1.21
	Caryophellene	1.49
Water melon peel	Chlorophyll	5.28
	Diosmetin	1.57
	Pheophytin	1.27
	Malvidin 3,5 diglycoside	1.23
Potato peel	Gallic acid	0.16
	Protocatecheic	1.84
	p-Hydroxybenzoic	0.26
	Caffeic acid	0.19
	Vanillic acid	0.04
	Chlorogenic acid	0.28
Orange peel	p-coumaric	1.02
	Ferulic acid	0.91
	Syringic acid	7.71
	Naritutin	1.21
	Nazingin	3.83
	Ascorbic acid	14.9
Olive leaves	Oleuropein	71.61
	Apigenin 7-glucoside	4.10
	Rutin	0.15
	Vanillin	0.15
	Vanillic acid	1.87
	Caffeic acid	1.02
	Hydroxytyrosel	3.29

Adapted from Zeyada et al. (2008)

and applications for nutritional and/or health claims under the EU Regulations (EC) No. 1924/2006 and 834/2007 (Perles et al. 2011).

Guava (*Psidium guajava* L.) seed powder (red guava cv. Paluma) obtained from guava pulp processing contained varying amounts of macronutrients and micronutrients, high content of total dietary fiber (64 g/100 g), protein (11.2 g/100 g), iron (13.8 mg/100 g), zinc (3.3 mg/100 g) and reduced calorie content (182 kcal/100 g). Their lipid profile showed a predominance of unsaturated fatty acids (87.1 per cent), especially linoleic acid and oleic acid. The powder obtained contained significant amounts of bioactive compounds, such as ascorbic acid (87.4 mg/100 g), total carotenoids (1.3 mg/100 g) and insoluble dietary fiber (63.6 g/100 g). With regard to their microbiological quality, the samples were found suitable for consumption. Thus, guava seeds can be a viable alternative to prevent various diseases and malnutrition in our country and to reduce the environmental impact of agricultural waste (Athayde et al. 2014).

Polygonum cuspidatum, used in important traditional Chinese medicine, is widely distributed in the world and many parts of the plant are used to treat hyperlipemia, inflammation, infection, cancer, etc. The roots of this plant are used as an effective agent in pre-clinical and clinical practice for regulating lipids, anti-endotoxic shock, anti-infection, anti-inflammation, anti-cancer and other diseases in China and Japan. Over 67 compounds including quinones, stilbenes, flavonoids, counmarins and ligans have been isolated and identified from this plant (Peng et al. 2013).

Olive mill wastewater is an agricultural waste material produced in high quantities in the Mediterranean basin. It is an inexpensive source of health-promoting phytochemicals with the potential economic value including many low molecular weight compounds, such as the bioaccessible verbascosides (Cardinali et al. 2011). The biophenolic fraction of olive mill waste in Australia contained higher phenol content, higher antioxidant activity and broad spectrum antibacterial activity as seen against *Staphylococcus aureus*, *Bacillus subtilis*, *Escherichia coli* and *Pseudomonas aeruginosa*; molluscicidal activity against *Isidorella newcombi* was also reported (Obied et al. 2007).

Parkia speciosa, known as stink bean, is popular in south-eastern Asia including Malaysia and north-eastern India. The seeds are normally separated from the pods before they are ready to sell. *P. speciosa* is believed by the locals to have medicinal properties; the pods have hypoglycaemic activity (Jamaluddin et al. 1995), seeds contain antibacterial cyclic polysulfides, also responsible for their strong pungent flavour and anticancer activity due to presence of thiazolidine-4-carboxylic acid (Pandeya 1972). The waste of fruits and beans, peels and pods, which were previously used as fertilisers, feeds and landfills, are now considered more precious for their highly antioxidant pectin-like polysaccharide, a functional carbohydrate in food application studies (Gan et al. 2010b).

Controlling the spread of food-borne pathogens in moist environments is an important microbial food safety issue. Isolation of compounds from agricultural waste (such as fruit peels) that would control spread of human pathogens was explored by Mahadwar et al. (2015) by using *Salmonella enterica* serovar *typhimurium* as a model organism. Pomegranate peels were found to have great potential as a bioactive repellent for pathogenic micro-organisms.

Encapsulation

Encapsulation is a process to entrap solid, liquid, or gaseous active agents (or core) within a carrier, wall or coating material (Fang and Bhandari 2010). It is a useful tool to improve delivery of bioactive molecules (e.g., antioxidants, minerals, vitamins, phytosterols, lutein, fatty acids, lycopene) and living cells (e.g., probiotics) into foods (Vos et al. 2010). The materials used for encapsulation must be food-grade, biodegradable and able to form a barrier between the internal phase and its surroundings, the most widely used material being polysaccharides. Among the polysaccharides, starch and their derivates—amylose, amylopectin, dextrins, maltodextrins, polydextrose, syrups and cellulose and their derivatives are commonly used. Plant exudates and extracts—gum Arabic, gum tragacanth, gum karaya, mesquite gum, galactomannans, pectins and soluble soybean polysaccharides are also used. Recently, marine extracts, such as carrageenans and alginate, and microbial and animal polysaccharides like dextran, chitosan, xanthan and gellan have also been exploited. Apart from natural and modified polysaccharides, proteins and lipids are also used as coatings for encapsulation. Examples of the most common milk and whey proteins are caseins, gelatine and gluten and the lipid materials are fatty acids and fatty alcohols, waxes (beeswax, carnauba wax, candellia wax), glycerides and phospholipids. Other materials employed are PVP, paraffin, shellac and inorganic materials (Wandrey et al. 2009).

The most extensively applied encapsulation technique in the food industry is spray drying, because it is economical, flexible and continuous, producing particles of size less than 40 μm (Zuidam and Heinrich 2009). Other encapsulation techniques include spray-chilling, freeze-drying, melt extrusion and melt injection. Encapsulation in cyclodextrins and liposomal vesicles are less preferred since they are more expensive technologies. The advantages of encapsulation of bioactive material include providing barriers between sensitive bioactive materials and the environment (oxygen or water), thus allowing differentiation of taste and aroma, masking bad taste (bitterness and astringency of polyphenols) or smell, stabilizing food ingredients (less evaporation and degradation of volatile actives such as aroma) or increasing their bioavailability. Encapsulation may also be used to immobilize cells or enzymes in food-processing applications, such as fermentation and metabolite production. There is an increasing demand to find suitable solutions that provide high productivity and quality of the final food products (Nedovic et al. 2011). Encapsulation facilitates packaging materials in small capsules that release their contents at controlled rates over prolonged periods and under specific conditions (Desai and Park 2005). Produced particles usually have diameters of a few nm to a few mm (Wandrey et al. 2009). Nanotechnology is the science of the miniature, wherein biological and non-biological structures smaller than 100 nm are now being exploited by agricultural producers and food manufacturers to encapsulate bioactive compounds, to benefit consumers in the long term, to foster a competitive and innovative domestic agricultural and food system and provide new methods to improve safety and nutritional value of food products (Weiss et al. 2006).

Conclusion

To conclude, agricultural wastes have increased in tune with increase in food production, consumption and processing, the sources being the farmer, retailer, food industries, consumer/household, with each source contributing different quantities, cumulatively adding up to a substantial quantum of waste. This not only adds to the environmental pollution burden, but also results in a massive waste of unutilized wealth. This chapter elaborates on the processes required for an effective utilisation of agro-wastes. Agro-wastes from a variety of residues like molasses, bagasse, oilseed cakes, milling by-products, etc., are a potential source of value-added products, such as food, fuel, feed and a variety of chemicals, nutraceuticals and functional foods. Valorization of the ubiquitous phenolics which are present in the peels and seeds of agri-wastes with diverse medicinal value, for development of nutraceuticals and functional foods is the most effective approach for their utilisation. Utilisation of agro-wastes begins with a process of pretreatment required to break down the structure of agro-residues using physical, chemical, biological and a combination of these methods. This is followed by extraction techniques to harvest the bioactive compounds from the matrix of the plant material adopting classical and/or more advanced non-classical methods. The classical extraction methods include Soxhlet extraction, maceration and hydrodistillation; the more efficient non-classical clean and green extraction methods are solvent extraction, pressurized liquid extraction, subcritical and supercritical fluid extraction, microwave and ultrasonic assisted extractions, pulsed electric field extraction, enzyme assisted extraction, fermentation methods, instant controlled pressure-drop extraction and many more. These extraction methods may be used in combination to increase the efficiency of extraction. The extracted bioactive chemicals must be identified for better understanding of their chemistry and biological activity before being taken forward for the development of functional foods, of which numerous products have already made it to the retail shelves. Encapsulation of bioactives further serves to improve the functionality of the food components. Thus this chapter details the processes by which agricultural waste can be turned to wealth.

References

Adam, F., M. Abert-Vian, G. Peltier and F. Chemat. 2012. Solvent-free ultrasound-assisted extraction of lipids from fresh microalgae cells: A green, clean and scalable process. Bioresour. Technol. 114: 457–465.

Aider, M. and C. Barbana. 2010. Canola proteins: Composition, extraction, functional properties, bioactivity, applications as a food ingredient and allergenicity—a practical and critical review. Trends Food Sci. Technol. 22(1): 21–39.

Ajila, C. M., S. K. Brar, M. Verma, R. D. Tyagi, S. Godbout and J. R. Valero. 2011. Extraction and analysis of polyphenols: Recent trends. Crit. Rev. Biotechnol. 31(3): 227–49.

Allaf, K. and P. Vidal. 1988. Feasibility study of a new process of drying/swelling by instantaneous decompression toward vacuum of vegetables in pieces in view of a rapid rehydration. Gradient Activity Plotting, University of Technology of Compiegne UTC No. CR/89/103. Compiegne, France.

Allaf, T., V. Tomao, K. Ruiz and F. Chemat. 2012. Instant controlled pressure drop technology and ultrasound assisted extraction for sequential extraction of essential oil and antioxidants. Ultrason Sonochem http://dx.doi.org/10.1016/j.ultsonch.2012.05.013.

Alonso, A. M., D. A. Guillén, C. G. Barroso, B. Puertas and A. García. 2002. Determination of antioxidant activity of wine byproducts and its correlation with polyphenolic content. J. Agrl. Food Chem. 50: 5832–5836.

Aludatt, M. H., T. M. Rababah, A. Johargy, S. Gammoh, K. Ereifej, M. N. Alhamad, S. M. Brewer, A. A. Saati, S. Kubow and M. Rawshdeh. 2016. Extraction, optimisation and characterisation of phenolics from *Thymus Vulgaris* L.: Phenolic content and profiles in relation to antioxidant, antidiabetic and antihypertensive properties. Intl. J. Food Sci. Tech. 51(3): 720–730.

Alvira, P., E. Tomaś-Pejo, M. Ballesteros and M. J. Negro. 2010. Pretreatment technologies for an efficient bioethanol production process based on enzymatic hydrolysis: A review. Bioresour. Technol. 101(13): 4851–4861.

Amro, B., T. Aburjai and S. Al-Khalil. 2002. Antioxidative and radical scavenging effects of olive cake extract. Fitoterapia 73: 456–461.

Anwar, Z., M. Gulfraz and M. Irshad. 2014. Agro-industrial lignocellulosic biomass a key to unlock the future bio-energy: A brief review. J. Radiation Res. Applied Sci. 7: 163–173.

Asgher, M., Z. Ahmad and H. M. N. Iqbal. 2013. Alkali and enzymatic delignification of sugarcane bagasse to expose cellulose polymers for saccharification and bio-ethanol production. Indl. Crops Prod. 44: 488–495.

Athayde, U. T. A. M., S. E. Castro, C. J. O. Beserra, M. S. M. De, L. A. De, M. C. Galvo, A. C. Duarte, F. B. A. Travassos, R. A. L. Moreira, R. S. Praciano, T. J. C. De Albuquerque, S. J. Do Nascimento and R. L. Lages. 2014. Chemical composition, fatty acid profile and bioactive compounds of guava seeds (*Psidium guajava* L.). Ciência e Tecnologia de Alimentos 34(3): 485–492.

Babbar, N., H. S. Oberoi, D. S. Uppal and R. T. Patil. 2011. Total phenolic content and antioxidant capacity of extracts obtained from six important fruit residues. Food Res. International 44: 391–396.

Babbar, N. and H. S. Oberoi. 2013. Potential of agro-residues as sources of bioactive compounds. pp. 261–298. *In*: Biotransformation of Waste Biomass: Fine Chemicals, Springer, USA.

Babbar, N., H. S. Oberoi and S. K. Sandhu. 2015. Therapeutic and nutraceutical potential of bioactive compounds extracted from fruit residues. Crit. Rev. Food Sci. Technol. 55: 319–337.

Ballard, T. S., P. Mallikarjunan, K. Zhou and S. O'Keefe. 2010. Microwave-assisted extraction of phenolic antioxidant compounds from peanut skins. Food Chem. 120(4): 1185–92.

Barbosa, E. S., D. Perrone, A. L. A. Vendramini and S. G. F. Leite. 2008. Vanillin production by *Phanerochaete chrysosporium* grown on green coconut agro-industrial husk in solid state fermentation. Biores. 3: 1042–1050.

Barrios-González, J., F. J. Fernández, A. Tomasini and A. Mejía. 2005. Secondary metabolites production by solid-state fermentation. Malays J. Microbiol. 1: 1–6.

Ben Amor, B. and K. Allaf. 2009. Impact of texturing using instant pressure drop treatment prior to solvent extraction of anthocyanins from Malaysian Roselle (*Hibiscus sabdariffa*). Food Chem. 115(3): 820–825.

Berka-Zougalia, B., A. Hassani, C. Besombes and K. Allaf. 2010. Extraction of essential oils from Algerian myrtle leaves using instant controlled pressure drop technology. J. Chromatogr. 1217(40): 6134–42.

Bernal, J., J. A. Mendiola, E. Ibáñez and A. Cifuentes. 2010. Advanced analysis of nutraceuticals. J. Pharm. Biomed. Anal. 55(4): 758–774.

Bhanja, T., S. Rout, R. Banerjee and B. C. Bhattacharyya. 2008. Studies on the performance of a new bioreactor for improving antioxidant potential of rice. LWT—Food Science and Technology 41(8): 1459–1465.

Bhavsar, N. H., B. V. Raol, S. S. Amin and G. G. Raol. 2015. Production, optimization and characterization of fungal cellulase for enzymatic saccharification of lignocellulosic agro-waste. Int. J. Curr. Microbiol. App. Sci. 4(3): 30–46.

Bonzanini, F., R. Bruni, G. Palla, N. Serlataite and A. Caligiani. 2009. Identification and distribution of lignans in *Punica granatum* L. fruit endocarp, pulp, seeds, wood knots and commercial juices by GC–MS. Food Chem. 117: 745–749.

Bryant, G. and J. Wolfe. 1987. Electromechanical stress produced in the plasma membranes of suspended cells by applied electrical fields. J. Membrane Bio. 96(2): 129–139.

Buetler, T. M., M. Renard, E. A. Offord, H. Schneider and U. T. Ruegg. 2002. Green tea extract decreases muscle necrosis in mdx mice and protects against reactive oxygen species. American J. Clin. Nut. 75: 749–753.

Burdock, G. A., I. G. Carabin and J. C. Griffiths. 2006. The importance of GRAS to the functional food and nutraceutical industries. Toxicol. 221(1): 17–27.

Cardinali, A., V. Linsalata, V. Lattanzio and M. G. Ferruzzi. 2011. Verbascosides from olive mill waste water: assessment of their bioaccessibility and intestinal uptake using an *in vitro* digestion/bCaco-2 model system. J. Food Sci. 76(2): H48–H54.

Carvalho, A. P., M. Mendes, M. M. Moreira, D. Cruz, J. M. C. S. Magalhães, M. F. Barroso, M. J. Ramalhosa, A. Duarte, L. F. Guido, A. M. Gomes and C. D. Matos. 2016. Microwave-assisted extraction in Goji berries: Effect on composition and bioactivity, evaluated through conventional and nonconventional methodologies. Intl. J. Food Sci. Tech. 51(6): 1401–1408.

Cavero, S., M. García-Risco, F. Marín, L. Jaime, S. Santoyo, F. J. Señoráns, G. Reglero and E. Ibañez. 2006. Supercritical fluid extraction of antioxidant compounds from oregano: Chemical and functional characterization via LC-MS and *in vitro* assays. J. Supercrit. Fluid 38(1): 62–69.

Chandini, S. K., L. J. Rao, M. K. Gowthaman, D. J. Haware and R. Subramanian. 2011. Enzymatic treatment to improve the quality of black tea extracts. Food Chem. 127(3): 1039–1045.

Chen, S., X. H. Xing, J. J. Huang and M. S. Xu. 2010. Enzyme-assisted extraction of flavonoids from *Ginkgo biloba* leaves: Improvement effect of flavonol transglycosylation catalyzed by *Penicillium decumbens* cellulase. Enzyme Microb. Technol. 48(1): 100–105.

Cheung, Y. C., K. C. Siu, Y. S. Liu and J. Y. Wu. 2012. Molecular properties and antioxidant activities of polysaccharide–protein complexes from selected mushrooms by ultrasound-assisted extraction. Process Biochem. 47(5): 892–895.

Chia, S. L., H. C. Boo, K. Muhamad, R. Sulaiman, F. Umanan and G. H. Chong. 2015. Effect of subcritical carbon dioxide extraction and bran stabilization methods on rice bran oil. J. Am. Oil Chemists' Soc. 92(3): 393–402.

Cho, K. M., S. Y. Hong, R. K. Math, J. H. Lee, D. M. Kambiranda, J. M. Kim, S. M. A. Islam, M. G. Yun, J. J. Cho, W. J. Lim and H. D. Yun. 2009. Biotransformation of phenolics (isoflavones, flavanols and phenolic acids) during the fermentation of cheonggukjang by *Bacillus pumilus* HY1. Food Chem. 114: 413–419.

Corralesa, M., S. Toepflb, P. Butza, D. Knorrc and B. Tauschera. 2008. Extraction of anthocyanins from grape by-products assisted by ultrasonics, high hydrostatic pressure or pulsed electric fields: A comparison. Innovative Food Sci. Emerging Tech. 9(1): 85–91.

Cortazar, E., L. Bartolomé, A. Delgado, N. Etxebarria, L. A. Fernández, A. Usobiaga and O. Zuloaga. 2005. Optimisation of microwave-assisted extraction for the determination of nonylphenols and phthalate esters in sediment samples and comparison with pressurised solvent extraction. Anal. Chim. Acta 534: 247–254.

Cuccioloni, M., M. Mozzicafreddo, L. Sparapani, M. Spina, A. M. Eleuteri, E. Fioretti and M. Angeletti. 2009. Pomegranate fruit components modulate human thrombin. Fitoterapia 80: 301–3305.

Curiel, J. A., L. Betancor, B. de las Rivas, R. Muñoz, J. M. Guisan and G. Fernández-Lorente. 2010. Hydrolysis of tannic acid catalyzed by immobilized-stabilized derivatives of tannase from *Lactobacillus plantarum*. J. Agric. Food Chem. 58(10): 6403–6409.

Daintree, L. S., A. Kordikowski and P. York. 2008. Separation processes for organic molecules using SCF Technologies. Adv. Drug Del. Rev. 60(3): 351–372.

Day, L., R. Seymour, K. Pitts, I. Konczak and L. Lundin. 2009. Incorporation of functional ingredients into foods. Trends Food Sci. Technol. 20(9): 388–3395.

de Vos, P., M. M. Faas, M. Spasojevic and J. Sikkema. 2010. Review: Encapsulation for preservation of functionality and targeted delivery of bioactive food components. Int. Dairy J. 20: 292–302.

Desai, K. G. H. and H. J. Park. 2005. Recent developments in microencapsulation of food ingredients. Drying Technol. 23: 1361–1394.

Di Luccio, M., F. Capra, N. P. Ribeiro, G. D. L. P. Vargas, D. M. G. Freire and D. Oliveira. 2004. Effect of temperature, moisture and carbon supplementation on lipase production by solidstate fermentation of soy cake by *Penicillium simplicissimum*. Appl. Biochem. Biotechnol. 113: 173–180.

Diplock, A. T., P. J. Aggett, M. Ashwell, F. Bornet, E. B. Fern and M. B. Roberfroid. 1999. Scientific concepts of functional foods in Europe: Consensus document. British J. Nut. 81(1): S1–S28.

Dobias, P., P. Pavlikova, M. Adam, A. Eisner, B. Benova and K. Ventura. 2010. Comparison of pressurised fluid and ultrasonic extraction methods for analysis of plant antioxidants and their antioxidant capacity. Cent. Eur. J. Chem. 8(1): 87–95.

Dorta, E., M. González, M. G. Lobo, C. Sánchez Moreno and B. De Ancos. 2014. Screening of phenolic compounds in by-product extracts from mangoes (*Mangifera indica* L.) by HPLC-ESI-QTOF-MS and multivariate analysis for use as a food ingredient. Food Res. Intl. 57: 51–60.

Du Pont, M. S., R. N. Bennett, F. A. Mellon and G. Williamson. 2002. Polyphenols from alcoholic apple cider are absorbed, metabolized and excreted by humans. J. Nut. 132: 172–175.

Dunford, N., S. Irmak and R. Jonnala. 2010. Pressurised solvent extraction of policosanol from wheat straw, germ and bran. Food Chem. 119(3): 1246–1249.

Eggeman, T. and R. T. Elander. 2005. Process economic analysis of pretreatment technologies. Biores. Tech. 96: 2019–2025.

Esclapez, M. D., J. V. García-Pérez, A. Mulet and J. A. Cárcel. 2011. Ultrasound-assisted extraction of natural products. Food Eng. Rev. 3: 108–120.

Fang, Z. and B. Bhandari. 2010. Encapsulation of polyphenols—A review. Trends Food Sci. Technol. 21: 510–523.

Farag, M. A., A. Porzel and L. A. Wessjohann. 2012. Comparative metabolite profiling and fingerprinting of medicinal licorice roots using a multiplex approach of GC–MS, LC–MS and 1D NMR Techniques. Phytochem. 76: 60–72.

Fincan, M. and P. Dejmek. 2002. *In situ* visualization of the effect of a pulsed electric field on plant tissue. J. Food Eng. 55(3): 223–230.

Fincan, M., F. De Vito and P. Dejmek. 2004. Pulsed electric field treatment for solid-liquid extraction of red beetroot pigment. J. Food Eng. 64(3): 381–388.

Fridovich, I. 1978. The biology of oxygen radicals. Science 201: 875–880.

Fu, Y. J., W. Liu, Y. G. Zu, M. H. Tong, S. M. Li, M. M. Yan, T. Efferth and H. Luo. 2008. Enzyme-assisted extraction of luteolin and apigenin from pigeonpea (*Cajanus cajan* (L.) Millsp.) leaves. Food Chem. 111(2): 508–512.

Fuhrman, B., N. Volkova, M. Rosenblat and M. Aviram. 2000. Lycopene synergistically inhibits LDL oxidation in combination with vitamin E, glabridin, rosmarinic acid, carnosic acid, or garlic. Antioxidant Redox Signal 2: 491–506.

Gan, C. Y., N. H. A. Manaf and A. A. Latiff. 2010. Physico-chemical properties of alcohol precipitate pectin-like polysaccharides from *Parkia speciosa* pod. Food Hydrocolloids 24: 471–478.

García-Marino, M., J. C. Rivas-Gonzalo, E. Ibáñez and C. García-Moreno. 2006. Recovery of catechins and proanthocyanidins from winery by-products using subcritical water extraction. Anal. Chim. Acta 563(1-2): 44–50.

Gil, M. I., F. A. Tomás-Barberán, B. Hess-Pierce, D. M. Holcroft and A. A. Kader. 2000. Antioxidant activity of pomegranate juice and its relationship with phenolic composition and processing. J. Agric. Food Chem. 48: 4581–4589.

Guderjan, M., S. Töpfl, A. Angersbach and D. Knorr. 2005. Impact of pulsed electric field treatment on the recovery and quality of plant oils. J. Food Eng. 67(3): 281–287.

Hammel, K. E., A. N. Kapich, K. A. Jensen Jr and Z. C. Ryan. 2002. Reactive oxygen species as agents of wood decay by fungi. Enz. Microb. Technol. 30(4): 445–453.

Harborne, J. B., H. Baxter and G. P. Moss. 1999. Phytochemical Dictionary: Handbook of Bioactive Compounds from Plants (2nd Ed.), Taylor & Francis, London.

Hasbay Adil, I., H. I. Çetin, M. E. Yener and A. Bayındırl. 2007. Subcritical (carbon dioxide+ethanol) extraction of polyphenols from apple and peach pomaces and determination of the antioxidant activities of the extracts. J. Supercritical Fluids 43(1): 55–63.

Hassas-Roudsari, M., P. Chang, R. Pegg and R. Tyler. 2009. Antioxidant capacity of bioactives extracted from canola meal by subcritical water, ethanolic and hot water extraction. Food Chem. 114(2): 717–726.

Hayat, K., S. Hussain, S. Abbas, U. Farooq, B. M. Ding, S. Q. Xia, C. S. Jia, X. M. Zhang and W. S. Xia. 2009. Optimized microwave-assisted extraction of phenolic acids from citrus mandarin peels and evaluation of antioxidant activity *in vitro*. Sep. Purif. Technol. 70(1): 63–70.

Heinz, V., S. Toepfl and D. Knorr. 2003. Impact of temperature on lethality and energy efficiency of apple juice pasteurization by pulsed electric fields treatment. Innovat. Food Sci. Emerg. Tech. 4(2): 167–175.

Hekmat, S., H. Soltani and G. Reid. 2009. Growth and survival of *Lactobacillus reuteri* RC-14 and *Lactobacillus rhamnosus* GR-1 in yogurt for use as a functional food. Innovat. Food Sci. Emerg. Tech. 10(2): 293–296.

Hernández, J. S., A. F. Aguilera-Carbó, R. Rodríguez Herrera, J. L. Martínez and C. N. Aguilar. 2008. Kinetic production of the antioxidant ellagic acid by fungal solid state culture. Proceedings of the 10th International Chemical and Biological Engineering Conference. Portugal: CHEMPOR. pp. 1849–1854.

Herrero, M., A. Cifuentes and E. Ibanez. 2006. Sub and supercritical fluid extraction of functional ingredients from different natural sources: plants, food-by-products, algae and microalgae: A review. Food Chem. 98(1): 136–148.

Herrero, M., J. A. Mendiola, A. Cifuentes and E. Ibanez. 2010. Supercritical fluid extraction: Recent advances and applications. J. Chromatogr. 1217(16): 2495–2511.

Holtzapple, M. T., J. E. Lundeen, R. Sturgis, J. E. Lewis and B. E. Dale. 1992. Pretreatment of lignocellulosic municipal solid-waste by ammonia fiber explosion (AFEX). Appl. Biochem. Biotechnol. 34(5): 5–21.

Hu, Z., Y. Wang and Z. Wen. 2008. Alkali (NaOH) pretreatment of switchgrass by radio frequency-based dielectric heating. Appl. Biochem. Biotech. 148: 71–81.

Huang, S. H. and L. T. Ng. 2012. Quantification of polyphenolic content and bioactive constituents of some commercial rice varieties in Taiwan. J. Food Compos. Anal. 26: 122–127.

Ichwan, M. and T. W. Son. 2011. Study on organosolv pulping methods of oil palm biomass. pp. 364–370. *In*: International Seminar on Chemistry.

Imeh, U. and S. Khokhar. 2002. Distribution of conjugated and free phenols in fruits, antioxidant activity and cultivar variations. J. Agrl. Food Chem. 50: 6301–6306.

Iqbal, H. M. N., I. Ahmed, M. A. Zia and M. Irfan. 2011. Purification and characterization of the kinetic parameters of cellulose produced from wheat straw by *Trichoderma viride* under SSF and its detergent compatibility. Adv. Biosci. Biotech. 2(3): 149–156.

Jamaluddin, F., S. Mohamed and M. N. Lajis. 1995. Hypoglycaemic effect of stigmast-4-en-3-one, from *Parkia speciosa* empty pods. Food Chem. 54: 9–13.

Jiménez-Escrig, A., L. O. Dragsted, B. Daneshvar, R. Pulido and F. Saura-Calixto. 2003. *In vitro* antioxidant activities of edible artichoke (*Cynara scolymus* L.) and effect on biomarkers of antioxidants in rats. J. Agrl. Food Chem. 51: 5540–5545.

Joana Gil-Chávez, G., A. Jose, J. Villa, Fernando Ayala-Zavala, J. Basilio Heredia, David Sepulveda, M. Elhadi Yahia and A. Gustavo González-Aguilar. 2013. Technologies for extraction and production of bioactive compounds to be used as nutraceuticals and food ingredients: An overview. Comp. Rev. Food Sci. Food Safety 12: 5–23.

Jostein, G. and E. Johnny. 1980. Cellulases: Microbial enzymes and bioconversions. pp. 283–327. *In*: A. H. Rose (ed.). Economic Microbiology. Academic Press, London.

Kähkönen, M. P., A. I. Hopia and M. Heinonen. 2001. Berry phenolics and their antioxidant activity. J. Agrl. Food Chem. 49: 4076–4082.

Kar, B., R. Banerjee and B. C. Bhattacharyya. 1999. Microbial production of gallic acid by modified solid state fermentation. J. Indl. Microbiol. Biotech. 23: 173–177.

Karki, B., B. P. Lamsal, S. Jung, J. H. Van Leeuwen, A. L. Pometto, D. Grewell and S. K. Khanal. 2010. Enhancing protein and sugar release from defatted soyflakes using ultrasound technology. J. Food Eng. 96(2): 270–278.

Katiyar, S. K. and C. A. Elmets. 2001. Green tea polyphenolic antioxidants and skin photoprotection. Intl. J. Oncol. 18: 1307–1313.

Kaur, U., H. S. Oberoi, V. K. Bhargav, R. S. Shivappa and S. S. Dhaliwal. 2012. Ethanol production from alkali- and ozone-treated cotton stalks using thermotolerant *Pichia kudriavzevii* HOP-1. Industrial Crops Products 37: 219–226.

Khandare, V., S. Walia, M. Singh and C. Kaur. 2010. Black carrot (*Daucus carota* ssp. sativus) juice: Processing effects on antioxidant composition and color. Food Bioprod. Process 89(4): 482–486.

Kim, T. H., J. S. Kim, C. Sunwoo and Y. Y. Lee. 2003. Pretreatment of corn stover by aqueous ammonia. Biores. Technol. 90(1): 39–47.

Kim, T. H., F. Taylor and K. B. Hicks. 2008. Bioethanol production from barley hull using SAA (soaking in aqueous ammonia) pretreatment. Biores. Technol. 99(13): 5694–5702.

Kris-Etherton, P. M., K. D. Hecker, A. Bonanome, S. M. Coval, A. E. Binkoski, K. F. Hilpert, A. E. Griel and T. D. Etherton. 2002. Bioactive compounds in foods: Their role in the prevention of cardiovascular disease and cancer. Am. J. Med. 113: 71S–88S.

Kristiawan, M., V. Sobolik and K. Allaf. 2008. Isolation of Indonesian cananga oil using multi-cycle pressure drop process. J. Chromatogr. 1192(2): 306–318.

Kujala, T., J. Loponen and K. Pihlaja. 2001. Betalains and phenolics in red beetroot (*Beta vulgaris*) peel extracts, extraction and characterisation. Zeitschrift für Naturforschung C 56: 343–348.

Kurilich, A. C., E. H. Jeffery, J. A. Juvik, M. A. Wallig and B. P. Klein. 2002. Antioxidant capacity of different broccoli (*Brassica oleracea*) genotypes using the oxygen radical absorbance capacity (ORAC) assay. J. Agrl. Food Chem. 50: 5053–5057.

Lattanzio, V., P. A. Kroon, V. Linsalata and A. Cardinali. 2009. Globe artichoke: A functional food and source of nutraceutical ingredients. J. Funct. Foods 1(2): 131–144.

Leighton, T. G. 2007. What is ultrasound? Prog. Biophys. Mol. Biol. 93(1-3): 3–83.

Li, H., Z. Deng, T. Wu, R. Liu, S. Loewen and R. Tsao. 2011. Microwave assisted extraction of phenolics with maximal antioxidant activities in tomatoes. Food Chem. 130(4): 928–936.

Liao, W., Y. Liu, Z. Wen, C. Frear and S. Chen. 2007. Studying the effects of reaction conditions on components of dairy manure and cellulose accumulation using dilute acid treatment. Biores. Technol. 98: 1992–1999.

Liza, M. S., R. Abdul Rahman, B. Mandana, S. Jinap, A. Rahmat, I. Zaidul and A. Hamid. 2010. Supercritical carbon dioxide extraction of bioactive flavonoid from strobilanthes crispus (*Pecah kaca*). Food Bioprod. Process 88(2-3): 319–326.

López, N., E. Puértolas, S. Condón, I. Álvarez and J. Raso. 2008. Effects of pulsed electric fields on the extraction of phenolic compounds during the fermentation of must of Tempranillo grapes. Innovative Food Sci. Emerging Tech. 9(4): 477–482.

López, N., E. Puértolas, S. Condón, J. Raso and I. Álvarez. 2009. Enhancement of the extraction of betanine from red beetroot by pulsed electric fields. J. Food Eng. 90(1): 60–66.

Lucas, M., S. K. Hanson, G. L. Wagner, D. B. Kimball and K. D. Rector. 2012. Evidence for room temperature delignification of wood using hydrogen peroxide and manganese acetate as a catalyst. Biores. Technol. 119: 174–180.

Mahadwar, G., K. R. Chauhan, G. V. Bhagavathy, C. Murphy, A. D. Smith and A. A. Bhagwat. 2015. Swarm motility of *Salmonella enterica* Serovar Typhimurium is inhibited by compounds from fruit peel extracts. Letters in Appl. Microbiol. 60(4): 334–340.

Makris, D. P., G. Boskou and N. K. Andrikopoulos. 2007. Polyphenolic content and *in vitro* antioxidant characteristics of wine industry and other agri-food solid waste extracts. J. Food Comp. Anal. 20: 125–132.

Mandal, V., Y. Mohan and S. Hemalatha. 2007. Microwave-assisted extraction-an innovative and promising extraction tool for medicinal plant research. Pharmacogn. Rev. 1(1): 7–18.

Marette, S., J. Roosen, S. Blanchemanche and E. Feinblatt-Meleze. 2010. Functional food, uncertainty and consumers' choices: A lab experiment with enriched yoghurts for lowering cholesterol. Food Policy 35(5): 419–428.

Markom, M., M. Hasan, W. R. W. Daud, H. Singh and J. M. Jahim. 2007. Extraction of hydrolysable tannins from *Phyllanthus niruri* Linn.: Effects of solvents and extraction methods. Sep. Purif. Technol. 52: 487–496.

Martins, S., S. I. Mussatto, G. Martínez-Avila, J. Montañez-Saenz, C. N. Aguilar and J. A. Teixeira. 2011. Bioactive phenolic compounds: Production and extraction by solid-state fermentation. A review. Biotech. Adv. 29: 365–373.

Miron, T., M. Plaza, G. Bahrim, E. Ibanez and M. Herrero. 2010. Chemical composition of bioactive pressurized extracts of Romanian aromatic plants. J. Chromatogr. 1218(30): 4918–4927.

Mitchell, D. A., N. Krieger, D. M. Stuart and A. Pandey. 2000. New developments in solid-state fermentation: II. Rational approaches to the design, operation and scale-up of bioreactors. Process Biochem. 35: 1211–1225.

Moreira, M. M., S. Morais, A. A. Barros, C. Delerue-Matos and L. F. Guido. 2012. A novel application of microwave-assisted extraction of polyphenols from brewer's spent grain with HPLC-DAD-MS analysis. Anal. Bioanal. Chem. 403(4): 1019–1029.

Mosier, N., C. Wyman, B. Dale, R. Elander, Y. Y. Lee, M. Holtzapple and M. Ladisch. 2005. Features of promising technologies for pretreatment of lignocellulosic biomass. Bioresour. Technol. 96(6): 673–686.

Mussatto, S. I., G. Dragone and I. C. Roberto. 2007. Ferulic and p-coumaric acids extraction by alkaline hydrolysis of brewer's spent grain. Ind. Crop Prod. 25: 231–237.

Nakamura, Y., M. Daidai and F. Kobayashi. 2004. Ozonolysis mechanism of lignin model compounds and microbial treatment of organic acids produced. Water Sci. Technol. 50(3): 167–172.

Naveena, B. M., A. R. Sen, S. Vaithiyanathan, Y. Babji and N. Kondaiah. 2008. Comparative efficacy of pomegranate juice, pomegranate rind powder extract and BHT as antioxidants in cooked chicken patties. Meat Sci. 80: 1304–1308.

Naz, T., J. Packer, P. Yin, J. J. Brophy, H. Wohlmuth, D. E. Renshaw, J. Smith, Yaegl Community Elders, S. R. Vemulpad and J. F. Jamie. 2016. Bioactivity and chemical characterisation of *Lophostemon suaveolens*—An endemic Australian Aboriginal traditional medicinal plant. Natural Prod. Res. 30(6): 693–696.

Nedovic, V., A. Kalusevica, V. Manojlovicb, S. Levica and B. Bugarski. 2011. An overview of encapsulation technologies for food Applications. Procedia Food Science 11th International Congress on Engineering and Food (ICEF 11) 1, Published by Elsevier B.V. pp. 1806–1815.

Nigam, P. S. 2009. Production of bioactive secondary metabolites. pp. 129–145. *In*: P. S. Nigam and A. Pandey (eds.). Biotechnology for Agro-industrial Residues Utilization (1st Ed.), Springer, Netherlands.

Nigam, P. S., N. Gupta and A. Anthwal. 2009. Pre-treatment of agro-industrial residues. Biotechnology for agro-industrial residues utilisation. Springer Netherlands, pp. 13–33.

Nobili, S., D. Lippi, E. Witort, M. Donnini, L. Bausi, E. Mini and S. Capaccioli. 2009. Natural compounds for cancer treatment and prevention. Pharmacol. Res. 59(6): 365–378.

Oberoi, H. S., P. V. Vadlani, K. Brijwani, V. K. Bhargav and R. T. Patil. 2010a. Enhanced ethanol production by fermentation of rice straw using hydrolysate adapted *Candida tropicalis* ATCC 13803. Process Biochem. 45: 1299–1306.

Oberoi, H. S., P. V. Vadlani, R. Madl, L. Saida and J. P. Abeykoon. 2010b. Ethanol production from orange peels by two stage hydrolysis followed by fermentation using parameters optimized through statistical experimental design. J. Agricul. Food Chem. 58: 3422–3429.

Oberoi, H. S., N. Babbar, S. K. Sandhu, S. S. Dhaliwal, U. Kaur, B. S. Chadha and V. K. Bhargav. 2011a. Ethanol production from alkali-treated rice straw *via* simultaneous saccharification and fermentation using newly isolated thermotolerant *Pichia kudriavzevii* HOP-1. J. Industrial Microbiol. Biotechnol. 39: 557–566.

Oberoi, H. S., P. V. Vadlani, A. Nanjundaswamy, S. Bansal, S. Singh, S. Kaur and N. Babbar. 2011b. Enhanced ethanol production from Kinnow mandarin (*Citrus reticulata*) waste *via* a statistically optimized simultaneous saccharification and fermentation process. Bioresour. Technol. 102: 1593–1601.

Oberoi, H. S., P. V. Vadlani, L. Saida, S. Bansal and J. D. Hughes. 2011c. Ethanol production from banana peels using statistically optimized simultaneous saccharification and fermentation process. Waste Management 31: 1576–1584.

Obied, H. K., D. R. Bedgood, P. D. Prenzler and K. Robards. 2007. Bioscreening of Australian olive mill waste extracts: Biophenol content, antioxidant, antimicrobial and molluscicidal activities. Food and Chem. Toxicol. 45: 1238–1248.

Ooshima, H., K. Aso, Y. Harano and T. Yamamoto. 1984. Microwave treatment of cellulosic materials for their enzymatic-hydrolysis. Biotechnol. Lett. 6: 289–294.

Pandey, A. 2003. Solid state fermentation. Biochem. Eng. J. 13: 81–84.

Pandeya, S. N. 1972. Role of sulphides (Thioethers) in biological systems. J. Scientific and Indl. Res. 31: 320–331.

Pasquet, V., J. R. Cherouvrier, F. Farhat, V. Thiery, J. M. Piot, J. B. Berard, R. Kaas, B. Serive, T. Patrice and J. P. Cadoret. 2011. Study on the microalgal pigments extraction process: Performance of microwave assisted extraction. Process Biochem. 46(1): 59–67.

Pedroza, M. A., M. Carmona, G. L. Alonso, M. R. Salinas and A. Zalacain. 2013. Pre-bottling use of dehydrated waste grape skins to improve colour, phenolic and aroma composition of red wines. Food Chem. 136(1): 224–236.

Peng, W., R. Qin, X. Li and H. Zhou. 2013. Botany, phytochemistry, pharmacology, and potential application of *Polygonum cuspidatum* Sieb.et Zucc.: A Review. J. Ethnopharmacol. 148(3): 729–745.

Perles, R. D., D. A. Moreno, M. Carvajal and C. G. Viguera. 2011. Composition and antioxidant capacity of a novel beverage produced with green tea and minimally-processed byproducts of broccoli. Innovative Food Sci. Emerging Technol. 12(3): 361–368.

Peschel, W., F. Sánchez-Rabaneda, W. Diekmann, A. Plescher, I. Gartzía, D. Jiménez, R. Lamuela-Raventós, S. Buxaderas and C. Codina. 2006. An industrial approach in the search of natural antioxidants from vegetable and fruit wastes. Food Chem. 97(1): 137–150.

Philippoussis, A. N. 2009. Production of mushrooms using agro-industrial residues as substrates. pp. 163–196. *In*: P. S. Nigam and A. Pandey (eds.). Biotechnology for Agro-industrial Residues Utilization. Ist Ed. Netherlands: Springer.

Pingret, D., A. S. Fabiano-Tixier, C. L. Bourvellec, C. M. G. C. Renard and F. Chemat. 2012. Lab and pilot-scale ultrasound-assisted water extraction of polyphenols from apple pomace. J. Food Eng. 11(1): 73–81.

Plaza, M., A. Cifuentes and E. Ibanez. 2008. In the search of new functional food ingredients from algae. Trends Food Sci. Technol. 19(1): 31–39.

Plaza, M., M. Amigo-Benavent, M. D. del Castillo, E. Ibanez and M. Herrero. 2010a. Neoformation of antioxidants in glycation model systems treated under subcritical water extraction conditions. Food Res. Int. 43(4): 1123–9.

Plaza, M., M. Amigo-Benavent, M. D. del Castillo, E. Ibanez and M. Herrero. 2010b. Facts about the formation of new antioxidants in natural samples after subcritical water extraction. Food Res. Int. 43(10): 2341–2348.

Plaza, M., S. Santoyo, L. Jaime, G. Garcıa-Blairsy Reina, M. Herrero, F. J. Señoráns and E. Ibanez. 2010c. Screening for bioactive compounds from algae. J. Pharm. Biomed. Anal. 51(2): 450–455.

Puértolas, E., N. López, G. Saldaña, I. Álvarez and J. Raso. 2010. Evaluation of phenolic extraction during fermentation of red grapes treated by a continuous pulsed electric fields process at pilot-plant scale. J. Food Eng. 119(3): 1063–1070.

Qu, W., H. Ma, J. Jia, R. He, L. Luo and Z. Pan. 2012. Enzymolysis kinetics and activities of ACE inhibitory peptides from wheat germ protein prepared with SFP ultrasound-assisted processing. Ultrason. Sonochem. 19(5): 1021–1026.

Rawat, R., N. Srivastava, B. S. Chadha and H. S. Oberoi. 2014. Generating fermentable sugars from rice straw using functionally active cellulolytic enzymes from *Aspergillus niger* HO. Energy and Fuels 28: 5067–5075.

Renard, C. M. C. G., Y. Rohou, C. Hubert, G. della Valle, J. F. Thibault and J. P. Savina. 1997. Bleaching of apple pomace by hydrogen peroxide in alkaline conditions: optimisation and characterisation of the products. Lebensmittel-Wissenschaft und-Technologie 30: 398–405.

Robards, K., P. D. Prenzler, G. Tucker, P. Swatsitang and W. Glover. 1999. Phenolic compounds and their role in oxidative processes in fruits. Food Chem. 66: 401–436.

Roberfroid, M. B. 2002. Functional foods: Concepts and application to inulin and oligofructose. British J. Nut. 87(2): S139–S143.

Robledo, A., A. Aguilera-Carbó, R. Rodríguez, J. L. Martinez, Y. Garza and C. N. Aguilar. 2008. Ellagic acid production by *Aspergillus niger* in solid state fermentation of pomegranate residues. J. Ind. Microbiol. Biotech. 35: 507–513.

Routray, W. and V. Orsat. 2011. Microwave-assisted extraction of flavonoids: A review. Food Bioprocess Technol. 5: 409–424.

Sánchez, C. 2009. Lignocellulosic residues: Biodegradation and bioconversion by fungi. Biotechnol. Adv. 27: 185–194.

Sanders, M. E. 1998. Overview on functional foods: Emphasis on probiotic bacteria. Int. Dairy J. 8: 341–347.

Sarkar, N., S. K. Ghosh, S. Bannerjee and K. Aikat. 2012. Bioethanol production from agricultural wastes: An overview. Renew. Energy 37(1): 19–27.

Saulnier, D. M. A., J. K. Spinler, G. R. Gibson and J. Versalovic. 2009. Mechanisms of probiosis and prebiosis: Considerations for enhanced functional foods. Curr. Opin. Biotechnol. 20(2): 135–141.

Schieber, A., F. C. Stintzing and R. Carle. 2001. By-products of plant food processing as a source of functional compounds—Recent developments. Trends Food Sci. Technol. 12: 401–413.

Sihvonen, M., E. Jarvenpaa, V. Hietaniemi and R. Huopalahti. 1999. Advances in supercritical carbon dioxide technologies. Trends Food Sci. Technol. 10(6-7): 217–222.

Sills, D. L. and J. M. Gossett. 2011. Assessment of commercial hemicellulases for saccharification of alkaline pretreated perennial biomass. Biores. Technol. 102: 1389–1398.

Silva, L. V., D. L. Nelson, M. F. B. Drummond, L. Dufossé and M. B. A. Glória. 2005. Comparison of hydrodistillation methods for the deodorization of turmeric. Food Res. Intl. 38(8-9): 1087–1096.

Siro, I., E. Kapolna, B. Kapolna and A. Lugasi. 2008. Functional food. Product development, marketing and consumer acceptance—A review. Appetite 51(3): 456–467.

Soria, A. C. and M. Villamiel. 2010. Effect of ultrasound on the technological properties and bioactivity of food: A review. Trends Food Sci. Technol. 21(7): 323–331.

Soxhlet, F. 1879. Die gewichts analytische Bestimmung des Milchfettes. Dingler's Polytechnisches J. 232: 461–465.

Starmans, D. and H. Nijhuis. 1996. Extraction of secondary metabolites from plant material: A review. Trends Food Sci. Technol. 7(6): 191–197.

Suttirak, W. and S. Manurakchinakorn. 2014. *In vitro* antioxidant properties of mangosteen peel extract. J. Food Sci. Technol. 51(12): 3546–3558.

Szentmihalyi, K., P. Vinkler, B. Lakatos, V. Illes and M. Then. 2002. Rose hip (*Rosa canina* L.) oil obtained from waste hip seeds by different extraction methods. Biores. Technol. 82(2): 195–201.

Szydlowska-Czerniak, A., G. Karlovits, G. Hellner, C. Dianoczki and E. Szlyk. 2010. Effect of enzymatic and hydrothermal treatments of rapeseeds on quality of the pressed rapeseed oils: Part I: Antioxidant capacity and antioxidant content. Process Biochem. 45(1): 7–17.

Teymouri, F., L. Laureano-Perez, H. Alizadeh and B. E. Dale. 2005. Optimization of the ammonia fiber explosion (AFEX) treatment parameters for enzymatic hydrolysis of corn stover. Biores. Technol. 96(18): 2014–2018.

Vankar, P. S. 2004. Essential oils and fragrances from natural sources. Resonance 9(4): 30–41.

Vattem, D. A. and K. Shetty. 2003. Ellagic acid production and phenolic antioxidant activity in cranberry pomace (*Vaccinium macrocarpon*) mediated by *Lentinus edodes* using a solid-state system. Process Biochem. 39: 367–379.

Vorobiev, E., A. B. Jemai, H. Bouzrara, N. I. Lebovka and M. I. Bazhal. 2005. Pulsed electric field assisted extraction of juice from food plants. pp. 105–130. *In*: G. Barbosa-Canovas, M. S. Tapia and M. P. Cano (eds.). Novel Food Processing Technologies, CRC Press, New York.

Wandrey, C., A. Bartkowiak and S. E. Harding. 2009. Materials for encapsulation. pp. 31–100. *In*: N. J. Zuidam and V. A. Nedovic (eds.). Encapsulation Technologies for Food Active Ingredients and Food Processing, Springer: Dordrecht, The Netherlands.

Wang, T., R. Jónsdóttir, H. G. Kristinsson, G. O. Hreggviðsson, J. O. Jónsson and G. Ólafsdóttir. 2010. Enzyme-enhanced extraction of antioxidant ingredients from red algae *Palmaria palmate*. LWT-Food Sci. Technol. 43(9): 1387–1393.

Wang, L. and C. L. Weller. 2006. Recent advances in extraction of nutraceuticals from plants. Trends Food Sci. Technol. 17(6): 300–312.

Wang, M., J. E. Simon, I. F. Aviles, K. He, Q. Y. Zheng and Y. Tadmor. 2003. Analysis of antioxidative phenolic compounds in artichoke (*Cynara scolymus* L.). J. Agrl. Food Chem. 51: 601–608.

Weiss, J., P. Takhistov and D. J. McClements. 2006. Functional materials in food nanotechnology. J. Food Sci. 71: R107–R116.

Wolfe, K. L., X. Wu and R. H. Liu. 2003. Antioxidant activity of apple peels. J. Agrl. Food Chem. 51: 609–614.

Wyman, C. E. 1999. Biomass ethanol: Technical progress, opportunities and commercial challenges. Ann. Rev. Energy Environment 24: 189–226.

Xiao, W., Y. Wang, S. Xia and P. Ma. 2012. The study of factors affecting the enzymatic hydrolysis of cellulose after ionic liquid pretreatment. Carbohydrate Polymers 87: 2019–2023.

Yanagida, A., T. Kanda, M. Tanabe, F. Matsudaira and J. G. O. Cordeiro. 2000. Inhibitory effects of apple polyphenols and related compounds on cariogenic factors of mutans streptococci. J. Agrl. Food Chem. 48: 5666–5671.

Yoo, J., S. Alavi, P. Vadlani and V. Amanor-Boadu. 2011. Thermo-mechanical extrusion pretreatment for conversion of soybean hulls to fermentable sugars. Bioresour. Technol. 102(16): 7583–7590.

Yu, X. W. and Y. Q. Li. 2008. Expression of *Aspergillus oryzae* tannase in *Pichia pastoris* and its application in the synthesis of propyl gallate in organic solvent. Food Technol. Biotechnol. 46(1): 80.

Zeyada, N. N., M. A. M. Zeitoum and O. M. Barbary. 2008. Utilization of some vegetables and fruit waste as natural antioxidants. Alex. J. Food Sci. Technol. 5: 1–11.

Zhang, H. F., X. H. Yang and Y. Wang. 2011. Microwave assisted extraction of secondary metabolites from plants: Current status and future directions. Trends Food Sci. Technol. 22(12): 672–688.

Zhao, L., G. Zhao, F. Chen, Z. Wang, J. Wu and X. Hu. 2006. Different effects of microwave and ultrasound on the stability of (all-E)-astaxanthin. J. Agric. Food Chem. 54(21): 8346–8351.

Zhao, X., K. Cheng and D. Liu. 2009. Organosolv pretreatment of lignocellulosic biomass for enzymatic hydrolysis. Appl. Microbiol. Biotechnol. 82(5): 815–827.

Zhao, Y., Y. Wang, J. Y. Zhu, A. Ragauskas and Y. Deng. 2008. Enhanced enzymatic hydrolysis of spruce by alkaline pretreatment at low temperature. Biotechnol. Bioeng. 99: 1320–1328.

Zheng, J. and L. Rehmann. 2014. Extrusion pretreatment of lignocellulosic biomass: A review. Intl. J. Molc. Sci. 15: 18967–18984.

Zheng, Z. and K. Shetty. 2000. Solid-state bioconversion of phenolics from cranberry pomace and role of *Lentinus edodes* β-glucosidase. J. Agric. Food Chem. 48: 895–900.

Zhu, J., C. Wan and Y. Li. 2010. Enhanced solid-state anaerobic digestion of corn stover by alkaline pretreatment. Biores. Technol. 101: 7523–7528.

Zill-e Huma, M., A. Vian, A. S. Fabiano-Tixier, M. Elmaataoui, O. Dangles and F. Chemat. 2011. A remarkable influence of microwave extraction: Enhancement of antioxidant activity of extracted onion varieties. Food Chem. 127(4): 1472–1480.

Zuidam, N. J. and J. Heinrich. 2009. Encapsulation of aroma. pp. 127–160. *In*: N. J. Zuidam and V. A. Nedovic (eds.). Encapsulation Technologies for Food Active Ingredients and Food Processing; Springer: Dordrecht, The Netherlands.

Zupko, I., J. Hohmann, D. Redei, G. Falkay, G. Janicsak and I. Mathe. 2001. Antioxidant activity of leaves of *Salvia* species in enzyme-dependent and enzyme-independent systems of lipid peroxidation and their phenolic constituents. Planta Medica 67: 366–368.

CHAPTER 6

Isolation, Purification and Encapsulation Techniques for Bioactive Compounds from Agricultural and Food Production Waste

Viktor A. Nedović,[1,*] *Fani Th Mantzouridou,*[2] *Verica B. Đorđević,*[3] *Ana M. Kalušević,*[1] *Nikolaos Nenadis*[2] and *Branko Bugarski*[3]

Introduction

Food related materials that are discharged worldwide are residues of high organic load generated through different stages of food's life cycle, from agricultural production up to postharvest handling, processing, retail and consumption. Typically, these substances are characterized as 'waste', in line with the definition given from the Waste Framework Directive (Commission Directive 2006/12/EC): "*Waste: a material which the holder discards or indented or is required to discard*". The global amount of food waste is enormous, accounting for about 1.3 billion tons per yr (Gustavsson et al. 2011). This situation has created concerns worldwide due to environmental and economical problems.

For a long time, proper food waste treatment was seen as a primary focus for most food waste management strategies. In recent years there has been a steady flow of EU policies that promote waste minimization (Mirabella et al. 2014, Tsakona et al. 2016). Moreover, the utilisation and recovery of resources from non-preventable wastes is

[1] Department of Food Technology and Biochemistry, Faculty of Agriculture, University of Belgrade, Nemanjina 6, 11080 Zemun, Belgrade, Serbia. E-mail: ana.kalusevic@agrif.bg.ac.rs
[2] Laboratory of Food Chemistry and Technology, School of Chemistry, Aristotle University of Thessaloniki, 54124 Thessaloniki, Greece. E-mail: fmantz@chem.auth.gr; niknen@chem.auth.gr
[3] Department of Chemical Engineering, Faculty of Technology and Metallurgy, University of Belgrade, Karnegijeva 4, 11000 Belgrade, Serbia. E-mail: vmanojlovic@tmf.bg.ac.rs; branko@tmf.bg.ac.rs
* Corresponding author: vnedovic@agrif.bg.ac.rs

associated with increasing EU target for eco-innovation, aimed at 'zero waste' society and circular economy where these fractions are used as raw material for new products and applications (COM/2014/0398, Tsakona et al. 2016). Considering that many of these food wastes contain considerable amounts of valuable compounds derived from the original material, they have the potential to be reused in other production systems (e.g., bio-refineries) for the production of higher value and marketable products rather than usual food waste processing to produce energy, feed and fertilizers (Lin et al. 2013).

Valorization of food waste for edible purposes represents a challenging field of research. In this direction, scientists focus mainly on the recovery of functional compounds derived from agricultural and food processing side streams and not the later stages of food's life cycle. This is attributed to the fact that the former are abundant, exist in concentrated industrial-urban areas and are less susceptible to biological deterioration as compared to the latter (Tornberg and Galanakis 2008, Galanakis 2012).

The current trend of research and development for utilisation of agricultural and food processing waste as a source for value-added bio-based product focuses on (1) recovery of compounds by using conventional and emerging technologies, (2) biotransformation of compounds by means of enzymes or whole cells to obtain analogues with higher value, and (3) development of bioprocesses using microorganisms for the production of bioactive compounds (Mèridas et al. 2012). Recovery/production of food additives, such as antioxidants and flavor-active compounds, is a good example of desirable added-value compounds. In the following sections characteristic examples of high value-added bio-based chemicals and corresponding waste streams are discussed with special focus on the conventional and emerging technologies applied from the source to the final product.

Phenolic Compounds

Phenolic compounds are a broad group of plant secondary metabolites, which arise biosynthetically from the shikimate or/and polyketide (acetate) pathways (Harborne 1989). These compounds are multifunctional since they contribute to plant growth and development, protection from environmental stress, viral, fungal and bacterial infestation, enzyme modulation, insect attraction or repulsion, coloration of flowers and fruits, etc. (Crozier et al. 2006). Due to the above, phenols are ubiquitous in plant tissues. Consequently, they can be found in a wide range of waste/by-products (liquid and solid) derived from the processing of raw materials or other steps in the line of production in the agricultural industries. Their natural abundance, as well as the fact that the dietary intake in phenols through daily consumption of plant-derived foods has been associated with various health benefits, such as prevention of cancer and cardiovascular diseases formation (Birt et al. 2001, Kris-Etherton et al. 2002), list them among the target substances to be recovered from these sources or even agricultural residues for the development of innovative products with high value-added (Naziri et al. 2014, Galanakis 2016). Towards this direction, various efforts have been made and the field of research is of great importance for the local and international economy (Tornberg and Galanakis 2008, Galanakis 2012, Naziri et al. 2014). The following paragraphs draw on the case study for the recovery of ferulic acid from agro-industrial by-products and waste, which is already exploited.

Recovery of ferulic acid from agro-industrial by-products and waste

Ferulic acid (4-hydroxy-3-methoxy phenyl-propenoic acid) is a member of hydroxycinnamic acids that is ubiquitously found in the plant kingdom. As a matter of fact it is the most prevalent one with a key role in plant architecture (Barberousse et al. 2008). It is biosynthesized from phenylalanine and tyrosine (Zhao and Moghadasian 2008) and is largely located in the bran of grains, the peel of fruits, roots and peel of vegetables, hulls, straws (Clifford 1999) justifying its presence in various agro-industrial residues. It is found in its free form but mainly bound or in the form of oligomers (Barberousse et al. 2008). Bounding is made to various organic acids, mono/di- and poly-saccharides, sterols or even inorganic elements such as silicon, lignin and proteins, usually through esterification, and less frequently through etherification, protecting the plant cell walls from biodegradation (Inanaga et al. 1995, Barberousse et al. 2008, Zhao and Moghadasian 2008). It is widely available in the diet with its dietary intake estimated to be in the range 150–250 mg/d (Zhao and Moghadasian 2008). The compound is of low toxicity, is absorbed by the small intestine and easily metabolized (Ou et al. 2004). Furthermore, various biological properties have been ascribed, including antioxidant activity (Graf 1992). Due to these properties, the phenolic acid has been widely used in the food and cosmetic industries. More specifically in Japan, it has been approved as an antioxidant in foods, beverages and cosmetics. Despite not being considered as GRASS (generally recognized as safe) in the USA and most European countries, several medical essences and natural extracts selected for their high content of ferulic acid are added to foods as an FDA (Food and Drug Administration)-approved antioxidant formulation. Other applications include its use as an ingredient in food gels and edible films due to its cross-linking capability with proteins and polysaccharides, in sports beverages as an ergogenic substance and skin protection formulation acting as antioxidant. The respective compound may also serve as a substrate for the production of vanillin, a flavoring compound widely used in food, cosmetic and pharmaceutical industries (Ou et al. 2004).

Food or cosmetic grade ferulic acid was produced in Japan through hydrolysis of γ-oryzanol (feroyl steryl ester), a constituent of rice bran oil (Graf 1992). In most Japanese companies, this is still the method of production (http://www.oryza.co.jp/html/english/). Rice bran seems to contain appreciable amounts (33 mg/g) of ferulic acid (Gopalan et al. 2015). However, the oil contains oryzanols in the range 2–3 per cent (Erickson 1990). Thus, rice bran is underexploited for ferulic acid recovery and the same applies for other agro-industrial residues containing comparable or lower, but still adequate levels. Characteristic examples are maize bran (~ 31.2 mg/g), sweet potato stems (~ 22 mg/g), corn fiber (~ 18 mg/g), Jojoba meal (~ 9.1 mg/g), sugar beet pulp (~ 6.4 mg/g), wheat (~ 4 mg/g) and triticale bran (~ 2.2 mg/g), rice hulls (~ 1.5–2.0 mg/g), sugarcane baggage (~ 1.4 mg/g) (Nenadis et al. 2013, Gopalan et al. 2015).

The first step is an adequate recovery/liberation from the plant material of the target compound. Taking into account the above, it is evident that for the case of ferulic acid suitable conditions for its liberation from the walls should be adopted.

Therefore, common organic solvents, such as water or alcohol and their mixtures, are not recommended. On the basis of various efforts, this can be feasible through enzymatic or chemical processes (Barberousse et al. 2008, Gopalan et al. 2015). Application of far infrared irradiation prior to extraction, claiming to liberate phenolic antioxidants, has not been extensively explored for such a purpose. The latter may be due to the fact that ferulic acid was not detected unexpectedly among the identified constituents after GC-MS analysis of silylated extracts (Lee et al. 2003).

Enzymatic release

The enzymatic release is based on the use of the so-called feruloyl esterases (FAEs) which can cleave the esteric bonds formed between ferulic acid and the various constituents already mentioned (Gopalan et al. 2015). These enzymes are classified into four groups. Type A are effective only when esterification is through O-5 and not O-2 of L-arabinofuranose, whereas types B-D hydrolyze both but present differences on the effect of release of various ferulic acid dimmers (Crepin et al. 2004, Topakas et al. 2007). The enzymes are produced by filamentous fungi or bacteria preferably by solid-state fermentation, which among other advantages over submerged fermentation, offers higher enzyme titres and stability (Gopalan et al. 2015). The respective approach is of increasing interest as a green means that does not produce further chemical waste nor destroy other valuable constituents due to the high specificity, making consequently the substrate further exploitable (recovery of amino acids, bio-polymers, platform chemicals, fuel compounds) and increase the profit (Gopalan et al. 2015).

Despite the high specificity, the effectiveness is not always adequate. As summarized by Gopalan et al. (2015), the efficiency is low (0.7–58 per cent) and only in a limited number of applications does the efficiency reach 98–100 per cent. The success may depend on the characteristics of the enzymes, as well as the substrate structure which may affect the access to targeted bonds. Furthermore, it is not known to what extent the liberated ferulic acid inhibits the activity of esterases. The latter was highlighted for the first time by Xiros et al. (2009), using as a substrate brewer's spent grain. The ways to increase esterase efficiency is to pre-treat the substrate or/and to co-add other enzymes (xylenases, arabinofuranosidases, proteases) that may modify the substrate and 'open the way' for esterases. Pre-treatments include extensive comminution and sieving (Yu et al. 2002) which result in partial disintegration of the structure or steam explosion (flash explosion) which can cause the formation of smaller oligosaccharides suitable as substrate for enzymes, through sudden hot steam decompression (Saulnier et al. 2001). In the first case, passing oat hulls through a 250 μm screen or smaller improved the release by *Aspergilus* esterase. Even so, the profit was negligible in comparison to the results obtained with the co-addition of *Trichoderma* xylenase (7.8–32,768 U/assay). In such a case, the release by the esterase maintained at 26 U/assay increased from ~ 0.6 (control), 2.8–104-fold. In the second example, the improvement of release from maize bran due to pre-treatment is depicted in Fig. 1. Critical was also the temperature. Thus, more than 60 per cent of solubilization was achieved after explosion at 190°C for 1 min, achieving a more than three-fold increase in the release. Still, however, a considerable amount remained bound.

An example highlighting the role of proteases is the work of Xiros et al. (2009), who used a crude enzyme from *Fusarium oxysporum* for the release of ferulic acid by using

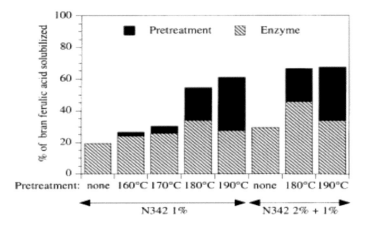

Figure 1. Total solubilisation of ferulic acid from maize bran by the combined action of flash-explosion and Novozym 342 (adopted with permission from Saulnier et al. 2001).

as a substrate brewer's spent grain. The release was ~ 2.5-fold (1 mg/g dry substrate) higher than that obtained by the synergistic action of an esterase (FoFaeC-12213) and *Trichoderma longibrachiatum* M3 xylanase. Such an achievement was related to the presence of alcalase and papain in the crude enzyme. The contribution of the latter was also verified by treatment of the substrate with a mixture of these two proteases, simulating the content in the crude enzyme.

An exceptional case was that reported by Shin et al. (2006), who discovered a fungal strain, namely *Neosartorya spinosa* NRRL185, which is able to produce an array of enzymes to fully liberate ferulic acid from corn bran and corn kernel skins (corn fibers). For corn fibers, the ~ 100 per cent release of ferulic acid was achieved after 24 hours, whereas for the corn bran 48 hours were required and a two-fold higher quantity of enzymes. The latter was considered reasonable due to the higher contentment in ferulic acid. The partial characterization of the extracellular proteome of the microbe revealed the presence of an array of cellulases, hemicellulases, iso-forms of xylanase and ferulic acid esterase. The recovery of ferulic acid was followed by a 76–100 per cent release of reducing sugars, implying that the field of application of enzymes derived from the respective microorganism can be broader.

Considering the above continuous characterization of the proteome of enzymes from various sources will permit efficient release of ferulic acid and other targeted bioactive phenols. Furthermore, taking into account that enzymatic activity requires the selection of appropriate parameters (e.g., enzyme quantities, temperature, incubation time), experimental design approaches-optimisation studies utilizing approaches such as response surface methodology can contribute further to a profitable process (Barberousse et al. 2009).

Chemical release

Chemical liberation of ferulic acid that is esterified to the various molecules in different substrates is rather facile to achieve. The breaking of bonds can be made only under alkaline conditions since acidic treatment may break glycosidic bonds, solubilizing, thus, the sugars but leaves the ester and ether bonds intact, whereas under intense conditions, they may even degrade phenolics (Krygier et al. 1982, Nenadis et al. 2013). A characteristic example is depicted for rice hulls treated either with NaOH or HCl acid solution of the same concentration under the same conditions of extraction in Fig. 2. Significant liberation of ferulic acid using alkaline conditions is possible even at room temperature. Nevertheless, application of higher temperature, especially more than 100°C affects positively the liberation and so does the use of higher concentration of alkali and higher incubation (Mussato et al. 2007, Fig. 2A).

Those commenting in favour of such an approach instead of the enzymatic one state that the application of enzymes is usually impractical due to the low production by the microorganisms and the long-time necessary to achieve the hydrolysis (Ou et al. 2007, Aarabi et al. 2016). After all, alkaline liberation is used as a reference to comment on the effectiveness of the enzymatic approach (Barberousse et al. 2008). At this point it should be stressed that the conditions employed varies (*vide infra*) and considering the fact that it may influence the quantitative findings, their selection should be made carefully for an accurate comparison with the enzymatic protocol developed.

Thus, production of solutions rich in ferulic acid from agricultural residues or by-products after alkaline extraction have been reported, using a wide range of NaOH concentrations (0.05–4 mol/L) at room temperature or with concomitant heating (50–180°C) and wide range (1.5–24 hours) of incubation time (Mussato et

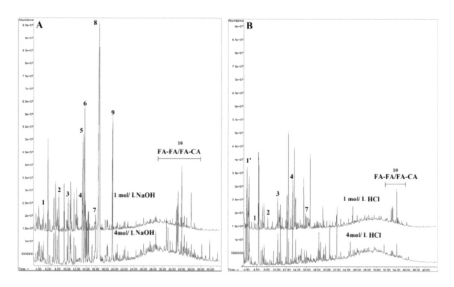

Figure 2. Total ion gas chromatogram of trimethylsilanol derivatives of hull extracts (Gladio variety) after treatment with (A) 1 and 4 mol/L NaOH or (B) 1 and 4 mol/L HCl solutions (120°C, 120 kPa, 2 h). 1': hydroxymethylfurfural; 1: p-hydoxybenzaldehyde; 2: vanillin; 3: p-hydroxybenzoic acid; 4: vanillic acid; 5: syringaldehyde; 6: cis-p-coumaric acid; 7: syringic acid; 8: trans-p-coumaric acid; 9: ferulic acid; 10: ferulic acid dimer (FA-FA) or ferulic acid-coniferyl alcohol (FA-CA) dimer (adapted with permission from Nenadis et al. 2013).

al. 2007, Ou et al. 2007, Tilay et al. 2008, Torre et al. 2008, Buranov and Mazza 2009, Nenadis et al. 2013, Zhao et al. 2014, Aarabi et al. 2016). These conditions were applied to substrates, such as sugarcane bagasse, brewer's spent grain, corn cobs, maize bran, flax shives, wheat bran, rice hulls and sugar-beet-pulp. An advantage of chemical hydrolysis using an alkali is that modern techniques of extraction making efficient release in a shorter time can be employed. For example, using a 0.5 mol/L NaOH solution with a pressurized extractor, similar levels of ferulic acid (25.1 mg/g) within 57 min were obtained from corn bran (T = 180°C, P = 5.2 MPa, flow rate = 5 mL/min), to those employing the same alkaline solution for 4 hours at 50°C. More characteristic is the application of microwaves—a technique that can be scaled up at industrial level (Li et al. 2013). Thus, utilisation of microwaves using 4 mol/L NaOH solution (170°C, 90 s, 1.1 MPa) provided extracts with almost the same ferulic acid content and antioxidant activity to that prepared within two-hour exposure to the same alkaline solution at 120°C 0.12 MPa (N.N., unpublished data). Taking into consideration the variability of parameters in the numerous studies, as already mentioned for the enzymatic approaches, response surface methodology could provide optimum conditions for each substrate. An example is the study of Tilay et al. (2008) who, applying such an approach for maize bran and considering as variables NaOH concentration, incubation time and temperature achieved a 1.3-fold increase in comparison to an un-optimized conventional extraction technique. The authors used in their experiment long incubation times (40 > t > 10 hours) and low temperatures (38 > t > 21°C). We believe that in case a high temperature and short incubation time could be selected, the optimum conditions would result in higher efficiency.

All in all, chemical treatment results in solutions of high content in ferulic acid. Nevertheless, as shown in Fig. 2A, other phenolics are liberated, consequently the approach is not very selective. In addition, depending on the conditions employed other valuable constituents may be damaged (Gopalan et al. 2015) rendering unprofitable the further exploitation of the substrate. Apart from these, a murky solution is obtained containing impurities and wax (Buranov and Mazza 2009), purification of which is a more challenging task than that of solutions obtained with enzymatic treatment.

Purification

Despite the fact that various studies have been carried out on the liberation of ferulic acid, less are the efforts regarding the purification in large-scale. Solvent extraction, the use of different resins and even powdered activated charcoal are some of the approaches examined.

In case of solvent extraction ethyl acetate, which is generally recognized as safe, is preferred to recover the dissolved ferulic acid from the hydrolysate. Following such an approach, the solution containing the liberated acid is extracted twice or more with equal volumes of ethyl acetate, the organic fractions are pooled together, dried over anhydrous sodium sulfate and concentrated, using reduced vacuum. Such a protocol employed by Shin et al. (2006) after the enzymatic liberation of ferulic acid from corn residues was adequate according to the HPLC chromatogram provided by the authors. Nevertheless, when alkaline treatment is applied, other phenolic compounds are also co-extracted with ethylacetate (Fig. 2A) and impurities; therefore, further purification is required. Another proposal to simplify purification involves exploitation of ferulic acid solubility in aqueous ethanol. More specifically, Buranov and Mazza (2009) carrying out solubility trials determined that 30–35 per cent of ethanol in water could recover the maximum amount of acid. The steps followed were neutralisation with HCl, addition of 95 per cent v/v ethanol to fix the alcohol concentration in the extract at 35 per cent v/v permitting high solubility of ferulic acid and removal of wax and glucomannans through centrifugation. The authors then employed ultrafiltration to minimize the use of ethanol and to remove polymeric hemicelluloses. Precipitation of oligomeric hemicelluloses was then achieved via ethanol addition, whereas finally pure ferulic acid could be recovered through precipitation and by vacuum evaporation of water. Even so, the authors did not provide the per cent purity but relied on the thin-layer chromatography and Fourier Infrared analysis of the isolate and pure standard. More recently, Aarabi et al. (2016) followed the above-recommended protocol, omitting the ultrafiltration part, to recover ferulic acid from the hydrolysate of sugar-beet pulp. These authors for further purification dissolved the isolate to anhydrous ethanol and centrifugation at 11000 x g. The liquid chromatogram provided showed traces of impurities; the corresponding one of the alkaline extract indicated that prior to purification the amount of impurities was low.

Adsorption through charge-transfer complexes and other physical mechanisms with activated carbons, by hydrophobic interactions (through the phenol ring) with polystyrenic resins (XAD) or hydrogen bonds (through the phenolic or carboxylic hydroxyl group) by polyvinylpolypyrrolidone (PVPP), have been examined by Couteau and Mathaly (1997) as a means to trap ferulic acid from a sugar-beet hydrolysate. The recovery could be obtained via elution with ethanol. Better adsorption was achieved when the solution was acidified at pH < pKa of frolic acid (4.5) so that the carboxylic group was maintained protonated. Chemical-activated carbon presented the highest affinity towards ferulic acid, followed by XAD-16 (the best among XAD-16, XAD-1180, XAD-2, XAD-4) and PVPP as shown by the calculated maximum specific capacity, which was ~ 22, 12 and 8 per cent (w/w ferulic acid/adsorbent) respectively. The carbon presented, however, higher equilibration time. Despite the fact that the desorbed ferulic acid was not pure, the authors claimed that an appreciable enrichment was achieved which should permit crystallization for a purified product. The aforementioned study was made on a solution produced after enzymatic treatment. According to Ou et al. (2007) the activated charcoal is not a feasible way to achieve the same goal in case an alkaline-hydrolysate is obtained. This is due to the fact that the color substances produced as a consequence of the chemical treatment can be washed out by using alcohol. Thus, washing with hot water and acetic ether was proposed for removing such impurities and then the recovery could be made with 0.2 mol/L NaOH. The authors made one step forward and proved that a high purity ferulic acid could be then obtained through treatment with an anion macroporous resin. Tilay et al. (2008) purified ferulic acid from an alkaline hydrolyzate of maize bran by column chromatography, followed by preparative high-performance thin-layer chromatography (HPTLC). The alkaline hydrolyzate extracted with ethyl acetate, after concentration was loaded on to a column filled with Amberlite XAD-16 and elution was carried out with ethanolic ammonium hydroxide (0.1 per cent). The richest fraction containing ~ 51 per cent ferulic acid was further purified with HPTLC up to 95.35 per cent. Recently, Zhao et al. (2014) proposed a suitable protocol that could provide ferulic acid crystal (8.47 g/kg corn bran) with 84.5 per cent purity from a difficult substrate, namely corn bran after alkaline extraction. The authors relied on membrane technology and in order to be able to apply ultrafiltration avoided blocking of polysaccharides the 0.25 mol/L NaOH solution contained also 50 per cent ethanol. As a second step, the authors applied nanofiltration and recovered the ferulic acid through crystallization after adjusting the pH value of the concentrate at 2.0. The corresponding technology accounted for 81 per cent recovery of ferulic acid, is considered simpler than anion exchange chromatography and produced less waste water and organic solvent (ethanol is recycled). In addition, the corn bran residue was easier to be digested by xylanse. Bearing in mind the above, it is interesting to stress that other techniques applied on an industrial scale could provide, after some appropriate pretreatment, high purity products. Towards this direction useful could be preparative HPLC (Sticher 2008), as well as preparative high-speed counter-current chromatography (HSCCC). The latter, which can be scaled up at industrial level (Sutherland 2007) was used by Liu et al. (2006) to isolate ferulic acid from extracts of *Radix Angelicae sinensis* with purity over 98 per cent.

Carotenoids

Carotenoids make up a diverse class of natural terpenoid pigments. Their major food use is as colouring agents (yellow to red). Noticeably, among the 50 carotenoids present in the human diet, only six are detectable in human plasma (α- and β-carotene, β-cryptoxanthin, lycopene, lutein, zeaxanthin) (Borel et al. 2007). The beneficial role of carotenoids on human health increased the interest in their commercialization as food and feed additives and as components in cosmetics and pharmaceuticals (http://ubic-consulting.com/wp-content/uploads/2015/08/The-World-Beta-Carotene-Ingredient Market.pdf). For example, β-carotene is the main precursor for vitamin A. Its role in the provision of an adequate supply of total vitamin A in both developed and developing countries has been noticed by researchers (Grune et al. 2010). Moreover, there is solid evidence to support protective effects of dietary carotenoid intake in widespread modern diseases related to oxidative stress (i.e., cancer, cardiovascular or photosensitivity disorders) in the context of carotenoids' unique antioxidative properties (Fiedor and Burda 2014). β-carotene and lycopene are two of the most important products economically available in the carotenoid market worldwide. Concerning β-carotene, its market size was valued at over $425 million in 2015 and is likely to exceed $500 million in 2023 (CAGR of more than 3 per cent from 2016 to 2023) (https://www.gminsights.com/industry-analysis/beta-carotene-market). Although lycopene has a smaller market share, the global lycopene market is expected to reach $84 million by 2018, up from $66 million in 2010 (Ciriminna et al. 2016).

The carotenoids used commercially are produced via chemical synthesis, extraction from natural sources, or microbial fermentation. Over time, most of the synthetic ones were prohibited due to their serious impact on health (Martins et al. 2016). The growing demand for more natural ingredients in foods has resulted in an increase in research concerning the microbial production and extraction of carotenoids from plant sources (Mezzomo and Ferreria 2016).

The fact that the market of natural carotenoids, especially for β-carotene and lycopene, is still in the developing stage reflects the relatively high price of the extracts from microorganisms and tomatoes. Well-developed technologies are involved in microbial production of carotenoids on a large scale, resulting in high yield and purity of the final product. Nevertheless, the cost effectiveness of such a process exists mainly due to the high cost of the raw materials. The high price of plant lycopene is a consequence of the low concentration of the pigment in the fruit, the complex purification process due to the numerous carotenoids present in the raw material and the use of pesticide-free fresh tomatoes grown using organic farming (Naziri et al. 2014, Ciriminna et al. 2016). Thus, scientific efforts are focused on the development of eco-inovative solutions by exploitation of agricultural and food processing waste for the recovery/bioproduction of natural carotenoids.

Recovery of plant-derived natural lycopene from tomato processing waste

Tomato (*Lycopersicon esculentum*) is the most important vegetable crop worldwide after potato, with 100 million tons produced annually. Tomatoes are produced either raw or after processing, providing a significant amount of carotenoids, mainly lycopene (80–90 per cent of total carotenoids) in the human diet (Mirabella et al. 2014). Beyond unprocessed fruit, the tomato processing waste (i.e., tomato pomace containing mainly peels and seeds), that account for 10–40 per cent of total processed tomatoes (Strati and Oreopoulou 2014) has been intensively researched as a cheap source of lycopene. Lycopene content in tomato peels is the highest (90 per cent of the total fruit concentration) among that found in the different fruit parts (Konar et al. 2012), depending on the growing season, variety, fruit stage of ripening and the industrial processing methods (Riggi and Avola 2008, Urbonaviciene et al. 2012). According to Kalogeropoulos et al. (2012), the carotenoid content of tomato-processing waste amounted to 413.7 and 149.8 μg/g dry weight for lycopene and β-carotene, respectively. In this context, the research interest increased over the years regarding the recovery of high quality lycopene products from tomato processing waste by conventional and emerging technologies (Strati and Oreopoulou 2014, Papaioannou et al. 2016).

Solvent extraction accomplished in a Soxhlet extractor or agitation vessel has been widely applied for the recovery of lycopene from tomato processing waste (Strati and Oreopoulou 2014). The optimum extraction conditions depend on the solvent type, solvent to solid ratio, particle size, temperature and extraction time (Strati and Oreopoulou 2011a). An array of polar and non-polar solvents (e.g., hexane, ethanol, acetone, ethyl acetate, ethyl lactate) as well as their mixtures has been used so far for the extraction of tomato lycopene (Strati and Oreopoulou 2011b). Owing to its lipophilic character, the mixtures of ethyl acetate with hexane was proposed as optimum by Strati and Oreopoulou (2011a) to increase lycopene solubilization and maximize the extraction yield (96 per cent of total extracted carotenoids) from tomato processing waste. Elevated temperature (up to 70°C) and sequential extraction steps increase the process yield. However, this is not a straightforward process as there are several factors (raw material, solvent and extraction conditions) that affect carotenoid isomerization/degradation as the temperature increases (Strati and Oreopoulou 2014). However, the majority of suitable solvent systems for efficient extraction of lycopene from tomato waste should be completely removed from the extracts before their reuse for food application due to their hazardous effects (Commission Directive 97/60/EC). Thus, the selection of appropriate and safe, food-grade extraction solvents is important to avoid the elimination problems of trace residual solvents (Papaioannou et al. 2016). In this direction, Strati and Oreopoulou (2011a) suggested ethyl lactate as an environment-friendly potent solvent to extract carotenoids from tomato waste in higher yields compared to ethyl acetate (243 vs. 46 mg/kg dry basis, respectively). This solvent was found to be efficient for the extraction of both trans- and cis-lycopene isomers (Ishida and Chapman 2009).

The limiting step for recovering lycopene from tomato waste is the penetration of solvent into the tomato tissues to solubilize the pigment, which is located in the chromoplast membrane structures (Lavecchia and Zuorro 2008). Enzyme treatment of plant material with mixtures of cell-wall degradating enzymes with pectinolytic, cellulolytic and hemicellulolytic activities has been successfully applied prior to conventional solvent extraction process in order to make the peel lose its structural integrity (e.g., Lavecchia and Zuorro 2008, Zuorro et al. 2011, Ranveer et al. 2013, Zuorro et al. 2014, Strati et al. 2015). This pretreatment step seems to enhance the yield and quality of the product in comparison to the conventional solvent extraction process which allows for extremely mild extraction conditions, such as low temperature and short extraction times (Strati and Oreopoulou 2014). The combined use of enzymatic treatment with another green chemistry tool, i.e., sonication, was found to improve further the extraction yield of lycopene from tomato peel with the concomitant reduction of enzyme amount and incubation time (Konwarh et al. 2012). It is well documented that ultrasonic waves create cavitation phenomena, causing cell lysis, and thus facilitating the release of intracellular compounds (i.e., lycopene) into the solvent. The possibilities of using organic-solvent-free extraction systems, based on the combined use of commercial enzyme preparation and surfactant-assisted or aqueous extraction, were also evaluated. In the first case, sequential enzymatic pretreatment using commercial pectinolytic enzymes and surfactant-assisted extraction with Span 20 (30 min for each step) led to a four-fold higher lycopene yield compared to enzymatic pretreatment for one hour and subsequent extraction with ethyl acetate (three extraction cycles, 2 hours for each cycle) (Papaioannou and Karabelas 2012). Recently, Rigi et al. (2016) proposed saponin as a renewable, natural, low-priced surfactant for enhanced lycopene extraction from tomato industrial waste. Also, the enzymatic-assisted aqueous extraction process was successfully applied for the recovery and purification of lycopene-containing chromoplasts with a 30-fold higher lycopene content compared with that in untreated peels (Cuccolini et al. 2013). The main steps of this novel process include (a) cell lysis with the combination of pH changes and hydrolytic enzyme treatments, (b) precipitation of the lycopene-containing chromoplasts by lowering the pH followed by centrifugation, and (c) product purification, using proteases. In the final product, lycopene accounts for 8–10 per cent (w/w, dry basis), which represents a 30-fold increase with respect to the lycopene concentration of the untreated peels. Beyond the avoidance of the organic solvent, the recovered lycopene-rich formulations are stable and easily applicable in the food and cosmetics industries.

The extraction of bioactive components from various plant materials is continuously updated and emerging technologies, such as microwave and pressure-accelerated extraction techniques that gained popularity in the last decade, also attracted the interest of scientists working on tomato lycopene. For example, microwave-assisted extraction has been successfully applied in the extraction of lycopene from tomato peel. As a result of the internal heat generated by microwaves, structural disruption of MAE-treated tomato peels was achieved, thus allowing for improved total lycopene extraction yield. At this point it should be stressed that conventional extraction under similar thermal conditions favored E–Z isomerization (Shi et al. 2008). A promising method by combining the ultrasound and microwave irradiation together has been proposed by Lianfu and Zelong (2008) for lycopene extraction from tomato paste. This combination resulted in an almost complete recovery of the target compound with a concomitant reduction of extraction time compared with single ultrasound-assisted extraction technique (97.4 per cent in 6.1 min and 89.4 per cent in 29.1 min, respectively). The other proposal to accelerate lycopene recovery from tomato peel is pressurized extraction accomplished in the Extractor Naviglio using water (Naviglio et al. 2008a, b). Quasi-crystalline solid form rich in lycopene is recovered and further purified by solid-phase extraction at more than 98 per cent chromatographic purity using a small amount of organic solvent.

Supercritical carbon dioxide ($SCCO_2$) extraction, one of the established extraction techniques that is suitable for the extraction of non-polar compounds, has been applied for the recovery of lycopene from tomato products and industrial tomato processing waste. Optimum extraction conditions range between 50–110°C for temperature, 0.5–8.0 h for extraction time and 300–400 bar for extraction pressure, and include the used of co-solvent, mainly ethanol (Konar et al. 2012). The higher mass transfer ability of $SCCO_2$ with regard to liquid solvents, allows efficient penetration into the solid matrix (Mohamed and Mansoori 2002) to release lycopene. Nevertheless, material pretreatment like drying, grinding and sieving is expected to increase the yields in lycopene recovery from tomato waste (Konar et al. 2012). Tomato seed oil has been also reported to improve the efficiency of the process. In particular, the yield of lycopene recovery from tomato peel increased to 56 per cent, by mixing peel and seed at a mass ratio of 37:63, compared with 18 per cent in the absence of seed (Machmudah et al. 2012).

Biotechnological production of carotenoids

Most of the commercially used carotenoids are synthetic or extracted from plant sources. Since 2001, biotechnologically produced β-carotene is approved as a colouring agent (Commission directive (EC) No. 50/2001). The zygomycete fungus *Blakeslea trispora* is a safe and efficient industrial source of β-carotene and lycopene (Commission directive (EC) No. 50/2001, Commission regulation (EC) No. 721/2006, Commission directive (EU) No. 3/2011). Other carotenogenic microorganisms are halotolerant algae of the genus *Dunaliella* which is grown autotrophically in high-saline media containing inorganic nutrients, unlike *B. trispora*, and yeasts like *Rhodotorula* spp. (Mata-Gómez et al. 2014).

In general, the selection of appropriate nutrient sources can influence dramatically the raw material costs. For this reason, scientific efforts are focused on the exploitation of agro-industrial waste/by-products that provide carbon, nitrogen and other elements appropriate for microbial metabolism with the concomitant reduction of the production costs (Naziri et al. 2014).

Focusing on highly produced carotenoid *B. trispora* cells, beet molasses and cheese whey have been considered as carbon sources of high potential for the production of bio-based carotenoids (Goksungur et al. 2004, Roukas et al. 2007,

Varzakakou and Roukas 2010). Also, research findings support the ability of *B. trispora* to valorize different origin agricultural solid waste, such as cabbage, watermelon husk and peach peels for carotenoids production by solid-state fermentation (SSF). HPLC analysis revealed that the bioprocesses allow for high total carotenoid volumetric production (230.49 ± 22.97 mg/L) and selectivity in terms of β-carotene production (over 76 per cent of total carotenoids). There are also a few examples on the use of hydrophobic substances for the production of carotenoids by this fungus. In particular, crude olive pomace oil, which is derived from olive pomace (the solid residue of olive oil production), was found to enhance fungal growth (20–26 g/L) and β-carotene volumetric yield (720 mg/L). Recently, Nanou and Roukas (2016) proposed waste cooking oil as a new substrate for carotene production by *B. trispora*. According to findings, the highest total carotene yield (~ 2000 mg/L) was obtained in the fermentation medium containing waste cooking oil (50.0 g/L) and corn steep liquor (80.0 g/L). In this study, the carotenes identified by HPLC were β-carotene (74.2 per cent), γ-carotene (23.2 per cent), and lycopene (2.6 per cent).

Various agro-industrial waste has been incorporated as the medium components for biotechnological production of carotenes by several red yeast strains using hydrolyzed mung bean waste flour (Tinoi et al. 2005), fermented radish brine (Malisorn and Suntornsuk 2009), whole stillage (Ananda and Vadlani 2010) and parboiled rice water (Valduga et al. 2014). Optimization of carotenoid production by microorganisms was achieved by altering the media components (i.e., type and amount of C sources, C/N ratio, nitrogen source, minerals) and physical parameters, such as temperature, aeration, pH and light. Most optimization studies relied on powerful statistical designs and response surface methodology (e.g., Mantzouridou et al. 2002a, b).

Extraction and recovery of carotenoids from microbial cells increases the process cost (Lee et al. 2016). Roukas and Mantzouridou (2001) proposed an improved method for the complete extraction of β-carotene from *B. trispora* cells by steaming the fermentation broth (121°C for 15 min) and direct removal of β-carotene from heat-treated cells by three repeated extraction steps using ethanol (ratio of 1:100 at 30°C, 2 hours for each step). This approach provides a solution to the problem associated with oxidation of the target carotenoid by drying of biomass. Recently, low-pressure steam explosion (at 0.2 MPa for 4.4 min) for cell disruption and ethyl lactate (ratio of 1:25 at 39°C for 45 min) for the extraction were proposed as a rapid and environment-friendly method for carotenoid extraction from *B. trispora* (Wang et al. 2015). Total carotenoid yield reached 95.6 per cent.

The potential of $SCCO_2$ extraction for the recovery of lycopene from the dried biomass of mated cultures of *B. trispora* was highlighted in the work of Choudhari and Singhal (2008). In this study, under optimized operation conditions (1.1 hour, 52°C and 349 bar) 92 per cent recovery of lycopene could be achieved.

Bioflavors

Biotechnology-derived production of natural flavours by microorganisms, plant cell cultures and enzymes was strengthened by the latest EC legislation (EU Regulation No. 1334 (2008) Official J EU 354/34). Of the approximately 6500 known flavors, only 50–100 are obtained by microbial fermentation, while the rest are mainly produced by chemical synthesis (Bicas et al. 2009). However, food additive flavors produced by bioconversion of food grade substrates are finding better consumer acceptance as compared to synthetically produced compounds. In addition, some bioflavors make favorable contribution to the promotion of consumer's health and well-being. For example, bioflavors show biological activity *in vitro* and *in vivo* against certain types of tumors (Bicas et al. 2010). Some of them can also be used to induce detoxifying enzymes for the inhibition of potato sprouting, as an anti-microbial agent, insect repellent, among other uses (De Carvalho and Da Fonseca 2006). Yeast, bacteria and fungi produce a large range of fruit-like or flower-like aromas, depending on the strain and culture conditions. Although several bacteria, yeasts and fungi produce aroma compounds (Table 1), a few species of yeasts and fungi are generally preferred and only a few of these find industrial application due to their GRAS (generally regarded as safe) status, such as *Kluyveromyces marxianus*.

The production of flavor compounds from agro-industrial waste with negligible or even no-cost is an interesting approach to build a sustainable and economically competitive microbial process. More than 60 per cent of plant biomass produced on earth is composed of lignocelluloses. This vast resource is a potential source of bioflavors, biofuels, biofertilizers, animal feed and chemical feedstocks. Waste from the fish industry has also been converted into flavors precursors and subsequently, flavour compounds. Recently, Peinado et al. (2016) used fish powder for the generation of fish flavor formulations (after protease biocatalysis and subsequent heating in the presence of glucose and/or fish oil). Main volatile products formed were 4-heptenal, 2,4-heptadienal and some pyrazines, also identified among the volatile profile of cooked seafood.

Biotechnological route of flavor in production is not yet an economically feasible alternative to chemical synthesis, mainly because the costs of extraction and purification of the products generally reduce the profit considered as economic. Thus, the selling prices of bioflavors are many times higher. Highlights are fruity/flowery flavors like peach, rose or vanilla, but also for banana, as a complex bioflavor with improved quality.

Bio-oxidation of terpenes

The terpenes are secondary metabolites of plants which are classified, on the basis of the number of carbons in the molecule. The simpler terpenes (mono- and sesquiterpenes) are the major constituents of essential oils and are widely used in the perfume industry. Carotenes are synthetized by bacteria, algae, fungi and green plants and comprise more than 600 known structures. Terpenes are a good starting material for synthesis of many fine chemicals due to their similar carbon skeleton. R-(+)limonene, for example, is the most abundant monocyclic monoterpene in Nature, and it represents more than 90 per cent of the orange peel oil. The oxygenated derivates of limonene (carveol, carvone, peryllil alcochol, mentol and α-terpineol) are recognized for their pleasant fragrances and some of them have even health-protection activity. The tetraterpene β-carotene, the orange pigment of tropical vegetables, is a precursor of norisoprenoid ionones, molecules responsible for desirable fruity and floral flavors. Bioxidation of terpenes, the process which can be catalysed by enzymes, integer cells, plant-cultured cells, fungi, yeast and bacteria is an interesting approach for the flavor industry (Bicas et al. 2009). Current optimization strategies for biocatalytic terpenoid production

Table 1. Examples of flavor active compound recovery from agro-industrial wastes.

Source	Method	Flavor compounds	Yield/ Production rate	Reference
Coffee husk	SSF by fungi *Ceratocystis*	ethyl acetate acetaldehyde, isopropanol, ethyl isobutyrate, isobutyl acetate, isoamyl acetate, ethyl-3-hexanoate	5.2–6.6 mmol/l per gram of total volatiles	Soares et al. 2000
Cassava bagasse, apple pomace	SF by four strains of *Rhizopus*	acetaldehyde, propanol, esters	–	Christen et al. 2000
Fish powder	Enzymatic hydrolysis	4-heptenal, 2,4-heptadienal, pyrazines, 1-octen-3-ol, 1-hepten-4-ol	–	Peinado et al. 2016
Orange peel waste	SSF by yeast (*S. cerevisiae*)	isoamyl acetate, ethyl dodecanoate, decanoate, octanoate, phenyl ethyl acetate	total aroma esters of ~ 250 mg/kg orange peel	Mantzouridou et al. 2015
Cassava bagasse, palm bran	SSF by *Kluyveromyces marxianu*	isoamylic alcohol, ethyl acetate, propyl acetate, butyl acetate, ethyl propionate, ethyl isobutyrate, isoamyl acetate	62–315 µmol/l/g 251–1395 µmol/l/g 32–35 µmol/l/g 64 µmol/l/g 7–28 µmol/l/g 7–18 µmol/l/g 43–74 µmol/l/g	Medeiros et al. 2000
Linseed cake, castor oil cake, olive press cake, sunflower cake	SSF by *Moniliella suaveolens*, *Trichoderma harzianum*, *Pityrosporum ovale*, *Ceratocytis oniliformis*	δ-and γ-decalactone	up to 1 g per kg dry matter	Laufenberg et al. 2003
Mixture of citric pulp and soya bran, sugarcane molasses, soya molasses	SSF by *Ceratocystis fimbriata*	ethyl acetate isoamyl acetate	99.60 µmol/L g of total volatiles	Rossi et al. 2009
Wheat bran	Three step process: (I) Enzymatic hydrolysis (II) fermentation by metabolically engineered *Escherichia coli* (III) Purification by ion exchange	biovanillin	up to 70 per cent	Di Gioia et al. 2007

Table 1 contd. ...

...Table 1 contd.

Source	Method	Flavor compounds	Yield/ Production rate	Reference
Mixed solid and liquid food industry wastes (i.e., cheese whey, molasses, brewer's spent grains, malt spent rootlets, orange and potato pulp)	SSF by *S. cerevisiae*, *K. marxianus*, Kefir culture	ε-pinene	4.2 g/kg with Kefir culture	Aggelopoulos et al. 2014
Sugarcane bagasse	SSF by *Trichoderma viride*	Coconut aroma (6-pentyl-α-pyrone δ-octalactone, γ-nonalactone, γ-undecalactone, γ-dodecalactone and δ-dodecalactone)	Total volatiles up to 17.5 mg/g of dry matter	Fadel et al. 2015
Kraft lignin	Three-step process: alkaline lignin oxidation, ultrafiltration and ion exchange separation	vanillin	10 per cent (w/w)	Da Silva et al. 2009
Rice bran oil	Fermentation by *Aspergillus niger* and *Pycnoporus cinnabarinus*	biovanillin	2.8 g/L	Zheng et al. 2007
Corn cob	Enzymatic hydrolysis followed by fermentation by metabolically engineered *Escherichia coli*	biovanillin	11 g/L	Torres et al. 2009
Citrus waste	Hydrolization with sulfuric acid, followed by flash evaporation and condenzation	limonene	800 t/year based on 100.000 t/ year of citrus waste*	Lohrasbi et al. 2010
Orange peels waste	Microwaves and ultrasounds assisted extraction	essential oils	4.2 per cent 346 g of essential oils from 100 kg of orange fruit	Boukroufa et al. 2015
Orange pulp	SF by *Saccharomyces cerevisiae*	ethyl hexanoate octanoate decanoate	1329 μg/L 9049 μg/L 9967 μg/L	Mantzouridou and Paraskevopoulou 2013

SSF - solid state fermentation; SF - submerged fermentation
*Based on economic analysis

are mainly focused on metabolic engineering of the biosynthesis pathway within an expression host. For example, recently Toogood et al. (2015) have demonstrated the successful biosynthesis of menthol isomers from pulegone by incorporating part of the peppermint biosynthetic pathway into *E. coli*; the best performing enzymes (according to factors, such as catalytic rate, enantio-specificity, soluble expression) were identified and assembled into a functional biocatalytic cascade.

Biovanillin production

A good example for biotechnologically-derived flavor is biovanillin, which can be produced from plant biomass waste containing ferulic acid or isoeugenol (biovanillin natural precursors) through microbial conversion rather than conventional chemical reagents. In fact, biovanillin is recognized as the most attractive and promising alternative for natural vanillin extracted from vanilla pods, which has a high price in the market, between \$1200 and \$4000 kg^{-1} (Zamzuri and Abd-Aziz 2013). The potential agro waste for biovanillin production that is highly reported include cereal bran, sugar beet pulp, rice bran oil and palm oil biomass (Priefert et al. 2001, Zamzuri and Abd-Aziz 2013). Strains of *Amycolatopsis* sp. or *Streptomyces setonii* have been utilized in a patented process (Rabenhorst and Hopp 2000). Lesage-Meessen et al. (2002) and later Zhang et al. (2007) reported a two-step process with white rot fungi, involving the transformation of ferulic acid into vanillic acid by using *Aspergillus niger*, and later, conversion into biovanillin by *Pycnoporus cinnabarinus* or *Phanerochaete chrysosporium*. A number of microbial strains and filamentus fungi, such as *Sporotrichum pulverulentum* (Ander et al. 1980), *Rhodococcus rhodochrous* (Chatterjee et al. 1999), *Pseudomonas species/Pseudomonas fluorescens* (Civolani et al. 2000), *Pseudomonas chlororaphis* (Ramesh et al. 2007), *Pseudomonas putida* (Furukawa et al. 2003) and *Bacillus* species (Rapid Sobti et al. 2000, Shimoni et al. 2000, Zhao et al. 2005, 2006) successfully converted ferulic acid or isoeugenol into biovanillin. However, the process development is difficult because of the slow growth of actinomycetes and high viscosity of broths fermented by them. Therefore, the construction of new recombinants strains of quickly-growing bacteria to overproduce vanillin is favored. For example, resting cells of metabolically-engineered *E. coli*, containing the genes responsible for conversion of ferulic acid into vanillin, were used to transform ferulic acid in the hydrolyzed waste mixture into vanillin (Yoon et al. 2005, Barghini et al. 2007, Di Gioia et al. 2007, Torres et al. 2009). Pre-cultivation conditions, such as medium heat sterilization, preliminary adaptation to the hydrolyzate and initial biomass level affected vanillin productivity (Torres et al. 2009).

SSF for production of bioflavors

Amongst the various types of microbial processes to convert solid waste into flavors, the use of SSF emerged as a means to improve cost effectiveness of these processes.

Crop residues (straw, corn by-products, bagasse, etc.) are particularly suitable lignocellulosic waste materials for SSF, since they are available in large quantities in processing facilities. The conditions of the SSF-process are especially well adapted to the requirements of fungi, which represent about 60 per cent of the micro-organisms used in flavor production (Laufenberg et al. 2003). Capital and operating costs are reduced when compared to conventional submerged fermentation as a result of lower working volumes per product yield and lower energy costs for sterilization and stirring. The exact content of the volatiles produced by fermentation of the solid waste depends on its composition and particle size, microbial strain used and the process parameters, such as initial substrate pH, cultivation temperature, initial substrate moisture and particle size. Generally, smaller substrate particles provide a larger surface area for the microbial attack, but substrate agglomeration may occur with poor microbial growth; in contrast, larger particles provide better aeration but a limited surface for microbial attack (Couto et al. 2006). The process usually requires a few additional nutrients, such as precursors, glucose, nitrogen sources, etc. In fact, in some cases, the aroma production was strongly inhibited in the absence of glucose as glucose was the main factor controlling the production (Medeiros et al. 2000). Adding a nitrogen source could enhance the formation of total volatiles by several times.

As examples, cassava bagasse, sugarcane bagasse, wheat bran, palm bran, linseed cake, olive oil cake, apple pomace, soya bran, coffee husk, orange peel and mixtures of different liquid and solid waste have been evaluated for this purpose by cultivating different microorganisms (Laufenberg et al. 2003, Rossi et al. 2009, Bicas et al. 2010, Aggelopoulos et al. 2014, Fadel et al. 2015, Mantzouridou et al. 2015). The use of mixed waste substrates has been scarcely reported. For example, Aggelopoulos et al. (2014) used mixed solid and liquid food industry waste for production of proteins, nutrient minerals, as well as for aroma; regarding volatile esters, ε–pinene was a main product with a yield of 4 kg/tn of solid-state fermentation which was achieved by mixed cell cultures of kefir.

Scale up, purification of end products and biomass estimation are the major challenges of SSF in practice. A number of bioreactors have been designed (tray, packed-bed, horizontal drum and fluidized bed) which could overcome the problems of scale up, along with on-line monitoring and control of several parameters (Singhania et al. 2009). The newer generations of bioreactors for SSF have better porosity of the solid substrate and improved mass and heat transfer, which is well described by Thomas et al. (2013).

Separation processes for bioflavors

The current industrial options for aroma isolation and purification from waste are distillation, partial condensation, solvent extraction, adsorption, or a combination thereof, while distillation has been mostly used in the past. For example, essential oils, highly concentrated flavoring agents are extracted from aromatic plants, mainly by steam- or water-distillation. During this process, the essential oil and the condensed steam (distillation waste water) are brought together in intimate contact for a long time. As a result, some of the essential oil components (mainly the oxygenated ones) are partitioned into the water phase. The new concepts for extraction are based on solvent-free processes using ultrasound and microwave techniques. For example, Boukroufa et al. (2015) developed a new bio-refinery concept of orange peel waste. It was based on optimized ultrasound- and microwave-assisted extraction process to recover essential oils from orange peel waste. Essential oil extracted from citrus by-product can be used in food as flavouring ingredients in drinks, sweets and other food products; in the pharmaceutical industry for its anti-inflammatory and antibacterial effect; and in preparation of toilet soaps, perfumes, cosmetics and other home care products. The yield of the novel process was similar to that of steam distillation (~ 4 per cent).

Novel solutions for flavor recovery from agro-industrial waste refer to combined highly concentrated production and on-line separation. Membrane-based separation systems include reverse osmosis, perstraction, pervaporation and membrane evaporation, as well as liquid/liquid extraction, adsorption and gas stripping.

The solid-phase extraction techniques play an important role in isolation and recovery of volatile food flavor components. It is a convenient method for flavor separation from aqueous waste streams, such as flavors drained with condensed water during the distillation of aromatic plants (Edris et al. 2003). The essential forces of adsorption are physical in nature, so that the physical and chemical properties of the sorbed compounds are not affected. Different types of sorbents are used in industrial applications, e.g., zeolite, alumina, silica and activated carbon.

Pervaporation is based on a selective transport through a dense membrane (polymeric or ceramic) linked with the recovery of the permeate from the vapor phase. A feed liquid mixture contacts one side of a membrane while the permeate is removed as a vapor from the other side. The vapor-pressure difference between the feed solution and the permeate vapor (vacuum) induces transport through the membrane. Pervaporation has been used for the extraction of compounds biotechnologically produced or recovered from perfumery waste where PDMS (polydimethylsiloxane) has been most often used as membrane material. Araujo et al. (2008) simulated this membrane separation process to recover ethyl butyrate from orange juice prior to concentration (heating exposal) in evaporators or even from the evaporation effluents (water essence phases or water condensates).

In general, all separation systems suffer from some engineering or operational problem. Thus, membrane systems have a high selectivity for solvents but clogging and fouling might occur; liquid/liquid extractions might form emulsions which reduce their effectiveness; upon gas stripping, there are residual solvents. Conventional adsorption materials are quite expensive, so the process with the use of bioadsorbents made of vegetable waste seems attractive. Most of the research work done with bioadsorbents made from agro-industrial wastes refers to removal of metal ions and textile dyes.

Encapsulation Technology for Waste-derived Bioactive Compounds

Most of the bioactives discussed in this chapter, like polyphenols, (pro)vitamins, essential oils and colorants are labile compounds prone to degradation in the presence of oxygen and light, and susceptible to pH changes and high temperatures—the conditions generally associated with processing, storage and digestion of food products. In addition, when implemented in food products, they may interact with other food constituents, which often result due to formation of unfavorable compounds. Apart from offering protection, encapsulated systems can enable release of actives in a controllable manner, for example, polyphenols could be delivered to lower parts of the gastrointestinal tract, so their bioavailability and bioaccessibility are improved. Masking unpleasant taste, converting liquid compounds into solid forms, precise dosage to food products are also some of the beneficial aspects of encapsulation technology in the food sector. The following part of the chapter refers to the processes for producing high-value products based on bioactives which are isolated from agro-industrial wastes (Fig. 3). Among numerous encapsulation methods used for food compounds, only those which are most suitable for the waste-derived bioactives according to the recent scientific work will be discussed.

Spray drying

Spray drying is an economical, flexible and continuous operation that produces microparticles of good stability and quality. Thus, it is the most widely used microencapsulation technique in the food industry for the preparation of dried, stabilized various food additives and flavors (Đorđević et al. 2015). Most of such microencapsulates have been produced with common carrier materials like maltodextrin, gum Arabic, whey and soy protein isolate, skim milk powder, pectin and gelatin. In addition, with the right choice of a polymer which has a pH-dependent solubility, it is possible to design gastro-resistant microcapsules. For example, Lauro et al. (2015) used cellulose acetate phthalate (CAP) to produce microsystems of citrus by-product extract (rich in flavanones), which were able to protect the extract in the gastric environment. Another example of such material is shellac, a biopolymer with

Figure 3. Illustration of the concept—from waste to high-value products: production of natural colorants by encapsulation of the extract obtained from vinification waste.

gastric-resistant properties, which has been registered for use in food (E 904). Due to the shift in pH from acidic to close to neutral or mildly alkaline during passage from the stomach to the intestine, the shellac coating dissolves and release actives (Oidtmann et al. 2012). Moreover, spray-drying combined with other encapsulation technologies can give complex delivery systems with superior properties—controlled release, stability at different temperatures and pH. For example, in a recent study of Gültekin-Özgüven et al. (2016), firstly liposomes containing black mulberry extract were produced and then coated with cationic chitosan by the layer-by-layer deposition method. Then, liposomes were spray dried upon addition of maltodextrin. Thus the formulations (1) were easily manageable to be implemented in a food product (chocolate) (at different conching temperatures and alkalization degrees) and (2) enhanced *in vitro* bioaccessability of anthocyanins.

Effects of various process parameters (inlet temperature, coating/core ratio, flow rate) on the encapsulation efficiency/yield and the main physical properties of the microcapsules have been vast studied (Goula and Adamopoulos 2012, Çam et al. 2014, Ranveer et al. 2015, Kaderides et al. 2015). An increase in inlet air temperature, drying air flow rate, feed solid concentration and carrier/core ratio generally lead to increase in encapsulation efficiency of active compounds. However, a precaution should be taken when increasing the drying temperature since most of bioactive compounds are prone to oxidation upon heating. In fact, the optimal inlet air temperature should be high enough to provide efficient encapsulation, but not to impair antioxidant activity of active compounds. The reported values of optimal inlet air temperature for spray-drying of waste-derived sensitive compounds mainly ranged between ~ 130°C (de Souza et al. 2014, 2015) to ~ 190°C (Goula and Adamopoulos 2012). Paini et al. (2015) determined that the polyphenols from olive pomace exhibited high antiradical power but only when lower inlet air temperature (130 versus 160°C) and higher feed flow (10 mL/min versus. 5 mL/min) were applied with maltodextrin as a coating agent (10 per cent w/w).

The encapsulation efficiency largely depends on the type of wall material used. Maltodextrin is the most exploited carrier material (Zhang et al. 2007, Burin et al. 2011, Rubilar et al. 2012, Paini et al. 2015, Çam et al. 2014, Laokuldilok and Kanha 2015, Kaderides et al. 2015, de Souza et al. 2015, Chang et al. 2016) for oils, anthocyanins and other polyphenolic compounds produced from agro-industrial waste (see Table 2).

It also provides good thermal stability to core compounds. For example, based on HPLC and thermogravimetric analysis, Paini et al. (2015) confirmed the thermal protection effect of maltodextrin for phenolic compounds of olive pomace. Mixtures of maltodextrin with other carbohydrate- or protein-based materials in most cases provided the best encapsulation property, for example, according to Kaderides et al. (2015), the encapsulation efficiency of phenolic compounds extracted from pomegranate peel varied from 69.8 per cent when maltodextrin was used as a coating agent, to 73.8 per cent for maltodextrin/gum Arabic, 91.5 per cent for skimmed milk powder, 97.2 per cent for maltodextrin/skimmed milk powder, and 98.6 per cent for maltodextrin/ whey protein isolate. In another example, degradation of roselle anthocyanins (natural food colorants of brilliant red color) upon heating (at 60 and 80°C) was the lowest for the combination of maltodextrin and gum arabic in comparison to one-component encapsulating agents (Idham et al. 2012).

Table 2. Encapsulation of bioactive compounds isolated from agro-industrial waste.

Source	Compounds	Carrier	Technique	Implementation	Reference
Blueberry pomace	Anthocyanins	Whey protein	Spray drying	–	Flores et al. 2014
Mulberry waste	Anthocyanins	Chitosan Maltodextrin	Spray drying	Dark chocolate	Gültekin-Özgüven et al. 2016
Bilberry Pomace	Anthocyanins	Whey protein Pectin Schellac Maltodextrin	Emulsification/ Heat gelation Extrusion Spray drying	–	Kropat et al. 2013 Oidtmann et al. 2012
Grape skin	Anthocyanins	Gum Arabic Maltodextrin μ-cyclodextrin	Spray drying	Soft drinks	Burin et al. 2011
Jabuticaba skin	Anthocyanins	Alginate Polyethyleneglycol	Ionic gelification Rapid expansion	–	Santos et al. 2013
Egg plant	Anthocyanins	Alginate	Ionic gelification	Jelly crystals	Chatterjee and Bhattacharjee 2015
Black glutinous rice bran	Anthocyanins	Maltodextrin	Spray drying Freeze drying	–	Laokuldilok and Kanha 2015
Black bean coats	Anthocyanins	β-cyclodextrins	Complex inclusion	Sport beverages	Aguilera et al. 2016
Black soybean seed coat	Anthocyanins	Carnauba wax	Melt solidification	–	Salević et al. 2015
Black soybean seed coat	Anthocyanins	Alginate Pectin	Internal gelation	–	Kalušević et al. 2016
Cherry pomace	Polyphenols, Anthocyanins	Wheyprotein Soy protein	Freeze drying	Cookies	Tumbas Šaponjac et al. 2016
Black current pomace	Polyphenols Anthocyanins	Maltodextin Inulin	Spray drying	–	Bakowska-Barczaka and Kolodziejczy 2011
Citrus byproduct	Flavonoids, Anthocyanins	Cellulose acetate phthalate	Spray drying	–	Lauro et al. 2015
Grape marc	Polyphenols Anthocyanins	Sunflower oil, palm oil Maltodextrin	Emulsification Spray drying	–	Sessa et al. 2013 Spigno et al. 2013
Wine production waste	(+)Catechin (−)Epicatechin	Alginate Alginate/chitosan	Vibration nozzle method	–	Aizpurua-Olaizola et al. 2016
Blueberry pomace	Procyanidin	White sorghum	Extrusion	–	Khanal et al. 2009
Grape seeds	Procyanidins	Gum Arabic Maltodextrin	Spray drying	–	Zhang et al. 2007

Table 2 contd. ...

...Table 2 contd.

Source	Compounds	Carrier	Technique	Implementation	Reference
Pomegranate peel	Phenolics	Maltodextrin	Spray drying	Ice cream	Çam et al. 2014
Pomegranate peel	Phenolics	Skim milk powder Maltodextrin Gum Arabic Whey protein isolate	Spray drying	Hazelnut paste	Kaderides et al. 2015
Sour cherry pomace	Phenolic compounds	Maltodextrin/gum Arabic	Freeze drying	–	Cilek et al. 2012
Star fruit (*Averrhoa carambola*) pomace	Polyphenols	Maltodextrin	Spray drying Freeze drying	–	Saikia et al. 2015
Rubus ulmifolius Schott flowers	Polyphenols	Alginate	Atomization/ coagulation	Yogurts	Martins et al. 2014
Olive leaves	Polyphenols	Chitosan	Spray drying	–	Kosaraju et al. 2006
Olive pomace	Polyphenols	Maltodextrin	Spray drying	–	Paini et al. 2015
Olive mill wastewater Pomegranate waste	Polyphenols Oil	Maltodextrin Skimmed milk powder, Maltodextrin/ skimmed milk powder Maltodextrin/ whey protein isolate Maltodextrin/gum Arabic	Spray drying	–	Goula and Lazarides 2015
Olive kernels and leaves	Polyphenols	Maltodextrin	Freeze drying	–	Chanioti et al. 2016
Winemaking byproducts vinification byproducts	Phenolic compound	Maltodextrin	Spray drying	–	de Souza et al. 2014, 2015
Apple pomace	Polyphenols Vitamin C	Alginate	Co-extrusion	Milk with probiotics	Shinde et al. 2014
Red dragon fruit peels	Betalains	Maltodextrin/gum Arabic Maltodextrin/ pectin	Freeze drying	–	Rodriguez et al. 2016
Grape seed	Oil	Alginate	Electrostatic extrusion	Fermented sausages	Stajić et al. 2014
Pomegranate seed	Oil	Skim milk powder	Emulsification Spray drying	–	Goula and Adamopoulos 2012

Table 2 contd. ...

...Table 2 contd.

Source	Compounds	Carrier	Technique	Implementation	Reference
Tomato processing waste	Lycopene	Gelatin Sucrose	Spray drying	–	Ranveer et al. 2013
Tomato processing waste	Lycopene	Gelatin and poly(ç-glutamic acid)	Freeze drying	–	Chiu et al. 2007
Opuntia fruits peel and epidermal layer	Indicaxanthin	Maltodextrin	Spray drying	–	Gandía-Herrero et al. 2010
Red beet and flower of *Lampranthus productus*	Miraxanthin V (betaxanthin) and betanidin (betacyanin)	Maltodextrin and chitosan	Spray drying	–	Gandía-Herrero et al. 2013
H. pluvialis	Oleoresin and asthaxantin	Modified starch	Spray drying	–	Bustamante et al. 2016

The physicochemical properties of spray-dried powders also depend on process conditions. Thus, higher inlet air temperature and lower feed flow usually imply lower moisture content, smoother particle surface, better powder dispersability and flowability, while increasing concentration of a wall material causes lower bulk density. The size of spray-dried powder particles generally varies between a few and 100 μm, depending on operating conditions (core/carrier ratio among others), material properties and pretreatment of a feed mixture (Ranveer et al. 2015, Kosaraju et al. 2006). Storage tests confirmed that encapsulated formulations are able to protect sensitive bioactives from degradation upon heating or photodegradation. Most studies report retarded decrease in phenolic content and antioxidant activity in comparison to control (un-encapsulated compounds). According to Jiménez-Aguilar et al. (2011), the rate of this decrease appeared to be linear for non-encapsulated control and non-linear for the encapsulated extracts, but mathematical information regarding the observed trends was not discussed. However, there are exceptions to this rule. Thus, Flores and coauthors (2014) revealed actual increase in phenolic content and antioxidant capacity of spray-dried powders made from blueberry pomace extract and whey proteins upon storage at different conditions (22, 37, 45°C and exposure to light), although the degradation of anthocyanins occurred according to the first-order kinetics. The authors explained that upon heating and prolonged storage, the anthocyanins possibly decomposed (to phloroglucinaldehyde and benzoic acid derivatives, such as syringic acid or 4-hydroxybenzoic acid) and also polymerized. In this way, thermally-produced phenolic compounds from degradation or polymerization may partially or fully compensate for loss in antioxidant activity arising from decreased monomeric anthocyanins (Fischer et al. 2013). In fact, the disadvantage of using whey protein as a wall material in spray drying arises from significantly higher rates of anthocyanin degradation compared with polysaccharide-based wall materials (Tonon et al. 2009, Idham et al. 2012). At low temperatures, bioactives which are encapsulated seem to be stable for long storage time (Paini et al. 2015, Çam et al. 2014, Ranveer et al. 2015, Kha et al. 2015). For example, Çam et al. (2014) reported constant phenolic content of pomegranate peel microcapsules made of maltodextrin for the storage period of 90 days at 4ºC. In another study, Ranveer et al. (2015) claimed more than 90 per cent retention of lycopene from tomato processing waste in microencapsulated sample (made of sucrose-gelatin mixture) in comparison to less than 5 per cent retention of un-encapsulated lycopene after 42 days of storage at various storage conditions (the presence and absence of air, sunlight, at room temperature and under refrigeration).

Spray-dried formulations of waste-derived bioactives have been tested in different food products, such as dark chocolate (Gültekin-Özgüven et al. 2016), hazelnut paste (Kaderides et al. 2015), ice cream (Çam et al. 2014), biscuits (Umesha et al. 2015), soup powder (Rubilar et al. 2012), yoghurt (Kha et al. 2015, Martins et al. 2014), pasteurised milk (Kha et al. 2015) and soft drinks (Burin et al. 2011) (Table 2). According to the sensory evaluations reported in literature, majority of the panelists accepted the phenolic-enriched food products. This lends support to commercial introduction of such products to the general public with the potential as functional foods.

Freeze-drying

Freeze-drying has been proposed as an alternative for spray drying of heat-sensitive bioactive food components, but it is economically sustainable technology only for high-price food products. By freeze-drying a high percentage of encapsulation efficiency is possible to achieve. However, the long period of processing and high energy input (Nedovic et al. 2011) in comparison to those of spray drying make it unfavorable for large-scale applications. In a recent study done by Laokuldilok and Kanha (2015), the two techniques, spray drying (at different inlet air temperatures 140, 160 and 180ºC and freeze drying at –45ºC), were compared with respect to powder properties of the black glutinous rice (*Oryza sativa* L.) bran anthocyanins produced by them. The freeze-dried anthocyanin powders made from maltodextrins (commercial and extracted from black glutinous rice) exhibited higher moisture content but higher flowability, higher process yield and better anthocyanin retention in comparison to spray-dried formulations. Similarly, Saikia et al. (2015) claim for higher encapsulating efficiency of *Averrhoa carambola* pomace extract by freeze-drying (78–97 per cent) than by spray drying (63–79 per cent) with maltodextrin as a carrier in both cases. The scientists are looking for optimal formulations regarding the carrier material and core-to-carrier ratio. Likewise for spray drying, maltodextrin is the most exploited carrier material for freeze drying of food compounds (Cilek et al. 2012, Yazicioglu et al. 2015, Saikia et al. 2015, Chanioti et al. 2016, Rodriguez et al. 2016). Gum Arabic contributed to more efficient encapsulation in maltodextrin-based particles of phenolic compounds extracted from sour cherry pomace due to emulsifying and stabilizing effects of gum Arabic during ultrasonication—the process which preceded freeze drying step (Cilek et al. 2012). In another study, maltodextrin/gum Arabic and maltodextrin/pectin formulations of betalains (rare red-purple and yellow pigments) from red dragon fruit exhibited antiinflammatory, antioxidative and antiangiogenic activities several times stronger than the un-encapsulated betalain extract (Rodriguez et al. 2016).

Lyophilized microcapsules of waste-derived polyphenols based on polysaccharide and protein materials were implemented in several food products, such as cake (Luca et al. 2014) and cookies (Tumbas Šaponjac et al. 2016). Encapsulation was effective in the masking of the flavor of the phenolic powders, but at elevated temperatures associated with the baking process (~ 200°C), a significant degradation or oxidation loss of primary polyphenol structure carrying antioxidant activity occurred along with reaction of phenolics with other nutritional components in cookie recipes, such as sugar fragments originated from caramelization and Maillard reaction. Ferulic acid presents low water solubility and a tendency to be oxidized. Such issues can be addressed via formation of inclusion complex with hydroxypropyl-β-cyclodextrin (HP-β-CD) by freeze-drying. The stoichiometry of the complex was 1:1, the water solubility of ferulic acid was significantly improved whereas a marked stability in aqueous solution towards irradiation was observed (Wang et al. 2011).

Extrusion methods

Extrusion methods are based on the dispersing of aqueous solutions of the polymer and an active compound into droplets dropping into a gelling bath. The dripping tool could be a pipette, syringe, vibrating or spraying nozzle, jet cutter, or an atomizing disk. The alginate is most often used as a carrier material for beads formation, where sodium alginate solution upon ion exchange with calcium cations from the gelling solution creates a dense matrix network. The polymer concentration, the force applied and content of the core material determine the size and other properties of the (micro)beads (Chatterjee and Bhattacharjee 2015). When chitosan is added to the gelling bath, a membrane is formed through ionic interactions between the carboxylic residues of the alginate and positively-charged amino groups of chitosan, reducing the active compounds from leaking during encapsulation. On the other hand, the microcapsules lose some water during this process, leading to a size reduction and therefore, to a lower loading efficiency (Aizpurua-Olaizola et al. 2016). The reported values of encapsulation efficiency are between 20 and nearly 100 per cent, depending on the extrusion method applied and operating parameters (Oidtmann et al. 2012, Santos et al. 2013, Shinde et al. 2014, Chatterjee and Bhattacharjee 2015, Aizpurua-Olaizola et al. 2016, Kalušević et al. 2016). Calcium alginate formulations provide long-term stability to bioactives. For example, the shelf-life of anthocyanins extracted from peels of eggplants was 16-fold enhanced by encapsulation within Ca-alginate in comparison to un-encapsulated pigments (Chatterjee and Bhattacharjee 2015). Calcium alginate formulations are stable in gastric environment, e.g., after 2 hours at 37°C and pH of 1.4, the Ca-alginate beads bearing extract from jabuticaba skins (*Myrciaria cauliflora*) released only 20 per cent of their anthocyanin content (Santos et al. 2013). Chitosan membrane of Ca-alginate beads provides additional protection to actives, e.g., Aizpurua-Olaizola et al. (2016) showed decreased degradation of (+) catechin and (–)epicatechin from wine waste extracts during storage under different conditions. Shinde et al. (2014) showed that co-extrusion of apple skin (a fruit waste) as a bioactive ingredient (containing polyphenols, pectin and vitamin C) with probiotic cells within calcium alginate matrix was beneficial for the preservation of their viability.

Other methods

Novel approaches for encapsulation of bioactives extracted from natural sources provide the answer to increasing demands imposed by the food sector for additives which should be on one hand, natural and at the same time, have superior properties. However, these are still under scrutiny as they are based on time- and cost-consuming methods unsuitable for applications on a large scale. Beta-cyclodextrin inclusion complexes (Nunes and Mercadante 2007, Aguilera et al. 2016), biopolymer nanoparticles (Silva et al. 2014) and micro- and nano-emulsions (Sessa et al. 2013, Naik et al. 2014, Khemakhem et al. 2016) are some of the rare examples found in the literature.

Future Perspectives

The challenge for agricultural and food production waste valorization into higher added-value products rather than feed, fertilizers and energy has attracted the interest of researchers and industry and urged the EU towards a zero-waste economy by 2025. To this direction, recent scientific findings provide opportunities for developing eco-innovative and emerging technologies for efficient reuse of these streams, leading to the recovery of new products and their recycling into the food chain. The problems that should be overcome are the scale up and technology transfer gaps and the long time required for successful commercialization.

References

Aarabi, A., M. Mizani, M. Honarvar, H. Faghihian and A. Gerami. 2016. Extraction of ferulic acid from sugar beet pulp by alkaline hydrolysis and organic solvent methods. J. Food Measur. Charact. 10: 42–47.

Aggelopoulos, T., K. Katsieris, A. Bekatorou, A. Pandey, I. M. Banat and A. A. Koutinas. 2014. Solid-state fermentation of food waste mixtures for single cell protein, aroma volatiles and fat production. Food Chem. 145: 710–716.

Aguilera, Y., L. Mojica, M. Rebollo-Hernanz, M. Berhow, E. G. de Mejía and M. A. Martín-Cabrejas. 2016. Black bean coats: New source of anthocyanins stabilized by β-cyclodextrin copigmentation in a sport beverage. Food Chem. 212: 561–570.

Aizpurua-Olaizola, O., P. Navarro, A. Vallejo, M. Olivares, N. Etxebarria and A. Usobiaga. 2016. Microencapsulation and storage stability of polyphenols from *Vitis vinifera* grape wastes. Food Chem. 190: 614–621.

Ananda, N. and P. V. Vadlani. 2010. Production and optimization of carotenoid-enriched dried distiller's grains with solubles by *Phaffia rhodozyma* and *Sporobolomyces roseus* fermentation of whole stillage. J. Ind. Microbiol. Biotechnol. 37: 1183–1192.

Ander, P., A. Hatakka and K. E. Eriksson. 1980. Vanillic acid metabolism by the white-rot fungus *Sporotrichum pulverulentum*. Arch. Microbiol. 25: 189–202.

Araujo, W. A., M. E. T. Alvarez, E. B. Moraes and M. R. Wolf-Maciel. 2008. Evaluation of Pervaporation Process for Recovering a Key Orange Juice Flavour Compound: Modeling and Simulation. 18th European Symposium on Computer Aided Process Engineering—ESCAPE 18 Bertrand Braunschweig and Xavier Joulia (Eds.). pp. 175–180.

Bakowska-Barczak, A. M. and P. P. Kolodziejczyk. 2011. Black currant polyphenols: Their storage stability and microencapsulation. Ind. Crops Prod. 34: 1301–1309.

Barberousse, H., O. Roiseux, C. Robert, M. Paquot, C. Deroanne and C. Blecker. 2008. Analytical methodologies for quantification of ferulic acid and its oligomers. J. Sci. Food Agric. 88: 1494–1511.

Barberousse, H., A. Kamoun, M. Chaabouni, J. M. Giet, O. Roiseux, M. Paquot, C. Deroanne and C. Blecker. 2009. Optimization of enzymatic extraction of ferulic acid from wheat bran, using response surface methodology and characterization of the resulting fractions. J. Sci. Food and Agric. 89: 1634–1641.

Barghini, P., D. Di Gioia, F. Fava and M. Ruzzi. 2007. Vanillin production using metabolically engineered *Escherichia coli* under nongrowing conditions. Microb. Cell. Fact. 6: 13–23.

Bicas, J. C., J. L. Silva, D. A. Paula and G. M. Pastore. 2010. Biotechnological production of bioflavors and functional sugars. Ciênc. Tecnol. Aliment. 30: 7–18.

Bicas, J. L., A. P. Dionisio and G. M. Pastore. 2009. Bio-oxidation of terpenes: An approach to flavour industry. Chem. Rev. 109: 4518–4531.

Birt, D. F., S. Hendrich and W. Wang. 2001. Dietary agents in cancer prevention: Flavonoids and isoflavonoids. Pharmacol. Therap. 90: 157–177.

Borel, P., M. Moussa, E. Reboul, B. Lyan, C. Defoort, S. Vincent-Baudry, M. Maillot, M. Gastaldi, M. Darmon, H. Portugal, R. Planells and D. Lairon. 2007. Human plasma levels of vitamin E and carotenoids are associated with genetic polymorphisms in genes involved in lipid metabolism. J. Nutr. 137: 2653–9.

Boukroufa, M., C. Boutekedjiret, L. Petigny, N. Rakotomanomana and F. Chemat. 2015. Bio-refinery of orange peels waste: A new concept based on integrated green and solvent free extraction processes using ultrasound and microwave techniques to obtain essential oil, polyphenols and pectin. Ultrasonics Sonochem. 24: 72–79.

Buranov, A. U. and G. Mazza. 2009. Extraction and purification of ferulic acid from flax shives, wheat and corn bran by alkaline hydrolysis and pressurised solvents. Food Chem. 115: 1542–1548.

Burin, V. M., P. N. Rossa, N. E. Ferreira-Lima, M. C. Hillmann and M. T. Boirdignon-Luiz. 2011. Anthocyanins: Optimisation of extraction from Cabernet Sauvignon grapes, microcapsulation and stability in soft drink. Int. J. Food Sci. Tech. 46: 186–193.

Bustamante, A., L. Masson, J. Velasco, J. M. del Valle and P. Robert. 2016. Microencapsulation of *H. pluvialis* oleoresins with different fatty acid composition: Kinetic stability of astaxanthin and alpha-tocopherol. Food Chem. 190: 1013–1021.

Çam, M., N. C. İçyer and F. Erdoğan. 2014. Pomegranate peel phenolics: Microencapsulation, storage stability and potential ingredient for functional food development. LWT—Food Sci. Tech. 55: 117–123.

Chang, C., N. Varankovich and M. T. Nickerson. 2016. Microencapsulation of canola oil by lentil protein isolate-based wall materials. Food Chem. 212: 264–273.

Chanioti, S., P. Siamandoura and C. Tzia. 2016. Evaluation of extracts prepared from olive oil by-products using microwave-assisted enzymatic extraction: Effect of encapsulation on the stability of final products. Waste Biomass Valorization 7: 831–842.

Chatterjee, D. and P. Bhattacharjee. 2015. Encapsulation of colour from peels of eggplant in calcium alginate matrix. Nutrafoods 14: 87–96.

Chatterjee, T., B. K. De and D. K. Bhattacharyya. 1999. Microbial conversion of isoeugenol to vanillin by *Rhodococcus rhodochrous*. Indian. J. Chem. B 38: 538–541.

Chiu, Y. T., C. P. Chiu, J. T. Chien, G. H. Ho, J. Yang and B. H. Chen. 2007. Encapsulation of lycopene extract from tomato pulp waste with gelatin and poly(γ-glutamic acid) as carrier. J. Agricult. Food Chem. 55: 5123–5130.

Choudhari, S. M. and R. S. Singhal. 2008. Supercritical carbon dioxide extraction of lycopene from mated cultures of *Blakeslea trispora* NRRL 2895 and 2896. J. Food Eng. 89: 349–354.

Christen, P., A. Bramorski, S. Revah and C. R. Soccol. 2000. Characterization of volatile compounds produced by *Rhizopus* strains grown on agro-industrial solid wastes. Bioresource Technol. 71: 211–215.

Cilek, B., A. Luca, V. Hasirci, S. Sahin and G. Sumnu. 2012. Microencapsulation of phenolic compounds extracted from sour cherry pomace: Effect of formulation, ultrasonication time and core to coating ratio. Eur. Food Res. Technol. 235: 587–596.

Ciriminna, R., A. Fidalgo, F. Meneguzzo, L. M. Ilharco and M. Pagliaro. 2016. Lycopene: Emerging production methods and applications of a valued carotenoid. ACS Sustainable Chem. Eng. 4: 643–650.

Civolani, C., P. Barghini, A. R. Roncetti, M. Ruzzi and A. Schiesser. 2000. Bioconversion of ferulic acid into vanillic acid by means of a vanillate-negativemutant of *Pseudomonas fluorescens* strain BF13. Appl. Environ. Microbiol. 66: 2311–2317.

Clifford, M. N. 1999. Chlorogenic acids and other cinnamates—Nature, occurrence and dietary burden. J. Sci. Food Agric. 79: 362–372.

Commission directive (EC) 97/60/EC of the European Parliament and of the Council amending for the third time Directive 88/344/EEC on the approximation of the laws of the Member States on extraction solvents used in the production of foodstuffs and food ingredients. Official Journal of the European Union 331: 7–9 (1997).

Commission directive (EC) No. 50/2001 of 3 July 2001 amending Directive 95/45/EC laying down specific purity criteria concerning colors for use in foodstuffs. Official Journal of the European Union L 190(14) (2001).

Commission directive 3/2011/EU of 17 January 2011 amending Directive 2008/128/EC laying down specific purity criteria on colours for use in foodstuffs. Official Journal of the European Union L 13/59 (2011).

Commission regulation (EC) No. 721/2006 of 23 October 2006 authorising the placing on the market of lycopene from *Blakeslea trispora* as a novel food ingredient under Regulation (EC) No. 258/97 of the European Parliament and of the Council (notified under document number C(2006) 4973). Official Journal of the European Union L296 (2006).

Couteau, D. and P. Mathaly. 1997. Purification of ferulic acid by adsorption after enzymic release from a sugar-beet pulp extract. Ind. Crop Prod. 6: 237–252.

Couto, S. R. and M. A. Sanroman. 2006. Application of solid-state fermentation to food industry—A review. J. Food Eng. 76: 291–302.

Crepin, V. F., C. B. Faulds and I. F. Connerton. 2004. Functional classification of the microbial feruloyl esterases. Appl. Microbiol. Biotechnol. 63: 647–652.

Crozier, A., I. B. Jaganath and M. N. Clifford. 2006. Phenols, polyphenols and tannins: An overview. pp. 1–24. *In*: A. Crozier, M. N. Clifford and H. Ashihara (eds.). Plant Secondary Metabolites: Occurrence, Structure and Role in the Human Diet. Blackwell Publishing Ltd., Oxford.

Cuccolini, S., A. Aldini, L. Visai, M. Daglia and D. Ferrari. 2013. Environmentally friendly lycopene purification from tomato peel waste: Enzymatic assisted aqueous extraction. J. Agric. Food Chem. 61: 1646−1651.

Da Silva, E. A. B., M. Zabkova, J. D. Araújo, C. A. Cateto, M. F. Barreiro, M. N. Belgacem and A. E. Rodrigues. 2009. An integrated process to produce vanillin and lignin-based polyurethanes from Kraft lignin. Chem. Eng. Res. Des. 87: 1276–1292.

De Carvalho, C. C. C. R. and M. M. R. Da Fonseca. 2006. Carvone: Why and how should one bother to produce this terpene. Food Chem. 95: 413–42.

de Souza, V. B., A. Fujita, M. Thomazini, E. R. da Silva, J. F. Lucon, M. I. Genovese and C. S. Favaro-Trindade. 2014. Functional properties and stability of spray-dried pigments from Bordo grape (*Vitis labrusca*) winemaking pomace. Food Chem. 164: 380–386.

de Souza, V. B., M. Thomazini, J. C. de Carvalho Balieiro and C. S. Fávaro-Trindade. 2015. Effect of spray drying on the physicochemical properties and color stability of the powdered pigment obtained from vinification byproducts of the Bordo grape (*Vitis labrusca*). Food Bioprod. Process. 93: 39–50.

Di Gioia, D., L. Sciubba, L. Setti, F. Luziatelli, M. Ruzzi, D. Zanichelli and F. Fava. 2007. Production of biovanillin from wheat bran. Enzyme Microb. Technol. 41: 498–505.

Directive 2006/12/EC of the European Parliament and of the Council of 5 April 2006 on waste. 2006. Official Journal of the European Union L 114/9.

Đorđević, V., B. Balanč, A. Belščak-Cvitanović, S. Lević, K. Trifković, A. Kalušević, B. Bugarski and V. Nedović. 2015. Trends in encapsulation technologies for delivery of food bioactive compounds. Food Eng. Rev. 7: 452–490.

Edris, A. E., S. B. Girgis and H. H. M. Fadel. 2003. Recovery of volatile aroma components from aqueous waste streams using an activated carbon column. Food Chem. 82: 195–202.

Erickson, D. R. 1990. Edible Fats and Oils Processing: Basic Principles and Modern Practices: World Conference Proceedings, AOCS Press, Campaign, Illinois.

European Commission (EC). Communication from the Commission to the European Parliament, the Council, the European Economic and Social Committee and the Committee of the Regions: Towards a Circular Economy: A Zero Waste Programme for Europe. COM (2014) 0398; European Commission (EC): Brussels, Belgium, 2014.

Fadel, H. H. M., M. G. Mahmoud, M. M. S. Asker and S. N. Lotfy. 2015. Characterization and evaluation of coconut aroma produced by *Trichoderma viride* EMCC-107 in solid state fermentation on sugarcane bagasse. Electron. J. Biotechnol. 18: 5–9.

Fiedor, J. and K. Burda. 2014. Potential role of carotenoids as antioxidants in human health and disease. Nutrients 6: 466–488.

Fischer, U. A., R. Carle and D. R. Kammerer. 2013. Thermal stability of anthocyanins and colourless phenolics in pomegranate (*Punica granatum* L.) juices and model solutions. Food Chem. 138: 1800–1809.

Flores, F. P., R. K. Singh and F. Kong. 2014. Physical and storage properties of spray-dried blueberry pomace extract with whey protein isolate as wall material. J. Food Eng. 137: 1–6.

Furukawa, H., H. Morita, T. Yoshida and T. Nagasawa. 2003. Conversion of isoeugenol into vanillic acid by *Pseudomonas putida* I58 cells exhibiting high isoeugenol-degrading activity. J. Biosci. Bioeng. 96: 401–403.

Galanakis, C. M. 2012. Recovery of high added-value components from food wastes: Conventional, emerging technologies and commercialized applications. Trends Food Sci. Tech. 26: 68–87.

Galanakis, C. M. 2016. Food Waste Recovery Processing Technologies and Industrial Techniques. Academic Press, London.

Gandía-Herrero, F., M. Jiménez-Atiénzar, J. Cabanes, F. García-Carmona and J. Escribano. 2010. Stabilization of the bioactive pigment of *Opuntia* fruits through maltodextrin encapsulation. J. Agri. Food Chem. 58: 10646–10652.

Gandía-Herrero, F., J. Cabanes, J. Escribano, F. García-Carmona and M. Jiménez-Atiénzar. 2013. Encapsulation of the most potent antioxidant betalains in edible matrixes as powders of different colors. J. Agric. Food Chem. 61: 4294–4302.

Goksungur, Y., F. Mantzouridou, T. Roukas and P. Kotzekidou. 2004. Production of b-carotene from beet molasses by *Blakeslea trispora* in stirred-tank and bubble column reactors: Development of a mathematical modeling. Appl. Biochem. Biotechnol. 112: 37–54.

Gopalan, N., L. V. Rodríguez-Duran, G. Saucedo-Castaneda and K. M. Nampoothiri. 2015. Review on technological and scientific aspects of feruloyl esterases: A versatile enzyme for biorefining of biomass. Biores. Technol. 193: 534–544.

Goula, A. M. and K. G. Adamopoulos. 2012. A method for pomegranate seed application in food industries: Seed oil encapsulation. Food Bioprod. Process. 90: 639–652.

Goula, A. M. and H. N. Lazarides. 2015. Integrated processes can turn industrial food waste into valuable food by-products and/or ingredients: The cases of olive mill and pomegranate wastes. J. Food Eng. 167: 45–50.

Graf, E. 1992. Antioxidant potential of ferulic acid. Free Radic. Biol. Med. 13: 435–448.

Grune, T., G. Lietz, A. Palou, A. C. Ross, W. Stahl, G. Tang, D. Thurnham, S. Yin and H. K. Biesalski. 2010. b-Carotene is an important vitamin a source for humans. J. Nutr. 140: 2268–2285.

Gültekin-Özgüven, M., A. Karadağ, Ş. Duman, B. Özkal and B. Özçelik. 2016. Fortification of dark chocolate with spray dried black mulberry (*Morus nigra*) waste extract encapsulated in chitosan-coated liposomes and bioaccessability studies. Food Chem. 201: 205–212.

Gustavsson, J., C. Cederberg, U. Sonesson, R. Van Otterdijk and A. Meybeck. 2011. Global food losses and food waste. Extend, causes and prevention. Rome: Food and agriculture organization of the United Nations. http://www.fao.org/fileadmin/user_upload/ags/publications/GFL_web.pdf.

Harborne, J. B. 1989. Methods in Plant Biochemistry. Volume 1. Plant Phenolics. Academic Press, London.

Idham, Z., I. I. Muhamad, S. H. M. Setapar and M. R. Sarmidi. 2012. Effect of thermal processes on roselle anthocyanins encapulated in different polymer matrices. J. Food Process. Preserv. 36: 176–184.

Inanaga, S., A. Okasaka and S. Tanaka. 1995. Does silicon exist in association with organic compounds in rice plant? Soil Sci. Plant Nutr. 41: 111–117.

Ishida, B. K. and M. H. Chapman. 2009. Carotenoid extraction from plants using a novel, environmentally friendly solvent. J. Agric. Food Chem. 57: 1051–1059.

Jiménez-Aguilar, D. M., A. E. Ortega-Regules, J. D. Lozada-Ramírez, M. C. I. Pérez-Pérez, E. J. Vernon-Carter and J. Welti-Chanes. 2011. Color and chemical stability of spray-dried blueberry extract using mesquite gum as wall material. J. Food Compos. Anal. 24: 889–894.

Kaderides, K., A. M. Goula and K. G. Adamopoulos. 2015. A process for turning pomegranate peels into a valuable food ingredient using ultrasound-assisted extraction and encapsulation. Innov. Food Sci. Emerg. 31: 204–215.

Kalogeropoulos, N., A. Chiou, V. Pyriochou, A. Peristeraki and V. T. Karathanos. 2012. Bioactive phytochemicals in industrial tomatoes and their processing byproducts. LWT—Food Sci. Technol. 49: 213–216.

Kalušević, A., A. Salević, S. Lević, B. Čalija, S. Žilić, J. Milić and V. Nedović. 2016. Encapsulation of bioactive compounds from black soybean seed coats by internal gelation. Works Fac. Agric. Food Sci., University of Sarajevo 16: 146–151.

Kha, T. C., M. H. Nguyen, P. D. Roach and C. E. Stathopoulos. 2015. A storage study of encapsulated gac (*Momordica cochinchinensis*) oil powder and its fortification into foods. Food Bioprod. Proces. 96: 113–125.

Khanal, R. C., L. R. Howard, C. R. Brownmiller and R. L. Prior. 2009. Influence of extrusion processing on procyanidin composition and total anthocyanin contents of blueberry pomace. J. Food Sci. 74: H52–H58.

Khemakhem, M., G. Sotiroudis, E. Mitsou, S. Avramiotis, T. G. Sotiroudis, N. Bouzouita and V. Papadimitriou. 2016. Melanin and humic acid-like polymer complex from olive mill waste waters. Part II. Surfactant properties and encapsulation in W/O microemulsions. J. Mol. Liq. 222: 480–486.

Konar, N., I. Haspolat, E. S. Poyrazoğlu, K. Demir and N. Artık. 2012. Review on supercritical fluid extraction (SFE) of lycopene from tomato and tomato products. Karaelmas Sci. Eng. J. 2: 69–75.

Konwarh, R., S. Pramanik, D. Kalita, C. L. Mahanta and N. Karak. 2012. Ultrasonication—A complementary 'green chemistry' tool to biocatalysis: A laboratory-scale study of lycopene extraction. Ultrason. Sonochem. 19: 292–299.

Kosaraju, S. L., L. D'ath and A. Lawrence. 2006. Preparation and characterisation of chitosan microspheres for antioxidant delivery. Carbohydr. Polym. 64: 163–167.

Kris-Etherton, P. M., K. D. Hecker, A. Bonanome, S. M. Coval, A. E. Binkoski, K. F. Hilpert, A. E. Griel and T. D. Etherton. 2002. Bioactive compounds in foods: Their role in the prevention of cardiovascular disease and cancer. Amer. J. Medic. 113: 71S–88S.

Kropat, C., M. Betz, U. Kulozik, S. Leick, H. Rehage, U. Boettler, N. Teller and D. Marko. 2013. Effect of microformulation on the bioactivity of an anthocyanin-rich bilberry pomace extract (*Vaccinium myrtillus* L.) *in vitro*. J. Agric. Food Chem. 61: 4873–4881.

Krygier, K., F. Sosulski and L. Hogge. 1982. Free, esterified, and insoluble-bound phenolic acids. 1. Extraction and purification procedure. J. Agric. Food Chem. 30: 330–334.

Laokuldilok, T. and N. Kanha. 2015. Effects of processing conditions on powder properties of black glutinous rice (*Oryza sativa* L.) bran anthocyanins produced by spray drying and freeze drying. LWT—Food Sci. Tech. 64: 405–411.

Laufenberg, G., B. Kunz and M. Nystroem. 2003. Transformation of vegetable waste into value added products: (A) the upgrading concept; (B) practical implementations. Bioresour. Technol. 87: 167–198.

Lauro, M. R., L. Crasci, C. Carbone, R. P. Aquino, A. M. Panico and G. Puglisi. 2015. Encapsulation of a citrus by-product extract: Development, characterization and stability studies of a nutraceutical with antioxidant and metalloproteinases inhibitory activity. LWT—Food Sci. Tech. 62: 169–176.

Lavecchia, R. and A. Zuorro. 2008. Improved lycopene extraction from tomato peels using cell-wall degrading enzymes. Eur. Food Res. Technol. 228: 153–158.

Lee, J. J. L. and W. W. N. Chen. 2016. The production, regulation and extraction of carotenoids from *Rhodosporidium toruloides*. J. Mol. Genet. Med. 10: 215.

Lee, S. C., J. H. Kim, S. M. Jeong, D. R. Kim, J. U. Ha, K. C. Nam and D. U. Ahn. 2003. Effect of far-infrared radiation on the antioxidant activity of rice hulls. J. Agric. Food Chem. 51: 4400–4403.

Lesage-Meessen, L., A. Lomascolo, E. Bonnin, J. F. Thibault, A. Buleon, M. Roller, M. Asther, E. Record, B. Colonna Ceccaldi and M. Asther. 2002. A biotechnological process involving filamentous fungi to produce natural crystalline vanillin from maize bran. Appl. Biochem. Biotechnol. 102: 141–153.

Li, Y., M. Radoiu, A. S. Fabiano-Tixier and F. Chemat. 2013. From laboratory to industry: Scale-up, quality, and safety consideration for microwave-assisted extraction. pp. 207–229. *In*: F. Chemat and G. Cravotto (eds.). Microwave-assisted Extraction for Bioactive Compounds: Theory and Practice. Food Engineering Series 4, Springer Science+Business Media, New York.

Lianfu, Z. and L. Zelong. 2008. Optimization and comparison of ultrasound/microwave assisted extraction (UMAE) and ultrasonic assisted extraction (UAE) of lycopene from tomatoes. Ultrason. Sonochem. 15: 731–737.

Lin, C. S. K., L. A. Pfaltzgraff, L. Herrero-Davila, E. B. Mubofu, S. Abderrahim, J. H. Clark, A. A. Koutinas, N. Kopsahelis, K. Stamatelatou, F. Dickson, S. Thankappan, Z. Mohamed, R. Brocklesby and R. Luque. 2013. Food waste as a valuable resource for the production of chemicals, materials and fuels. Current situation and global perspective. Energy Environ. Sci. 6: 426–464.

Liu, Z., J. Wang, P. Shen, C. Wang and Y. Shen. 2006. Microwave-assisted extraction and high-speed counter-current chromatography purification of ferulic acid from Radix *Angelicae sinensis*. Sep. Purif. Technol. 52: 18–21.

Lohrasbi, M., M. Pourbafrani, C. Niklasson and M. J. Taherzadeh. 2010. Process design and economic analysis of a citrus waste biorefinery with biofuels and limonene as products. Bioresource Technol. 101: 7382–7388.

Luca, A., B. Cilek, V. Hasirci, S. Sahin and G. Sumnu. 2014. Storage and baking stability of encapsulated sour cherry phenolic compounds prepared from micro- and nano-suspensions. Food Bioprocess. Technol. 7: 204–211.

Machmudah, S., S. Zakaria, S. Winardi, M. Sasaki, M. Goto, N. Kusumoto and K. Hayakawa. 2012. Lycopene extraction from tomato peel by-product containing tomato seed using supercritical carbon dioxide. J. Food Eng. 108: 290–296.

Malisorn, C. and W. Suntornsuk. 2009. Improved b-carotene production of *Rhodotorula glutinis* in fermented radish brine by continuous cultivation. Biochem. Eng. J. 43: 27–32.

Mantzouridou, F., T. Roukas, P. Kotzekidou and M. Liakopoulou. 2002a. Optimization of β-carotene production from synthetic medium by *Blakeslea trispora*. A mathematical modeling. Appl. Biochem. Biotechnol. 101: 153–175.

Mantzouridou, F., T. Roukas and P. Kotzekidou. 2002b. Effect of the aeration rate and agitation speed on β-carotene production and morphology of *Blakeslea trispora* in a stirred tank reactor: Mathematical modeling. Biochem. Eng. J. 10: 123–135.

Mantzouridou, F. and A. Paraskevopoulou. 2013. Volatile bio-ester production from orange pulp-containing medium using *Saccharomyces cerevisiae*. Food Bioprocess Technol. 6: 3326–3334.

Mantzouridou, F. T., A. Paraskevopoulou and S. Lalou. 2015. Yeast flavor production by solid state fermentation of orange peel waste Biochem. Eng. J. 101: 1–8.

Martins, A., L. Barros, A. M. Carvalho, C. Santos-Buelga, I. P. Fernandes, F. Barreiro and I. C. Ferreira. 2014. Phenolic extracts of *Rubus ulmifolius* Schott flowers: Characterization, microencapsulation and incorporation into yogurts as nutraceutical sources. Food Funct. 5: 1091–1100.

Martins, N., C. L. Roriz, P. Morales, L. Barros and I. C. F. R. Ferreira. 2016. Food colorants: Challenges, opportunities and current desires of agroindustries to ensure consumer expectations and regulatory practices. Trends Food Sci. Tech. 52: 1–15.

Mata-Gómez, L. C., J. C. Montañez, A. Méndez-Zavala and C. N. Aguilar. 2014. Biotechnological production of carotenoids by yeasts: An overview. Microb. Cell Fact. 13: 1–12.

Medeiros, A., A. Pandey, R. J. S. De Freitas, P. Christen and C. R. Soccol. 2000. Optimization of the production of aroma compounds by *Kluyveromyces marxianus* in solid-state fermentation using factorial design and response surface methodology. Biochem. Eng. J. 6: 33–39.

Mèridas, O. S., A. G. Coloma and R. S. Vioque. 2012. Agricultural residues as a source of bioactive natural products. Phytochem. Rev. 11: 447–466.

Mezzomo, N. and S. R. S. Ferreira. 2016. Carotenoids functionality, sources and processing by supercritical technology: A Review. J. Chem. Article ID 3164312, 16 pages.

Mirabella, N., V. Castellani and S. Sala. 2014. Current options for the valorization of food manufacturing waste: A review. J. Clean. Prod. 65: 28–41.

Mohamed, R. S. and G. A. Mansoori. The use of supercritical fluid extraction technology in food processing. *In*: Featured Article—Food Technology Magazine, June 2002. The World Markets Research Centre, London, UK.

Mussatto, S. I., G. Dragone and I. C. Roberto. 2007. Ferulic and *p*-coumaric acids extraction by alkaline hydrolysis of brewer's spent grain. Ind. Crop. Prod. 25: 231–237.

Naik, A., V. Meda and S. S. Lele. 2014. Freeze drying for microencapsulation of α-linolenic acid rich oil: A functional ingredient from *Lepidium sativum* seeds. Eur. J. Lipid Sci. Technol. 116: 837–846.

Nanou, K. and T. Roukas. 2016. Waste cooking oil: A new substrate for carotene production by *Blakeslea trispora* in submerged fermentation. Bioresource Technol. 203: 198–203.

Naviglio, D., F. Pizzolongo, L. Ferrara, A. Aragón and A. Santini. 2008a. Extraction of pure lycopene from industrial tomato by-product in water using a new high-pressure process. J. Sci. Food Agric. 88: 2414–2420.

Naviglio, D., T. Caruso, P. Iannece, A. Aragón and A. Santini. 2008b. Characterization of high purity lycopene from tomato wastes using a new pressurized extraction approach. J. Agric. Food Chem. 56: 6227–6231.

Naziri, E., N. Nenadis, F. Th. Mantzouridou and M. Z. Tsimidou. 2014. Valorization of the major agrifood industrial by-products and waste from Central Macedonia (Greece) for the recovery of compounds for food applications. Food Res. Int. 65: 350–358.

Nedovic, V., A. Kalusevic, V. Manojlovic, S. Levic and B. Bugarski. 2011. An overview of encapsulation technologies for food applications. Procedia Food Sci. 1: 1806–1815.

Nenadis, N., A. Kyriakoudi and M. Z. Tsimidou. 2013. Impact of alkaline or acid digestion to antioxidant activity, phenolic content and composition of rice hull extracts. LWT—Food Sci. Technol. 54: 207–215.

Nunes, I. L. and A. Z. Mercadante. 2007. Encapsulation of lycopene using spray-drying and molecular inclusion processes. Braz. Arch. Biol. Technol. 50: 893–900.

Oidtmann, J., M. Schantz, K. Mäder, M. Baum, S. Berg, M. Betz, U. Kulozik, S. Leick, H. Rehage, K. Schwarz and E. Richling. 2012. Preparation and comparative release characteristics of three anthocyanin encapsulation systems. J. Agric. Food Chem. 60: 844–851.

Oryza Oil and Fat Chemical CO., Ltd. (http://www.oryza.co.jp/html/english/) date of access 21-8-2016.

Ou, S. and K. C. Kwok. 2004. Ferulic acid: Pharmaceutical functions, preparation and applications in foods. J. Sci. Food Agric. 84: 1261–1269.

Ou, S., Y. Luo, F. Xue, C. Huang, N. Zhang and Z. Liu. 2007. Seperation and purification of ferulic acid in alkaline-hydrolysate from sugarcane bagasse by activated charcoal adsorption/anion macroporous resin exchange chromatography. J. Food Eng. 78: 1298–1304.

Paini, M., B. Aliakbarian, A. A. Casazza, A. Lagazzo, R. Botter and P. Perego. 2015. Microencapsulation of phenolic compounds from olive pomace using spray drying: A study of operative parameters. LWT—Food Sci. Tech. 62: 177–186.

Papaioannou, E. H. and A. J. Karabelas. 2012. Tomato peel lycopene recovery under mild conditions assisted by enzymatic pre-treatment and non-ionic surfactants. Acta Biochim. Polon. 59: 71–74.

Papaioannou, E. H., M. Liakopoulou-Kyriakides and A. J. Karabelas. 2016. Natural origin lycopene and its "green" downstream processing. Crit. Rev. Food Sci. Nutr. 56: 686–709.

Peinado, I., G. Koutsidis and J. Ames. 2016. Production of seafood flavour formulations from enzymatichydrolysates of fish by-products LWT—Food Sci. Technol. 66: 444–452.

Priefert, H., J. Rabenhorst and A. Steinbuchel. 2001. Minireview: Biotechnological production of vanillin. Appl. Microbiol. Biotechnol. 56: 296–314.

Rabenhorst, J. and R. Hopp. 2000. Process for the Preparation of Vanillin and Suitable Microorganisms. U.S. Patent #6133003.

Ramesh, C. K., K. S. Upendra, S. Nandini and K. S. Arun. 2007. Isolation and identification of a novel strain of *Pseudomonas chlororaphis* capable of transforming isoeugenol to vanillin. Curr. Microbiol. 54: 457–461.

Ranveer, R. C., S. N. Patil and A. K. Sahoo. 2013. Effect of different parameters on enzyme-assisted extraction of lycopene from tomato processing waste. Food Bioprod. Process. 91: 370–375.

Ranveer, R. C., A. A. Gatade, H. A. Kamble and A. K. Sahoo. 2015. Microencapsulation and storage stability of lycopene extracted from tomato processing waste. Braz. Arch. Biol. Technol. 58: 953–960.

Rapid Sobti, R. C., K. G. Gupta, P. Sharma, B. Karmakar, R. M. Vohra and H. Nandanwar. 2000. Degradation of ferulic acid via 4-vinylguiacol and vanillin by a newly isolated strain of *Bacillus coagulans*. J. Biotechnol. 80: 195–202.

Riggi, E. and G. Avola. 2008. Fresh tomato packinghouses waste as high added-value biosource. Resour. Conserv. Recy. 53: 96–106.

Rigi, A. A. and S. Abbasi. 2016. Microemulsion-based lycopene extraction: Effect of surfactants, co-surfactants and pretreatments. Food Chem. 197: 1002–1007.

Rodriguez, E. B., M. L. P. Vidallon, D. J. R. Mendoza and C. T. Reyes. 2016. Health-promoting bioactivities of betalains from red dragon fruit (*Hylocereus polyrhizus* (Weber) Britton and Rose) peels as affected by carbohydrate encapsulation. J. Sci. Food Agr. DOI: 10.1002/jsfa.7681.

Rossi, S. C., L. P. S. Vandenberghe, B. M. P. Pereira, F. D. Gago, J. A. Rizzolo, A. Pandey, C. R. Soccol and A. B. P. Medeiros. 2009. Improving fruity aroma production by fungi in SSF using citric pulp. Food Res. Int. 42: 48–484.

Roukas, T. and F. Mantzouridou. 2001. An improved method for the extraction of b-carotene from *Blakeslea trispora*. Appl. Biochem. Biotechnol. 90: 37–45.

Roukas, T., F. Mantzouridou, T. Boumpa, A. Vafiadou and Y. Goksungur. 2007. Production of b-carotene from beet molasses and deproteinized whey by *Blakeslea trispora*. Food Biotechnol. 21: 195–196.

Rubilar, M., E. Morales, K. Contreras, C. Ceballos, F. Acevedo, M. Villarroe and C. Shene. 2012. Development of a soup powder enriched with microencapsulated linseed oil as a source of omega-3 fatty acids. Eur. J. Lipid Sci. Technol. 114: 423–433.

Saikia, S., N. K. Mahnot and C. L. Mahanta. 2015. Optimisation of phenolic extraction from *Averrhoa carambola* pomace by response surface methodology and its microencapsulation by spray and freeze drying. Food Chem. 171: 144–152.

Salević, A., A. Kalušević, S. Lević, B. Čalija, J. Milić, S. Žilić and V. Nedović. 2015. Encapsulation of Black Soybean Seed Coats Anthocyanins by Melt Solidification. Proc. 23th Int. Conf. Bioencapsulation. pp. 162–163.

Santos, D. T., J. Q. Albarelli, M. M. Beppu and M. A. Meireles. 2013. Stabilization of anthocyanin extract from jabuticaba skins by encapsulation using supercritical CO_2 as solvent. Food Res. Int. 50: 617–624.

Saulnier, L., C. Marot, M. Elgorriaga, E. Bonnin and J. F. Thibault. 2001. Thermal and enzymatic treatments for the release of free ferulic acid from maize bran. Carbohydr. Polym. 45: 269–275.

Sessa, M., A. A. Casazza, P. Perego, R. Tsao, G. Ferrari and F. Donsì. 2013. Exploitation of polyphenolic extracts from grape marc as natural antioxidants by encapsulation in lipid-based nanodelivery systems. Food Bioprocess Tech. 6: 2609–2620.

Shi, J., Y. Dai, Y. Kakuda, G. Mittal and S. J. Xue. 2008. Effect of heating and exposure to light on the stability of lycopene in tomato puree. Food Control 19: 514⁻520.

Shimoni, E., U. Ravid and Y. Shoham. 2000. Isolation of a *Bacillus* sp. Capable of transforming isoeugenol to vanillin. J. Biotechnol. 78: 1–9.

Shin, H. D., S. McClendon, T. Le, F. Taylor and R. R. Chen. 2006. A complete enzymatic recovery of ferulic acid from corn residues with extracellular enzymes from Neosartorya spinosa NRRL185. Biotechnol. Bioeng. 95: 1108–1115.

Shinde, T., D. Sun-Waterhouse and J. Brooks. 2014. Co-extrusion encapsulation of probiotic *Lactobacillus acidophilus* alone or together with apple skin polyphenols: An aqueous and value-added delivery system using alginate. Food Bioprocess Technol. 7: 1581–1596.

Silva, L. M., L. E. Hill, E. Figueiredo and C. L. Gomes. 2014. Delivery of phytochemicals of tropical fruit by-products using poly (DL-lactide-co-glycolide) (PLGA) nanoparticles: Synthesis, characterization, and antimicrobial activity. Food Chem. 165: 362–370.

Singhania, R. R., A. K. Patel, C. R. Soccol and A. Pandey. 2009. Recent advances in solid-state fermentation. Biochem. Eng. J. 44: 13–18.

Soares, M., P. Christen, A. Pandey and C. R. Soccol. 2000. Fruity flavor production by *Ceratocystis fimbriata* grown on coffee husk in solid-state fermentation. Process Biochem. 35: 857–861.

Spigno, G., F. Donsì, D. Amendola, M. Sessa, G. Ferrari and D. M. De Faveri. 2013. Nanoencapsulation systems to improve solubility and antioxidant efficiency of a grape marc extract into hazelnut paste. J. Food Eng. 114: 207–214.

Stajić, S., D. Živković, V. Tomović, V. Nedović, M. Perunović, N. Kovjanić, S. Lević and N. Stanišić. 2014. The utilisation of grapeseed oil in improving the quality of dry fermented sausages. Int. J. Food Sci. Tech. 49: 2356–2363.

Sticher, O. 2008. Natural product isolation. Nat. Prod. Reports 25: 517–554.

Strati, I. F. and V. Oreopoulou. 2011a. Process optimisation for recovery of carotenoids from tomato waste. Food Chem. 129: 747–752.

Strati, I. F. and V. Oreopoulou. 2011b. Effect of extraction parameters on the carotenoid recovery from tomato waste. Int. J. Food Sci. Tech. 46: 23–29.

Strati, I. F. and V. Oreopoulou. 2014. Recovery of carotenoids from tomato processing by-products–a review. Food Res. Int. 65: 311–321.

Strati, I. F., E. Gogou and V. Oreopoulou. 2015. Enzyme and high pressure assisted extraction of carotenoids from tomato waste. Food Bioprod. Process 94: 668–674.

Sutherland, I. A. 2007. Recent progress on the industrial scale-up of counter-current chromatography. J. Chromatogr. A 1151 1-2: 6–13.

Thomas, L., C. Larroche and A. Pandey. 2013. Current developments in solid-state fermentation. Biochem. Eng. J. 81: 146–161.

Tilay, A., M. Bule, J. Kishenkumar and U. Annapure. 2008. Preparation of ferulic acid from agricultural wastes: Its improved extraction and purification. J. Agric. Food Chem. 56: 7644–7648.

Tinoi, J., N. Rakariyatham and R. L. Deming. 2005. Simplex optimization of carotenoid production by *Rhodotorula glutinis* using hydrolyzed mung bean waste flour as substrate. Process Biochem. 40: 2551–2557.

Tonon, R. V., C. Brabet, D. Pallet, P. Brat and M. D. Hubinger. 2009. Physicochemical and morphological characterisation of açaí (*Euterpe oleracea* Mart.) powder produced with different carrier agents. Int. J. Food Sci. Technol. 44: 1950–1958.

Toogood, H. S., A. N. Cheallaigh, S. Tait, D. J. Mansell, A. A. Jervis, A. Lygidakis, L. Humphreys, E. Takano, J. M. Gardiner and N. S. Scrutton. 2015. Enzymatic menthol production: One-pot approach using engineered *Escherichia coli*. ACS Synth. Biol. 4: 1112−1123.

Topakas, E., C. Vafiadi and P. Christakopoulos. 2007. Microbial production, characterization and applications of feruloyl esterases. Process Biochem. 42: 497–509.

Tornberg, E. and C. Galanakis. 2008. Olive Waste Recovery. WO Patent # 082343.

Torre, P., B. Aliakbarian, B. Rivas, J. M. Domínguez and A. Converti. 2008. Release of ferulic acid from corn cobs by alkaline hydrolysis. Biochem. Eng. J. 40: 500–506.

Torres, B. R., B. Aliakbarian, P. Torre, P. Perego, J. M. Domínguez, M. Zilli and A. Converti. 2009. Vanillin bioproduction from alkaline hydrolyzate of corn cob by *Escherichia coli* JM109/pBB1. Enzyme Microb. Technol. 44: 154–158.

Tsakona, S., A. G. Skiadaresis, N. Kopsahelis, A. Chatzifragkou, S. Papanikolaou, I. K. Kookos and A. A. Koutinas. 2016. Valorisation of side streams from wheat milling and confectionery industries for consolidated production and extraction of microbial lipids. Food Chem. 198: 85–92.

Tumbas Šaponjac, V., G. Ćetković, J. Čanadanović-Brunet, B. Pajin, S. Djilas, J. Petrović, I. Lončarević, S. Stajčić and J. Vulić. 2016. Sour cherry pomace extract encapsulated in whey and soy proteins: Incorporation in cookies. Food Chem. 207: 27–33.

Umesha, S. S., R. Sai Manohar, A. R. Indiramma, S. Akshitha and K. Akhilender Naidu. 2015. Enrichment of biscuits with microencapsulated omega-3 fatty acid (Alpha-linolenic acid) rich Garden cress (*Lepidium sativum*) seed oil: Physical, sensory and storage quality characteristics of biscuits. LWT—Food Sci. Tech. 62: 654–661.

Urbonaviciene, D., P. Viskelis, J. Viskelis and J. Jankauskiene. 2012. Lycopene and b-carotene in non-blanched and blanched tomatoes. J. Food Agri. Environ. 10: 142–146.

Valduga, E., A. H. R. Ribeiro, K. Cence, R. Colet, L. Tiggemann, J. Zeni and G. Toniazzo. 2014. Carotenoids production from a newly isolated *Sporidiobolus pararoseus* strain using agroindustrial substrates. Biocatal. Agric. Biotechnol. 3: 207–213.

Varzakakou, M. and T. Roukas. 2010. Identification of carotenoids produced from cheese whey by *Blakeslea trispora* in submerged fermentation. Prep. Biochem. Biotechnol. 40: 76–82.

Wang, H. D., L. W. Zhang, J. Luo and L. J. Yu. 2015. Rapid and environmentally-friendly extraction of carotenoids from *Blakeslea trispora*. Biotechnol. Lett. 37: 2173–2178.

Wang, J., Y. Cao, B. Sun and C. Wang. 2011. Characterisation of inclusion complex of *trans*-ferulic acid and hydroxypropyl-β-cyclodextrin. Food Chem. 124: 1069–1075.

Xiros, C., M. Moukouli, E. Topakas and P. Christakopoulos. 2009. Factors affecting ferulic acid release from Brewer's spent grain by *Fusarium oxysporum* enzymatic system. Bioresour. Technol. 100: 5917–5921.

Yazicioglu, B., S. Sahin and G. Sumnu. 2015. Microencapsulation of wheat germ oil. J. Food Sci. Technol. 52: 3590–3597.

Yoon, S. H., C. Li, J. E. Kim, S. H. Lee, J. Y. Yoon, M. S. Choi, W. T. Seo, J. K. Yang, J. Y. Kim and S. W. Kim. 2005. Production of vanillin by metabolically engineered *Escherichia coli*. Biotechnology 27: 1829–1832.

Yu, P., D. D. Maenz, J. J. McKinnon, V. J. Racz and D. A. Christensen. 2002. Release of ferulic acid from oat hulls by *Aspergillus* ferulic acid esterase and *Trichoderma xylanase*. J. Agric. Food Chem. 50: 1625–1630.

Zamzuri, N. A. and S. Abd-Aziz. 2013. Biovanillin from agro wastes as an alternative food flavour. J. Sci. Food Agric. 93: 429–438.

Zhang. L., D. Mou and Y. Du. 2007. Procyanidins: Extraction and micro-encapsulation. J. Sci. Food Agric. 87: 2192–2197.

Zhao, L. Q., Z. H. Sun, P. Zheng and L. L. Zhu. 2005. Biotransformation of isoeugenol to vanillin by a novel strain of *Bacillus fusiformis*. Biotechnol. Lett. 27: 1505–1509.

Zhao, L. Q,. Z. H. Sun, P. Zheng and J. Y. He. 2006. Biotransformation of isoeugenol to vanillin by Bacillus fusiformis CGMCC1347 with addition of resin HD-8. Process Biochem. 41: 1673–1676.

Zhao, S., S. Yao, S. Ou, J. Lin, Y. Wang, X. Peng, A. Li and B. Yu. 2014. Preparation of ferulic acid from corn bran: Its improved extraction and purification by membrane separation. Food Bioprod. Process. 92: 309–313.

Zhao, Z. and M. H. Moghadasian. 2008. Chemistry, natural sources, dietary intake and pharmacokinetic properties of ferulic acid: A review. Food Chem. 109: 691–702.

Zheng, L., P. Zheng, Z. Sun, Y. Bai, J. Wang and X. Guo. 2007. Production of vanillin from waste residue of rice bran oil by *Aspergillus niger* and *Pycnoporus cinnabarinus*. Bioresour. Technol. 98: 1115–1119.

Zuorro, A., M. Fidaleo and R. Lavecchia. 2011. Enzyme-assisted extraction of lycopene from tomato processing waste. Enzyme Microb. Technol. 49: 567–573.

Zuorro, A., R. Lavecchia, F. Medici and L. Piga. 2014. Use of cell wall degrading enzymes for the production of high-quality functional products from tomato processing waste. Chem. Eng. Trans. 38: 355–360.

CHAPTER 7

Extraction, Isolation and Utilisation of Bioactive Compounds from Rice Waste

Binh T. Ho[1],* and *Khang N. Tran*[2]

Introduction

Rice is a staple food in most Asian countries and a source of energy for 30 per cent of the world's population (Müller-Fischer 2013). For human consumption, rice is usually processed by being milled and polished into white grain. Rice processing typically yields 56–58 per cent milled white rice (head rice), 10–12 per cent broken grain, 18–20 per cent husk and 10–12 per cent bran (Kahlon and Chow 2001). Head rice is the main commercial product, while the remaining categories constitute rice waste or by-products. Straw is also considered rice waste and its yield per ratio to grain is 1:6 (Schiere et al. 2004). Straw and husk are mainly used for non-food purposes, such as animal feed, craft art, chemical industries and amino acids (Matano et al. 2014, Van Soest 2006). Bran contains the majority of functional substances of rice grain and is unstable due to its oil and lipase content (Kahlon and Chow 2001). This paper provides a review of the bioactive compounds in rice waste and their extraction, isolation and utilisation.

Bioactive Compounds in Rice Waste

Dietary fiber

Crude fiber is one of the major components of rice waste. In straw and husks, it is mostly in the form of crystalline cellulose fiber, which is a good source of processed ethanol (Alemdar and Sain 2008). In rice bran, crude fiber is reported to be a good source of

[1] Department of Food Technology, Faculty of Agriculture and Natural Resources, An Giang University, Vietnam. 18 Ung Van Khiem, Dong Xuyen Ward, Long Xuyen City, An Giang, Vietnam.
[2] Australian Research Council - Industrial Transformation Training Centres (ARC - ITTC) - School of Agriculture and Food Sciences, The University of Queensland, Australia, St. Lucia, QLD 4072, Australia.
* Corresponding author: hothanhbinhvn@yahoo.com

dietary fiber (DF) (Alemdar and Sain 2008). The Codex Alimentarius Commission Dietary (2013) defines DF as a carbohydrate polymer with ten or more monomeric units that are not hydrolyzed by endogenous enzymes in the small intestine and which have physiological beneficial effects on human health. Cellulose, hemicellulose, pectin, resistant starch, lignin and gums are considered to be DFs in plants (Bao 2012). DF is classified according to its solubility in water. The ratio of insoluble DF to soluble DF has an effect on both its properties and biological functions (Figuerola et al. 2005). Cellulose, hemicellulose and lignin are the main insoluble fibers while soluble fibers comprise mostly pectin and gums (Gao and Yue 2012). Due to their better taste characteristics, soluble DFs are more practical and acceptable as food than crude fibers and insoluble DFs (Wan et al. 2014).

Brown and white rice contain 2.9–4.0 per cent and less than 2.3 per cent DFs, respectively (Abdul-Hamid and Luan 2000). Bran contains 17–29 per cent DFs and is a suitable material for food and pharmaceutical applications. Cellulose, hemicellulose, lignin and wax are abundant in the outer layers and cell walls of rice bran (Luh 1991). These cellulosic components have been reported as zero energy components in food by Macagnan et al. (2016). About 90 per cent of dietary fiber in defatted rice bran is insoluble fiber, with the remainder being water soluble dietary fiber (Kahlon et al. 1990). Compared to other grains, rice bran is an abundant and cheap material, but it is underestimated as a source of DF in the food industry. The benefits of rice bran DF have been reported by a large number of research studies.

Functions

Both soluble and insoluble DFs in rice bran have good physical and bioactive properties. Various studies show their potential health benefits. Fiber can reduce chronic health syndromes including diabetes, obesity and cancer. Recommended daily DFs intakes are 25 g per day for women and 38 g per day for men (Panel on Macronutrients 2005).

Rice bran dietary fiber and diabetes

The consumption of high fiber diets based on rice bran DF can reduce the risk of diabetes and significantly reduce the serum glucose levels in both Type 1 and Type 2 diabetic patients (Silva et al. 2005). Qureshi et al. (2002) also reported that by consuming stabilized rice bran for 60 days insulin-dependence can be reduced, as well as glycosylated haemoglobin levels. Similarly, 14 type 2 diabetic subjects significantly reduced their serum glucoses level after the intake of stabilized rice bran for 12 weeks (Cheng et al. 2010). When consumed, soluble DFs transform to gel and become highly viscos, which could delay their passage through the intestinal tract and slow down the formation of glucose in serum (Jones et al. 1996). Soluble DFs could also modify the gastrointestinal myoelectrical activity, causing a delay in the small bowel transit time. Lastly, the high viscosity of soluble DF fiber might prevent the transportation of glucose through the unstirred water layer in the intestine (Gao and Yue 2012). Insoluble DFs could increase the insulin sensitivity of the human body, causing a decrease in the serum insulin. This mechanism was explained by Bäckhed et al. (2007) as insoluble DFs reducing the number of gram-negative bacteria in the gut, which in turn increases insulin sensitivity. This evidence shows that both soluble and insoluble DFs could help prevent or reduce the risk of diabetes. Rice bran is therefore a good source of beneficial natural DFs for diabetic patients.

Rice bran dietary fiber and obesity

DFs have been reported to play important roles in reducing the risk of obesity and lowering cholesterol. A study on nearly 3,000 adults over 10 years showed that people who consumed the most DF gained less weight than those consuming the least DF at any level of fat intake (Ludwig et al. 1999). Howarth et al. (2001) also reported that consumption of high-fiber diets reduces the risk of obesity. This author indicated that the consumption of an additional 14 g/day DF could decrease body weight by more than 1.9 kg over 3.8 months. A study on feeding rice bran (10 per cent total DF) to hamsters for 21 days resulted in a significantly lower plasma, liver cholesterol and triglycerides than cholesterol-feed hamsters (Kahlon et al. 1992). Another study on feeding hamsters with rice bran (10 per cent total DF, 0.5 per cent cholesterol) showed a similar result, with the lowering of liver cholesterol and liver lipid content in hamsters (Kahlon and Chow 2000).

The mechanisms for preventing obesity and lowering cholesterol have been reported by Gao and Yue (2012). Only 40 per cent of the fiber could be absorbed and used as energy. Therefore, rice bran fibres contain lower energy density. In addition, since the fiber can bind and trap a large amount of water, the consumption of fiber could reduce the energy-to-weight ratio in food, thereby extending satiety and decreasing energy intake. Finally, the high bulk and viscosity of fiber could cover other nutrients, causing a delay in digestion and energy absorbtion. A study showed that the consumption of 48 g DF per day could decrease energy absorption by 8 per cent in comparison with consuming 20 g DF per day (Wisker et al. 1988).

Rice bran DFs and gastrointestinal health

Studies show positive effects of rice bran DFs on the human gastrointestinal system. Consuming up to 28 g bran DF per day shows an increase in fecal weight and fecal volume with a resulting increase in the intestinal regularity (Miyoshi et al. 1986, Silva et al. 2005, Tomlin and Read 1988). DFs in rice bran, including hemicellulose, saccharide or α-glycan, have been reported as the main inhibitors of carcinogenicity or cytotoxicity of carcinogens (The Cosmetic Ingredient Review Expert Panel 2006). Similar reports have indicated that the DFs in rice bran as anticancer compounds can reduce the risk of colon cancer, esophagel cancer and stomach cancer (Gao and Yue 2012). The DFs may absorb the carcinogens which are then excreted into the faeces, and also act as a substrate for bacterial fermentation to release short chain fatty acids. These have anticarcinogenic effects *in vitro* (Harris et al. 1996). In addition, as the DFs can increase the fecal weight and reduce the transit time, they bring many advantages in preventing colon cancer (Gao and Yue 2012). Another study by Soler et al. (2001) reported that the addition of DFs to daily meals can prevent esophageal cancer by reducing hyperinsulinemia and the formation of insulin-like growth factors. DFs in grain have also been reported as anti-stomach cancer factors, although the mechanism has remained unclear (Jansen et al. 1999). Rice bran is therefore a good source of DFs for food. The health benefits have been reported in a range of studies.

Protein

Rice bran is reported to be a commercially attractive source of protein (Champagne 2004, Kahlon 2009, Kennedy and Burlingame 2003, Shih 2003). The protein is located in the aleurone cells, embryo and starchy endosperm. Its content in rice bran ranges from 12 to 20 per cent (Lasztity 1996, Shih 2003). Most nitrogen in rice bran is protein nitrogen (Luh 1991).

Protein can be fractionated based on its solubility in different solvents; the Osborne method is one of the most well-known methods (Betschart et al. 1977). Within this method, proteins are fractionated in a sequence of different solvents, including water, salt, alkali and alcohol. The water soluble protein fraction, salt water soluble protein fraction, ethanol soluble protein fraction and alkaline soluble protein fraction are named albumin, globulin, prolamin and glutelin, respectively (Hettiarachchy et al. 2012).

In rice kernels, most albumin and globulin are distributed in the embryo and outer layers of the kernels (Hattori et al. 2011), while prolamin and glutelin (oryzenin) are concentrated mostly in the endosperm (Lasztity 1996, Shih 2004). Albumin, globulin, prolamin and glutelin are distributed in the BRM at concentrations of about 31 per cent (including germ 13 per cent), 40 per cent (germ 10 per cent), 21 per cent (germ 10 per cent) and 5 per cent, respectively (Champagne et al. 2004, Saunders 1985, Shih 2003, 2004). Landers and Hamaker (1994) reported that 69 per cent of extractable protein in rice bran comprises albumins and globulins. Although albumins dissolve in water and globulins are salt-water soluble, the minerals in rice bran protein can quickly dissolve in the water during extraction and dilute the globulin fractions at the same time (Lasztity 1996). Therefore, the majority of protein in rice bran can be easily extracted by water.

Health benefit and properties of rice bran protein

Proteins from rice and rice bran can be classified as hypoallergenic, lactose free, soy free and gluten free (Friedman 2013, Shoji et al. 2001, The Cosmetic Ingredient Review Expert Panel 2006, Tsuji et al. 2001, Van Hooser and Crawford 1989, Wang et al. 1999). The nutritional properties of rice and rice bran protein also have advantages compared with other protein sources. Rice bran protein has a very high digestibility, more than 90 per cent (Wang et al. 1999). The protein efficiency ratio (PER) of RMBP is 2.0 to 2.2, which is close to the PER value of milk (2.5) (Kahlon 2009).

Many reports agree that rice and rice protein are hypoallergenic (Champagne 2004, Landers and Hamaker 1994, The Cosmetic Ingredient Review Expert Panel 2006, Van Hooser and Crawford 1989). Although Matsuda (1991) reported that 16 kDa rice protein showed a positive reaction to serum class E immunoglobulins, there are very few reports related to allergies from cooked rice (The Cosmetic Ingredient Review Expert Panel 2006). According to Van Hooser and Crawford (1989) rice is the only grain allowed for widespread consumption without allergy testing. In addition, rice bran protein has high lysine content and consequently has one of the highest nutritional value among cereal proteins (Lasztity 1996, The Cosmetic Ingredient Review Expert Panel 2006). Rice bran protein and its amino acid content are reported to have a desirable amino acid profile and to be suitable sources for consumption by infants and children (Helm and Burks 1996, Khan 2011, Wang et al. 1999).

The use of rice bran proteins in food and cosmetics has been reported in a number of studies. Yadav and Chaudhary (2011) increased the nutritional quality of biscuits by adding 15 per cent BRM protein. In addition, the addition of 1–5 per cent rice bran protein to bread helps reduce weight loss and microbial count (Harper et al. 2005). Rice bran proteins are used to upgrade the nutritional value of milk and drinks for children (Cho and Samuel 2009, Watchararuji et al. 2008). Hydrolyzed rice bran proteins have been used as hair conditioning and skin-conditioning agents because they retain moisture. Rice bran protein is also found in hair sprays, shampoos and moisturizing products (The Cosmetic Ingredient Review Expert Panel 2006). Moreover, rice bran protein is reported to have functional properties that are suitable for various food formulations including weaning foods, dry mixes, baked foods, whipped toppings, salad dressings, soups, sauces and gravies (Chandi and Sogi 2007, Fabian and Ju 2011). The properties of rice bran protein and some common protein sources are summarised in Table 1. Rice bran protein isolate has good foaming capacity, close to that of egg white (Wang et al. 1999). The water and oil absorption capacity of rice bran-isolated proteins are high, especially under high sugar and salt concentration and therefore, could be a good emulsifier in food (Chandi and Sogi 2007).

Rice bran is generally considered a safe protein source for human consumption on account of its high quality nutritional and functional properties, and thus could potentially be a valuable protein source for the food industry (Jiamyangyuen et al. 2005, Saunders 1985). However, it is not yet commercially available in this form to date (Gupta et al. 2008, Wang et al. 1999) due to its inefficiency in extraction and a lack of economic value as a commercial product (Fabian and Ju 2011). A suitable extraction method for an inexpensive and high quality protein product of protein from rice bran is needed to improve the new value-added opportunities for rice bran, which is an abundant by-product in the processing of high value milled rice.

Table 1. Properties of rice bran (RB) protein isolate and some other food proteins.

	RB protein isolate	Egg albumin	Soy protein	Casein
Nitrogen solubility index (NSI) at pH 7 (%)	60[1], 35[4], 65[5], 73[6]		35[4]	27.44[6]
Bulk density (g/cm^3)	0.4[1][3], 0.12[6]			0.89[6]
Water absorption capacity (mL/g)	3.5[1], 2.9[3], 4.04[6]			2.48[6]
Oil absorption capacity (mL/g)	3.0[1], 2.3[3], 8.14[6]			1.72[6]
Foaming capacity (mL)	10.4[1], 18.9[2], 8.1[6]	20.5[2]	47.6[4]	14.25[6]
Foaming stability (min)	71[1], 108[2], 90.6[3]	120[2]	78.3[4]	
Emulsifying stability (min)	4.1[1], 17.3[4]		27.5[4]	

[1] Khan et al. (2011) [2] Fabian and Ju (2011) [3] Yadav et al. (2011) [4] Tang et al. (2003) [5] Wang et al. (1999) [6] Chandi and Sogi (2007)

Antioxidant Compounds

An antioxidant is defined as any nutrient or chemical compound that can react with and neutralise free radicals to prevent the oxidative cell damage which leads to cancer (Schwab 2009). In rice, six classes of antioxidants have been identified, including γ-oryzanol, tocopherols and tocotrienols, phenolic acids, phytic acid, flavonoid and anthocyanins and pro-anthocyanidins (Goufo and Trindade 2014). The concentrations of these antioxidants are different, depending on the rice grain fraction (bran, embryo, endosperm and the husk), grain colour (brown, purple, black and red) and rice varieties. These components are mostly concentrated in the outer layers and the embryo of rice grain. Phenolic acids, however, have their highest concentration in the rice husk (Goufo and Trindade 2014). This section discusses the distribution and functions of antioxidants in rice bran and rice husk.

γ-oryzanol (steryl ferulate)

γ-oryzanol is concentrated mostly in the bran layer (3500–31640 mg/kg) and germ (532–1750 mg/kg) rather than in the rice endosperm (19–406.5 mg/kg) (Goufo and Trindade 2014). γ-oryzanol is a combination of ferulate ester and plant sterols. Cycloartenyl ferulate, 24-methylenecycloartanyl ferulate and campesteryl ferulate account for 80 per cent of γ-oryzanol. γ-oryzanol is reported as a highly effective antioxidant which decreases cholesterol absorption. The ability of treating hyperlipidemia and disorders of menopause of γ-oryzanol has also been studied. It can improve muscle mass and act as a skin-protecting agent (Patel and Naik 2004).

Antioxidant properties: The antioxidant capacity of γ-oryzanol is four times higher than vitamin E, because it has a ferulic acid structure which is a phenolic acid antioxidant (Hiramitsu and Armstrong 1991). The antioxidant activity of γ-oryzanol can be improved when it is mixed with linoleic acid at molar ratios of 1:100 or 1:250 (Xu and Godber 2001). 24-methylenecycloartanyl ferulate is the highest antioxidant component among the other components belonging to the γ-oryzanol group (Xu et al. 2001).

Cholesterol lowering: γ-oryzanol has been studied for its cholesterol lowering properties in both humans and animals. A study involving 66 persons who consumed brown rice extract containing mainly γ-amino butyric acid (GABA) and γ-oryzanol showed a significant decrease in total cholesterol level by 200 mg/dL (Patel and Naik 2004). Other research on cholesterolemic hamsters fed with 1 per cent γ-oryzanol showed a decrease in total cholesterol by 28 per cent (Rong et al. 1997). γ-oryzanol also has effects on cholesterol absorption by increasing the cholesterol conversion process that turns cholesterol into fecal bile acids and sterols. γ-oryzanol can lower the low-density lipoprotein cholesterol and total serum cholesterol (Kahlon et al. 1992, Nicolosi et al. 1991).

Other bioactivities: 300 mg γ-oryzanol was added to the diets of 40 peri-menopausal women for 4 to 8 weeks. Ninety per cent of them confirmed a menopausal reduction (Patel and Naik 2004). γ-oryzanol can also increase endorphin release and aid muscle development when used as a sports supplement.

Bioactivities of phenolic acids in rice bran

Phenolic compound, a substance containing a phenolic ring and an organic carboxylic acid function, is widely available in rice grain (Goufo et al. 2014). Twelve phenolic acids have been identified in rice, with most of them being present in the rice bran (177.6–319.9 mg/100 g) and rice husk (477.6 mg/100 g) (Goufo and Trindade 2014, Zhou et al. 2004). Ferulic acids account for up to 77 per cent of phenolic acids in rice bran, while p-coumaric acids are the most abundant phenolic acid in rice husk. Phenolic acids in rice can be in water-soluble forms, soluble conjugates and insoluble forms. They are in plant cell vacuoles (water-soluble form), sugar or low molecular components esterification (soluble conjugates) or covalently linked to cell-wall components, such as cellulose, lignin and protein (insoluble phenolic acids) (Adom et al. 2003, Arranz et al. 2010). In rice bran, up to 74 per cent of phenolic compounds are present in an insoluble form, strongly binding to the cell wall (Adom et al. 2003). These bound phenolics can be extracted by alkali, acid or enzymatic treatment. Alkali extraction (2M) has been reported as the most suitable method for extracting the bound phenolic compounds (Su et al. 2014). The soluble phenolic acids, however, remain in the rice bran after the defatting process due to their polar properties (Fabian et al. 2010). These compounds can be extracted by organic solvents for further food use. In pigmented rice varieties, such as red, purple or black rice, flavonoids are present in high levels and are recognized as high value bioactive phenolic components (Goufo and Trindade 2014).

Phenolic compounds in rice bran have been recognized as antioxidants by various researchers (Arab et al. 2011, Byungrok et al. 2011, Chen et al. 2012, Pitija et al. 2013). The antioxidant ability of phenolic compounds is four times higher than α-tocopherol but lower than that for anthocyanins (Yawadio et al. 2007). In a study by Kondo et al. (2011), ethanol extraction from jasmine rice bran, which contains a high ratio of soluble phenolic acids, showed the highest level of antibacterial activity.

Bioactivities of phosphorus compounds in rice waste

Phosphorus compounds can be in the form of phytate phosphorus, cellular phosphorus and inorganic phosphorus (Frontela et al. 2008, Goufo and Trindade 2014, Liu et al. 2005, Ren et al. 2007). In rice bran, phytate phosphorus compounds account for 60–70 per cent of total phosphorus, followed by cellular phosphorus (20–30 per cent) and inorganic phosphorus (2–4 per cent). In the rice endosperm, 60 per cent of total phosphorus is present as cellular phosphorus, while phytate phosphorus accounts for 37 per cent. In rice husks, 82 per cent of phosphorus is phytate phosphorus, 14 per cent is inorganic phosphorus and the rest is cellular phosphorus (Goufo and Trindade 2014).

In rice bran, 65–73 per cent of the total phosphorus is in the form of phytic acid, which is also known as myo-inositol 1,2,3,4,5,6-hexakis-dihydrogen phosphate or IP6 (Goufo and Trindade 2014). Phytic acid occurs at a level of 5–6 per cent in rice bran, depending on cultivation conditions and rice variety (Kasim and Edwards 1998, Liu et al. 2005). In the original pH of rice bran, phytic acid is negatively charged that, allows positive binding to metal cations, protein and starch (Fuh and Chiang 2001, Rimbach et al. 2008). Forming complexes with phytic acid and protein, mineral and starch could alter their solubility, functionality, digestibility and absorption. Phytic acid, therefore, is recognized as an anti-nutrient factor (Rickard and Thompson 1997). However, recent studies reveal the potential beneficial effects of phytic acid on human health as a bioactive compound (Canan et al. 2011).

The antioxidant ability of phytic acid has also been recognized in a number of studies. Norazalina et al. (2010) reported that 71 per cent of carcinogen-treated rats supplied with drinking water containing 0.5 per cent rice bran phytic acid were found to have no signs of colon cancer after 8 weeks. Phytic acid can inhibit the formation of hydroxyl radicals by binding with Fe^{2+} to form the iron chelate that suppresses the oxidative reaction. This mechanism has been used to explain the ability of phytic acid in reducing the risk of colon cancer (Graf and Eaton 1990). Diabetic KK mice fed with a dietry supplement of 1 per cent of phytic acid over 8 weeks were found to have a reduction in blood glucose (Lee et al. 2006). However, the mechanism by which phytic acid lowers blood glucose has been remained unclear. The strong antioxidant ability of phytic acid can reduce neurodegeneration, which is associated with Parkinson's disease (Xu et al. 2008). As it could prevent lipid oxidation, phytic acid has been used to inhibit the development of an over-warm flavour of beef and chicken meat (Lee et al. 1998, Soares et al. 2004).

Extraction and Isolation of Bioactive Compounds from Rice Waste

Extraction method for dietary fiber (DF)

As fiber includes a wide range of complex compounds, many different extraction methods are potentially applicable (Codex Alimentarius International Food Standard 2013, Dhingra et al. 2012). DFs can be extracted as whole total fiber, soluble or insoluble fiber using selected extraction solvents (Daou and Zhang 2012, Maphosa and Jideani 2016). The selection of an appropriate extraction method and solvent types depends on the end products desired and the intended applications (Daou and Zhang 2012). The extraction usually starts with elimination of undesirable components, mostly starch and protein, followed by collection of the remaining DFs through precipitation and centrifugation.

Acidic extraction

The DFs can be extracted by soaking ground raw material in sulphurous acid. The soluble protein in the sulphurous solvent is separated by pH precipitation or ultrafiltration. The starch and other contaminations are removed by gravimetric separation (Ramírez et al. 2009) and the soluble DF is finally collected. This method has been reported to be time consuming, costly and environment-unfriendly (Maphosa and Jideani 2016).

Enzymatic and alkali extraction

Alkali solvents can be used to extract DFs by using sodium hydroxide and calcium hydroxide (Daou and Zhang 2013, Wan et al. 2014). Defatted rice bran was treated with 0.15N NaOH for 64.3 minutes in a ratio of 1:5. Starch was hydrolyzed by amylaze and amyl glucosidase. The soluble fiber was collected by ethanol precipitation and centrifugation, with the residue being insoluble fiber. The highest yields of total DFs and soluble DF were 31 per cent and 2.69 per cent, respectively (Daou and Zhang 2013). Soluble DF (7.89 per cent) in rice bran was extracted using a 1:30 ratio of 2 per cent Ca(OH) and defatted rice bran at 60°C over 4 hours (Wan et al. 2014). The starch was removed by amylase at 90°C for 15 minutes before submergence in 2 per cent $Ca(OH)_2$. Another study extracted fiber from rice husk by using NaOH (4 per cent), followed by a bleaching process using NaCl 1.7 wt per cent in acetic acid buffer and an acid hydrolysis (H_2SO_4 10 M) (Johar et al. 2012). The concentration of final fiber reached 96 per cent, but the severe conditions of this method could limit the potential uses of the extracted fiber. The combination of enzymes and chemicals in DF extraction could increase the extractability and shorten the extraction time (Maphosa and Jideani 2016). This method is more environmental friendly than the chemical method.

Physical method

Insoluble DF can be isolated based on its swelling property. At room temperature, fiber has a higher swelling property than starch. Insoluble DF and starch can therefore be fractionated by washing through 53 to 90 μm sieves. This method could potentially increase the yield of insoluble DF by 10–15 per cent (Dalgetty and Baik 2003). However, this method requires the use of additional chemicals or enzymes for the extraction of protein.

The extraction method is ideal as it is cost effective, easy to perform and environment-friendly. Thus, water is usually preferred in combination with enzymes for DF extractions. However, the effectiveness of this extraction method depends on the properties of the raw material and the intended application of the DFs. Recently, enzymatic and alkali extractions have been the most popular methods for both soluble and insoluble DFs extraction in rice bran.

Extraction of rice bran protein

A number of methodologies have been developed for protein extraction, including physical, alkaline and enzymatic methods. However, the process still faces technical challenges. As rice bran is a complex mixture of four protein fractions, four different solvents for extraction are used (Betschart et al. 1977, Hamada 1997). This results in a complicated process with low efficiency in large-scale extraction (Shih 2003). Hamada (1997) reported that rice bran protein may have poor solubility. The majority of rice bran proteins are distributed inside the cells. Thus the cell walls need to be disrupted to maximize the extraction of protein (Fabian and Ju 2011). The most efficient methods for extraction of protein from rice bran are based on the use of highly alkaline and dissociating solvents, but this approach could potentially have negative effects on the nutritional properties of the extracted protein (Hamada 1997, Phillips and Finley 1989).

Physical methods

In food processing, physical methods are preferred to chemical or enzymatic methods due to their freedom from the use of additional chemical compounds and minimal unnatural alterations of the protein during the extraction process. The physical forces can disrupt cell walls and release the protein for extraction. Early research has showed that 38 per cent of protein can be extracted from rice bran by using colloid milling, followed by homogenization at room temperature (Anderson and Guraya 2001). Tang (2002) applied freeze-thawing, sonication (150–750 W, 5 minutes), high-speed blending (soaking for 16 hours, blending 5 minutes) and high pressure (200–800 MPa), to extract protein from heat-stabilized BRM with distilled water at room temperature. It is evident that the physical methods can extract rice bran protein but the protein yield is quite low (9.9–14.3 per cent of total protein content).

Alkali methods

Alkaline extraction is the most common and convenient method for protein extraction from BRM (Fabian and Ju 2011, Prakash and Ramaswamy 1996, Saunders 1985, Shih 2003). Fabian (2011) showed that 80 per cent of protein from rice bran was extracted at pH 12 and 72 per cent was extracted at pH 11 (Prakash and Ramanatham 1994). However, in seed protein extraction, there are a number of reports that alkali solvent at high pH (10–12.2) could negatively change nutritional properties of protein and create toxic compounds, such as lysinoalanine (Fabian and Ju 2011, Fennema 1996). This drawback limits the use of highly alkaline conditions in protein extraction for food. Using mild alkaline conditions to extract BRM protein, however, results in low extraction yield. For example, a study by Gupta (2008) showed that only 21.2 per cent of protein was extracted from rice bran at pH 9.5. Thus, it is necessary to study and develop other effective methods to extract rice bran protein, or combine other methods with low alkaline extraction.

Enzymatic methods

The enzymatic extraction method is preferred to chemical extraction because it requires milder conditions and is legally acceptable for food (Shih 2003). Rice bran can be treated with single or multiple enzymes, with the protein then being extracted by water or mild alkali solvent, followed by isoelectric precipitation (Fabian and Ju 2011). Hamada (1999) reported that the yield of extracted protein at pH 8.0 and 50°C increased from 1–20 per cent to 63–90 per cent following treatment with protease. However, the recovered proteins from this method are hydrolyzed types which do not maintain their original structure or functional properties (Shih 2003). Wang (1999) used phytase and xylanase to hydrolyze phytic acid and xylan in rice bran to increase the subsequent extraction at pH 10. Phytic acid can reduce protein solubility, while xylan is a common cell wall component. The treatment increases the protein yield from 20–40 per cent to 52–74 per cent (Wang et al. 1999). The maximum protein yield achieved is comparable to that achieved with the high alkaline extraction method, although this study treated the rice bran under moderately alkaline conditions (pH 10) for one hour at room temperature. Enzymes were used to hydrolyze the starch and other components in the cell wall to improve the protein extraction. Tang (2002) treated rice bran with amylase at pH 7 and 45°C and increased the protein yield slightly from 10–42 per cent to 10–52 per cent. Other carbohydrases, such as cellulase or glucanase, were also tested, but the protein extraction yield only increased by between 2 and 15 per cent (Tang et al. 2003).

Currently, subcritical water is used to extract rice bran protein and give a higher protein yield than physical methods. Under supercritical conditions, water contains H^+ and OH^- turn it into a better solvent, dissolving the protein more readily. Between 70 and 96 per cent of protein can be extracted from rice bran by using subcritical water (Fabian and Ju 2011). However, due to a high temperature treatment (200–250°C) during this process, the protein is denatured. In addition, this method also requires specific and costly equipment.

Due to the natural combination between protein and other components, such as xylan, starch, cellulose and phytate, the protein extraction yield from rice bran is improved by hydrolyzing these substances. However, although the protein solubility could be driven by interactions between protein and other rice bran components, the mechanism involved remain unclear. It is, therefore, important to conduct further studies on the factors that impact the solubility of rice bran protein so as to develop effective extraction methods.

Extraction of Other Antioxidants from Rice Waste

γ-oryzanol extraction and purification

Since γ-oryzanol is a high value bioactive compound, the research on extraction and isolation of γ-oryzanol is of great interest to food and pharmaceutical specialists. Narayan et al. reported that more than 40 patents on γ-oryzanol isolation have been globally granted (Narayan et al. 2006). γ-oryzanol can generally be extracted directly from rice bran by supercritical CO_2 or be isolated from crude rice bran oil by a solvent (Imsanguan et al. 2008). Zhimin Xu and Godber (2000) reported that 5.39 mg γ-oryzanol per gram of rice bran was extracted in supercritical fluid extraction at 50°C and 680 atm for 25 min. γ-oryzanol can be extracted using a solvent mixture of hexane and isopropanol (1:1) at 60°C for 45–60 min, but this method has given only one quarter of the yield of that obtained from supercritical fluid extraction.

γ-oryzanol can be extracted from rice bran oil using an alkali followed by distilled HCL hydrolysis. The residue is then refluxed, using NaOH in methanol and then filtered. The residue is then dissolved in diethyl ether and treated with disodium trioxide to remove fatty acids. γ-oryzanol is finally washed with water and dried (Patel and Naik 2004). In the rice bran oil industry, γ-oryzanol can be isolated from the by-product of oil refining—soapstock. γ-oryzanol (85 per cent) in soapstock can be recovered by a two-step leaching process. CO_2 is used in the first leaching followed by methanol and ethanol to wash out the weak alkali salts (Narayan and Barhate 2006). The extracted oryzanol can be purified using ketones and/or alcohol to provide highly concentrated crystallised oryzanol (Patel and Naik 2004).

Extraction of Phenolic Compounds from Rice Bran

Phenolic compounds can be extracted by a solvent. Extraction yields differ depending on the type of solvent, solvent concentration time of extraction and physical treatment. Phenolic acids can be extracted by ethanol, acetone, methanol, HCl, distilled water or NaOH. Soluble phenolic acids can be extracted by using 40 per cent acetone, followed by 40 per cent ethanol and methanol for the highest extractability (21.6

per cent) (Jun et al. 2012). Subcritical water was used to extract phenolic acids from rice bran with 30 minutes of heating and holding for 5 minutes at 175ºC, but the yield was low (0.2 per cent) (Fabian et al. 2010). Insoluble phenolic compounds can be extracted by a mixture of solvent including NaOH, HCl and ethyl acetate. Rice bran can be hydrolyzed by NaOH followed by HCL neutralisation. Ethyl acetate is then used to extract the insoluble phenolic acids. However, this method requires intensive separating steps to isolate the phenolic compounds, as the supernatant contains a mixture of different compounds, such as protein, fiber, lipid and phosphorus substances (Wang et al. 2015). Goufo and Trindade (2014) reported that a mixture of acetone, water and acetic acid in the ratio of 70-29.5-0.5, respectively, is often applied to extract the proanthocyanidin in the pigment of rice bran. Physical treatments have a strong effect on the extractability of phenolic compounds in rice bran. Rice bran can be processed by vortex extraction, ultrasonic extraction, high pressured extraction and high temperature extraction to increase the extractability of phenolic compounds (Goufo et al. 2014).

Extraction of Phytic Acids from Rice Bran

Phytic acids have been reported as antibiotic compounds since 1990. However, studies on the extraction and purification of these components are limited. The recent phytic acids extraction process is mostly based on qualitative testing (Goufo et al. 2014). As with other bioactive substances in rice grains, phytic acid is usually isolated by solvent extraction followed by precipitation. Phytic acids can also be isolated by shaking with 1.8 per cent (w/v) HCl for 2 hours. The extracted phytic acids are then precipitated by 0.2 per cent $FeCl_3$, followed by washing with NaOH, HCl and de-ionized water. Concentrated phytic acids in rice bran are in the range of 38–60 mg/g (Wang et al. 2011). Other research extracted and isolated phytic acids in rice bran using HCl, $Ca(OH)_2$ and NaOH. Rice bran is firstly submerged in 0.1 M HCl for 30 min and then treated with microwave and ultrasonic treatment. $Ca(OH)_2$ (10 per cent) was then added along with 1 M NaOH, to precipitate the calcium phytic acids. This compound was then acidified with 001 x 7 resin to produce phytic acid. The maximum extraction rate of this method was reported to be 6.75 per cent (Xiong et al. 2012). Phytic acids have also been extracted by using HCl (5 per cent w/w) at pH 1.0 by shaking, followed by centrifugation. Freeze drying has been used to concentrate extracted phytic acids (Norazalina et al. 2010). However, studies on the extraction and purification of phytic acids in rice grain have been limited and need to be the focus of further research.

Conclusion

Rice waste contains valuable functional compounds. Protein, dietary fiber and antioxidants are distributed mostly in rice bran and show their potential health benefits in reference to diabetes, obesity, gastronomy and cancer. γ-oryzanol, phenolic compounds and phytic acids are among the most valuable functional substances in human health benefits. The potentially useful bioactive compounds can be extracted from rice waste based on physical, chemical and biological principles. Although various extraction methods have been developed, their efficiency and yield need to be improved.

References

Abdul-Hamid, A. and Y. S. Luan. 2000. Functional properties of dietary fiber prepared from defatted rice bran. Food Chemistry 68(1): 15–19. doi:10.1016/S0308-8146(99)00145-4.

Adom, K. K., R. H. Liu and M. E. Sorrells. 2003. Phytochemical profiles and antioxidant activity of wheat varieties. Journal of Agricultural and Food Chemistry 51(26): 7825–7834. doi:10.1021/jf0304041.

Alemdar, A. and M. Sain. 2008. Biocomposites from wheat straw nanofibers: Morphology, thermal and mechanical properties. Composites Science and Technology 68(2): 557–565. doi:10.1016/j.compscitech.2007.05.044.

Anderson, A. K. and H. S. Guraya. 2001. Extractability of protein in physically processed rice bran. Journal of the American Oil Chemists' Society 78(9): 969–972. doi:10.1007/s11746-001-0373-1.

Arab, F., I. Alemzadeh and V. Maghsoudi. 2011. Determination of antioxidant component and activity of rice bran extract. Scientia Iranica 18(6): 1402–1406. doi:http://dx.doi.org/10.1016/j.scient.2011.09.014.

Arranz, S., J. M. Silván and F. Saura-calixto. 2010. Nonextractable polyphenols, usually ignored, are the major part of dietary polyphenols: A study on the Spanish diet. Molecular Nutrition & Food Research 54(11): 1646–1658. doi:10.1002/mnfr.200900580.

Bäckhed, F., J. K. Manchester, C. F. Semenkovich and J. I. Gordon. 2007. Mechanisms underlying the resistance to diet-induced obesity in germ-free mice. Proceedings of the National Academy of Sciences of the United States of America 104(3): 979–984. doi:10.1073/pnas.0605374104.

Bao, J. 2012. Nutraceutical Properties and Health Benefits of Rice Cereals and Pulses. pp. 37–64. Wiley-Blackwell.

Betschart, A. A., R. Y. Fong and R. M. Saunders. 1977. Rice by-products: Comparative extraction and precipitation of nitrogen from U.S. and spanish bran and germ. Journal of Food Science 42(4): 1088–1093. doi:10.1111/j.1365-2621.1977.tb12673.x.

Byungrok, M., A. M. McClung and C. Ming-Hsuan. 2011. Phytochemicals and antioxidant capacities in rice brans of different color. (Report). Journal of Food Science 76(1): c117–c126.

Canan, C., F. T. L. Cruz, F. Delaroza, R. Casagrande, C. P. M. Sarmento, M. Shimokomaki and E. I. Ida. 2011. Studies on the extraction and purification of phytic acid from rice bran. Journal of Food Composition and Analysis 24(7): 1057–1063. doi:http://dx.doi.org/10.1016/j.jfca.2010.12.014.

Champagne, E. T. 2004. Rice: Chemistry and Technology. St. Paul, Minn: American Association of Cereal Chemists.

Champagne, E. T., D. F. Wood, B. O. Juliano and D. B. Bechtel. 2004. The rice grain and its gross composition. *In*: E. T. Champange (ed.). Rice Chemistry and Technology. USA: American Association of Cereal Chemists.

Chandi, G. K. and D. S. Sogi. 2007. Functional properties of rice bran protein concentrates. Journal of Food Engineering 79(2): 592–597. doi:10.1016/j.jfoodeng.2006.02.018.

Chen, X. Q., N. Nagao, T. Itani and K. Irifune. 2012. Anti-oxidative analysis, and identification and quantification of anthocyanin pigments in different coloured rice. Food Chemistry 135(4): 2783–2788. doi:http://dx.doi.org/10.1016/j.foodchem.2012.06.098.

Cheng, H. -H., H. -Y. Huang, Y. -Y. Chen, C. -L. Huang, C. -J. Chang, H. -L. Chen and M. -H. Lai. 2010. Ameliorative effects of stabilized rice bran on type 2 diabetes patients. Annals of Nutrition and Metabolism 56(1): 45–51. doi:10.1159/000265850.

Cho, S. and P. Samuel. 2009. Fiber Ingredients: Food Applications and Health Benefits. Boca Raton: CRC Press.

Codex Alimentarius International Food Standard. 2013. Guidelines on Nutrition Labelling (Vol. CAC/GL 2-1985). Rome: FAO.

Dalgetty, D. D. and B. -K. Baik. 2003. Isolation and characterization of cotyledon fibers from peas, lentils and chickpeas. Cereal Chemistry Journal 80(3): 310–315. doi:10.1094/CCHEM.2003.80.3.310.

Daou, C. and H. Zhang. 2012. Study on functional properties of physically modified dietary fibers derived from defatted rice bran. 4(9). doi:10.5539/jas.v4n9p85.

Daou, C. and H. Zhang. 2013. Optimization of processing parameters for extraction of total, insoluble and soluble dieatary fibers of deffatted rice bran. Food Science and Nutrition 28(8): 13.

Dhingra, D., M. Michael, H. Rajput and R. T. Patil. 2012. Dietary fiber in foods: A review. Journal of Food Science and Technology 49(3): 255–266. doi:10.1007/s13197-011-0365-5.

Fabian, C., N. Y. Tran-thi, N. S. Kasim and Y. H. Ju. 2010. Release of phenolic acids from defatted rice bran by subcritical water treatment. Journal of the Science of Food and Agriculture 90(15): 2576–2581. doi:10.1002/jsfa.4123.

Fabian, C. and Y. -H. Ju. 2011. A review on rice bran protein: Its properties and extraction methods. Critical Reviews in Food Science and Nutrition 51(9): 816–827. doi:10.1080/10408398.2010. 482678.

Fennema, O. R. 1996. Food Chemistry Vol. 76. New York: Marcel Dekker.

Figuerola, F., M. A. L. Hurtado, A. M. A. Estévez, I. Chiffelle and F. Asenjo. 2005. Fiber concentrates from apple pomace and citrus peel as potential fiber sources for food enrichment. Food Chemistry 91(3): 395–401. doi:10.1016/j.foodchem.2004.04.036.

Friedman, M. 2013. Rice brans, rice bran oils, and rice hulls: Composition, food and industrial uses, and bioactivities in humans, animals, and cells. Journal of Agricultural and Food Chemistry 61(45): 10626–10641. doi:10.1021/jf403635v.

Frontela, C., F. J. García-Alonso, G. Ros and C. Martínez. 2008. Phytic acid and inositol phosphates in raw flours and infant cereals: The effect of processing. Journal of Food Composition and Analysis 21(4): 343–350. doi:http://dx.doi.org/10.1016/j.jfca.2008.02.003.

Fuh, W. S. and B. H. Chiang. 2001. Dephytinisation of rice bran and manufacturing a new food ingredient. Journal of the Science of Food and Agriculture 81(15): 1419–1425. doi:10.1002/ jsfa.962.

Gao, Y. and J. Yue. 2012. Dietary Fiber and Human Health. Cereals and Pulses. pp. 261–271. Wiley-Blackwell.

Goufo, P., J. Pereira, J. Moutinho-Pereira, C. M. Correia, N. Figueiredo, C. Carranca and H. Trindade. 2014. Rice (*Oryza sativa* L.) phenolic compounds under elevated carbon dioxide (CO2) concentration. Environmental and Experimental Botany 99: 28–37. doi:http://dx.doi. org/10.1016/j.envexpbot.2013.10.021.

Goufo, P. and H. Trindade. 2014. Rice antioxidants: Phenolic acids, flavonoids, anthocyanins, proanthocyanidins, tocopherols, tocotrienols, γ-oryzanol and phytic acid Vol. 2. pp. 75–104.

Graf, E. and J. W. Eaton. 1990. Antioxidant functions of phytic acid. Free Radical Biology and Medicine 8(1): 61–69. doi:10.1016/0891-5849(90)90146-A.

Gupta, S., G. K. Chandi and D. S. Sogi. 2008. Effect of extraction temperature on functional properties of rice bran protein concentrates. International Journal of Food Engineering 4(2): 8–19. doi:10.2202/1556-3758.1165.

Hamada, J. S. 1997. Characterization of protein fractions of rice bran to devise effective methods of protein solubilization. Cereal Chemistry 74(5): 662–668. doi:10.1094/CCHEM.1997.74.5.662.

Hamada, J. S. 1999. Use of proteases to enhance solubilization of rice bran proteins. Journal of Food Biochemistry 23(3): 307–321. doi:10.1111/j.1745-4514.1999.tb00022.

Harper, W. J., S. Jiamyangyuen and V. Srijesdaruk. 2005. Extraction of rice bran protein concentrate and its application in bread. Songklanakarin Journal of Science and Technology 27(1): 55–64.

Harris, P. J., C. M. Triggs, A. M. Roberton, M. E. Watson and L. R. Ferguson. 1996. The adsorption of heterocyclic aromatic amines by model dietary fibers with contrasting compositions. Chemico-Biological Interactions 100(1): 13–25. doi:http://dx.doi.org/10.1016/0009-2797(95)03682-2.

Hattori, Y., K. Nagai and M. Ashikari. 2011. Rice growth adapting to deepwater. Current Opinion in Plant Biology 14(1): 100–105. doi:10.1016/j.pbi.2010.09.008.

Helm, R. M. and A. W. Burks. 1996. Hypoallergenicity of rice protein. Cereal Foods World 41(11): 839–843.

Hettiarachchy, N. S., K. Sato, M. R. Marshall and A. Kannan. 2012. Food Proteins and Peptides: Chemistry, Functionality, Interactions, and Commercialization. Boca Raton, FL: CRC Press.

Hiramitsu, T. and D. Armstrong. 1991. Preventive effect of antioxidants on lipid peroxidation in the retina. Ophthalmic Research 23(4): 196–203. doi:10.1159/000267103.

Howarth, N. C., E. Saltzman and S. B. Roberts. 2001. Dietary fiber and weight regulation. Nutrition Reviews 59(5): 129–139. doi:10.1111/j.1753-4887.2001.tb07001.

Imsanguan, P., A. Roaysubtawee, R. Borirak, S. Pongamphai, S. Douglas and P. L. Douglas. 2008. Extraction of α-tocopherol and γ-oryzanol from rice bran. LWT—Food Science and Technology 41(8): 1417–1424. doi:http://dx.doi.org/10.1016/j.lwt.2007.08.028.

Jansen, M. C. J. F., H. B. Bueno-de-Mesquita, L. Räsänen, F. Fidanza, A. Menotti, A. Nissinen and D. Kromhout. 1999. Consumption of plant foods and stomach cancer mortality in the seven countries study. Is grain consumption a risk factor? Nutrition and Cancer 34(1): 49–55. doi:10.1207/S15327914NC340107.

Jiamyangyuen, S., W. J. Harper, V. Srijesdaruk and K. Kumthonglang. 2005. Study of extraction and functional properties of rice bran protein concentrate. Milchwissenschaft-milk Science International 60(2): 192–195.

Johar, N., I. Ahmad and A. Dufresne. 2012. Extraction, preparation and characterization of cellulose fibers and nanocrystals from rice husk. Industrial Crops and Products 37(1): 93–99. doi:http://dx.doi.org/10.1016/j.indcrop.2011.12.016.

Jones, K. L., M. Horowitz, B. I. Carney, J. M. Wishart, S. Guha and L. Green. 1996. Gastric emptying in early noninsulin-dependent diabetes mellitus. The Journal of Nuclear Medicine 37(10): 1643–1648.

Jun, H. i., G. s. Song, E. i. Yang, Y. Youn and Y. s. Kim. 2012. Antioxidant activities and phenolic compounds of pigmented rice bran extracts. Journal of Food Science 77(7): C759–C764. doi:10.1111/j.1750-3841.2012.02763.

Kahlon, T. S. 2009. Rice bran: Production, composition, functionality and food applications, physiological benefit. *In*: S. S. Cho and P. Samuel (eds.). Fiber Ingredients: Food Applications and Health Benefits. Boca Raton: CRC press.

Kahlon, T. S., R. M. Saunders, F. I. Chow, M. M. Chiu and A. A. Betschart. 1990. Influence of rice bran, oat bran, and wheat bran on cholesterol and triglycerides in hamsters. Cereal Chemistry 67(5): 439–443.

Kahlon, T. S., R. M. Saunders, R. N. Sayre, F. I. Chow, M. M. Chiu and A. A. Betschart. 1992. Cholesterol-lowering effects of rice bran and rice bran oil fractions in hypercholesterolemic hamsters. Cereal Chemistry 69(5): 485–489.

Kahlon, T. S. and F. I. Chow. 2000. Lipidemic response of hamsters to rice bran, uncooked or processed white and brown rice, and processed corn starch. Cereal Chemistry 77(5): 673–678.

Kahlon, T. S. and F. I. Chow. 2001. Rice bran—production, composition, availability, healthful properties, safety, and food applications. *In*: S. S. Cho and M. L. Dreher (eds.). Handbook of Dietary Fiber New York: Taylor & Francis.

Kasim, A. B. and H. M. Edwards. 1998. The analysis for inositol phosphate forms in feed ingredients. Journal of the Science of Food and Agriculture 76(1): 1–9. doi:10.1002/(SICI)1097-0010(199801)76:1<1::AID-JSFA922>3.0.CO 2-9.

Kennedy, G. and B. Burlingame. 2003. Analysis of food composition data on rice from a plant genetic resources perspective. Food Chemistry 80(4) 589–596. doi:10.1016/S0308-8146(02)00507-1.

Khan, S. H. 2011. Quality evaluation of rice bran protein isolate-based weaning food for preschoolers. International Journal of Food Sciences and Nutrition 62(3): 280–288. doi:10.3109/09637486.2010.529802.

Khan, S. H., M. S. Butt, M. K. Sharif, A. Sameen, S. Mumtaz and M. T. Sultan. 2011. Functional properties of protein isolates extracted from stabilized rice bran by microwave, dry heat, and parboiling. Journal of Agricultural and Food Chemistry 59(6): 2416–2420. doi:10.1021/jf104177.

Kondo, S., R. Teongtip, D. Sri Chana and A. Itharat. 2011. Antimicrobial activity of rice bran extracts for diarrheal disease. Journal of the Medical Association of Thailand 94(7): 117–138.

Landers, P. S. and B. R. Hamaker. 1994. Antigenic properties of albumin, globulin, and protein-concentrate fractions from rice bran. Cereal Chemistry 71(5): 409–411.

Lasztity, R. 1996. The Chemistry of Cereal Proteins. Boca Raton: CRC Press.

Lee, B. J., D. G. Hendricks and D. P. Cornforth. 1998. Antioxidant effects of carnosine and phytic acid in a model beef system. Journal of Food Science 63(3): 394–398. doi:10.1111/j.1365-2621.1998.tb15750.

Lee, S. -H., H. -J. Park, H. -K. Chun, S. -Y. Cho, S. -M. Cho and H. S. Lillehoj. 2006. Dietary phytic acid lowers the blood glucose level in diabetic KK mice. Nutrition Research 26(9): 474–479. doi:http://dx.doi.org/10.1016/j.nutres.2006.06.017.

Liu, Z., F. Cheng and G. Zhang. 2005. Grain phytic acid content in japonica rice as affected by cultivar and environment and its relation to protein content. Food Chemistry 89(1): 49–52. doi:http://dx.doi.org/10.1016/j.foodchem.2004.01.081.

Ludwig, D. S., M. A. Pereira, C. H. Kroenke, J. E. Hilner, L. Van Horn, M. L. Slattery and J. D. R. Jacobs. 1999. Dietary fiber, weight gain, and cardiovascular disease risk factors in young adults. JAMA 282(16): 1539–1546. doi:10.1001/jama.282.16.1539.

Luh, B. S. 1991. Rice. New York, NY: Van Nostrand Reinhold.

Macagnan, F. T., L. P. da Silva and L. H. Hecktheuer. 2016. Dietary fibre: The scientific search for an ideal definition and methodology of analysis, and its physiological importance as a carrier of bioactive compounds. Food Research International 85: 144–154. doi:10.1016/j.foodres.2016.04.032.

Maphosa, Y. and V. A. Jideani. 2016. Dietary fiber extraction for human nutrition—A review. Food Reviews International 32(1): 98–115. doi:10.1080/87559129.2015.1057840.

Matano, C., T. M. Meiswinkel and V. F. Wendisch. 2014. Amino acid production from rice straw hydrolyzates. In: R. R. Watson, V. R. Preedy and S. Zibadi (eds.). Wheat and Rice in Disease Prevention and Health. Amsterdam: Elsevier Inc.

Matsuda, T., R. Nomura, M. Sugiyama and R. Nakamura. 1991. Immunochemical studies on rice allergenic proteins. Agricultural and Biological Chemistry 55(2): 509–513.

Miyoshi, H., T. Okuda, Y. Oi and H. Koishi. 1986. Effects of rice fiber on fecal weight, apparent digestibility of energy, nitrogen and fat, and degradation of neutral detergent fiber in young men. Journal of Nutritional Science and Vitaminology 32(6): 581–589.

Müller-Fischer, N. 2013. Nutrient-focused Processing of Rice Agricultural Sustainability. Bühler, Uzwil, Switzerland: Elsevier Inc.

Narayan, A. and R. Barhate. 2006. Extraction and purification of oryzanol from rice bran oil and rice bran oil soapstock. JAOCS, Journal of the American Oil Chemists' Society 83(8): 663–670.

Narayan, A. V., R. S. Barhate and K. S. M. S. Raghavarao. 2006. Extraction and purification of oryzanol from rice bran oil and rice bran oil soapstock. Journal of the American Oil Chemists' Society 83(8): 663–670. doi:10.1007/s11746-006-5021-2.

Nicolosi, R. J., L. M. Austrian and D. M. Hegsted. 1991. Rice bran oil lowers serum total and low density lipoprotein cholesterol and apo B levels in nonhuman primates. Atherosclerosis 88(2): 133–142. doi:10.1016/0021-9150(91)90075-E.

Norazalina, S., M. E. Norhaizan, I. Hairuszah and M. S. Norashareena. 2010. Anticarcinogenic efficacy of phytic acid extracted from rice bran on azoxymethane-induced colon carcinogenesis in rats. Experimental and Toxicologic Pathology 62(3): 259–268. doi:10.1016/j.etp.2009.04.002.

Panel on Macronutrients. 2005. Dietary Reference Intakes for Energy, Carbohydrate, Fiber, Fat, Fatty Acids, Cholesterol, Protein and Amino Acids (Macronutrients). Retrieved from Washington, D.C.: http://www.nap.edu/catalog/10490/dietary-reference-intakes-for-energy-carbohydrate-fiber-fat-fatty-acids-cholesterol-protein-and-amino-acids-macronutrients.

Patel, M. and S. N. Naik. 2004. Gamma-oryzanol from rice bran oil—A review. Journal of Scientific & Industrial Research 63(7): 9.

Phillips, R. D. and J. W. Finley. 1989. Protein Quality and the Effects of Processing. Vol. 29. New York: M. Dekker.

Pitija, K., M. Nakornriab, T. Sriseadka, A. Vanavichit and S. Wongpornchai. 2013. Anthocyanin content and antioxidant capacity in bran extracts of some Thai black rice varieties. International Journal of Food Science & Technology 48(2): 300–308. doi:10.1111/j.1365-2621.2012.03187.

Prakash, J. and G. Ramanatham. 1994. Effect of stabilization of rice bran on the extractability and recovery of proteins. Nahrung Food 38(1): 87–95.

Prakash, J. and H. S. Ramaswamy. 1996. Rice bran proteins: Properties and food uses. Critical Reviews in Food Science and Nutrition 36(6): 537–552. doi:10.1080/10408399609527738.

Qureshi, A. A., S. A. Sami and F. A. Khan. 2002. Effects of stabilized rice bran, its soluble and fiber fractions on blood glucose levels and serum lipid parameters in humans with diabetes mellitus Types I and II. The Journal of Nutritional Biochemistry 13(3): 175–187. doi:10.1016/S0955-2863(01)00211.

Ramírez, E. C., D. B. Johnston, A. J. McAloon and V. Singh. 2009. Enzymatic corn wet milling: Engineering process and cost model. Biotechnology for Biofuels 2(1): 1–9. doi:10.1186/1754-6834-2-2.

Ren, X. -L., Q. -L. Liu, H. -W. Fu, D. -x. Wu and Q. -Y. Shu. 2007. Density alteration of nutrient elements in rice grains of a low phytate mutant. Food Chemistry 102(4): 1400–1406. doi:http://dx.doi.org/10.1016/j.foodchem.2006.05.065.

Rickard, S. E. and L. U. Thompson. 1997. Interactions and biological effects of phytic acid. Interactions and Biological Effects of Phytic Acid. pp. 294–312. Washington, DC (USA): American Chemical Society.

Rimbach, G., J. Moehring, J. Pallauf, K. Kraemer and A. M. Minihane. 2008. Effect of dietary phytate and microbial phytase on mineral and trace element bioavailability—A literature review. Current Topics in Nutraceutical Research 6(3): 131–144.

Rong, N., L. Ausman and R. Nicolosi. 1997. Oryzanol decreases cholesterol absorption and aortic fatty streaks in hamsters. Lipids 32(3): 303–309. doi:10.1007/s11745-997-0037-9.

Saunders, R. M. 1985. Rice bran: Composition and potential food uses. Food Reviews International 1(3): 465–495. doi:10.1080/87559128509540780.

Schiere, J. B., A. L. Joshi, A. Seetharam, S. J. Oosting, A. V. Goodchild, B. Deinum and H. Van keulen. 2004. Grain and straw for whole plant value: Implications for crop management and Genetic improvement strategies. Expl. Agric. 40: 277–294.

Schwab, M. 2009. Encyclopedia of Cancer: Springer Berlin Heidelberg.

Shih, F. F. 2003. An update on the processing of high-protein rice products. Die Nahrung 47(6): 420–424. doi:10.1002/food.200390093.

Shih, F. F. 2004. Rice proteins. In: E. T. Champagne (ed.). Rice: Chemistry and Technology. Minnesota USA: American Association of Cereal Chemist, Inc.

Shoji, Y., T. Mita, M. Isemura, T. Mega, S. Hase, S. Isemura and Y. Aoyagi. 2001. A fibronectin-binding protein from rice bran with cell adhesion activity for animal tumor cells. Bioscience, Biotechnology and Biochemistry 65(5): 1181–1186. doi:10.1271/bbb.65.1181.

Silva, C. R., J. E. D. De Oliveira, R. A. H. G. De Souza and H. C. Silva. 2005. Effect of a rice bran fiber diet on serum glucose levels of diabetic patients in Brazil. Archivos Latinoamericanos de Nutricion 55(1): 23–27.

Soares, A. L., R. Olivo, M. Shimokomaki and E. I. Ida. 2004. Synergism between dietary vitamin E and exogenous phytic acid in prevention of warmed-over flavour development in chicken breast meat, Pectoralis major M. Synergism between dietary vitamin E and exogenous phytic acid in prevention of warmed-over flavour development in chicken breast meat, Pectoralis major M. doi:10.1590/S1516-89132004000100008.

Soler, M., C. Bosetti, S. Franceschi, E. Negri, P. Zambon, R. Talamini and C. La Vecchia. 2001. Fiber intake and the risk of oral, pharyngeal and esophageal cancer. International Journal of Cancer 91(3): 283–287. doi:10.1002/1097-0215(200002)9999:9999<::AID-IJC1047>3.0.CO;2-I.

Su, D., R. Zhang, F. Hou, M. Zhang, J. Guo, F. Huang and Z. Wei. 2014. Comparison of the free and bound phenolic profiles and cellular antioxidant activities of litchi pulp extracts from different solvents. BMC Complementary and Alternative Medicine 14: 9–9. doi:10.1186/1472-6882-14-9.

Tang, S., N. S. Hettiarachchy and T. H. Shellhammer. 2002. Protein extraction from heat-stabilized defatted rice bran. Physical processing and enzyme treatments. Journal of Agricultural and Food Chemistry 50(25): 7444–7448. doi:10.1021/jf025771.

Tang, S., N. S. Hettiarachchy, S. Eswaranandam and P. Crandall. 2003. Protein extraction from heat-stabilized defatted rice bran: II. The role of amylase, celluclast, and viscozyme. Journal of Food Science 68(2): 471.

Tang, S., N. S. Hettiarachchy, R. Horax and S. Eswaranandam. 2003. Physicochemical properties and functionality of rice bran protein hydrolyzate prepared from heat-stabilized defatted rice bran with the aid of enzymes. Journal of Food Science 68(1): 152–157. doi:10.1111/j.1365-2621.2003.tb14132.

The Cosmetic Ingredient Review Expert Panel. 2006. Amended final report on the safety assessment of Oryza Sativa (rice) Bran Oil, Oryza Sativa (rice) Germ Oil, Rice Bran Acid, Oryza Sativa (rice) Bran Wax, Hydrogenated Rice Bran Wax, Oryza Sativa (rice) Bran Extract, Oryza Sativa (rice) Extract, Oryza Sativa (rice) Germ Powder, Oryza Sativa (rice) Starch, Oryza Sativa (rice) Bran, Hydrolyzed Rice Bran Extract, Hydrolyzed Rice Bran Protein, Hydrolyzed Rice Extract, and Hydrolyzed Rice Protein (1091-5818). Retrieved from United States.

Tomlin, J. and N. W. Read. 1988. Comparison of the effects on colonic function caused by feeding rice bran and wheat bran. European Journal of Clinical Nutrition 42(10): 857–861.

Tsuji, H., M. Kimoto and Y. Natori. 2001. Allergens in major crops. Nutrition Research 21(6): 925–934. doi:10.1016/S0271-5317(01)00291-3.

Van Hooser, B. and L. V. Crawford. 1989. Allergy diets for infants and children. Comprehensive Therapy 15(10): 38–47.

Van Soest, P. J. 2006. Rice straw, the role of silica and treatments to improve quality. Animal Feed Science and Technology (130): 137–171.

Wan, Y., L. A. Espinoza Rodezno, K. M. Solval, J. Li and S. Sathivel. 2014. Optimization of soluble dietary fiber extraction from defatted rice bran using response surface methodology: Soluble dietary fiber extraction. Journal of Food Processing and Preservation 38(1): 441–448. doi:10.11 11/j.1745-4549.2012.00792.

Wang, K. M., J. G. Wu, G. Li, D. P. Zhang, Z. W. Yang and C. H. Shi. 2011. Distribution of phytic acid and mineral elements in three indica rice (*Oryza sativa* L.) cultivars. Journal of Cereal Science 54(1): 116–121. doi:http://dx.doi.org/10.1016/j.jcs.2011.03.002.

Wang, M., N. S. Hettiarachchy, M. Qi, W. Burks and T. Siebenmorgen. 1999. Preparation and functional properties of rice bran protein isolate. Journal of Agricultural and Food Chemistry 47(2): 411–416. doi:10.1021/jf9806964.

Wang, W., J. Guo, J. Zhang, J. Peng, T. Liu and Z. Xin. 2015. Isolation, identification and antioxidant activity of bound phenolic compounds present in rice bran. Food Chemistry 171: 40–49. doi:http://dx.doi.org/10.1016/j.foodchem.2014.08.095.

Watchararuji, K., M. Goto, M. Sasaki and A. Shotipruk. 2008. Value-added subcritical water hydrolysate from rice bran and soybean meal. Bioresource Technology 99(14): 6207–6213. doi:http://dx.doi.org/10.1016/j.biortech.2007.12.021.

Wisker, E., A. Maltz and W. Feldheim. 1988. Metabolizable energy of diets low or high in dietary fiber from cereals when eaten by humans. The Journal of Nutrition 118(8): 945–952.

Xiong, J., W. Zhang, H. Gu, L. Xi, Y. Zhang, Y. Hu and T. Zhang. 2012. Preparation of phytic acid and its characteristics as copper inhibitor. Energy Procedia 17: 1641–1647. doi:http://dx.doi. org/10.1016/j.egypro.2012.02.292.

Xu, Q., A. G. Kanthasamy and M. B. Reddy 2008. Neuroprotective effect of the natural iron chelator, phytic acid in a cell culture model of Parkinson's disease. Toxicology 245(1): 101–108. doi:10.1016/j.tox.2007.12.017.

Xu, Z. and J. Godber. 2000. Comparison of supercritical fluid and solvent extraction methods in extracting γ-oryzanol from rice bran. Journal of the American Oil Chemists' Society 77(5): 547–551. doi:10.1007/s11746-000-0087-4.

Xu, Z. and J. Godber. 2001. Antioxidant activities of major components of γ-oryzanol from rice bran using a linoleic acid model. Journal of the American Oil Chemists' Society 78(6): 645–649. doi:10.1007/s11746-001-0320-1.

Xu, Z., N. Hua and J. S. Godber. 2001. Antioxidant activity of tocopherols, tocotrienols, and gamma-oryzanol components from rice bran against cholesterol oxidation accelerated by 2,2'-azobis(2-methylpropionamidine) dihydrochloride. Journal of Agricultural and Food Chemistry 49(4): 2077.

Yadav, R. B., B. S. Yadav and D. Chaudhary. 2011. Extraction, characterization and utilization of rice bran protein concentrate for biscuit making. British Food Journal 113(9): 1173–1182. doi:10.1108/00070701111174596.

Yawadio, R., S. Tanimori and N. Morita. 2007. Identification of phenolic compounds isolated from pigmented rices and their aldose reductase inhibitory activities. Food Chemistry 101(4): 1616–1625. doi:http://dx.doi.org/10.1016/j.foodchem.2006.04.016.

Zhou, Z., K. Robards, S. Helliwell and C. Blanchard. 2004. The distribution of phenolic acids in rice. Food Chemistry 87(3): 401–406. doi:http://dx.doi.org/10.1016/j.foodchem.2003.12.015.

CHAPTER 8

Extraction, Characterization and Utilisation of Bioactive Compounds from Wine Industry Waste

Ariel R. Fontana,* Andrea Antoniolli and *Rubén Bottini*

Introduction

Viticulture is one of the most important agro economic activities in the world with more than 60 million tons produced globally every year. This production is partially directed at fresh consumption of table fruit and juice, or raisins. However, about 80 per cent of the world-wide produced grape is used in wine production (Fontana et al. 2013, http://www.fao.org 2010). Therefore, the overall winemaking industry produces huge amounts of by-products like organic wastes, wastewater, greenhouse gases and inorganic residues (Teixeira et al. 2014). Only considering the winemaking process, about 20 per cent of the weight of processed grapes remains as pomace and stems after grape juice extraction. Currently, these residues are not valued as profitable, being mainly directed to composting or discarded in open areas, thus causing an ecological and economical waste management issue. Therefore, the demand for greener industrial production along with the challenge of minimizing the by-product treatment costs has prompted the search for strategies for utilisation of these residues (Teixeira et al. 2014). Similarly, the increase in consumer awareness about the use of additives in food products and the attention that functional foods have attained in the last few years, create a need for the identification of alternative natural and safer sources of food antioxidants (Fontana et al. 2013). Several compounds present in winemaking waste, such as polyphenols and dietary fiber, have health-promoting effects and other biological properties mostly related to their antioxidant characteristics. The valorization of waste will provide alternatives to reduce the environmental impact of

Laboratorio de Bioquímica Vegetal, Instituto de Biología Agrícola de Mendoza, Consejo Nacional de Investigaciones Científicas y Técnicas-Universidad Nacional de Cuyo, Almirante Brown 500, M5528AHB Chacras de Coria, Argentina.

* Corresponding author: afontana@mendoza-conicet.gob.ar; fontana_ariel@yahoo.com.ar winemaking activities and to add value through the commercialization of extracts rich in bioactive compounds. In this sense, the attention should be focused on enriching the overall waste chain, thus optimizing extraction approaches for pilot-scale applications. The isolation and characterization of bioactive compounds will be a necessary step to increase the value and justify the commercial application in pharmaceutical, cosmetic and food industries.

Waste from Wine Industry

Grape pomace

Grape pomace (GP) is the left-over from winemaking process immediately after fermentation, composed mainly of berry skins and seeds (Fontana et al. 2013). In winemaking of white grapes, the seeds and skins are removed before fermentation. For red-grape winemaking, grapes are subjected to a gentle but prolonged extraction with a hydro-ethanolic mixture that provides the red wine with a variable content of polyphenols. Seeds and skins are removed after maceration in contact with fermenting must. After fermentation, a residue mainly constituted of skins and seeds is separated from the wine. Since winemaking is not an exhaustive process, relatively high levels of bioactive compounds remain in this by-product. Grape seeds are rich in extractable phenolic antioxidants, such as phenolic acids, flavonoids, procyanidins and resveratrol between other stilbenes, while grape skins contain abundant anthocyanins (Yu and Ahmedna 2013). In other words, a profuse amount of different bioactive compounds are present in the winemaking by-products, which present the chance of turning a problem into an opportunity for improving the sustainability of the industry.

Grape stems and canes

Aside from GPs, other types of solid waste are generated by the winemaking and viticultural practices, such as stems and canes.

Grape stems (which include rachis and pedicels) are produced in large amounts during the winemaking process, representing approximately 5 per cent of the grape cluster weight and having no commercial value although sometimes it is used as animal food or soil fertilizer (Maria Anastasiadi et al. 2012, Apostolou et al. 2013). This by-product is removed before the vinification process with the aim to avoid its negative contribution on the organoleptic characteristics of wine due to its astringency (Teixeira et al. 2014).

Grape cane is the lignocellulosic by-product resulting during the annual pruning of grapevines that is usually incorporated into the vineyard soil and transformed to charcoal or burned (Vergara et al. 2012). Both residues have been scarcely studied and there are only few reports on their polyphenol content, but their industrial applications are very limited in comparison to GP (Anastasiadi et al. 2012, Apostolou et al. 2013). Some studies discuss the high potential of grape stems and canes for the recovery of bioactive compounds. These by-products have high levels of stilbenes derivatives as well as total polyphenolic and flavonoid contents (Anastasiadi et al. 2012). In fact, the presence of various non-colored bioactive phenolics in grape stems, such as stilbenes (resveratrol and derivatives), flavan-3-ols (particularly (+)-catechin), flavonols and phenolic acids in extracts of different native grape varieties from Greece has been assessed (Apostolou et al. 2013), thereby establishing that grape stems constitute a rich source of bioactive polyphenols with antioxidant properties. Vergara et al. (2012) reported high levels of trans-resveratrol and other stilbenes, higher than those previously reported in the literature, in grape cane samples from vineyards in the Bio-Bio region of Chile.

Besides the grape stems' potential as a source of polyphenols, the content of complex carbohydrates makes this by-product a rich source of good quality dietary fiber (Barros et al. 2015). Additionally, the waste contains complex polysaccharides, valuable as a source of cellulose for paper production, in food industries and as an additive in optical and pharmaceutical industries (Barros et al. 2015).

Bioactive Compounds Composition

Phenolic compounds

Phenolic compounds represent an assorted group of secondary metabolites present in plants, and conspicuously in grapes. The phenolic content and composition of grape by-products are importantly influenced by technological practices to which grapes are exposed during winemaking. Commencement with the first 'French paradox' report, interest in phenolic compounds has definitely increased because of their recognized health benefits (Kammerer et al. 2014). During the last two decades, polyphenols have been associated with a multitude of beneficial effects on health, purportedly preventing damages and diseases caused by oxidative stress. Polyphenols are unequally distributed in grapevine tissues, and this is consistent with their biological function. In berries, UV-absorbing flavonoids and phenolic acids, attractant anthocyanins and deterrent tannins are mainly deposited in the outer layers of the skin, aleurone cells, seed coats and hulls, respectively (Kammerer et al. 2005). On the other hand, seeds are rich in astringent polyphenols.

Polyphenols are characterized by their high structural diversity, including not only an ample variety of molecules with the polyphenol structure (i.e., several hydroxyl groups on aromatic rings) but also molecules with one phenol ring, such as phenolic acids and phenolic alcohols. They could be subdivided into several classes: phenolic acids (mainly benzoic and hydroxycinnamic acids), flavonoids (flavones, flavonols, flavanones, flavononols, flavanes, flavanols, anthocyanidins and anthocyanins, chalcones and dihydrochalcones), tannins (hydrolyzable and non-hydrolyzable or condensed tannins), stilbenes, coumarins, phenylethanol derivatives, lignans and neolignans (Garrido and Borges 2013).

Dietary fiber

According to the new Codex definition, dietary fiber consists of carbohydrate polymers which are resistant to hydrolysis (digestion) by endogenous human enzymes (Cummings et al. 2009). Dhingra et al. (2012) defined dietary fiber as that part of plant material in the diet which is resistant to enzymatic digestion. This classification includes cellulose type compounds, non-cellulosic polysaccharides as xyloglucans, pectins, gums, mucilages and components that are not carbohydrates, such as lignins

(Dhingra et al. 2012). However, Monro and Burlingame (1996) pointed at the difficulty of defining dietary fiber and do not refer the term to specific chemical species but to overlapping polymer blends, defined both by their solubility under certain conditions as chemical identity. The Association of Official Analytical Chemists (AOAC) defines dietary fiber as polysaccharides and remnants resistant to hydrolysis (digestion) of plant material by human digestive enzymes (AOAC 1995). Therefore, dietary fiber includes many complex substances which possess distinctive chemical and physical properties (Yu and Ahmedna 2013). Thus the term dietary fiber describes chemical, physiological and nutritional properties of compounds in which the grouping is based. Furthermore, Saura-Calixo et al. state that certain insoluble proanthocyanidins, mainly tannins, are present in dietary fiber due to its high molecular weight and/or because they are bounded to the cell wall. For this reason, non-extractable polyphenols are misidentified as lignins or polysaccharides, even if the physiological and chemical properties of these compounds are markedly different (Saura-Calixto 1998). Therefore, non-extractable polyphenols could be included on an extended definition of dietary fiber (Saura-Calixto 2012).

The traditional source of dietary fiber used as functional ingredient has been mainly from the cereal industry (Fuentes-Alventosa et al. 2009). However, recent studies suggest that by-products obtained from processing berries and culturing vines respectively, could be used as potential sources of dietary fiber (Beres et al. 2016, Deng et al. 2011, González-Centeno et al. 2010, Llobera and Cañellas 2007, 2008). Dietary fiber from GP represents a matrix rich in phenolic compounds and, therefore, is a supplement that combines benefits inherent in both the components, which may be useful in prevention of cancer and cardiovascular diseases (Pozuelo et al. 2012).

Beres et al. (2016) recovered dietary fiber with antioxidant properties from GP cv. Pinot noir. González-Centeno et al. analyzed stems and GP of ten different cultivars, finding that in GP, the main cell wall polymers were pectin with low methyl-esterification, whereas in stems cellulose predominated. These authors highlighted the potential of wine by-products as raw materials for the production of dietary fiber concentrates. The development of commercial fiber-rich products with added nutritional value constitutes an interesting option to be promoted due to the relevance of a suitable consumption of dietary fiber in human daily diet.

Extraction Techniques for Recovery of Bioactive Compounds from Wine Industry Waste

Techniques, such as solid–liquid or Soxhlet extractions, have been used for many decades, but they are time-consuming and require relatively large quantities of solvents. Also, due to the common extractive steps used by these techniques, including heating, boiling, or refluxing, a loss of polyphenols due to ionization, hydrolysis, and oxidation occurs during the procedure (Fontana et al. 2013). Besides, factors such as extraction solvent and sample/solvent ratio are relevant to achieving good recoveries, especially considering the polar nature of most compounds. In recent years, with the development of miniaturization/simplification, different extraction techniques, such as ultrasound-assisted, microwave-assisted, supercritical fluid and accelerated solvent extraction, have been developed for the recovery of bioactive compounds from wine-industry waste. These techniques are primarily focused on shortening the extraction time and decreasing the solvent consumption—the facts associated with an increase in sustainability without modifying (or even improving) the recovery of compounds.

Solid-liquid extraction

The solvent extraction technique, also known as solid-liquid extraction (SLE), comprises a mass transport phenomenon in which migration of compounds through a solid matrix towards a liquid solvent system occurs (Corrales et al. 2009, Ignat et al. 2011). SLE is a popular method for bioactive plant food constituents (Amendola et al. 2010) as it is a cheap and simple way to recover target compounds from a solid matrix, like GP (Brazinha et al. 2014). The efficiency of mass transport is related to fluctuation in concentration gradients, diffusion coefficients, or boundary layer and thus depends on the extraction method, solvent, particle size, temperature and extraction time as well as the presence of interfering substances in the matrix (Corrales et al. 2009, Ignat et al. 2011). The solvent separates the soluble from insoluble fractions. So the choice of an appropriate solvent is a relevant step (Cacace and Mazza 2003, Pinelo et al. 2006). The efficiency of extraction is related to how fast the compound is dissolved and the equilibrium in the liquid is reached. Polyphenols are polar compounds; hence, they are easily solubilized in polar mediums, like hydro-alcoholic solutions. Phenolic fractions could then be obtained by varying alcohol concentration in mixtures with increasing concentration of low-polar solvents (Galanakis 2012). The solvents mixture of water with methanol, ethanol, propanol, acetone, ethyl acetate or dimethylformamide, result in different efficiencies in extraction of phenolics (Bucić-Kojić et al. 2009). To know phenolic extraction kinetics is useful for process and scaling-up designs, although there are few reports about optimization of SLE approaches (Bucić-Kojić et al. 2009, Spigno et al. 2007). The solubility and the diffusion coefficient is enhanced when the temperature increases, thus favoring extraction efficiency (Spigno et al. 2007, Pinelo et al. 2005). Some authors reported that above 50°C, phenolic stability decreases and denaturation of membranes may occur. So this factor cannot be increased indefinitely (Pinelo et al. 2005). Spigno et al. compared the simultaneous effect of extraction temperature and time, demonstrating that both exert a significant impact and may be regulated after an economical appraisal of the energy cost. To achieve high recoveries of compounds and to minimize energy costs, extraction time and temperature are important parameters to be optimized (Fontana et al. 2013). Yilmaz and Toledo analyzed the effect of aqueous solutions of ethanol, methanol or acetone and single-compound solvent systems on the extraction of total phenols from grapeseed powder and found that extraction with aqueous solutions was better than the single-compound solvent system. Bucić-Kojić et al. reported that the extraction solvent and temperature had a significant influence on the extraction of polyphenolics, the most efficient being a mix of 50 per cent of aqueous ethanol.

Supercritical fluid extraction

Supercritical fluid extraction (SFE) is a novel technique to extract compounds from solid matrices employing properties of supercritical fluids. A supercritical fluid is a substance above its critical temperature and pressure with a good solvating power, high diffusivity, low viscosity and marginal surface tension (Wells 2003). The critical temperature is the highest temperature at which a gas can be converted to liquid by an increase in pressure (Cavalcanti and Meireles 2012). The critical pressure is the highest pressure at which a liquid can be converted into a gas by an increase in temperature (Cavalcanti and Meireles 2012). Due to these characteristics, SFE has considerable advantages over SLE. SFE permits a rapid mass transfer in the supercritical phase and an improved ability to penetrate the pores in the sample matrix, thereby achieving a fast and efficient extraction. SFE is an environment-friendly alternative to the organic solvent extraction because it avoids the use of large amounts of solvents, is rapid, automatable and selective. Similarly, the absence of light and air during extraction reduces the degradation processes that may occur during organic solvent extraction.

The SFE of solid matrices is separated into two phases: extraction and separation. During extraction, the solvent is fed into the extractor and evenly dispersed inside the fixed bed. The supercritical fluid flows through the fixed bed of solid particles, dissolving and extracting the compounds. Successively, the mixture (solvent and compounds) leaves the extractor and runs into the separator (Cavalcanti and Meireles 2012). In separation, a reduction in pressure is produced, resulting in the vaporization of the solvent that re-circulates and precipitation of the solute that is collected and analyzed.

During the development of SFE methods, some factors must be considered, such as selection of supercritical fluids, sample preparation, use of modifiers and extraction conditions. For extraction of phenolics from GP, the used solvent is supercritical carbon dioxide (SC-CO$_2$) (Casas et al. 2010, Fiori 2010, Oliveira et al. 2013, Otero-Pareja et al. 2015, Palenzuela et al. 2004, Vatai et al. 2009). Although SC-CO$_2$ is not very suitable for extraction of polar analytes, the addition of modifiers (ethanol or methanol) to a supercritical fluid (SCF) can change its polarity, obtaining a more selective extraction power (Casas et al. 2010). Extraction of phenolics from GP shows that if a co-solvent is not added, higher temperatures are needed for better extraction. Conversely, using a co-solvent like ethanol avoids heating the system over 35°C, which is positive considering the stability of some compounds (Casas et al. 2010). So, the SC-CO$_2$ extraction uses an extraction temperature of 30°C, avoiding degradation of the analytes. Additionally, the methods based on SC-CO$_2$ have the advantage of being environment-friendly, obtaining extracts free of residual toxic solvents, which is preferred for natural products to be used as functional foods and nutraceuticals.

Accelerated solvent extraction

Accelerated solvent extraction (ASE, also known as pressurized liquid extraction, pressurized solvent extraction and pressurized fluid extraction) works well for the recovery of many compounds from different matrices. ASE uses conventional solvents at elevated temperatures (100–180°C) and high pressures (1500–2000 psi) to enhance the extraction of organic analytes from solid samples (Richter and Raynie 2012). Conditions of elevated pressure and temperature exert changes in the solvent, the sample and the interactions among them. Under high pressure, the solvent boiling point increases; consequently the extraction is conducted at higher temperatures. The high pressure allows the solvent to penetrate deeper into the sample matrix, increasing the extraction of analytes confined to the matrix pores (Richter and Raynie 2012). This is crucial for complex samples like grape by-products where many of the compounds have interactions and different chemical nature. The solubility of analytes increases and mass transfer goes faster at elevated temperatures. The high temperature weakens the van der Waals forces, hydrogen bonding, London forces and dipole attractions from solute-matrix. In addition, at high temperatures the solvent viscosity and surface tension get reduced, enhancing the solvent penetration. All of these factors lead to faster extraction, higher recoveries of analytes and less solvent consumption in comparison with conventional SLE (Fontana et al. 2013).

SLE is used for extraction of antioxidants from winemaking residues and other natural products. So far, different solvent mixtures, temperatures and pressure combinations have been used (Monrad et al. 2010a, b). Soural et al. (2015) studied different factors, like drying, grinding, temperature, extraction time, solvent in solid-liquid extraction versus others methods in grape cane. They obtained more stilbenes using ASE in methanol, but yields were comparable with extraction at 50°C for methanol and acetone; microwave-assisted extraction; extraction by fluidized-bed extraction with methanol; and, extraction with methanol heated to reflux. Besides, higher yields were obtained from powdered than with cut material. Multiple extractions gave higher yields compared with single-step extraction at higher temperature.

A modification of SLE using subcritical water as solvent was reported (Aliakbarian et al. 2012), with recoveries of bioactive compounds from GP at 140°C, similar to those obtained by using organic solvents, but in a time significantly shorter. In this method no organic solvent is used, so the product is free of residual solvents. Elimination of organic solvents is desirable to produce functional foods and nutraceuticals. The SLE using subcritical water can be considered a cost effective and benign process to obtain antioxidants from winery by-products. Efforts should be devoted to developing pilot plants to scale up this method for industrial application.

Other authors reported a differential extraction yield for anthocyanin and for tannins and tannin-anthocyanin adducts at different temperatures, which highlights the importance to optimize the extraction according to the compounds of interest (Vergara-Salinas et al. 2013).

Alternative physical treatments

The application of Ultrasound Radiation (US) is a suitable alternative to conventional SLE with stirring because it increases extraction efficiencies of compounds. US effects are related with cavitation, which involves the implosion of bubbles formed in the liquid medium. Bubble implosion generates compression of gases and vapors within the bubbles or cavities and, as a consequence, high temperature and pressure are produced (Luque de Castro and Priego-Capote 2006). Increasing pressure favors the penetration of the solvent into the matrix, while improving the transport between the solid matrix and the liquid phase. This leads to an increment in analyte solubility and diffusivity from the sample matrix to the solvent, which is the limiting step of mass transfer (Luque de Castro and Priego-Capote 2006). Reports of antioxidant extraction from winery waste showed that US is as effective as any other extraction process (Ghafoor et al. 2009). US greatly reduce the extraction time without the need for high temperatures that may affect phenol stability. The efficiency of US extraction relays on that sonication simultaneously, enhancing the hydration and fragmentation process while facilitating the mass transfer of solutes (Ghafoor et al. 2009).

High-voltage electric discharge (HVED) is another extraction technology recently introduced. It is based on the application of high voltage between two electrodes, so that the electrons are accelerated and reach sufficient energy to excite water molecules. Then, 'a flood' of electrons, called streamer, is created. If the applied electric field is intense enough, the streamer propagates from the positive to the negative electrode. When one streamer attains the negative electrode, electrical breakdown occurs. So high-amplitude pressure shock waves, bubble cavitation and liquid turbulence are created. These result in particle fragmentation and cell structure damage that accelerate the extraction of intracellular compounds (Boussetta et al. 2009). The optimum conditions for recovery of phenols and the possibility of applying these results on a pilot scale have been established (Boussetta et al. 2009, Boussetta et al. 2009, Boussetta et al. 2011, Boussetta et al. 2012). The initial development points to accelerate the extraction of total soluble matter and polyphenols from GP into distilled water. Also there is a synergistic effect on polyphenol extraction when HVED is combined with freezing or elevated temperatures. This is explained as an impact on cell membrane permeability that increases because heating is like any freezing process (Boussetta et al. 2009). The results show that comparatively higher energy is required to obtain equivalent extraction rates on the pilot scale when compared with laboratory scale (Boussetta et al. 2012). Thus, further studies are necessary to determine the energy costs of HVED related to other extraction techniques to obtain a sustainable alternative to traditional extraction approaches.

Pulsed Ohmic Heating (POH) combines electric fields and temperatures for the extraction of cell compounds. El Darra et al. studied the effects of electric fields of 100–800 V/cm and 0–50 per cent ethanol in water on phenolics extraction. Cell membranes denaturation occurred due to POH in a dose-response manner. A synergy was observed when POH was combined at 50°C and 30 per cent ethanol. POH pretreatment accelerates the extraction of total phenolics without the need for raised temperatures. POH-accelerated extraction is promising for application in wastes from agricultural and food origin without hydroalcoholic solvent use.

Barba et al. (2015) compared the effect of US, HVED and POH on the selective recovery of compounds from GP. They compared their efficiency for phenolic extraction, particularly for anthocyanins, by evaluation of cell disintegration indexes. HVED was the best technique for recovery of higher phenolics with lower energy requirement. However, HVED was less selective than PEF and US, regarding the recovery of anthocyanins. This fact might be due to the value of cell disintegration (evaluated as Cell Disintegration Index Z) increases significantly with the increase of energy input. This phenomenon reflects the extraction of ionic intracellular components from damaged cells to modify the electrical conductivity of the media. Consequently, Z index could be a useful indicator to evaluate the recovery of anthocyanins.

Isolation and chemical characterization

Due to the different properties of phenolics, obtaining concentrated extracts of specific bioactive components from GP is relevant from both industrial and analytical aspects. Extracts from extraction processes have many bioactive compounds, so isolation and/ or concentration of specific chemical groups sometimes is required because individual polyphenols exhibit different functionalities and chemical activities Fontana et al.

2013. Taking into consideration the complexity of the extracts makes mandatory the application of some sample preparation step, such as purification to isolate different groups of compounds prior to instrumental analysis.

The isolation, fractionation and/or pre-concentration step simplify the identification and quantification of analytes. Purification removes interfering compounds from complex matrixes being a decisive step of any method. Some authors used solid-phase extraction (SPE) to increase selectivity of the chromatographic techniques for identification of phenolics. Different sorbent materials have been used for the fractionation between phenolic acids, flavonoids and anthocyanins present in GP. The most common sorbent used is reverse-phase octadecyl silane (C_{18}) which has good affinity for phenolic compounds. Kammerer et al. (2004) developed a method for phenolics based on the adsorption of such compounds onto C_{18} mini-columns followed by sequential elution with acidified water (hydroxybenzoic and hydroxycinnamic acids), ethyl acetate (flavanols, flavonols and stilbenes) and acidified methanol (anthocyanins). Anthocyanins were analyzed directly in the acidified extract of GP previous SPE. Preceding the SPE step as well, they extracted non-anthocyanins from GP with ethyl acetate. The purpose was to reduce the abundance of anthocyanins, by increasing the retention capacity of C_{18} cartridges (Kammerer et al. 2005). The method reported by Kammerer et al. was efficient in terms of the number of compounds reported in GP from many different samples. Others used different variations of the SPE C_{18} technique to profile anthocyanins or anthocyanidins in GP with good results in terms of purification and selectivity prior to the analysis by HPLC (Bonilla et al. 1999, Negro et al. 2003, Wang et al. 2010). Yi et al. (2005) proposed the use of hydrophilic-lipophilic-balanced (HLB) reversed-phase sorbent for the initial separation of two fractions: (1) phenolic acids and (2) anthocyanins and other flavonoids. Fraction 2 is loaded on to a Sephadex LH20 cartridge to elute first the anthocyanins and flavonols by using methanol acidified with formic acid and then washing with 70 per cent acetone for the elution of tannins and proanthocyanidins. After freeze-drying, the anthocyanin and flavonol fraction is solubilized in 5 per cent formic acid and applied to a second HLB cartridge. The cartridge is washed with 5 per cent formic acid, followed by ethyl acetate and then 10 per cent formic acid in methanol. The ethyl acetate eluted the flavonols and the acidified methanol eluted the anthocyanins. The different fractions are evaluated against viability and apoptosis in cancer cells. This procedure, however, has several steps that augment sample manipulation and therefore, may lessen reproducibility of the method. It is also time consuming as compared with the method reported by Kammerer et al. Altogether, these techniques are effective for extraction of polyphenols, but they are time-consuming and require high volumes of organic solvents. Taking into account these drawbacks, a new method has been recently applied for the extraction and concentration of phenolics from GP based on the QuEChERS (quick, easy, cheap, effective, rugged and safe) extraction technique (Antoniolli et al. 2015, Fontana et al. 2016, Fontana and Bottini 2014). This procedure involves an initial single-phase extraction of compounds of interest with acetonitrile followed by a salting-out extraction/partitioning step by adding a combination of salts. Subsequently, the clean-up is performed using dispersive-solid-phase extraction (d-SPE), which is based on the addition of the sorbent material into an aliquot of the extract to remove the matrix interferences. The d-SPE clean-up avoids passing the extracts through SPE cartridges, requiring much smaller quantities of sorbent and solvent. The principal advantages of QuEChERS method are its simplicity, repeatability, low cost, speed and wide applicability to different types of samples and analytes. This method was successfully used for characterization and quantification of at least 20 polyphenols representatives of different chemical classes (phenolic acids, flavanols, flavonols, stilbenes and phenylethanol analogs) in GPs.

Different spectrophotometric methods for quantification of total phenolics have been developed. These procedures are based on different principles and are used to determine various structural groups present in phenolic molecules. Some of these methods include the measurement of absorption at 280 nm and the Folin-Ciocalteu assay. Several authors have studied the properties of GP by determining the antioxidant power using different techniques as well as the total phenolic content of extracts or different isolated fractions. Because synergistic and antagonistic effects have been observed in *in vitro* tests, similar interactions are expected *in vivo*. Therefore, the spectrophotometric methods that give broad-spectrum information about the extract might be complemented by chromatographic techniques coupled with different detection systems to identify and quantify the individual phenolics present in each fraction. These data are needed to support the technological application of recovered bioactive compounds in diverse industries. The analytical methods most frequently used to establish correlations between the properties of extracts (or their fractions) with each individual phenolic profiled have been recently reviewed by Fontana et al.

Potential Use of Bioactive Compounds of Waste from Winemaking Industry

Biological properties: health beneficial effects

Oxidative stress has been defined as the state in which reactive oxygen species (ROS) levels are higher than the redox cell capacity (García et al. 2003), thus resulting in an oxidative injury to molecules as DNA, lipids and proteins (Prior and Cao 1999) and cellular structural damage (Wang et al. 2010). As a consequence ROS have been implicated in hundreds of human diseases including arthritis, atherosclerosis, senility, Alzheimer and Parkinson diseases, intestinal dysfunction, tumor promotion and carcinogenesis, among others (Alonso et al. 2002).

The health-promoting and anti-aging effects of polyphenolics are associated with their capacity to capture ROS involved in conditions ranging from inflammatory-immune injury to myocardial infarction and cancer. So they behave as powerful antioxidant and metal chelators (Fontana et al. 2013, Middleton et al. 2000, Šeruga et al. 2011, Wu et al. 2010). In this sense, grape waste extracts are a rich source of bioactive compounds, such as (+)-catechin, (−)-epicatechin and (−)-epicatechin-3-O-gallate, *trans* resveratrol as well as dimeric, trimeric and tetrameric procyanidins, among others.

A study of the individual impact of wine phenolic constituents on to the catalytic activity of cyclooxygenase-1 and 2 (COX-1 and COX-2), and 5-lipoxygenase (LOX) was performed by Kutil et al. (2014). COX-1, COX-2 and LOX are pro-inflammatory enzymes which affect platelet aggregation, vasoconstriction, vasodilatation and then atherosclerosis development. The effect of piceatannol, trans-resveratrol, quercetin and myricetin was evaluated and the authors observed that piceatannol was the strongest inhibitor of LOX, followed by kaempherol, quercetin and myricetin. Trans-resveratrol was identified as the most potent inhibitor of COX-1 and COX-2 enzymes. In another study (Choi et al. 2016), the activity of beta-amiloide (Aß), the main pathogenic molecule in Alzheimer's disease, was reduced by high concentration of oxy-resveratrol and resveratrol. With piceatannol, however, this inhibition was independent of concentration and cellular viability was maintained. Therefore piceatannol has a greater capacity than resveratrol and oxy-resveratrol to reduce Aß levels without causing cell death.

Rodríguez Lanzi et al. (2016) tested protective health effects of lyophilized whole GP and GP extracts supplied to the food of rats with high-fat-fructose diet-induced metabolic syndrome, by monitoring reduction of adiposity and improvement of insulin signaling. It was observed that whole GP was more effective in preventing the increased systolic blood pressure and triglycerides plasma levels than GP extracts, and the differences might be due to the dietary fiber content of whole GP. Rodriguez-Rodriguez et al. (2012) evaluated the effects of GP extracts in the aorta tissue of rats and found that the extracts induce endothelium-dependent vasodilatation and attenuates vascular contraction through a NO-dependent mechanism. Modifications of enzyme activities, as well as modifications in lipid peroxidation in rats with GP diet were reported (Chidambara Murthy et al. 2002). The authors indicate that the activities of hepatic enzymes, which play important roles in combating the ROS, would be protected by GP. Hogan et al. (2010) studied the application of GP in mice with an obesity-induced diet and reported that supplementation with GP produced anti-inflammatory activity but not a reduction in oxidative stress (Hogan et al. 2010). Terra et al. (2009) observed that commercial extracts of grape seeds prevented low-grade inflammation in rats fed with a high-fat diet by adjusting adipose tissue cytokine imbalance, enhancing anti-inflammatory molecules and diminishing pro-inflammatory ones.

Apostolou et al. (2013) also showed prevention from ROS-induced DNA damage and inhibitory activity against liver and cervical cancer cell growth by applying extracts of grape stems. They observed the antioxidant activity of extracts and its effect on protection against ROS-induced DNA damage and inhibition at low concentrations of the growth of liver and cervical cancer cells. They concluded that the activities of grape stem extracts were comparable to those of seed extracts.

Food Systems and Antimicrobial Effects

In the last few years, the use of antimicrobial compounds of synthetic origin has been discussed due to environmental and health concerns. Industrial residues are an inexpensive potential source of natural by-products with antioxidant properties and thus the interest to obtain these compounds from natural sources has increased (Fontana et al. 2013). Dias et al. (2015) studied the *in vitro* antibacterial effects of grape stem extracts on Gram (+) (*Listeria monocytogenes*, *Staphylococcus aureus* and *Enterococcus faecalis*) and Gram (–) (*Pseudomonas aeruginosa*, *Escherichia coli* and *Klebsiella pneumoniae*) intestinal pathogens, observing that the former strains were more sensible than the latter. The authors suggest that the use of grape stem extracts would contribute to reduce the risk of intestinal microbial disturbances and the results encourage *in vivo* studies.

Tseng and Zhao (2012) applied GP and grape skin extracts to study the minimum inhibitory concentration (MIC) as indicator of antibacterial activity. Their results indicate that the extracts exhibited antibacterial effects and the GP extract was more effective against *Listeria innocua* as compared to the *E. coli* strain assessed. In a similar study, Jayaprakasha et al. (2003) also found that grape seed extracts were more efficient inhibiting Gram (+) than Gram (–) bacteria. Anastasiadi et al. (2009) evaluated the antibacterial effect of GP extract against *Listeria monocytogenes*, with results that justify the extracts incorporation in food systems to prevent growth of these bacteria. Other authors evaluated antibacterial and antifungal activities of supercritical fluid (CO_2) extracts of Merlot and Syrah GP against *Staphylococcus aureus*, *Bacillus cereus*, *Escherichia coli* and *Pseudomonas aeruginosa* and fungi (*Candida albicans*, *Candida parapsilosis*, *Candida krusei*). It was observed that the *Merlot* extracts were more active against Gram (+) bacteria and the Syrah extracts were less efficient against all the microorganisms tested (Oliveira et al. 2013). Delgado Adámez et al. (2012) reported that grape seed extract showed inhibitory effects against Gram (+) (*Listeria innocua*, *Brochothrix thermosphacta*, *Staphylococcus aureus* subsp. *aureus*) and Gram (–) (*Pseudomonas aeruginosa*, *Salmonella enterica* subsp. *enteric* and *Escherichia coli*) bacteria. Due to antibacterial and antioxidant properties of grape seed extracts, the authors suggest that these are a feasible alternative to synthetic compounds as antibacterial and antioxidant agents to prevent spoilage of stored food by bacteria and oxidation.

The interest on application of winemaking waste or its extracts to enhance food nutritional and functional value for improving the shelf-life of food systems have been increasing. Scientists focus on the antioxidant and antimicrobial activities as well as the potential health benefits of the waste components. Although many *in vitro* studies have been conducted successfully, applications on the food system are scarcely reported.

Ruiz-Moreno et al. (2015) evaluated the feasibility to replace and/or reduce SO_2 use in winemaking by utilizing an extract of grape stems due to its antioxidant and antimicrobial activities; they also appraised its potential effect on wine aroma. These authors establish that grape stems are a rich source of compounds with a low sourcing cost, high antioxidant activity and good antimicrobial properties. Grape stem extracts therefore represent an economic alternative to SO_2 in winemaking and its addition will help the industry's sustainability.

The feasibility to produce antioxidant-rich yoghurt through fortification with grape seed extract has been reported by Chouchouly et al. (2013). The authors observed that grape seed addition did not affect yoghurt pH and *Lactobacilli* counts, neither did it produce major defects in consistency, color or flavor as compared to controls. Polyphenol levels, antiradical and antioxidant activity along 3–4 weeks of storage were higher than control.

Chitosan has been studied as an edible biopolymer in films and coatings. Incorporation of GP extract on chitosan to improve the shelf-life of food products has been proposed (Ferreira et al. 2014). The chitosan films supplemented with different GP extracts (aqueous, berry skin wax and grape seed oil) modulated its mechanical characteristics, resulting in an enhancement of its antioxidant properties. Therefore, the authors suggest that these films can prove a promising alternative to synthetic material and a vehicle for functional compounds, promoting biological applications, including the increase in food's shelf-life.

The utilisation of GP from white grapes as an additive to wheat flour to improve the nutritional characteristics of wheat biscuits has been proposed (Mildner-Szkudlarz et al. 2013). Addition levels up to 10 per cent showed considerably higher dietary fiber than control samples and were characterized by significantly higher antioxidant activities associated with their phenolic contents.

Conclusion

The recovery of bioactive compounds from winemaking waste represents a front-facing challenge for scientists and industries. The implementation of sustainable alternatives to recover compounds and add values to these residues is a complex approach depending on several parameters. Researchers should be able to develop scaling-up processes without affecting the functional properties of the recovered compounds, develop a product that meets the high quality standards for safety and organoleptic characteristics that consumers demand. Further research in the field of biochemical and physiological activities of bioactive compounds is needed. Many studies justifying the positive health effects associated with the known antioxidative properties of GP have been presented till today. In this regard, characterization of the chemical composition of extracts will help to evaluate effects and safety during technological application in systems industries. Research should focus on extraction-emerging technologies, particularly by using simplified techniques and non-toxic (or none) solvents that meet the environmental safety status. These will give more sustainable processes for the recovery of bioactive compounds, increasing the safety of products coming from grape waste and justifying the recycling process from both technological and environmental points of view.

References

Aliakbarian, B., A. Fathi, P. Perego and F. Dehghani. 2012. Extraction of antioxidants from winery wastes using subcritical water. J. Supercrit. Fluids 65: 18–24.

Alonso, Á. M., D. A. Guillén, C. G. Barroso, B. Puertas and A. García. 2002. Determination of antioxidant activity of wine by-products and its correlation with polyphenolic content. J. Agric. Food Chem. 50: 5832–5836.

Amendola, D., D. M. De Faveri and G. Spigno. 2010. Grape marc phenolics: Extraction kinetics, quality and stability of extracts. J. Food Eng. 97: 384–392.

Anastasiadi, M., N. G. Chorianopoulos, G. J. E. Nychas and S. A. Karoutounian. 2009. Antilisterial activities of polyphenol-rich extracts of grapes and vinification by-products. J. Agr. Food Chem. 57: 457–463.

Anastasiadi, M., H. Pratsinis, D. Kletsas, A. -L. Skaltsounis and S. A. Haroutounian. 2012. Grape stem extracts: Polyphenolic content and assessment of their *in vitro* antioxidant properties. LWT—Food Sci. Technol. 48: 316–322.

Anastasiadi, M., H. Pratsinis, D. Kletsas, A. L. Skaltsounis and S. A. Haroutounian. 2012. Grape stem extracts: Polyphenolic content and assessment of their *in vitro* antioxidant properties. LWT—Food Sci. Technol. 48: 316–322.

Antoniolli, A., A. R. Fontana, P. Piccoli and R. Bottini. 2015. Characterization of polyphenols and evaluation of antioxidant capacity in grape pomace of the cv. Malbec. Food Chem. 178: 172–178.

AOAC. 1995. Official Methods of Analysis of the Association of Official Analystical Chemists. 16th Edition: 7–9.

Apostolou, A., D. Stagos, E. Galitsiou, A. Spyrou, S. Haroutounian, N. Portesis, I. Trizoglou, A. Wallace Hayes, A. M. Tsatsakis and D. Kouretas. 2013. Assessment of polyphenolic content, antioxidant activity, protection against ROS-induced DNA damage and anticancer activity of *Vitis vinifera* stem extracts. Food Chem. Toxicol. 61: 60–68.

Barba, F. J., S. Brianceau, M. Turk, N. Boussetta and E. Vorobiev. 2015. Effect of alternative physical treatments (Ultrasounds, pulsed electric fields, and high-voltage electrical discharges) on selective recovery of bio-compounds from fermented grape pomace. Food Bioprocess Technol. 8: 1139–1148.

Barros, A., A. Gironés-Vilaplana, A. Texeira, N. Baenas and R. Domínguez-Perles. 2015. Grape stems as a source of bioactive compounds: Application towards added-value commodities and significance for human health. Phytochem. Rev. 14: 921–931.

Beres, C., F. F. Simas-Tosin, I. Cabezudo, S. P. Freitas, M. Iacomini, C. Mellinger-Silva and L. M. C. Cabral. 2016. Antioxidant dietary fibre recovery from Brazilian Pinot noir grape pomace. Food Chem. 201: 145–152.

Bonilla, F., M. Mayen, J. Merida and M. Medina. 1999. Extraction of phenolic compounds from red grape marc for use as food lipid antioxidants. Food Chem. 66: 209–215.

Boussetta, N., A. De Ferron, T. Reess, L. Pecastaing, J. L. Lanoisellé and E. Vorobiev. 2009. Improvement of polyphenols extraction from grape pomace using pulsed arc electro-hydraulic discharges. *In*: PPC2009—17th IEEE International Pulsed Power Conference. pp. 1088–1093.

Boussetta, N., J. L. Lanoisellé, C. Bedel-Cloutour and E. Vorobiev. 2009. Extraction of soluble matter from grape pomace by high voltage electrical discharges for polyphenol recovery: Effect of sulphur dioxide and thermal treatments. J. Food Eng. 95: 192–198.

Boussetta, N., E. Vorobiev, V. Deloison, F. Pochez, A. Falcimaigne-Cordin and J. L. Lanoisellé. 2011. Valorisation of grape pomace by the extraction of phenolic antioxidants: Application of high voltage electrical discharges. Food Chem. 128: 364–370.

Boussetta, N., E. Vorobiev, T. Reess, A. De Ferron, L. Pecastaing, R. Ruscassié and J. L. Lanoisellé. 2012. Scale-up of high voltage electrical discharges for polyphenols extraction from grape pomace: Effect of the dynamic shock waves. Innov. Food Sci. Emerg. Technol. 16: 129–136.

Brazinha, C., M. Cadima and J. G. Crespo. 2014. Optimization of extraction of bioactive compounds from different types of grape pomace produced at wineries and distilleries. J. Food Sci. 79: E1142–E1149.

Bucić-Kojić, A., M. Planinić, S. Tomas, L. Jakobek and M. Šeruga. 2009. Influence of solvent and temperature on extraction of phenolic compounds from grape seed, antioxidant activity and colour of extract. Int. J. Food Sci. Tech. 44: 2394–2401.

Cacace, J. E. and G. Mazza. 2003. Mass transfer process during extraction of phenolic compounds from milled berries. J. Food Eng. 59: 379–389.

Casas, L., C. Mantell, M. Rodríguez, E. J. M. D. I. Ossa, A. Roldán, I. D. Ory, I. Caro and A. Blandino. 2010. Extraction of resveratrol from the pomace of Palomino fino grapes by supercritical carbon dioxide. J. Food Eng. 96: 304–308.

Cavalcanti, R. N. and M. A. A. Meireles. 2012. Fundamentals of supercritical fluid extraction—Pawliszyn, Janusz. *In*: Comprehensive Sampling and Sample Preparation. pp. 117–133. Oxford: Academic Press.

Corrales, M., A. F. García, P. Butz and B. Tauscher. 2009. Extraction of anthocyanins from grape skins assisted by high hydrostatic pressure. J. Food Eng. 90: 415–421.

Cummings, J., J. Mann, C. Nishida and H. Vorster. 2009. Dietary fibre: An agreed definition. Lancet 373: 365–366.

Chidambara Murthy, K. N., R. P. Singh and G. K. Jayaprakasha. 2002. Antioxidant activities of grape (*Vitis vinifera*) pomace extracts. J. Agr. Food Chem. 50: 5909–5914.

Choi, B., S. Kim, B. G. Jang and M. J. Kim. 2016. Piceatannol, a natural analogue of resveratrol, effectively reduces beta-amyloid levels via activation of alpha-secretase and matrix metalloproteinase-9. J. Funct. Foods 23: 124–134.

Chouchouli, V., N. Kalogeropoulos, S. J. Konteles, E. Karvela, D. P. Makris and V. T. Karathanos. 2013. Fortification of yoghurts with grape (*Vitis vinifera*) seed extracts. LWT—Food Sci. Technol. 53: 522–529.

Delgado Adámez, J., E. Gamero Samino, E. Valdés Sánchez and D. González-Gómez. 2012. *In vitro* estimation of the antibacterial activity and antioxidant capacity of aqueous extracts from grape-seeds (*Vitis vinifera* L.). Food Control 24: 136–141.

Deng, Q., M. H. Penner and Y. Zhao. 2011. Chemical composition of dietary fiber and polyphenols of five different varieties of wine grape pomace skins. Food Res. Int. 44: 2712–2720.

Dhingra, D., M. Michael, H. Rajput and R. Patil. 2012. Dietary fibre in foods: A review. J. Food Sci. Technol. 49: 255–266.

Dias, C., R. Domínguez-Perles, A. Aires, A. Teixeira, E. Rosa, A. Barros and M. J. Saavedra. 2015. Phytochemistry and activity against digestive pathogens of grape (*Vitis vinifera* L.) stem's (poly) phenolic extracts 61: 25–32.

Ferreira, A. S., C. Nunes, A. Castro, P. Ferreira and M. A. Coimbra. 2014. Influence of grape pomace extract incorporation on chitosan films properties 113: 490–499.

Fiori, L. 2010. Supercritical extraction of grape seed oil at industrial-scale: Plant and process design, modeling, economic feasibility. Chem. Eng. Process 49: 866–872.

Fontana, A. R., A. Antoniolli and R. Bottini. 2013. Grape pomace as a sustainable source of bioactive compounds: Extraction, characterization, and biotechnological applications of phenolics. J. Agr. Food Chem. 61: 8987–9003.

Fontana, A. R. and R. Bottini. 2014. High-throughput method based on quick, easy, cheap, effective, rugged and safe followed by liquid chromatography-multi-wavelength detection for the quantification of multiclass polyphenols in wines. J. Chromatogr. A 1342: 44–53.

Fontana, A. R., A. Antoniolli and R. Bottini. 2016. Development of a high-performance liquid chromatography method based on a core–shell column approach for the rapid determination of multiclass polyphenols in grape pomaces. Food Chem. 192: 1–8.

Fuentes-Alventosa, J. M., G. Rodríguez-Gutiérrez, S. Jaramillo-Carmona, J. Espejo-Calvo, R. Rodríguez-Arcos, J. Fernández-Bolaños, R. Guillén-Bejarano and A. Jiménez-Araujo. 2009. Effect of extraction method on chemical composition and functional characteristics of high dietary fibre powders obtained from asparagus by-products. Food Chem. 113: 665–671.

Galanakis, C. M. 2012. Recovery of high added-value components from food wastes: Conventional, emerging technologies and commercialized applications. Trens Food Sci. Technol. 26: 68–87.

García, J., R. Pereira, C. Marques, M. Marques, I. Carolino and M. Oliveira Sousa. 2003. Biología Cellular y Molecular.

Garrido, J. and F. Borges. 2013. Wine and grape polyphenols—A chemical perspective. Food Res. Int. 54: 1844–1858.

Ghafoor, K., Y. H. Choi, J. Y. Jeon and I. H. Jo. 2009. Optimization of ultrasound-assisted extraction of phenolic compounds, antioxidants, and anthocyanins from grape (*Vitis vinifera*) seeds. J. Agr. Food Chem. 57: 4988–4994.

González-Centeno, M. R., C. Rosselló, S. Simal, M. C. Garau, F. López and A. Femenia. 2010. Physico-chemical properties of cell wall materials obtained from ten grape varieties and their by-products: Grape pomaces and stems. LWT—Food Sci. Technol. 43: 1580–1586.

Hogan, S., C. Canning, S. Sun, X. Sun and K. Zhou. 2010. Effects of grape pomace antioxidant extract on oxidative stress and inflammation in diet induced obese mice. J. Agr. Food Chem. 58: 11250–11256.

Hogan, S., L. Zhang, J. Li, S. Sun, C. Canning and K. Zhou. 2010. Antioxidant rich grape pomace extract suppresses postprandial hyperglycemia in diabetic mice by specifically inhibiting alpha-glucosidase. Nutr. Metab. 7: 1–9.

http://www.fao.org, F. -F. S. D. 2010. FAOSTAT-FAO Statistical Database.

Ignat, I., A. Stingu, I. Volf and V. I. Popa. 2011. Characterization of grape seed aqueous extract and possible applications in biological systems. Cellul. Chem. Technol. 45: 205–209.

Ignat, I., I. Volf and V. I. Popa. 2011. A critical review of methods for characterisation of polyphenolic compounds in fruits and vegetables. Food Chem. 126: 1821–1835.

Jayaprakasha, G. K., T. Selvi and K. K. Sakariah. 2003. Antibacterial and antioxidant activities of grape (*Vitis vinifera*) seed extracts. Food Res. Int. 36: 117–122.

Kammerer, D., A. Claus, R. Carle and A. Schieber. 2004. Polyphenol screening of pomace from red and white grape varieties (*Vitis vinifera* L.) by HPLC-DAD-MS/MS. J. Agr. Food Chem. 52: 4360–4367.

Kammerer, D., A. Claus, A. Schieber and R. Carle. 2005. A novel process for the recovery of polyphenols from grape (*Vitis vinifera* L.) pomace. J. Food Sci. 70: C157–C163.

Kammerer, D. R., A. Schieber and R. Carle. 2005. Characterization and recovery of phenolic compounds from grape pomace—A review. J. Appl. Bot. Food Qual. 79: 189–196.

Kammerer, D. R., J. Kammerer, R. Valet and R. Carle. 2014. Recovery of polyphenols from the by-products of plant food processing and application as valuable food ingredients. Food Res. Int. 65: 2–12.

Kutil, Z., V. Temml, D. Maghradze, M. Pribylova, M. Dvorakova, D. Schuster, T. Vanek and P. Landa. 2014. Impact of wines and wine constituents on cyclooxygenase-1, cyclooxygenase-2, and 5-lipoxygenase catalytic activity. Mediators of Inflammation.

Luque de Castro, M. D. and F. Priego-Capote. 2006. In Analytical Applications of Ultrasound. Ultrasound Assistance to Analytical Heterogeneous Liquid-Liquid Systems 26.

Llobera, A. and J. Cañellas. 2007. Dietary fibre content and antioxidant activity of Manto Negro red grape (*Vitis vinifera*): Pomace and stem. Food Chem. 101: 659–666.

Llobera, A. and J. Cañellas. 2008. Antioxidant activity and dietary fibre of Prensal Blanc white grape (*Vitis vinifera*) by-products. Int. J. Food Sci. Technol. 43: 1953–1959.

Middleton Jr, E., C. Kandaswami and T. C. Theoharides. 2000. The effects of plant flavonoids on mammalian cells: Implications for inflammation, heart disease, and cancer. Pharmacol. Rev. 52: 673–751.

Mildner-Szkudlarz, S., J. Bajerska, R. Zawirska-Wojtasiak and D. Górecka. 2013. White grape pomace as a source of dietary fibre and polyphenols and its effect on physical and nutraceutical characteristics of wheat biscuits. J. Sci. Food Agr. 93: 389–395.

Monrad, J. K., L. R. Howard, J. W. King, K. Srinivas and A. Mauromoustakos. 2010a. Subcritical solvent extraction of anthocyanins from dried red grape pomace. J. Agric. Food Chem. 58: 2862–2868.

Monrad, J. K., L. R. Howard, J. W. King, K. Srinivas and A. Mauromoustakos. 2010b. Subcritical solvent extraction of procyanidins from dried red grape pomace. J. Agr. Food Chem. 58: 4014–4021.

Monro, J. and B. Burlingame. 1996. Carbohydrates and related food components: INFOODS tagnames, meanings, and uses. J. Food Compos. Anal. 9: 100–118.

Negro, C., L. Tommasi and A. Miceli. 2003. Phenolic compounds and antioxidant activity from red grape marc extracts. Bioresource Technol. 87: 41–44.

Oliveira, D. A., A. A. Salvador, A. Smânia, E. F. A. Smânia, M. Maraschin and S. R. S. Ferreira. 2013. Antimicrobial activity and composition profile of grape (*Vitis vinifera*) pomace extracts obtained by supercritical fluids. J. Biotechnol. 164: 423–432.

Otero-Pareja, M. J., L. Casas, M. T. Fernández-Ponce, C. Mantell, E. J. M. De La Ossa. 2015. Green extraction of antioxidants from different varieties of red grape pomace. Molecules 20: 9686–9702.

Palenzuela, B., L. Arce, A. MacHo, E. Muñoz, A. Ríos and M. Valcárcel. 2004. Bioguided extraction of polyphenols from grape marc by using an alternative supercritical-fluid extraction method based on a liquid solvent trap. Anal. Bioanal. Chem. 378: 2021–2027.

Pinelo, M., M. Rubilar, M. Jerez, J. Sineiro and M. J. Núñez. 2005. Effect of solvent, temperature, and solvent-to-solid ratio on the total phenolic content and antiradical activity of extracts from different components of grape pomace. J. Agric. Food Chem. 53: 2111–2117.

Pinelo, M., J. Sineiro and M. A. J. Núñez. 2006. Mass transfer during continuous solid–liquid extraction of antioxidants from grape by-products. J. Food Eng. 77: 57–63.

Pozuelo, M. J., A. Agis-Torres, D. Hervert-Hernández, M. E. López-Oliva, E. Muñoz-Martínez, R. Rotger and I. Goñi. 2012. Grape antioxidant dietary fiber stimulates lactobacillus growth in rat cecum. J. Food Sci. 77: H59–H62.

Prior, R. L. and G. Cao. 1999. Antioxidant capacity and polyphenolic components of teas: Implications for altering in vivo antioxidant status 220: 255–261.

Richter, B. E. and D. Raynie. 2012. Accelerated solvent extraction (ASE) and high-temperature water extraction A2—Pawliszyn, Janusz. pp. 105–115. *In*: Comprehensive Sampling and Sample Preparation. Oxford: Academic Press.

Rodriguez-Rodriguez, R., M. L. Justo, C. M. Claro, E. Vila, J. Parrado, M. D. Herrera and M. Alvarez De Sotomayor. 2012. Endothelium-dependent vasodilator and antioxidant properties of a novel enzymatic extract of grape pomace from wine industrial waste. Food Chem. 135: 1044–1051.

Rodriguez Lanzi, C., D. J. Perdicaro, A. Antoniolli, A. R. Fontana, R. M. Miatello, R. Bottini and M. A. Vazquez Prieto. 2016. Grape pomace and grape pomace extract improve insulin signaling in high-fat-fructose fed rat-induced metabolic syndrome. Food Funct. 7: 1544–1553.

Ruiz-Moreno, M. J., R. Raposo, J. M. Cayuela, P. Zafrilla, Z. Piñeiro, J. M. Moreno-Rojas, J. Mulero, B. Puertas, F. Giron, R. F. Guerrero and E. Cantos-Villar. 2015. Valorization of grape stems 63: 152–157.

Saura-Calixto, F. 1998. Antioxidant dietary fiber product: A new concept and a potential food ingredient. J. Agric. Food Chem. 46: 4303–4306.

Saura-Calixto, F. 2012. Concept and health-related properties of nonextractable polyphenols: The missing dietary polyphenols. J. Agric. Food Chem. 60: 11195–11200.

Šeruga, M., I. Novak and L. Jakobek. 2011. Determination of polyphenols content and antioxidant activity of some red wines by differential pulse voltammetry, HPLC and spectrophotometric methods. Food Chem. 124: 1208–1216.

Soural, I., N. Vrchotová, J. Tříska, J. Balík, Š. Horník, P. Cuřínová and J. Sýkora. 2015. Various Extraction methods for obtaining stilbenes from grape cane of *Vitis vinifera* L. Molecules 20: 6093–6112.

Spigno, G., L. Tramelli and D. M. De Faveri. 2007. Effects of extraction time, temperature and solvent on concentration and antioxidant activity of grape marc phenolics. J. Food Eng. 81: 200–208.

Teixeira, A., N. Baenas, R. Dominguez-Perles, A. Barros, E. Rosa, D. A. Moreno and C. Garcia-Viguera. 2014. Natural bioactive compounds from winery by-products as health promoters: A review 15: 15638–15678.

Terra, X., G. Montagut, M. Bustos, N. Llopiz, A. Ardèvol, C. Bladé, J. Fernández-Larrea, G. Pujadas, J. Salvadó, L. Arola and M. Blay. 2009. Grape-seed procyanidins prevent low-grade inflammation by modulating cytokine expression in rats fed a high-fat diet. J. Nutr. Biochem. 20: 210–218.

Tseng, A. and Y. Zhao. 2012. Effect of different drying methods and storage time on the retention of bioactive compounds and antibacterial activity of wine grape pomace (Pinot Noir and Merlot). J. Food Sci. 77: H192–H201.

Vatai, T., M. Škerget and Z. Knez. 2009. Extraction of phenolic compounds from elder berry and different grape marc varieties using organic solvents and/or supercritical carbon dioxide. J. Food Eng. 90: 246–254.

Vergara-Salinas, J. R., P. Bulnes, M. C. Zúñiga, J. Pérez-Jiménez, J. L. Torres, M. L. Mateos-Martín, E. Agosin and J. R. Pérez-Correa. 2013. Effect of pressurized hot water extraction on antioxidants from grape pomace before and after enological fermentation. J. Agric. Food Chem. 61: 6929–6936.

Vergara, C., D. Von Baer, C. Mardones, A. Wilkens, K. Wernekinck, A. Damm, S. MacKe, T. Gorena and P. Winterhalter. 2012. Stilbene levels in grape cane of different cultivars in southern Chile: Determination by HPLC-DAD-MS/MS method. J. Agric. Food Chem. 60: 929–933.

Wang, X., H. Tong, F. Chen and J. D. Gangemi. 2010. Chemical characterization and antioxidant evaluation of muscadine grape pomace extract. Food Chem. 123: 1156–1162.

Wells, M. J. M. 2003. Sample preparation techniques in analytical chemistry. Chapter 2 Principles of extraction and the extraction of semivolatile organics from liquids. *In*: J. D. Winefordner (ed.). Sample Preparation Techniques in Analytical Chemistry, Vol. 162. Hoboken, New Jersey: John Wiley & Sons.

Wu, T. H., J. H. Liao, F. L. Hsu, H. R. Wu, C. K. Shen, J. M. P. Yuann and S. T. Chen. 2010. Grape seed proanthocyanidin extract chelates iron and attenuates the toxic effects of 6-hydroxydopamine: Implications for Parkinson's disease. J. Food Biochem. 34: 244–262.

Yi, W., J. Fischer and C. C. Akoh. 2005. Study of anticancer activities of muscadine grape phenolics *in vitro*. J. Agr. Food Chem. 53: 8804–8812.

Yu, J. and M. Ahmedna. 2013. Functional components of grape pomace: Their composition, biological properties and potential applications. Int. J. Food Sci. Technol. 48: 221–237.

CHAPTER 9

Extraction, Isolation and Utilisation of Bioactive Compounds from Waste Generated by the Olive Oil Industry

J. Lozano-Sánchez,[1,2,*] *I. Cea Pavez,*[3] *E. González-Cáceres,*[3]
H. Núñez Kalasic,[4] *P. Robert Canales*[3] and *A. Segura Carretero*[1,2]

Introduction

Extra virgin olive oil (EVOO) has acquired great importance due to its healthy properties as recently proved by various nutritional studies (Covas 2008). Among the bioactive properties, literature reveals that EVOO is antioxidant, anti-inflammatory, gastroprotective, cardioprotective, neuroprotective, hepato-protective, anti-diabetes, anti-obesity and anticancerous (Hassen et al. 2015, Scoditti et al. 2014, Whayne 2014). Indeed, EVOO can be regarded as functional food because it has beneficial effect on one or more target functions in the body, beyond providing nutrition, thus improving one's health and well-being or reducing the risk of disease.

These reported health properties are related to olive oil composition. EVOO compounds can be divided into major and minor fractions, based on their chemical composition. The major components, which include triacylglycerols, represent more than 98 per cent of the total oil weight. Minor components, which are present in very low amounts (about 2 per cent of oil weight), include more than 300 different chemical lipophilic and hydrophilic compounds, such as hydrocarbons, triterpenes, pigments, tocopherols, and phenols. Evidence from several studies point to the bioactivity of olive

[1] Department of Analytical Chemistry, University of Granada. Avda. Fuentenueva s/n, 18071 Granada (Spain).
[2] Research and Development of Functional Food Centre (CIDAF), PTS Granada, Avda. del Conocimiento s/n, Edificio Bioregión, 18016 Granada, Spain.
[3] Department of Food Sci. and Chem. Technology, University of Chile.
[4] Department of Agroindustry and Enology, University of Chile.
* Corresponding author: jesusls@cidaf.es

oil compounds belonging to major and minor fractions. The bioactive composition of EVOO is influenced by complex multivariate interactions of genotype, agricultural, environmental and technological factors (Frankel 2013, Lozano-Sánchez 2010).

Concerning technological factors, the production process includes collecting, washing, crushing of olives, malaxation of olive paste, centrifugation, filtration, and storage. These steps influence the final qualitative and quantitative bioactive composition of the oil. The loss of specific compounds due to major and minor fractions during the elaborate process is related to its partitioning between oil and by-products. Consequently, EVOO production is associated with generation of large quantities of waste, which could be an alternative source of bioactive compounds. The first by-product consists of olive leaves and twigs, resulting from the washing of the fruit. For three-phases plants, the main by-products are a liquid waste known as olive-mill wastewater, vegetation water, or 'alpechin' and a solid waste called pomace or *orujo*. The use of two-phase processing technique generates a by-product called *alperujo* or olive pomace, which includes a combination of liquid and solid waste. Other olive-oil by-products, generated by the storage and filtration of olive oil, are composed of solid and liquid sediments and cakes used for filtration.

The loss of healthy components in these by-products has prompted multidisciplinary research on recovering bioactive compounds. The present chapter focuses on different technologies aimed at utilizing olive residue through promoting their overall sustainable management and conversion into an affordable source of bioactive compounds with particular emphasis on phenolic compounds.

Waste Generated by Olive Oil Industry

Olive oil production is an agro-industrial activity of vital economic importance for Mediterranean countries. However, the production of olive oil generates large volumes of olive by-products (Papaphilippou et al. 2013). During olive processing, three main residual products are generated—twigs and olive leaves, olive pomace (*orujo* or *alperujo*, so-called depending on the centrifuge system) and olive mill waste water (OMWW) (Kapellakis et al. 2008). Other by-products generated by storage and filtration of olive oil are composed of solid and liquid storage waste and cakes used for filtering EVOO (Frankel et al. 2013). Figure 1 shows the flow chart of the by-products generated over the olive oil elaboration process.

Twigs and olive leaves

Twigs and olive leaves are generated during the pruning of olive trees (about 25 kg of branches and leaves per tree annually) (Herrero et al. 2011). Furthermore, these by-products are generated in large amounts during olive harvesting carried out by hand or using different device systems in order to increase the effectiveness of the harvesting: limb shaking devices, trunk shakers, impact shakers, double or single sided picking head mechanisms and straddle type. In olive oil industries, twigs and olive leaves are separated from the fruit before processing (about 10 per cent of the weight of olives) (Herrero et al. 2011). These leaves do not have any commercial value and are used as fuel after drying or eliminated in the orchard, through burning, with a negative environmental impact (Kapellakis et al. 2008, Ranalli et al. 2006, Sanchez et al. 2002).

The chemical composition of olive leaves varies depending on several conditions, such as origin, proportion of branches on the tree, storage, climatic conditions and moisture content among others (El Sedef and Karakaya 2009). Table 1 shows the chemical composition of this by-product (García 2003).

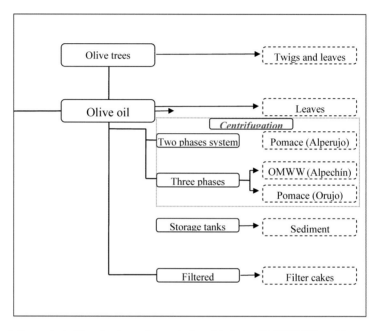

Figure 1. By-product of the olive tree culture and olive oil industry (OMWW: Olive mill waste water).

Table 1. Chemical composition of olive leaves.

Measurements	Olive leaves
Dry matter, g/kg fresh matter	586
g/kg dry matter	
Organic matter	838
Crude fat	32.1
Neutral detergent fiber	413
Acid detergent fiber	333
Acid detergent lignin	190
Crude protein (total N × 6.25)	70.0
g/kg dry matter	
Total extractable polyphenols	25.3
Total extractable tannins	1.00
Total extractable condensed tannins	0.75
Total condensed tannins	8.30

OMWW and olive pomace

After leaf removal and washing, olives are subjected to crushing and malaxation. Crushing is normally carried out by using a traditional stone mill or by means of a disk or hammer crusher. During malaxation, olive paste is subjected to a slow continuous kneading, aimed at breaking off the emulsions formed during the crushing process and facilitating adequate coalescence. Separation of the liquid and solid phases is carried out by centrifugal decanters. This separation technique can be divided into two different systems—three- and two-phase decanter. The three-phase decanter allows the separation of three flows of matter—the olive oil, olive pomace (solid remains of olive) and wastewater by adding water. Centrifugal force moves the heavier solid materials to the outside; a lighter water layer is formed in the middle, with the lightest oil layer on the inside. Two-phase decanter enables the oil phase to be separated from the malaxed olive paste without addition of water. This decanter has two exits—producing oil and olive pomace (Amirante et al. 2010, Rubio-Senet et al. 2012).

Concerning the pomace, several authors have reported the chemical composition of this by-product generated by three- and two-phase decanter (Table 2).

Olive pomace, which consist of olive pulp, skin, stone and water, is an inexpensive biomass that is generated in large quantities in Mediterranean countries. Its commercial value depends on its oil and water content. The olive pomace obtained through two-phase extraction procedure may be differentiated by the higher moisture and lower oil content than from the three-phase centrifugation procedure (Kapellakis et al. 2008). For this reason, the two-phase decanter is more efficient and environmental-friendly process than compared to the three-phase system (Frankel et al. 2013). Studies have been aimed at reducing the environmental impact of olive pomace and/or harnessing its potential economic value. Olive pomace is used as fuel, fertilizer or animal feed. Nevertheless, the profitability of olive pomace treatment plants is still in doubt because these activities represent a very small percentage of the olive pomace produced (Frankel et al. 2013).

OMWW is called *aqua reflue* in Italy, *alpechin* in Spain, *katsigaros* in Greece, *zebar* in Arab countries (Kapellakis et al. 2008). This waste is composed of vegetable

Table 2. Quantity and characteristics of olive pomace obtained with different extraction systems for olive oil (Pattara et al. 2010).

Measurements	Three-phase decanter	Two-phase decanter
Quantity (kg/ton olives)	450–550	800–850
Moisture (%)	45–55	65–75
Oil (% on fresh pomace)	3.5–4.5	3–4
Pulp (%)	15–25	10–15
Stones (%)	20–28	12–18
Ash (%)	2–4	3–4
Nitrogen (mg/100 g)	200–300	250–350
Phosphorus (mg/100 g)	30–40	40–50
Potassium (mg/100 g)	100–150	150–250
Total phenolic compounds (mg/100 g)	200–300	400–600

water from the fruit plus the water used at different stages of oil extraction. The OMWW physical-chemical properties consist in a dark red to black-coloured liquid with a strong smell of olive oil and is mildly acidic liquid of high conductivity (Kapellakis et al. 2008). This by-product contains olive pulp, mucilage, pectin, oil and other suspended components in a relatively stable emulsion. Their chemical composition may be divided into two fractions—water (83–92 per cent by weight) and organic matter, including readily fermentable proteins, organic acids (acetic, fumaric, glyceric and oxalic acid), sugars, oil, phenols, waxy and resinous substances, vitamins and traces of pesticides (Table 3). Their chemical composition is variable depending on olive variety, growing technique, harvesting period and especially the technology used for oil extraction (Roig et al. 2006, Frankel et al. 2013).

Storage By-products and Filter Cake

Storage by-products are generated during oil storage in the mill companies. Indeed, in the Mediterranean area, the production of olive oil is carried out from September to February and the oil storage is prolonged until its commercialization. During this storage time, two different by-products are generated—solid and aqueous wastes, which are deposited at the bottom of the storage tanks. These wastes are characterized

Table 3. Composition of olive mill waste water from pressure and centrifugation system (Kapellakis et al. 2008).

Constituents	Processing system	
	Pressure	Centrifugation
pH	4.5–5	4.7–5.2
Soluble solids (%)	0.1	0.9
Total soluble (%)	12	6
Organic constituents (%)		
Sugars	2.0–8.0	0.5–2.6
Nitrogen compounds	0.5–2.0	0.1–0.3
Organics acids	0.1–1.5	0.2–0.5
Polyalcohols	1.0–1.5	0.3–0.5
Pentoses, tannins	1.0–1.5	0.2–0.5
Polyphenols	2.0–2.4	0.3–0.8
Lipids	0.03–1.0	0.5–2.3
Inorganic constituents (%)		
P	0.11	0.03
K	0.72	0.27
Ca	0.07	0.02
Mg	0.04	0.01
Na	0.09	0.03
Cl	0.03	0.01

by olive pulp, skin, stone and water. With regard to the minor fraction, phenolic compounds belonging to different groups have been identified in these by-products (Lozano-Sánchez et al. 2011a).

Before commercialization of EVOO, the oil is filtered. The filtration process is carried out by using in organic or inorganic materials in conjunction with a variety of filtration equipment to enhance or conduct suspended solids and water-oil separations. As a result, the suspended solids are removed and the moisture is reduced. Consequently, this step generates a brilliant EVOO and a waste, named filter cake. The composition of the filter cake depends of the nature of the material used as filter aids. Filter aids for precoat filtration are produced from a wide variety of raw materials. Commonly utilized filter aids include diatomite with different particle sizes, different permeabilities and organic fibrous material (cellulose fibers) (Lozano-Sánchez et al. 2010). In addition, filter cakes are composed of water, polyphenols and other impurities retained in the filter. The phenolic composition of inorganic and organic filter aids generated after filtering olive oil have also been reported in relevant literature (Lozano-Sánchez et al. 2011a). This scientific report establishes that phenolic compounds are partitioned between olive oil and filter cakes.

Among these by-products, the most abundant ones is OMWW; approximately 3×10^7 m^3 of these effluents are produced annually in Mediterranean countries. OMWW is characterized by high chemical oxygen demand (COD: 40–220 g/L) and biochemical oxygen demand (BOD: 35–110 g/L), a pH of 3.0–6.0, together with a high phenolic content (0.5–24 g/L) (Niaounakis and Halvadakis 2006). This waste is a severe environmental problem mainly due to the high concentrations of phenolic compounds. Olive pomace is generated in large quantities in Mediterranean countries (450–550 and 800–850 kg/t olives generated by 3-phase and 2-phase decanters, respectively). The total phenolic compounds present in this by-product range from 200 to 600 mg/100 g of olive pomace and consequently, pose serious environmental problems (Pattara et al. 2010). Storage by-products are characterized by a high phenolic content (1.5 g per kg of solid waste or litres of aqueous wastes). The phenolic composition of filter cakes is 20-fold higher than that in olive oil (Férnandez-Gutiérrez et al. 2012).

The composition and amount of these by-products create serious environmental problems in the Mediterranean areas. The discharge of large quantities of the pollutants in the physical surroundings is not possible without any treatment. Furthermore, the pharmaceutical interest in olive by-products due to their phenolic composition is well-known and has stimulated multidisciplinary research on utilizing olive residues, either through promoting their overall sustainable management by green chemistry or conversion into an affordable source of natural antioxidant.

Bioactive Compounds from Waste Generated by Olive Oil Industry

Over the ages, olive and olive oils have been employed in traditional medicine for the treatment of diseases, in both humans and animals (Frankel 2011). This ability to treat disease has been associated to the compounds they contain, which give rise to a wide variety of biological activities. Nevertheless, the elaboration process of EVOO generates the loss of healthy components. The pharmaceutical interest in these compounds has stimulated multidisciplinary research on the composition of olive by-products as alternative source of bioactive compounds (Lozano-Sánchez et al. 2010). The biological activity of by-product extracts enriched on bioactive compounds might be assessed by: (a) nutritional epidemiology, a field of medical research that studies the relationship between nutrition and health; (b) *in vitro* studies which are performed with microorganisms, cells or biological molecules outside their normal biological context; and (c) *in vivo* studies conducted on animals, including humans and whole plants. This section summarizes the scientific reports on the chemical analysis of olive oil by-products addressed to bioactive compounds with particular emphasis on hydrocarbons, triterpenes and phenolic compounds.

Concerning hydrocarbons, squalene is a useful compound of olive pomace and OMWW. This hydrocarbon is a naturally occurring polyprenyl compound primarily known for its key role as an intermediate in cholerestol synthesis. *In vitro* and *in vivo* studies report that squalene administration decreased reactive oxygen species in lipoprotein fractions and caused a specific increase in high density lipoprotein (HDL)-cholesterol levels in animals (Gabás-Rivera et al. 2014). Furthermore, squalene is considered to be an important component due to its chemopreventative potential against cancer (Ghanbari et al. 2012).

With regard to triterpenes, two classes of triterpenic compounds are characterized in olive leaves—dialcohols (uvaol and erythrodiol) and triterpenic acids (oleanolic, ursolic and maslinic acids). Several studies show that these compounds possess healthy properties, which are anti-inflammatory and vasodilatory (Rodríguez-Rodríguez et al. 2007).

Phenolic fraction has aroused interest regarding its heath-promoting properties. Phenolic compounds, known for many years as 'polyphenols', are natural substances that generally present an aromatic ring with one or more hydroxyl substituents. They are divided into simple phenols, which are formed by one phenolic structure, and polyphenols, which include more than one in their structure. Phenolic compounds can be normally conjugated to one or more sugar residues linked to their hydroxyl groups, or even to the aromatic backbone. These compounds arise biogenetically from the shikimate/phenylpropanoid and mevalonate pathway (Quideau et al. 2011).

In olive by-products, the phenolic fraction consists of a heterogeneous mixture of compounds belonging to several families with different chemical structures—simple phenols (phenolic acids and phenolic alcohols), fenilpropanoids, flavonoids, lignans, oleosides and secoiridoids. Oleuropein is described as the major compound in olive leaves and twigs, which also contain high concentrations of glucosylated flavones, such as luteolin-7-glucoside and apigenin-7-glucoside (Lujan et al. 2008). Twigs are also characterized to an elevated amount of verbascoside (Fu et al. 2010, Paiva-Martin and Pinto 2008). OMWW is characterized by a high content of phenolic alcohols (hydroxytyrosol and tyrosol) (Agalias et al. 2007). However, oleoside derivatives, such as caffeoyl-6-oleoside, *p*-coumaroyl derivatives of oleoside/secologanoside and 6-O-[(2E)-2,6-dimethyl-8-hydroxy-2-octenoyloxy] secologanoside have also been found in olive leaves, OMWW, pomace and thermally treated olive-oil by-products (Rubio-Senent et al. 2013a, b, Kanakis et al. 2013, Quirantes-Piné et al. 2013). The chemical characterization of filter cake and storage by-products points to the presence of secoiridoids—decarboxymethyl elenolic acid linked to hydroxytyrosol (oleacein) and decarboxymethyl elenolic acid linked to tyrosol (oleocanthal) (Lozano-Sánchez et al. 2011b).

Phenolic compounds identified in olive by-products can be regarded as bioactive compounds as they affect beneficially one or more target functions in the body in a way that improves the health or reduces the risk of disease. Olive phenolics play an essential role in slowing down the oxidative process. The antioxidant activity of phenolic extracts obtained from olive leave, OMWW and pomace have been evaluated (El Sedef and Karakaya 2009, Alu'datt et al. 2010, Frankel et al. 2013). The results showed that the antioxidant activity of phenolic compounds in olive pomace provides a cheap source of natural antioxidants in concentrations up to 100 times higher than in EVOO. On the other hand, phenolic extracts from OMWW can be used as natural alternatives to commercial synthetic antioxidants with applications in food as well as in the development of nutraceutical and medical products (Frankel et al. 2013).

Furthermore, these compounds carry out biological activities, that are not confined to being antioxidant and free radical scavenging. In an attempt to establish the molecular mechanisms underlying these beneficial effects, several studies were carried out *in vitro* and *in vivo* experimental designs that shed light on such hypothesis (Ghanbari et al. 2012). Indeed, the distribution of phenolic compounds in rat tissues after the administration of a phenolic extract from olive pomace was demonstrated (Serra et al. 2012). The analysis of rat tissues showed a wide distribution of phenolic compounds and their metabolites, with the main detoxification route going through the kidneys. The free forms of some phenolic compounds, such as oleuropein derivative, were quantitated in plasma and brain, luteolin in kidney, testicle, and heart, and hydroxytyrosol in plasma, kidney and testicles. In this way, these compounds are available to produce different target functions in the body.

Concerning olive leaves, hypoglycemic, antihypertensive, antimicrobial and antiatherosclerotic effects of these samples were reported in various studies. These properties are attributed to the olive leaf phenolic compounds (El Sedef and Karakaya 2009). Phenolic extracts from OMWW establish the relationship between blood lipid profile and phenolic compounds. As is well known, high levels of total cholesterol and low-density lipoprotein cholesterol (LDL-C) are recognized risk markers for atherosclerosis. Oxidation of LDL damages the vascular wall, stimulating macrophage uptake and the formation of foam cells, which in turn result in the formation of plaque within the arterial wall. This is considered to be the primary cause of cardiovascular diseases, while high-density lipoprotein cholesterol (HDL-C) is assumed to be protective. *In vitro* studies show that hydroxytyrosol is one of the most promising compounds as a potent inhibitor of copper and peroxyl radical-induced oxidation of low-density lipoprotein (LDL) (Leger et al. 2000). In another study, individual verbascoside from OMWW was active as a scavenger of reactive oxygen species and as a chemopreventive agent protecting LDL from oxidative damage (Cardinali et al. 2012). Moreover, the administration of OMWW extract fractions and purified hydroxytyrosol to diabetic rats points to a decrease in the glucose level in the plasma (Hamden et al. 2009).

In addition, many authors have established the relation between phenolic compounds identified in olive by-products and biological effects. In fact, oleocanthal, belonging to the secoiridoid group, was showed to inhibit cyclooxygenase-1 (COX-1) and cyclooxygenase-2 (COX-2) activity (both involved in the inflammatory process) in the same way as the anti-inflammatory drug, ibuprofen, does (Beauchamp et al. 2005). In addition, hydroxytyrosol was reported to have significant anti-inflammatory properties that help to decrease pro-inflammatory cytokines, interleukin 1β (IL-1β) and tumour necrosis factor-alpha (TNF-α) in rats (Gong et al. 2009). This alcoholic phenol is related to the platelet aggregability. This phenomenon is proxy to thrombogenic potential and an important marker of cardiovascular risk. Administration of pure hydroxytyrosol to human volunteers lowered thromboxane B_2 (TXB$_2$) production in a time-dependent manner (Cicerale et al. 2010). Finally, phenolic compounds identified in olive by-products have anti-promotion and anti-progression capacity as they are able to inhibit the proliferation and induce apoptosis in cancer cells (Corona et al. 2009). Figure 2 shows some of the mechanisms of action and their respective biological activities on the by-product phenolic compounds.

Extraction and Isolation of Bioactive Compounds from Waste Generate by Olive Oil Industry

Synthetic procedures for the production of phenolic compounds are expensive and/or produce low yields. Taking into account the chemical characterization of olive waste, the by-products generated during olive oil production process can be used as valuable sources of components for nutraceuticals, food and pharmaceutical preparations or in the cosmetics industry. Recovering bioactive compounds from olive by-products requires isolation of a specific fraction or target compound using different procedures and their combinations, such as filtration, centrifugation, concentration, derivation, clean up, etc. Literature reports show different extraction systems for recovering

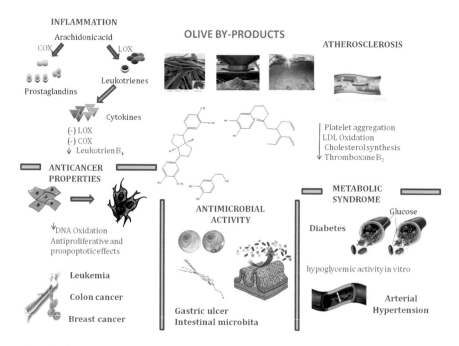

Figure 2. Biological properties of phenolic compounds from olive by-products.

polyphenols from samples under study—olive leaves, olive pomace, OMWW, storage by-products and filter cake. The extraction systems may be classified as conventional extraction, purification technologies and new advanced extraction techniques.

Conventional extraction technologies

With regard to conventional extraction technologies, several methodologies have been reported to provide phenolic enriched extracts, mainly in hydroxytyrosol, from olive by-products. The main systems for recovering the phenolic compounds include resin chromatography, membrane technologies (filtration and reverse osmosis) and solid-liquid or liquid-liquid solvent extractions. Among these processes, resin chromatography has proved a convenient and effective way for isolating phenolic compounds from aqueous solutions. This technique was applied by using different kinds of polymeric adsorbents. A patented system proposed to purify hydroxytyrosol from OMWW includes two-step chromatographic treatment. The process uses a non-activated ion-exchange-resin chromatographic method, followed by a second treatment on an XAD-type absorbent non-ionic resin which concentrates and completely purifies the hydroxytyrosol by means of elution with a methanol or ethanol: water dissolution (from 30–33 per cent). The final concentration in the obtained solution is at least 75 per cent of hydroxytyrosol, followed by removal of the polar organic solvent to produce a solid containing 95 per cent by weight of hydroxytyrosol. The patented system may also be applied to two-phase pomaces, three-phase pomaces and stones when subjected to a steam explosion process (Fernandez-Bolanos et al. 2002).

Membrane-separation processes have also been used for totally recovering the chemical components of OMWW. Indeed, a patented process is claimed for totally recovering the phenolic compounds in OMWW to reuse the concentrate residues in the production of fertilizers, biogas and highly purified aqueous products that may also be used as a basic component of beverages. Briefly, the process includes, in sequence, the following operations: (a) adjusting the pH of the collected OMWW within an acidic range; (b) treating the OMWW by enzymatic hydrolysis; (c) separating out the particles and suspended components, and a partially clarified liquid product; (d) treating the said liquid product resulting from operation by tangential microfiltration and providing a retentate phase and a permeate phase; (e) treating the permeate coming by tangential nanofiltration and providing for a new retentate phase and a permeate phase; (f) treating the permeate coming from tangential nanofiltration by reverse osmosis and providing a retentate phase rich in purified polyphenols and a permeate composed of purified water (Pizzichini 2005).

The combination of resin chromatography and membrane technologies was incorporated in the OMWW treatment. In this way, a pilot scale system has been developed for the recovery of valuable phenolic compounds (Agalias et al. 2007). The treatment consists of four steps: (a) successive filtration stages to gradually reduce the amount of suspended solids and decolorize water, (b) passage of the filtered wastewater through adsorbent resins to deodorize and decolorize the wastewater and recover the polyphenol and lactone components, (c) thermal evaporation and recovery of the organic solvent mixtures used to regenerate the resin, and (d) separation of the polyphenols and other organic substances by fast centrifuge partition chromatography. This procedure reduces 99.99 per cent of the phenolic compounds. In another pilot plant study, the application of hydrothermal treatment of olive pomace led to a final liquid phase that contained a high concentration of simple phenols (Rubio-Senent et al. 2012).

A new filtration process using native starch as main filter aid has been developed to improve the conventional filtration systems (Fernández-Gutiérrez et al. 2012). The filter aid can be used in both filtration process to remove suspended solid and to eliminate humidity and make the oil brilliant. Before filtering olive oil, a precoat layer is normally deposited in the filter tank. To achieve this, native starch is mixed with EVOO. In the next step, the mixture is passed through the filter tank. The native starch is deposited in the filtration equipment and a precoat layer is formed. Finally, filtration is carried out under a constant flow and increasing differential pressure. After the filter cake has completely been filled by suspended solid, the system reaches a maximum pressure and the filtration cycle completes. The filter cake, containing mainly native starch, removes suspended solids, moisture and makes the olive oil more brilliant in simply one filtration cycle. The filter cake generated as a by-product is rich in phenolic compounds coming from olive and which can be used in the development of functional food, nutraceuticals or ingredients in the food industry. The amount of hydroxytyrosol in native starch filter cakes is 20–30 times higher than unfiltered EVOO. In addition, the hydroxytyrosol retention power is 15–20 times higher than inorganic filter aids and twice more than other organic filter aids.

New advanced extraction techniques

Concerning new advanced extraction techniques, pressurized liquid extraction (PLE) and supercritical fluid extraction (SFE) have aroused growing interest as a green technology to replace the use of organic solvents, being more and more restricted due to their health and environment-related issues (Rosello-Soto et al. 2015). These extraction processes possess several advantages as compared to the conventional ones because of higher selectivity, shorter extraction time and lower consumption of toxic organic solvents. Pressurized liquid extraction (PLE), also known as accelerated solvent extraction, need particular extraction conditions in which the extracting solvents are heated at high temperatures and maintained at high pressures in order to keep their liquid state during the whole extraction procedure. In this way, the solubility of the analytes is enhanced, water diffusion improves and water viscosity decreases, allowing a better penetration into the matrix and an increase of mass transfer rate.

This extraction system is applied to the treatment of olive leaves, olive pomace and filter cake. Regarding olive leaves, several traditional methods have been used to extract phenolic compounds from this by-product, such as the pressing of leaves (Farag et al. 2007a, b) or conventional solvent extraction (Farag et al. 2003, Paiva-Martins et al. 2007). However, the yield of these procedures is low. For this reason, the capabilities of PLE using food-grade solvents, such as water and ethanol, to obtain antioxidant extracts rich on phenolic compounds from olive leaves have been evaluated (Herrero et al. 2011, Quirantes-Piné et al. 2013). In these contributions, different extraction conditions are tested. Extractions have been done using different combinations between solvent composition and temperatures to cover a wide range of dielectric constants. Thus, percentages of ethanol in the mixture ethanol:water from 0–85 per cent and temperatures above 120–200°C and below 40–63°C the boiling point of both solvent have been tested. The extraction of phenolic compounds is performed according at 1500 psi and 20 min as the pressure and extraction time. Using these parameters, a total of 25 different phenolic compounds are identified, including phenolic acids, secoiridoids, hydroxycinnamic acid derivatives, flavonols and flavones. Among them, hydroxytyrosol, oleuropein and luteolin-glucoside are the main phenolic compounds. Another study reports the effect of each parameter in the phenolic compounds recovered by central composite design (Xynos et al. 2014). Extraction parameters with statistical significance were ethanol content of the solvent mixture, temperature of the extraction and consecutive cycles of extractions. From the point of view of oleuropein content, a mixture of water/ethanol 43:57 in 190°C for one extraction cycle provided the optimal results.

Olive pomace obtained by the two-phase centrifugation system was used to obtain phenolic extracts by PLE. After a suitable separation of solid and liquid fractions, the solid waste was treated in an accelerated solvent extractor, using ethanol/water (80:20) at 80°C (Súarez et al. 2009). Concerning the olive oil filter cake, the PLE procedure considered mixtures of two solvents (ethanol and water) at temperatures ranging between 40–175°C (Lozano-Sánchez et al. 2014). Under these working conditions, the best isolation procedure to extract the phenolic fraction from the filter cake is accomplished using ethanol and water (50:50, v/v) at 120°C. The main phenolic compounds identified in the samples were characterized as phenolic alcohols or derivatives (hydroxytyrosol and its oxidation product), secoiridoids (decarboxymethylated and hydroxylated forms of oleuropein and ligstroside aglycones), flavones (luteolin and apigenin) and elenolic acid derivatives.

Supercritical Fluid Extraction (SFE) is a novel extraction technique that utilizes the solvent properties of fluids near their thermodynamic critical points. Increasing the pressure and temperature of the liquid/gas above the critical point generates supercritical fluids. These fluids have liquid-like solvent power and gas-like diffusivity, giving them great properties as ideal clean solvents for valuable compounds recovery from plant matrices. The physicochemical properties of supercritical fluids, such as the density, diffusivity, viscosity and dielectric constant can be controlled by varying the operating conditions of pressure and temperature or both in combination. Many supercritical fluids may be used, such as carbon dioxide, ethane, propane, butane, pentane, ethylene, ammonia, sulfur dioxide, water, chlorodifluoromethane, etc. However, CO_2 is the most desirable supercritical fluid for extracting phenolic compounds due to its availability, no toxicity and flammability, low cost (compared to liquid organic solvents), tunable solvent properties and easy to handle critical temperature and pressure (31.1°C and 7.38 MPa). Adjusting the pressure and temperature, altering thus the density, can change the extraction selectivity of supercritical fluids. Adding a co-solvent, such as ethanol, increases the polarity, which can also change the extraction selectivity.

In the olive oil industry, the main application of this technique is to recover phenolic compounds from olive leaves (Taamalli et al. 2012). Extraction of phenolic compounds from different varieties ('*Oueslati*', '*Chetoui*', '*Chemlali*', '*El Hor*', '*Jarboui*' and '*Chemchali*') has been carried out at 150 bar and 40°C using a mixture of CO_2 plus 6.6 per cent of ethanol. Despite that, the abundance of phenolic compounds was dependent on the variety with the main secoiridoid identified in all extracts being oleuropein. In addition, different flavonoids have been extracted using the above-mentioned experimental conditions—diosmetin, luteolin and apigenin.

In a recent study, the potential use of the combination of super/subcritical fluids to produce extracts enriched in oleuropein from olive leaves was tested. The phenolic compounds from olive leaves were extracted with PLE and SFE. The effect of the techniques was studied upon the yield, oleuropein content and scavenging radical activity. The serial combination of SFE-CO_2 modified by 5 per cent ethanol and subcritical water afforded high extract yield (44.1 per cent), high recovery of oleuropein (4.6 per cent) and good antioxidant activity. It was suggested that the removal of non-polar compounds with SFE-CO_2 resulted in enrichment of the residue of oleuropein, which was then extracted with PLE by subcritical water (Xynos et al. 2012). Table 4 summarizes the extraction methods applied to different olive by-products.

Trends in the Development of Functional Food and Nutraceutical using Bioactive Compounds from Waste Generated by Olive Oil Industry

The increasing interest in finding natural sources of bioactive compounds for use in food processing and nutraceutics has heightened research activities to replace synthetic by natural additives particularly derived from plants and agro-industry by-products (Farag et al. 2007, Jaber et al. 2012). In this way, phenolic compounds obtained from olive by-products find several food applications, like improvement of the nutritional profile or the stability of the food matrices. Enrichment of food with olive by-product extracts and the development of nutraceuticals is a technological challenge for food and pharmaceutical industry that searches for innovative, healthier and safe new ingredients as per market demands (Nunes et al. 2016).

There is no definition of functional foods and/or nutraceuticals as it varies across countries and markets. Functional foods are generally considered to go beyond the provision of basic nutrients to potentially offer additional benefits, such as reducing the risk of disease and/or promoting optimal health for the consumer (Aryee and Boye 2015). On the other hand, the term nutraceutical was coined in 1989 by Stephen Defelice, founder and chairman of the foundation for innovation in medicine, an organization which encourages medical health. According to him, "A nutraceutical is any substance that is a food or a part of food and provides medical or health benefits, including the prevention and treatment of disease" (Pandey et al. 2010). Taking into account the information summarized along this chapter, olive by-products could be used as a natural source for obtaining functional ingredients. Figure 3 shows a general diagram for developing ingredients from waste generated by the olive oil industry with food and nutraceutical applications.

Food and nutraceutical applications of phenolic compounds derived from olive leaf extract

The potential use of olive leaf phenolics as natural additives and functional ingredients has been evaluated in different food matrices. Concerning preservative applications, scientific reports show that complex olive leaves phenolic extracts express higher antioxidant capacity than isolated compounds (Benavente-García et al. 2000). Thus the application of complex phenolic extracts to increase the oxidative stability includes

Table 4. Extraction techniques for isolation of bioactive compounds from waste generated by olive oil industry.

Compounds extracted	Olive oil by-products	Technique	Extraction conditions	References
Hydroxytyrosol	OMWW	XAD-type absorbent non-ionic resin	Elution with methanol/ethanol: water	Fernandez-Bolanos et al. 2002
Hydroxytyrosol	OMWW	Membrane separation processes	Adjusting pH Tangential micro- and nano-filtration Reverse osmosis	Pizzichini 2005
Phenolic compounds	OMWW	Resin chromatography and membrane separation technology	Combination of successive filtration and absorbent resins	Agalias et al. 2007
Hydroxytyrosol	Filter cake	New filtration systems	Native starch as filter aids	Fernández-Gutiérrrez et al. 2012
Simple phenols	OMWW	Hydrothermal treatment		Rubio-Senent et al. 2012
Phenolic compounds	Olive leaves	PLE	Ethanol:water 40–200°C, 1500 psi, 20 min	Herrero et al. 2011
Oleuropein	Olive leaves	PLE	Ethanol:water 57:43 (v:v) 190°C, 1500 psi, 20 min	Xynos et al. 2014
Phenolic compounds	Olive pomace	PLE	Ethanol:water 80:20 (v:v) 80°C, 1500 psi, 20 min	Suarez et al. 2009
Phenolic compounds	Filter cake	PLE	Ethanol:water 50:50 (v:v) 120°C, 1500 psi, 20 min	Lozano-Sánchez et al. 2014
Oleuropein and flavonoids	Olive leave	SFE	150 bar, 40°C CO_2 + 6.6% ethanol	Taamalli et al. 2012
Oleuropein	Olive leave	SFE + PLE		Xynos et al. 2012

n-3 enriched eggs, edible oils, meat, fruit and dairy products. Botsoglou et al. studied the effect of olive leaf supplementation on oxidative stability and fatty acid profile of n-3 enriched eggs over the shelf-life (Botsoglou et al. 2012). Phenolic compounds showed a high protective effect of fatty acid against the oxidation process. In other studies, the protective effect of these compounds against the oxidative phenomenon of refined olive and sunflower oils was evaluated. These were found to be superior to

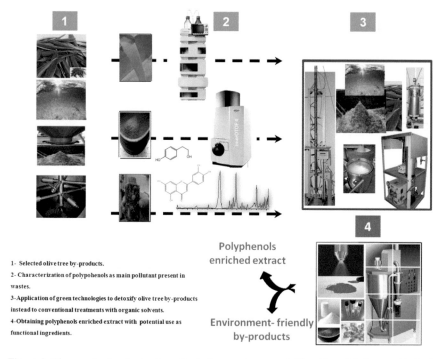

1- Selected olive tree by-products.

2- Characterization of polypohenols as main pollutant present in wastes.

3-Application of green technologies to detoxify olive tree by-products instead to conventional treatments with organic solvents.

4-Obtaining polyphenols enriched extract with potential use as functional ingredients.

Polyphenols enriched extract

Environment- friendly by-products

Figure 3. Diagram for developing ingredients from waste generated by olive oil industry with food and nutraceutical applications.

some synthetic antioxidants (Keceli and Harp 2014, Malheiro et al. 2013, Rafiee et al. 2012, Taghvaei and Jafari 2015). These results were in agreement with those reported by Chiou et al. who supplemented frying oils with olive leaf extracts. This study concluded that the use of fortified oils enhances the bioactive compounds content in fried potatoes, when compared to non-supplemented oils (Chiou et al. 2007). The same frying procedure was carried out in another work and the oleuropein content was monitored in the final product for eight successive frying sessions. In each session a gradual decrease in oleuropein content (in oil and food) was noticed (Chiou et al. 2013).

Hayes et al. (2010) studied the antimicrobial effect as well as the capacity to inhibit the lipid oxidation of lutein, sesamol, ellagic acid and olive leaves extract on meat. Under aerobic conditions, the olive leaf extract was found to reduce lipid oxidation in the range from 53–78 per cent. When this experiment was carried out under modified conditions, this extract reduced the lipid oxidation in the range from 76–84 per cent. The same effect was reported in the oxidative quality of cold stored beef cubes (Aytul et al. 2004, Hayes et al. 2010). However, the results obtained in these works showed that the protein-polyphenol binding phenomenon could be responsible for the changes in several protein properties (gelation and emulsification). With regard to fruits, the applications of olive leaf extracts to impregnate dehydrated apple cubes showed improvement in the fruit stability. In addition, the composition of the final product revealed high oleuropein content (1928 mg/100 g dry weight) (Ahmad-Qasem et al. 2015). Olive leaf extracts are also used in dairy products. Zoidou et al. (2014) added pure oleuropein in cow's milk and yogurt. It was found that heat treatment of milk did not affect the oleuropein because it was not hydrolyzed by acids produced during the fermentation process. Furthermore, it was not metabolized by lactic acid bacteria and nor did it inhibit the growth of bacteria and stability in the final product. The new product had the same flavor, color and texture as the conventional one.

Regarding nutraceutical formulations, scientific studies were addressed to develop new formulations using new advances in encapsulation technology. Olive leaf extracts were encapsulated, using inclusion complexation and spray drying. Molecular inclusion is generally achieved by using cyclodextrins as the encapsulating materials. The inclusion of this by-product, with high content of oleuropein, has been carried out using ß-cyclodextrins. The resultant cyclodextrin-encapsulated systems have increased their antioxidant activity due to the protection of the phenolic compounds against rapid oxidation by free radicals (Mourtzinos et al. 2007).

Spray drying is one of the most common techniques used to obtain a dried encapsulated powder, due to low operative costs in respect to other methods and high flexibility. Despite that, spray drying has been used in the food industry since the late 1950s (Fanga and Bhandaria 2010). It is a relatively new technique in encapsulation purpose of the industry by-products. The core material for encapsulation is homogenized with the wall material. The mixture is then fed into a spray dryer and atomized with a nozzle or spinning wheel. Water is evaporated by hot air coming in contact with the atomized material. The capsules are then collected after they fall to the bottom of the drier (Fanga and Bhandaria 2010). Different polymers may be used as the wall materials: chitosan, modified starch, maltodextrin, gum or other substances. Among these polymers, chitosan has been selected as a wall material and the encapsulation efficiency was evaluated by FTIR spectroscopy. Results indicate that the majority of the olive leaf extract was physically encapsulated in the chitosan matrix (Kosaraju et al. 2006). Another wall material successfully used for encapsulation of this by-product was protein-lipid (sodium caseinate-soy lecithin) emulsion (Kosaraju et al. 2008).

Although there is no recommended human consumption daily dose for oleuropein, in the market there are several commercial nutraceutical extracts based on this by-product (Benolea®, Nutrafur®, Solgar®, Nature's Way® and Olympian Labs® among others) with different contents of oleuropein, generally in the range of 6–22 per cent of oleuropein.

Food and nutraceutical applications of phenolic compounds derived from olive pomace

Regarding the use of phenolic compounds derived from the olive pomace, nutraceutical formulations have been developed by combining high pressure and temperature extraction followed by encapsulation by using spray dried with different ratios of gum arabic and maltodextrin as coating agents (0:100, 20:80, 40:60, 60:40, 80:20 and 100:0 per cent w/w). The total amount of coating agent was maintained constant and equal to 10 per cent w/v (100 g/L) (Aliakbarian et al. 2015). Inlet temperature, feed flow and aspiration rate were 160°C, 5 mL/min and 30 m/h, respectively. For all products, moisture content, water solubility index, total polyphenol content, antiradical power and microencapsulation yield were determined. The results of this study show that the coating agent and process conditions led to the production of microencapsulated powders with improved water dissolution rate and a minimal loss in phenolics during the drying phase. The obtained microparticles can have industrial applications as functional components for foods or nutraceuticals purposes.

Food and nutraceutical applications of phenolic compounds derived from OMWW

The potential use of OMWW in food and beverages has been evaluated by several authors. The impact of processing and storage on its composition, safety and the bioavailability has been described by Zbakh and El Abbassi (2012). Their contribution provides an overview of key research describing the phenolic profile of OMWW associated with health-promoting activities, including the monitoring of these compounds along with treatment and preparation of functional beverages. These results point that phenolic compounds recovered from OMWW are highly bioavailable and safe. The inclusion of OMWW phenolic extract in beverage preparations may have a significant impact on the health of population through reduction in the incidence of cardiovascular and chronic degenerative diseases.

Functional milk beverages (FMB) have been formulated using phenolic compounds from OMWW (100 and 200 mg/liter, being FMB 100 and FMB 200, respectively) (Servili et al. 2011). After addition of phenolic extracts, milk beverages were fermented with γ-amino butyric acid (GABA)-producing microorganism (Lactobacillus plantarum C48) as well as autochthonous human gastro-intestinal lactic acid bacteria (Lactobacillus paracasei 15N). Concerning the viability of microorganims, Lb. plantarum showed a decrease of Log 2.52–2.24 cfu/ml during storage. The cell density of functional Lb. paracasei remained above the value of Log 8.0 cfu/ml. All samples showed the same trends in the acidification kinetic, except for a longer latency phase of FMB 200. During fermentation, the total concentration of free amino acids markedly increased. The concentration of GABA also increased during storage. For both parameters, there was no significant difference between the beverages. After fermentation, FMB 100 and FMB 200 showed the same phenolic composition of the phenol extract from olive vegetable water but a different ratio between hydroxytyrosol and decarboxymethyl oleuropein aglycone. During storage, the concentrations of decarboxymethyl oleuropein aglycone, tyrosol and verbascoside decreased in both FMB 100 and FMB 200. However, the concentration of hydroxytyrosol was increased. In addition, sensory analyses based on triangle and paired comparison tests showed that phenolic compounds at concentrations of 100 or 200 mg/l were suitable for addition to functional milk beverages.

An integrated approach for complete utilisation of OMWW was developed by Goula and Lazarides 2015. The developed strategies are based on recovering valuable by-products and/or ingredients using designed fermentation, spray drying and encapsulation technologies. Some of the valuable by-products are spreadable olive paste or olive powder (to be included in food formulations) and encapsulated polyphenols.

Conclusion

The interests in the healthy benefits of EVOO polyphenols have recently increased due to different studies supporting their biological effects in reducing oxidative stress as well as hypoglycemic, antihypertensive, antimicrobial, antiatherosclerotic, antidiabetic and anticancer properties, among others. Therefore, research on finding the best way of concentrating phenolic compounds in EVOO and recovering them from its by-products could be very promising. To achieve this goal, many studies about the volume of waste as well as their chemical characterization were summarized to establish the best conditions to recover bioactive compounds from EVOO by-products.

Taking into account the literature used in this chapter, different pilot plant and industrial processes employed to recover phenolic compounds from olive oil by-products have been widely reviewed. OMWW and olive pomace are considered as natural sources of hydroxytyrosol. This alcoholic phenol has been related to antiinflammatory and anticancer properties. The main technique used to recover this compound are resin chromatography and membrane separation processes. However, new advanced extraction techniques, such as pressurized liquid extraction and supercritical fluid extraction have aroused growing interest as a green technology replacing the use of organic solvents and of being more and more restricted due to their health- and environmental-related issues. These new processes have been applied for extracting phenolic compounds from solid waste—olive leaves, olive pomace and olive filter cake. The phenolic extracts obtained are characterized by a high content of oleuropein, oleacein and oleocanthal. All of these compounds are endowed with beneficial properties.

Choosing the best system for recovering phenolic compounds from olive by-products may only be done after careful consideration of all the relevant parameter processes. The most important is the extrapolation of the laboratory experiments to industrial scale. To establish which extraction system is more suitable, it should be continuously and critically scrutinised with important criteria, such as high retention capability, high flow rate, problem-free handling and environment-friendly as well as physiological safety.

Finally, new trends in the development of functional food and nutraceutical using bioactive compounds from waste generated by olive oil industry have also been reviewed. Increasing research is being conducted on the application of phenolic compounds as natural additives to improve the food stability. In addition, new processes based on encapsulation technologies have been developed. The main procedures are based on molecular inclusion and spray drying. The advances in this area can promote the development of functional food and nutraceuticals with a significant impact on the health of the population through reduction in the incidents of cardiovascular and chronic degenerative diseases.

References

Agalias, A., P. Magiatis, A. Skaltsounis, E. Mikros, A. Tsarbopoulos, E. Gikas, I. Spanos and T. Manios. 2007. A new process for the management of olive oil mill waste water and recovery of natural antioxidants. Journal of Agricultural and Food Chemistry 55: 2671–2676.

Ahmad-Qasem, M., J. Santacatalina, E. Barrajón-Catalán, V. Micol, J. Cárcel and J. García-Pérez. 2015. Influence of drying on the retention of olive leaf polyphenols infused into dried apple. Food and Bioprocess Technology 8: 120–133.

Aliakbarian, B., A. A. Casazza and P. Perego. 2015. Kinetic and ishotermal modeling of the absortion of phenolic compounds from olive mill wastewater onto activated carbon. Food Technology and Biotechnology 53: 207–214.

Alu'datt, M. H., I. Alli, K. Ereifej, M. Alhamad, A. R. Al-Tawaha and T. Rababah. 2010. Optimisation, characterisation and quantification of phenolic compounds in olive cake. Food Chemistry 123: 117–122.

Amirante, P., M. L. Clodoveo, A. Leone, A. Tamborrino and V. B. Patel. 2010. Influence of different centrifugal extraction systems on antioxidant content and stability of virgin olive oil. Elsevier. 85–93.

Aryee, A. N. and J. I. Boye. 2015. Current and emerging trends in the formulation and manufacture of nutraceuticals and functional food products. pp. 1–52. *In*: J. I. Boye (ed.). Nutraceutical and Functional Food Processing Technology. Wiley, Chichester, UK.

Aytul, K. K., F. Korel, D. K. Arserim-Uçar, I. Uysal and O. Bayraktar. 2004. Efficacy of olive leaf extract for enhancing quality of beef cubes. *In*: Proceedings of the 54th International Congress of Meat Science and Technology. pp. 51. South Africa: Cape Town, 51.

Beauchamp, G. K., R. S. J. Keast, D. Morel, J. Lin, J. Pika, Q. Han, C. -H. Lee, A. B. Smith and P. A. S. Breslin. 2005. Phytochemistry: Ibuprofen-like activity in extra-virgin olive oil. Nature 437: 45–46.

Benavente-Garcia, O., J. Castillo, J. Lorente, A. Ortuno and J. A. Del Rio. 2000. Antioxidant activity of phenolics extracted from *Olea europaea* L. leaves. Food Chemistry 68: 457–462.

Botsoglou, E., A. Govaris, D. Fletouris and N. Botsoglou. 2012. Lipid oxidation of stored eggs enriched with very long chain n-3 fatty acids, as affected by dietary olive leaves (*Olea europea* L.) or α-tocopheryl acetate supplementation. Food Chemistry 134: 1059–1068.

Cardinali, A., S. Pati, F. Minervini, I. D'Antuono, V. Linsalata and V. Lattanzio. 2012. Verbascoside, Isoverbascoside, and their derivatives recovered from olive mill wastewater as possible food antioxidants. Jorunal of Agricultural and Food Chemistry 60: 1822–1829.

Chiou, A., F. N. Salta, N. Kalogeropoulos, A. Mylona, I. Ntalla and N. K. Andrikopoulos. 2007. Retention and distribution of polyphenols after pan-frying of French fries in oils enriched with olive leaf extract. Journal of Food Science 72: S574–S584.

Chiou, A., N. Kalogeropoulos, P. Efstathiou, M. Papoutsi and N. K. Andrikopoulos. 2013. French fries oleuropein content during the successive deep frying in oils enriched with an olive leaf extract. International Journal of Food Science & Technology 48: 1165–1171.

Cicerale, S., L. Lucas and R. Keast. 2010. Biological activities of phenolic compounds present in virgin olive oil. International Journal of Molecular Science 11: 458–479.

Corona, G., M. Deiana, A. Incani, D. Vauzour, M. A. Dessì and J. P. E. Spencer. 2009. Hydroxytyrosol inhibits the proliferation of human colon adenocarcinoma cells through inhibition of ERK1/2 and cyclin D1. Molecular Nutrition and Food Research 53: 897–903.

Covas, M. 2008. Bioactive effects of olive oil phenolic compounds in humans: Reduction of heart disease factors and oxidative damage. Inflammopharmacology 16: 216–218.

El Sedef, N. and S. Karakaya. 2009. Olive tree (*Olea europaea*) leaves: Potential beneficial effects on human health. Nutrition Reviews 11: 632–638.

Fanga, Z. and B. Bhandaria. 2010. Encapsulation of polyphenols. Trends in Food Science and Technology 21: 510–523.

Farag, R., G. Baroty and A. Basuny. 2003. The influence of phenolic extracts obtained from the olive plant (cvs. Picual and Koroneiki), on the stability of sunflower oil. International Journal of Food Science and Technology 38: 81–87.

Farag, R. S., E. A. Mahmoud and A. M. Basuny. 2007a. Use crude olive leaf juice as a natural antioxidant for the stability of sunflower oil during heating. International Journal of Food Science & Technology 42: 107–115.

Farag, R., E. Mahmoud and A. Basuny. 2007b. Use crude olive leaf juice as a natural antioxidant for the stability of sunflower oil during heating. Journal of Food Science and Technoly 42: 107–115.

Fernandez-Bolanos, J., A. Heredia, G. Rodriguez, R. Rodriguez, R. Guillen and A. Jimenez. 2002. Method for obtaining purified hydroxytyrosol from products and by-products derived from the olive tree. Patent WO02064537, Aug 2002.

Fernández-Gutiérrez, A., A. Segura-Carretero and J. Lozano-Sánchez. 2012. Procedimiento para filtrar aceite usando tortas filtrantes a base de almidón nativo. Patent ES-201101121, Feb 2012.

Frankel, E. N. 2011. Nutritional and biological properties of extra virgin olive oil. Journal of Agricultural and Food Chemistry 59: 785–792.

Frankel, E., A. Bakhouche, J. Lozano-Sánchez, A. Segura-Carretero and A. Fernández-Gutiérrez. 2013. Literature review on production process to obtain extra virgin olive oil enriched in bioactive compounds. Potential use of byproducts as alternative sources of polyphenols. Journal of Agricultural and Food Chemistry 61: 5179–5188.

Fu, S., D. Arraez-Roman, A. Segura-Carretero, J. A. Menendez, M. P. Menendez-Gutierrez, V. Micol and A. Fernandez-Gutierrez. 2010. Qualitative screening of phenolic compounds in olive leaf extracts by hyphenated liquid chromatography and preliminary evaluation of cytotoxic activity against human breast cancer cells. Analytical and Bioanalytical Chemistry 397: 643–654.

Gabás-Rivera, C., C. Barranquero, R. Martínez-Beamonte, M. A. Navarro, J. C. Surra and J. Osada. 2014. Dietary squalene increases high density lipoprotein-cholesterol and paraoxonase 1 and decreases oxidative stress in mice. PLoS One 12: 104224.

García, A. M., A. Moumen, D. Y. Ruiz and E. M. Alcaide. 2003. Chemical composition and nutrients availability for goats and sheep of two-stage olive cake and olive leaves. Animal Feed Science and Technology 107: 61–74.

Ghanbari, R., F. Anwar, K. M. Alkharfy, A. H. Gilani and N. Saari. 2012. Valuable nutrients and functional bioactives in different parts of olive (Olea europaea L.). A review. International Journal of Molecular Science 13: 3291−3340.

Gong, D., C. Geng, L. Jiang, J. Cao, H. Yoshimura and L. Zhong. 2009. Effects of hydroxytyrosol-20 on carrageenan-induced acute inflammation and hyperalgesia in rats. Phytherapy Research 23: 646–650.

Goula, A. M. and H. N. Lazaride. 2015. Integrated processes can turn industrial food waste into valuable food by-products and/or ingredients: the cases of olive mill and pomegranate wastes. Journal of Food Engineering 167: 45–50.

Hamden, K., N. Allouche, M. Damak and A. Elfeki. 2009. Hypoglycemic and antioxidant effects of phenolic extracts and purified hydroxytyrosolfrom olive mill waste *in vitro* and in rats. Chem.-Biol. Interact. 180: 421−432.

Hassen, I., H. Casabianca and K. Hosni. 2015. Biological activities of the natural antioxidant oleuropein: Exceeding the expectation—A mini-review. Journal of Functional Foods 18: 926–940.

Hayes, J. E., V. Stepanyan, P. Allen, M. N. O'Grady and J. P. Kerry. 2010. Effect of lutein, sesamol, ellagic acid and olive leaf extract on the quality and shelf-life stability of packaged raw minced beef patties. Meat Science 84: 613–620.

Herrero, M., T. N. Temirzoda, A. Segura-Carretero, R. Quirantes, M. Plaza and E. Ibañez. 2011. New possibilities for the valorization of olive oil by-products. Journal of Chromatography A 1218: 7511–7520.

Jaber, H., M. Ayadi, J. Makni, G. Rigane, S. Sayadi and M. Bouaziz. 2012. Stabilization of refined olive oil by enrichment with chlorophyll pigments extracted from Chemlali olive leaves. European Journal of Lipid Science and Technology 114: 1274–1283.

Kanakis, P., A. Termentzi, T. Michel, E. Gikas, M. Halabalaki and A. Skaltsounis. 2013. From olive drupes to olive OilAn HPLC-orbitrap-based qualitative and quantitative exploration of olive key metabolites. Planta Medica 79: 1576–1587.

Kapellakis, I. E., K. P. Tsagarakis and J. C. Crowther. 2008. Olive oil history, production and by-product management. Reviews in Environmental Science and BioTechnology 7: 1–26.

Keceli, T. and F. Harp. 2014. The effect of olive leaves and their harvest time on radical scavenging activity and oxidative stability of refined olive oil. Quality Assurance and Safety of Crops & Foods 6: 141–149.

Kosaraju, S. L., L. D'ath and A. Lawrence. 2006. Preparation and characterisation of chitosan microspheres for antioxidant delivery. Carbohydrate Polymers 64: 163–167.

Kosaraju, S. L., D. Labbett, M. Emin, I. Konczak and L. Lundin. 2008. Delivering polyphenols for healthy ageing. Nutrition & Dietetics 65: S48–S52.

Leger, C. L., N. Kadiri-Hassani and B. Descomps. 2000. Decreased superoxide anion production in cultured human promonocyte cells (THP-1) due to polyphenol mixtures from olive oil processing wastewaters. Journal of Agricultural and Food Chemistry 48: 5061−5067.

Lozano-Sánchez, J., L. Cerretani, A. Bendini, A. Segura-Carretero and A. Fernández-Gutierrez. 2010. Filtration process of extra virgin olive oil: Effect on minor components, oxidative stability and sensorial and physicochemical characteristics. Trends in Food Science and Technology 21: 201–211.

Lozano-Sánchez, J., A. Segura-Carretero and A. Fernández-Gutierrez. 2011a. Characterization of the phenolic compounds retained in different organic and inorganic filter aids used for filtration of extra virgin olive oil. Food Chemistry 124: 1146–1150.

Lozano-Sánchez, J., E. Giambanelli, R. Quirantes-Piné, L. Carretani, A. Bendini, A. Segura-Carretero and A. Fernández-Gutiérrez. 2011b. Wastes generated during the storage of extra virgin oilve oil as a natural source of phenolic compounds. Journal of Agricultural and Food Chemistry 59: 11491–11500.

Lozano-Sánchez, J., M. Castro-Puyana, J. A. Mendiola, A. Segura-Carretero, A. Cifuentes and E. Ibáñez. 2014. Recovering bioactive compounds from olive oil filter cake by advanced extraction techniques. International Journal of Molecular Science 15: 16270–16283.

Lujan, R. J., F. P. Capote, A. Marinas and M. D. L. de Castro. 2008. Liquid chromatography/triple quadrupole tandem mass spectrometry with multiple reaction monitoring for optimal selection of transitions to evaluate nutraceuticals from olive-tree materials. Rapid Communication Mass Spectrometry 22: 855–864.

Malheiro, R., S. Casal, H. Teixeira, A. Bento and J. Pereira. 2013. Effect of olive leaves addition during the extraction process of overmature fruits on olive oil quality. Food and Bioprocess Technology 6: 509–521.

Mourtzinos, I., F. Salta, K. Yannakopoulou, A. Chiou and V. T. Karathanos. 2007. Encapsulation of olive leaf extract in becyclodextrin. Journal of Agricultural and Food Chemistry 55: 8088–8094.

Niaounakis, M. and C. P. Halvadakis. 2006. Olive processing waste management: Literature review and patent survey, 2nd Ed. Waste Manage. 5: 514.

Nunes, M. A., F. B. Pimentel, A. S. Costa, R. C. Alves and M. B. P. Oliveira. 2016. Olive by-products for functional and food applications: Challenging opportunities to face environmental constraints. Innovative Food Science & Emerging Technologies 35: 139–148.

Paiva-Martins, F., R. Correia, S. Felix, P. Ferreira and M. Gordon. 2007. Effects of enrichment of refined olive oil with phenolic compounds from olive leaves. Journal of Agricultural and Food Chemistry 55: 4139–4143.

Paiva-Martins, F. and M. Pinto. 2008. Isolation and characterization of a new hydroxytyrosol derivative from olive (*Olea europaea*) leaves. Journal of Agricultural and Food Chemistry 56: 5582–5588.

Pandey, M., R. K. Verma and S. A. Saraf. 2010. Nutraceuticals: New era of medicine and health. Asian J. Pharm. Clin. Res. 3: 11–15.

Papaphilipou, P., C. Yiannapas, M. Politi, M. Daskalaki, Vasileia, C. Michael, N. Kalogerakis, Mantzavinos and D. Fatta-Kasinos. 2013. Sequential coagulation–flocculation, solvent extraction and photo-Fenton oxidation for the valorization and treatment of olive mill effluent. Chemical Enginnering Journal 224: 82–88.

Pattara, C., G. M. Cappelletti and A. Cichelli. 2010. Recovery and use of olive stones: Commodity, environmental and economic assessment. Renewable Sustainable Energy Review 14: 1484–1489.

Pizzichini, M. 2005. Process for Recovering the Components of Olive Mill Wastewater with Membrane Technologies. Patent WO2005/ 123603, Dec 2005.

Quideau, S., D. Deffieux, C. Douat-Casassus and L. Pouységu. 2011. Plant polyphenols: Chemical properties, biological activities, and synthesis. Angewandte Chemie International Edition. 50: 586–621.

Quirantes-Piné, R., J. Lozano-Sánchez, M. Herrero, E. Ibáñez, A. Segura-Carretero and A. Fernández-Gutiérrez. 2013. HPLC–ESI–QTOF–MS as a powerful analytical tool for characterising phenolic compounds in olive-leaf extracts. Phytochemical Analysis 24: 213–223.

Rafiee, Z., S. M. Jafari, M. Alami and M. Khomeiri. 2012. Antioxidant effect of microwave assisted extracts of olive leaves on sunflower oil. Journal of Agricultural Science and Technology 14: 1497–1509.

Ranalli, A., S. Contento, L. Lucera, M. Di Febo, D. Marchegiani and V. Di Fonzo. 2006. Factors affecting the contents of iridoid oleuropein in olive leaves (*Olea europaea* L.). Journal of Agricultural and Food Chemistry 54: 434–440.

Rodriguez-Rodriguez, R. 1., M. D. Herrera, M. A. de Sotomayor and V. Ruiz-Gutierrez. 2007. Pomace olive oil improves endothelial function in spontaneously hypertensive rats by increasing endothelial nitric oxide synthase expression. American Journal of Hypertension 20: 728–734.

Roig, A., M. L. Cayuela and M. A. Sánchez-Monedero. 2006. A review on olive mil wastes and their valorisation methods. Waste Managenent 26: 960–969.

Rosello-Soto, E., M. Koubaa, A. P. Moubarik, R. Lopes, J. A. Saraiva, N. Boussetta, N. Grimi and F. J. Barba. 2015. Emerging opportunities for the effective valorization of wastes and by-products generated during olive oil production process: Nonconventional methods for the recovery of high-added value compounds. Trends in Food Science and Technology 45: 296–310.

Rubio-Senent, F., G. Rodríguez-Gutierrez, A. Lama-Muñoz and J. Fernández-Bolaños. 2012. New phenolic compounds hydrothermally extracted from the olive oil by-product alperujo and their antioxidative activities. Journal of Agricultural and Food Chemistry 60: 1175–1186.

Rubio-Senent, F., G. Rodríguez-Gutiérrez, A. Lama-Muñoz and J. Fernández-Bolaños. 2013a. Phenolic extract obtained from steam-treated olive oil waste: Characterization and antioxidant activity. LWT—Food Science and Technology 54: 114–12427.

Rubio-Senent, F., A. Lama-Muñoz, G. Rodríguez-Gutiérrez and J. Fernández-Bolaños. 2013b. Isolation and identification of phenolic glucosides from thermally treated olive oil byproducts. Journal of Agricultural and Food Chemistry 61: 1235–1248.

Sánchez, S., A. J. Moya, M. Moya, I. Romero, R. Torrero, V. Bravo and M. P. S. Miguel. 2002. Aprovechamiento del residuo de poda del olivar. Ingeniería Química (Madrid) 34: 194–203.

Scoditti, E., C. Capurso, A. Capurso and M. Massaro. 2014. Vascular effects of the Mediterranean diet-Part II: Role of omega-3 fatty acids and olive oil polyphenols. Vascular Pharmacology 63: 127–134.

Serra, A., L. Rubio, X. Borras, A. Maciá and M. P. Romero. 2012. Distribution of olive oil phenolic compounds in rat tissues after administration of a phenolic extract from olive cake. Mol. Nutr. Food Res. 56: 486–496.

Servili, M., C. G. Rizzello, A. Taticchi, S. Esposto, S. Urbani, F. Mazzacane, I. Di Maio, R. Selvaggini, M. Gobbetti and R. Di Cagno. 2011. Functional milk beverage fortified with phenolic compounds extracted from olive vegetation water and fermented with functional lactic acid bacteria. International Journal of Food Microbiology 147: 45–52.

Súarez, M., M. P. Romero, T. Ramo, A. Maciá and M. J. Motilva. 2009. Methods for preparing phenolic extracts from olive cake for potential application as food antioxidants. Journal of Agricultural and Food Chemistry 57: 1463–1472.

Taamalli, A., D. Arráez-Román, E. Barrajón-Catalán, V. Ruiz-Torres, A. Pérez-Sánchez, M. Herrero et al. 2012. Use of advanced techniques for the extraction of phenolic compounds from Tunisian olive leaves: Phenolic composition and cytotoxicity against human breast cancer cells. Food and Chemical Toxicology 50: 1817–1825.

Taghvaei, M. and S. Jafari. 2015. Application and stability of natural antioxidants in edible oils in order to substitute synthetic additives. Journal of Food Science and Technology 52: 1272–1282.

Whayne Jr., T. F. 2014. Ischemic heart disease and the mediterranean diet. Current Cardiology Reports 16(6).

Xynos, N., G. Papaefstathiou, M. Psychis, A. Argyropoulou, N. Aligiannis and A. Skaltsounis. 2012. Development of a green extraction procedure with super/subcritical fluids to produce extracts enriched in oleuropein from olive leaves. Journal of Supercritical Fluids 67: 89–93.

Xynos, N., G. Papaefstathiou, E. Gikas, A. Argyropoulou, N. Aligiannis and A. Skaltsounis. 2014. Design optimization study of the extraction of olive leaves performed with pressurized liquid extraction using response surface methodology. Separation and Purification Technology 122: 323–330.

Zbakh, H. and A. El Abbassi. 2012. Potential use of olive mill wastewater in the preparation of functional beverages: A review. Journal of Functional Foods 4: 53–65.

Zoidou, E., P. Magiatis, E. Melliou, M. Constantinou, S. Haroutounian and A. L. Skaltsounis. 2014. Oleuropein as a bioactive constituent added in milk and yogurt. Food Chemistry 158: 319–324.

Extraction, Isolation and Utilisation of Bioactive Compounds from Fresh Fruit and Vegetable Waste

Narashans Alok Sagar, Sunil Sharma and *Sunil Pareek**

Introduction

Fruits and vegetables are an integrated part of our diet. Due to the growing population, demand for horticultural crops has intensively increased to meet human needs (Schieber et al. 2001). Global production of fruits and vegetables has increased manifold, rendering processing of major fruit and vegetable types reach new levels of productivity. Examples include pineapple (19.14 MT), mango and guava (39.98 MT), grapes (67.32 MT), apple (70.30 MT), banana (100.00 MT) and citrus (119.70 MT), peas (15.97 MT), cauliflower (18.93 MT), carrot and turnip (33.79 MT), cabbage (59.55 MT), tomato (150.51 MT) and potato (330.05 MT) (FAO 2012). As a result, the food processing industry has grown tremendously, resulting in huge wastage. Numerous attempts have been made to minimize waste and to maximize waste utilisation. It has been determined that horticultural waste is an ample source of crucial bioactive compounds and dietary fibers. Waste generated from fruit and vegetable processing ranks second after the waste generated in households. The Food and Agriculture Organisation (FAO) report (2013) reveals that processing, packaging, distribution and consumption of fruit and vegetables in parts of China, India, Philippines and the United States of America produced approximately 55 million tonnes of fruit and vegetable waste.

Horticultural by-products were considered undervalued waste. Now fruit and vegetable by-products are used for the recovery of highly-valued biomolecules. These can be further utilized in the food, cosmetic and pharmaceutical industries. Horticultural by-products are a crucial source of organic acids, sugar derivatives, minerals, phenolics

Department of Agriculture and Environmental Sciences, National Institute of Food Technology Entrepreneurship and Management, Kundli, Sonepat, Haryana 131 028 India. E-mail: sunil_ciah@yahoo.co.in
* Corresponding author: spareekhort@gmail.com

and dietary fibers. These have been linked with various biological properties, such as antiviral, antibacterial, antitumoral, antimutagenic and cardioprotective properties (Dilas et al. 2009). Many fruits and vegetables like apple, orange, peach, potato, carrot, green pea, onion, artichoke, asparagus are used for production of juice and other processed foods. A large amount of waste is generated from this type of food production (Rodriguez et al. 1997, 2006). By-products from these sources can be used for extraction and isolation of the potential phytochemicals for further utilisation in the food, cosmetic and pharmaceutical industries. Utilisation of these by-products can increase profits, reduce liability where there is an obligation to treat waste, lower waste treatment cost and can lead to healthier products (WSDE 1994). The proper use of by-products materials acquired from horticultural crops may indeed mitigate many environmental problems and establish an initiative for sustainable development of the food industry.

Waste from Fruits and Vegetables

Waste refers to any unconsumed or unused parts of the fruits, vegetables and other food items, including by-products of horticultural commodities which are discarded after processing. Horticultural waste may differ in form from commodity to commodity, but largely skin, seeds, stone, pomace, etc. are the main components of waste material. Some fruits and vegetables can generate 25–30 per cent waste, which includes seeds and skin, and such waste has limited further use (Ajila et al. 2007, 2010). Many fruits and vegetables are not consumed in raw form by humans and this means that they need to be processed first for a desired product. High quantity of waste can be generated from processing steps (Ayala-Zavala et al. 2010). For example, coffee, mango, papaya and macadamia are processed and generate waste containing many bioactive compounds (Miljkovic and Bignami 2002).

As far as wastage is concerned, sliced apples generate 10.91 per cent seed and pulp as by-products and 89.09 per cent as final product. Dicing of papaya produces 8.47 per cent peel, 6.51 per cent seeds, 32.06 per cent futile pulp (because of imperfection in cubes) and 52.96 per cent final product. The peeling of mandarins generates 16.05 per cent peels and 83.95 per cent finished goods. Pineapple processing yields 13.48 per cent peels, 9.12 per cent core, 14.49 per cent pulp, 14.87 per cent top and 48.04 per cent final product. Processing of mangoes produces 11 per cent peels, 13.5 per cent seeds, 17.94 per cent inoperable pulp and 57.56 per cent finished product (Ayala-Zavala et al. 2010, Joshi et al. 2012). The production, nature of waste, typical losses and waste quantities of main commodities are given in Table 1. The use of fruits and vegetables in food manufacturing along with its waste is given in Table 2.

The overall production of temperate fruits is less in comparison with that of tropical fruits because of their restricted growing regions. The production and consumption of tropical fruits are exponentially high. Consequently, a substantial amount of waste in the form of skin and seeds has been generated as shown in Table 3.

Bioactive compounds from fruit and vegetable waste

Fruits and vegetables wastes have been investigated thoroughly for the extraction of dietary fiber, polyphenols and other bioactive compounds (Galanakis et al. 2012).

Table 1. Production, waste nature, typical losses and potential waste quantities of some fruits and vegetables.

Commodities	Waste part	Production (tonnes)	Typical losses (%)	Waste quantities (tonnes)
Apple	Pomace, peel, seed	1,376.0	–	412.0
Banana	Peel	2,378.0	35	832.30
Citrus	Rag, peel and seeds	1,211.9	50	606.0
Grapes	Skin, stem and seeds	565.0	20	–
Guavas	Peel, core and seeds	565.0	10	–
Mango	Peel, stone	6,789.7	45	3,144.4
Onion	Outer leaves	112.0	–	–
Peas	Shell	107.7	40	68.3
Pineapple	Core, skin	75.7	33	24.7
Potato	Peel	2,769.0	15	415.3
Tomato	Core, skin and seeds	464.5	20	90.3

Source: Gupta and Joshi (2000)

Table 2. Total fruits and vegetables production along with waste and food manufacture (million tonnes).

Product	Production	Total wastage	Manufacturing of Food
Total fruits	503.3	42.4	1.4
Total vegetables	865.8	70.2	0.5

Source: Laufenberg et al. (2009)

Table 3. Proportion of flesh, skin and seeds in tropical fruits.

Fruit types	Pulp/Flesh (%)	Skin/Rind (%)	Seeds (%)	References
Dragon fruit	54–74	22–44	2–4	Cheok et al. (2016)
Durian	20–35	55–66	5–15	Siriphanich (2011)
Jackfruit	30–35	55–62	8–10	Saxena et al. (2011)
Mango	60–75	11–18	14–22	Mitra et al. (2013)
Mangosteen	25–29	60–65	6–11	Ketsa and Paull (2011), Chen et al. (2011)
Papaya	80–90	10–20	10–20	Lee et al. (2011), Parni and Verma (2014)
Passion fruit	44–54	45–52	1–4	Almeida et al. (2015), Arjona et al. (1991), Esquivel et al. (2007)
Pineapple	60–71	29–40	----	Ketnawa et al. (2012), Choonut et al. (2014)
Rambutan	34–54	37–62	4–9	Sirisompong et al. (2011), Issara et al. (2014)

Although, only pulp or flesh is consumed as edible parts of fruit and vegetable studies have found that high amounts of essential phytochemicals and nutrients are present in the inedible peels and seeds as compared to the inner edible tissues (Rudra et al. 2015). For example, peels of oranges, lemons and grapes and seeds of avocados, longans, jackfruit and mangoes contain approximately 15 per cent higher phenolic content than that of their pulps (Gorinstein et al. 2001, Soong and Barlow 2004). Thus, fruit and vegetable waste are valuable and important sources of polyphenols, dietary fiber, enzymes and flavouring agents.

Polyphenols

Polyphenols are secondary metabolites of plants which are responsible for the nutritional quality and sensory attributes of fruits and vegetables (Tomás-Barberán et al. 2000, Lapornik et al. 2005). These compounds cover the largest portion of all phytochemical groups with various biological functions (Popa et al. 2008, Ignat et al. 2011). Phenolic compounds contain aromatic ring along with one or two hydroxyl groups (Balasundram et al. 2006), which are the determinants of antioxidant activity (Heim et al. 2002). Polyphenolic compounds can be divided into different classes, such as flavonoids (sub classes; flavonols, flavanonols, flavanones, flavones, isoflavones, flavanols and anthocyanidins), phenolic acids, tannins, stilbenes and lignans (Porter 1989, Harborne et al. 1999, Hollman and Katan 1999, Robbins 2003, Bavaresco 2003).

Previous studies reveal that the peels, rinds and seeds originating from fruits and vegetables contain high levels of phenolics. The citrus industry generates high amounts of seeds and peel residue which constitute about 50 per cent of total fruit weight (Bocco et al. 1998, Ignat et al. 2011). Citrus waste is a rich source of phenolic components. The citrus peel has higher quantities of polyphenols compared to the edible parts of the citrus fruit (Balasundram et al. 2006). The peels of other fruits have also been found to have higher amounts of phenolics than the edible portions of the fruit. For example, Gorinstein et al. (2001) found that the peels of peaches, apples and pears contained double levels of polyphenols compared to those found in the peeled fruits. Someya et al. (2002) reported that the banana pulp has 232 mg/100 g dw of phenolics and this level only accounted for about 25 per cent of that in the peel. Similarly, pomegranate peels were found to contain 249.4 mg/g phenolic compounds, which equates to approximately 10 times that in the pulp (24.4 mg/g) (Li et al. 2006). The eight selected cultivars of Clingstone peaches were investigated and it was found that their peels contained phenolics 2–2.5 times higher than those of the edible flesh (Chang et al. 2000). Apple peels were reported to be a rich source of phenolics with phenolic content of 300 mg/100 g dw (Wolfe and Liu 2003). Grape skin and seeds, by-products of juice and wine industries, are also a rich source of mono-, oligo- and polymeric proanthocyanidins (phenolics) (Shrikhande 2000, Torres and Bobet 2001).

In the case of vegetable waste, potato peel was reported as a good source of phenolic compounds as it contains 50 per cent of the whole phenolic content (Friedmen 1997). Analysis of cucumber peel revealed that it is a cheap source of flavonoids for industrial purposes (Agarwal et al. 2012). Zeyeda et al. (2008) reported that vegetable waste is rich in phenolic content in an increasing order from olive leaves, tomato peel, cucumber peel, watermelon peel and potato peel. In addition to this, the seed extracts of five vegetables *Cucumis sativus* (cucumber), *Cucurbita pepo* (pumpkin), *Momordica charantia* (bitter gourd), *Lagenaria siceraria* (bottle gourd) and *Praecitrullus fistulosus* (round gourd) have been found to be highly effective against several microbes, like *Eischericia coli*, *Fusarium oxysporium*, *Streptococcus thermophilous*, *Serratia marcescens*, and *Trichoderma reesei* (Sood et al. 2012, Sonia et al. 2016). Phenolic compounds possess antioxidant properties and thus can be used as food preservatives (Valenzuela et al. 1992, Naczk and Shahidi 2006). The health benefits of phenolic compounds include antimicrobial, antifungal and anti-carcinogenic effects (Gordon 1996, Lutfullah et al. 2014). The types of different phenolic compounds in fruit and vegetable by-products are illustrated in Table 4.

Dietary fibers

Many definitions have been given for dietary fibers but the term of dietary fiber is still controversial. A chemical definition includes a non-starchy polysaccharide. The most used and common definition of dietary fiber is 'dietary fiber is oligosaccharides, polysaccharides and the (hydrophilic) derivatives which cannot be digested by the human digestive enzymes to absorbable components in the upper alimentary tract. This includes lignins (Thebaudin et al. 1997)'. Dietary fiber is classified into two categories, i.e., soluble (mucilages, pectins, gums) and insoluble (lignins, celluloses, some hemicelluloses).

Fruit and vegetable by-products, such as pomace, peel, seeds, stems, oil cakes, pods and husk are the potential sources of dietary fiber (Sharma et al. 2016). According to Larrauri (1999) (quoted by Kunzek et al. 2003) 'perfect fiber' should have the following qualities:

- It should not contain negative color, odor, texture and taste
- It must have balanced soluble and insoluble fibers with the presence of bioactive contents
- It should not have any anti-nutritional compounds
- It must not react with the food that it is being added to; moreover, it should have a long shelf-life
- It must have the desired physiological effects
- It must have a positive image for consumers
- It should be reasonable in price

The amount of total dietary fibers (soluble and insoluble) of fruit and vegetable by-products is described below:

Fruit by-products

Apple (*Malus domestica*): The dietary fiber content of whole apple (pulp and peel) was analyzed and it was found that the peel of apple had high amounts of total dietary fibers, which accounted for 0.91 per cent fresh weight (FW). The percentage of insoluble and soluble dietary fibers was (0.46 per cent FW) and (0.43 per cent FW), respectively; these fibres are a good proportion for health (Gorinstein et al. 2001).

Grapes (*Vitis vinifera* L.): Grape pomace is a rich source of dietary fibers, i.e., hemi-celluloses, cellulose and small proportions of pectins (Kammerer et al. 2005). González-Centeno et al. (2010) investigated the by-products (stems, pomace) of ten

Table 4. Types of phenolic compounds in by-products of fruits and vegetables.

Commodities	Waste	Phenolic compounds	References
Fruits			
Apple	Pomace	Catechins, hydroxycinnamates, phloretin glycosides, uercetin glycosides, and procyanidins	Foo and Lu (1999), Lommen et al. (2000), Lu and Foo (1997, 1998), Schieber et al. (2001)
Banana	Bract	Anthocyanidins (delphinidin, cyanidin, pelargonidin, peonidin, petunidin and malvidin)	Pazmino-Duran et al. (2001)
	Peel	Carotenoids (xanthophylls, laurate, palmitate or caprate)	Subagio et al. (1996)
Citrus fruits	Peel and solid residues	Hesperidin and eriocitrin	Coll et al. (1998)
Grapes	Seeds	Procyanidins	Fuleki and Ricardo (1997), Jayaprakasha et al. (2001), Kallithraka et al. (1995), Saito et al. (1998)
	Pomace	Anthocyanins, catechins, flavonol glycosides, stilbenes	Schieber et al. (2001)
	Skin	Catechin, epicatechin, picatechin gallate and epigallocatechin	Souquet et al. (1996)
Mango	Seed kernel	Gallic and ellagic acids, and gallates, gallotannins	Arogba (2000), Schieber et al. (2001)
	Peel	Flavonol glycosides	Schieber et al. (2000)
Vegetables			
Carrot	Pomace	Carotene (α and β)	Schieber et al. (2001)
Olive	Waste water	Hydroxytyrosol derivatives and oleuropein	Visioli et al. (1998, 1999)
Onion	Skin	Quercetin 3,40-O-diglucoside and quercetin 40-O-monoglucoside	Price and Rhodes (1997)
Potato	Peel	Chlorogenic, gallic, protocatechuic and caffeic acids	Onyeneho and Hettiarachchy (1993), Rodriguez et al. (1994)
Red beet	Peel	Tryptophane, p-coumaric and ferulic acids, cyclodopa glucoside derivatives	Kujala et al. (2001)
Tomato	Skin	Lycopene	Sharma and Maguer (1996)

varieties of grapes. 'Tempranillo', the red grape cultivar, had the highest dietary fibers content in the pomace (36.9 g/100 g FW), followed by the stem (34.8 g/100 g FW) and grape (5.1 g/100 g FW). 'Manto Negro': Red grape pomace was studied for its dietary fiber and total dietary content was found to be 77.2 per cent of dry matter (DM). In this, the soluble fiber (3.77 per cent DM) was less than the insoluble fiber (73.5 per cent DM) (Llobera and Cañellas 2007). The total dietary content of red grape pomace was found the same as the white grape pomace (71.56 per cent DM, 'Prensal Blanc' cultivar). Among the total dietary fiber, the soluble fraction (10.33 per cent DM) was found to be less than the insoluble fraction (61.26 per cent DM) of the white grape pomace (Llobera and Canellas 2008).

Mango (*Mangifera indica*): The by-products of mango have been found to contain high amounts of dietary fiber. An investigation revealed that 51.2 per cent DM content of total dietary fiber (32 per cent DM insoluble fiber and 19 per cent DM soluble fiber) was found in mango peel (Ajila et al. 2008, 2009). Vergara-Valencia et al. (2007) reported that 'Tommy Atkins' mangoes contained 28.05 per cent DM of dietary fiber content; of which 13.80 per cent DM was insoluble fiber and 14.25 per cent was DM soluble fiber.

Lemon (*Citrus lemon*): Many studies have revealed the health benefits of the dietary fiber obtained from lemon by-products. The dietary fiber level in lemon peel was found to be 14 g/100 g DM which is much higher than that of the peeled lemon 7.34 g/100 g DM (Gorinstein et al. 2001). Of total dietary fiber, the amount of insoluble fiber and soluble fiber was 9.04 g/100 g DM and 4.93 g/100 g DM, respectively.

Peach (*Prunus persica*): 'Sudanell' peaches were reported to contain 30.7–36.1 per cent total dietary fiber. This constituted of 12.3 per cent DM soluble fiber and 23.8 per cent DM insoluble fiber. Kurz et al. (2008) investigated polysaccharides in the cell wall of peach fruit and found pectin was the main polysaccharide. Little difference was seen between unpeeled and peeled peaches in the content of hemi-cellulose (HC), lignin and cellulose (C). The peeled peach contained 17.0 g/100 g AIR (alcohol-insoluble residue) and 13.1 g/100 g AIR amount of C and HC. The unpeeled peach contained 16.4 g/100 g AIR and 12.9 g/100 g AIR.

Orange (*Citrus sinensis*): The peel of 'Liucheng' cultivar of orange was investigated for dietary fiber components and the peel was found to contain 57 per cent DW of total dietary fibers (9.41 per cent DW was the soluble fraction and 47.6 per cent DW was insoluble). The main components of the fiber were cellulose and pectin polysaccharides (Chau and Huang 2003).

Vegetable by-products

The by-products of vegetables have also been found to contain a high content of dietary fiber. Some vegetable examples are discussed below:

Onion (*Allium cepa* L.): Dietary fiber is found in all layers of the onion but in different ratio. Jaime et al. (2002) investigated the entire onion, including skin and inner layers of three different varieties of onions, to determine dietary fiber content. They found the highest amount of total dietary fiber in skin 68.3 per cent DM of the onion ('Grano de Oro' variety) and lowest (11.6 per cent DM) in the inner part of the onion. Insoluble fibers were also reported in higher amounts in the skin of the onion (66.6 per cent DM 'Grano de Oro' cv.) compared with those in the inner parts.

Potato (*Solanum tuberosum* L.): Three varieties of potato and their peels were investigated and there was no significant difference found in the total fiber concentration

(5.6 per cent, Karnico cultivar). The potato also contained 72.4 per cent DM of total starch (Liu et al. 2007). Thed and Phillips (1995) analyzed the impact of cooking on the total dietary fibers content. They worked on baking, boiling, microwaving and deep frying of potato and found significant increase in total dietary fiber content from 7.60 per cent DM (control) to 8.92 per cent DM by deep-fat frying and 9.08 per cent DM by microwave heating.

Cauliflower (*Brassica oleracea* L. *var botrytis*): By-products (stem and floret) of cauliflower were analyzed for non-starch polysaccharides (NSP) and it was found that the stem contained higher amounts 3.11 per cent FW in comparison with floret NSP 2.31 per cent FW. In both types of waste, the insoluble fibers were higher than soluble fibers. Of those, pectic polysaccharides were the main components of NSP (Femenia et al. 1997, 1998).

Carrot (*Daucus carrota* L.): The carrot pomace was analyzed for dietary fiber content and it was found to contain 63.6 per cent DM total dietary fibers. Of this, the concentration of insoluble fibers was 50.1 per cent DM and the soluble fibers was 13.5 per cent DM (Chau et al. 2004). Fresh carrot peels were also studied for dietary fibers and the effect of blanching on these was analyzed. The content of total dietary fibers was significantly increased from 45.45 per cent DM to 73.32 per cent DM after blanching (Chantaro et al. 2008).

Tomato (*Solanum lycopersicum* L.): Of waste, tomato pomace contains up to 50 per cent fibers on dry basis (Del Valle et al. 2006). García Herrera et al. (2010) studied levels of fibers in tomato peel by fusion of ground and dried tomato peel and found that the total dietary fiber content was 82.7 per cent of FW, insoluble fibers were found to be more than soluble fibers.

Apart from commodities described above, dietary fibers in the waste of other fruits are summarized in Table 5.

Table 5. Concentration of total dietary fibers in the by-products of fruits and vegetables.

Commodities	Type of waste	Total dietary fibers (TDF) (%)	Reference
Apricot	Seeds	27–35	Seker et al. (2010)
Banana	Peel	50 TDF (IDF and SDF ratio 5.46:1)	Wachirasiri et al. (2009)
Cranberry	Seeds	45.93 IDF, 5.13 SDF	C.S.F. (2013)
Date	Seeds	57.87–92.4 TDF	Almana and Mahmoud (1994), Elleuch et al. (2008), Al-Farsi and Lee (2008)
Grape	Seeds	40	Bagchi et al. (2002)
Green Chilli	Peel and seeds	80.41	Matsuda (1997), Mckee and Latner (2000)
Pumpkin	Pomace	76.94	Turksoy and Ozkaya (2011)
Raspberry	Pomace	77.5 TDF (75 IDF, 2.5 SDF)	Gorecka et al. (2010)

TDF = Total dietary fibers, IDF = Insoluble dietary fiber, SDF = Soluble dietary fiber

Enzymes and flavoring agents

Various wastes of fruit and vegetables are used to obtain different enzymes and flavoring agents with the help of micro-organisms. This bioconversion is achieved by solid-state fermentation. Vanillin (4-hydroxy-3-methoxybenzaldehyde) is the most important and a highly used flavour in food, cosmetic, pharmaceutical and detergent industries or sectors. The extraction of natural vanillin is achieved from the fermented pods of vanilla orchids (*Vanilla planifolia*) (Panouillé et al. 2007). Pineapple flavour (ethyl butyrate) is produced with the help of the microorganism, *Ceratocystis fimbriata* from apple pomace. Coconut flavour (δ-decalactone) is also achieved through bioconversion with the help of the micro-organism, *Ceratocystis moniliformis* from olive press cake (Lanza and Palmer 1976).

Amylases have been widely used in food processing industries in various products, including fruit juices, starch syrup, moist cakes, chocolate cakes. They have been also used in processes such as brewing, preparation of digestive aids and baking (Laufenberg et al. 2009). Banana stalk is the best substrate for the production of amylases because it contains more carbohydrate derivatives. Pectinase degrades pectic elements of fruits and vegetables, such as the cell wall structural component. Pectate lyase and pectin lyase can break down the long carbon chains by breaking the glycosidic bonds, while pectin esterase works on methoxyl groups. The production of pectinase is done by SSF from grape pomace using *Aspergillus awamori* yeast (Botella et al. 2005). Along with these enzymes, tannase (EC{enzyme commission} 3.1.1.20, tannin-acyl-hydrolase), xylanases (EC 3.2.1.8), laccase (EC 1.10.3.2) and proteases (EC 3.4.21.19) are some more enzymes which are produced the same as others by Solid-State Fermentation (SSF) (Rodríguez Couto 2008). The potential products are illustrated in Table 6.

Besides these bioactive compounds, several fruit and vegetable wastes are used to extract different natural colors and oils.

Extraction of Bioactive Compounds

Extraction is the most critical step to obtain bioactive compounds from fruit and vegetable wastes (Khoddami et al. 2013). Extraction method may vary with respect to the targeted bioactive compounds. Bioactive components can be characterized after identification from stem, flower, leaves and fruit of the plants. Many factors, such as temperature, time, plant part, pressure and solvent may affect the extraction process (Hernández et al. 2009). Sample preparation is also one of the crucial factors to determine the amount of bioactive compounds. For example, Dorta et al. (2012) applied three different combinations of dehydration methods, i.e., freeze drying, oven drying with static air at 70°C and oven drying with forced air at 70°C with ethanol, ethanol; water and acetone; water as solvents for the extraction of bioactive compounds from the mango peel and seed. Results revealed extraction of freeze-dried mango peel and seed with ethanol: water contained the highest amount of phenolics and anthocyanins.

The extraction techniques can be divided into two broad categories—conventional techniques and advanced techniques. The conventional extraction techniques include Soxhlet extraction, hydrodistillation and maceration (Khoddami et al. 2013). Soxhlet

Table 6. Flavours and enzymes with their substrates and used microorganisms.

Waste/substrate	Flavour/enzymes	Microorganism used	References
Flavor			
Apple pomace	Pineapple (Ethyl butyrate)	*Ceratocystis fimbriata*	
Carrot pomace	Vanillin	*Pycnoporus cinnabarius*	Asther et al. (1996) Bonnin et al. (1999)
Olive press cake	δ-decalactone (coconut) γ-decalactone	*Ceratocystis moniliformis Pityrosporum ovale*	Lanza et al. (1976)
Sugar beet pulp	Vanillin	*Aspergillus niger*	Lesage-meessen et al. (1999)
Enzymes			
Apple and strawberry pomace	Pectinase	*Lentinus edodes*	Shah et al. (2005)
Banana waste	Laccase, Xylanase	*Aspergillus* MPS-002, *Phylostica* spp. MPS-001	Krishna and Chandrasekaran (1995)
Cabbage waste	Cellulase, amylase	*Pseudomonas* spp.	Kunamneni et al. 2005
Citrus waste	Pectinase	*Trichoderma viride*	Ramachandran et al. (2004)
Coconut oil cake	Amylase	*Aspergillus oryzae*	Rosales et al. (2005)
Grape pomace	Pectinase, cellulase, xylanase	*Aspergillus awamori*	Botella et al. (2005), Thed and Phillips (1995)
Kiwi waste, orange Peels	Laccase	*Trametes hirsute*	Rosales et al. (2007), Botella et al. (2007)
Mango peel	Cellulase	*Aspergillus niger*	Bakir et al. (2001)
Melon peel, apple pomace, hazelnut shell	Xylanase	*Trichoderma harzianum* 1073 D3	Seyis (2005), Sabu et al. (2005)
Tamarind seed powder	Tannase	*Aspergillus niger* ATCC 16620	Pandey et al. (2000)

extraction is a very popular and widely used technique, developed by German scientist, French Ritter Von Soxhlet (1879). It is very useful for extracting phytochemicals from plant parts. Hydrodistillation is also a classical technique for extracting crucial oils and various bioactive compounds from plant sources. Three types of hydrodistillation include water distillation, water and steam distillation and direct steam distillation (Vankar 2004). Maceration method has been used for a long time for the preparation of tonic beverage at home and it has been used for extraction of bioactive compounds from plants because it is inexpensive.

Conventional techniques have some major limitations, such as requirement of high purity and costly solvents, longer time of extraction, degradation of heat-sensitive compounds and low extraction selectivity (De Castro and García-Ayuso 1998). To overcome these limitations of conventional techniques, novel and emerging techniques have been introduced. There are several novel and emerging techniques, which have been applied for the extraction process, include microwave-assisted extraction (MAE), pulsed electric field (PEF), enzyme-assisted extraction (EAE), liquid-liquid extraction (LLE) and solid-liquid extraction (LSE).

The extraction yields of caffeine and polyphenols using MAE from the leaves of green tea were found to be higher in comparison with other methods of extraction (Pan et al. 2003). The extraction for ginsenosides from ginseng root achieved in 15 min by MAE technique was much higher in comparison with conventional solvent extraction for 10 hours (Shu et al. 2003). Likewise, Dorta et al. (2013) compared MAE and traditional solvent extraction (TE) for the extraction of antioxidants from mango peel and found MAE was more effective as it extracted 1.5–6.0 times more phytochemicals and antioxidants than the traditional method.

The PEF technique has been applied for the extraction of phytosterols from maize and isoflavonoids (daidzein and genistein) from soybeans, recovery was increased by 32.4 per cent for maize and 20–21 per cent for soybean (Guderjan et al. 2005). In comparison to classical processes, PEF treatment on Merlot skin gave higher amounts of anthocyanins and polyphenols (Delsart et al. 2012).

Isolation, identification and potential use of bioactive compounds from fruit and vegetable waste

Isolation is very crucial for the recovery of potential and beneficial bioactive compounds from horticulture waste. Mango peel is a high-magnitude waste obtained from processing of mango into diverse products or consumption by humans directly after ripening. The bioactive compounds were extracted by using acetone (80 per cent) from raw mango and ripe mango fruits followed by their acid hydrolysis. A high performance liquid chromatography (HPLC) coupled with a reverse phase C18 column (Shimadzu, Model LC-10A) and a diode array detector (operating at 280 nm and 320 nm) was used to identify the phenolic substances present in the peel. The results revealed the presence of gallic acid, protocatechuic acid and gentisic acid in the peel extract. Besides this Liquid Chromatography and Mass Spectrometry (LC-MS) have been applied to identify the diverse range of phenolic acid derivatives obtained from acetone-based extract from the peel of ripened and raw mango and other compounds, such as mangiferin, ellagic acid, gentisyl-protocatechuic acid, syringic acid, gallic acid and quercetin. In addition, glycosylated iriflophenone, derivative of maclurin, were found in raw mango peel samples. Furthermore, β-carotene, violaxanthin and lutein were also identified in the mango peel (Ajila et al. 2010). Mango peel also contains 40 per cent (w/v) reducing sugars. Therefore, supplementation of peel medium with diverse nutrients viz. yeast (*Saccharomyces cerevisiae*), peptone and wheat bran extract can be used for ethanol production. Direct fermentation of mango peel extract and the nutrient-supplemented mango peel medium could lead to the production of 5.13 per cent (w/v) and 7.14 per cent (w/v) of ethanol, respectively (Reddy et al. 2011).

Banana (*Musa acuminata* Colla AAA) peels contain the diverse compounds with potent antioxidant properties. Various solvents, which were used for the extraction of bioactive compounds, include ethanol, methanol, water, acetone, mixtures of ethanol and water, methanol and water or acetone and water. Out of these, the mixture of acetone and water was the most effective for extraction of compounds from the peels in comparison to other solvents. The extracted compounds were mainly phenolics, anthocyanin compounds. Dopamine, catecholamines and L-dopa were found in large amounts in the banana peel and confer significant antioxidant activity to it. There is no report for the presence of tocopherols or phytosterols and ascorbic acid in the banana peel. The anti-radical scavenging activity of peel extract was confirmed by 1, 1-diphenyl-2-picryl hydrazyl (DPPH) and 2, 2'-azino-bis 3-ethylbenzothiazoline-6-sulphonic acid (ABTS) method. Besides this, the extract was also reported to have anti-lipid peroxidation effect. Antioxidant activity in the peels of different cultivars of banana is reported to be similar (Gonzalez-Montelongo et al. 2010).

The ethanolic rind extract obtained from rambutan (*Nephelium lappaceum*) showed anti-hyper glycemic properties. The major bioactive compound found in the rind was geraniin which possessed high antioxidant activity. The analysis of geraniin was done by HPLC-LCMS/MS and NMR (Palanisamy et al. 2011). Geraniin, an ellagitannin, could be obtained through the crude ethanolic extract of rind of rambutan using reverse-phase C18 column chromatography, but only 21 per cent yield of geraniin could be obtained. The prominent impurities, like corilagin and elaeocarpus, were primarily identified by LC-MS and these were reported to possess similar bioactive properties as geraniin (Perera et al. 2012).

Generally after the extraction of pectin from the fruits, there is ample scope to extract substances having antioxidant and antimicrobial properties from the leftover waste. The waste from *Citrus limetta, Solanum lycopersicum, Musa* sp., *Psidium* sp. and *Citrullus lanatus* was used for extracting total soluble proteins (TSP) and heat soluble proteins (HSP). HSP from *S. lycopersicum* was found to prevent the growth of *Escherichia coli*. HSP obtained from *C. limetta* and *Musa* sp. could inhibit *Pseudomonas* sp. *Fusarium oxysporum*, a fungal pathogen. There was no impact of HSP from *C. lanatus* on any of the pathogens. The fruit residues of *Musa* sp. and *Psidium* sp. also showed antioxidant activity alongwith high phenolic content. Thus, bioactive compounds from these residues are of great use in nutraceutical and pharmaceutical sectors (Farha et al. 2012).

In pomegranate, the pith and carpellary membrane constitute around 13 per cent of fruit weight and are a rich source of bioactive molecules. Column chromatography and HPLC are potential techniques to isolate bioactive molecules, which possess high radical scavenging activity. Punicalagin was found in a large concentration in the pith and the carpellary membrane of the pomegranate fruit. Punicalagin was reported to have high antioxidant activity. The ultraviolet (UV), infra-red (IR), 2 dimension-nuclear magnetic resonance (2D NMR), gas chromatography and mass (GC-Mass) and Matrix-assisted laser desorption/ionization (MALDI-Mass) are used to identify puncalagin (Kulkarni et al. 2004).

The quantification of bioactive molecules from the twelve native tropical fruits, viz., tamarind, sapodilla, Surinam cherry, passion fruit, soursop, mango, papaya, cashew apple, monbin, guava, acerola and pineapple from Brazil was carried out. The bioactive molecules which were studied were resveratrol and coumarins. By-products like peels, seeds and pulp leftovers made up part of the study. For isolating resveratrol, coumarin and gallic acid, HPLC analysis was performed. Resveratrol was mainly identified in the by-product of guava and Surinam cherry, while coumarins were reported to be found in the by-products of guava, passion fruit and Surinam cherry. Therefore, these by-products might be used for recovery of bioactive molecules for further utilisation (Da Silva et al. 2014).

Olive mill waste (OMW) is one of the potential sources of biophenols. Biophenols present in OMW accounted for 98 per cent of total biophenols present in the whole fruit. OMW was reported to be a reservoir of antioxidants and antimicrobials. HPLC or HPLC-MS-MS was found to be a potential technique in isolating the bioactive compounds from olive mill waste (Obied et al. 2005). In general, various extraction and isolation techniques can be applied to the bioactive compounds from the waste. These compounds have potent biological properties and thus can be applied in the food, cosmetic and pharmaceutical industries (Table 7). Therefore, bio-waste from vegetables and fruits can be further studied for their potential use as a rich source for recovery of bioactive compounds.

Table 7. Bioactive compounds obtained from the bio-waste of fruits and vegetables and their potential use.

Fruit/ vegetable	Waste	Bioactive compounds	Uses	Reference
Apple	Pomace	Polyphenols, antioxidants	Nutraceutical, pharmaceutical	Wijngaard and Brunton (2009)
Banana	Peel	Phenolic compounds, anthocyanin, dopamine, catecholamines and L-Dopa	Nutraceutical, pharmaceutical	Gonzalez-Montelongo et al. (2010)
Grapes	Seeds	High phenolic content	Food and pharmaceutical	Babbar et al. (2011)
Kinnow	Seeds, peel	Phenols	Food, pharmaceutical	Babbar et al. (2011)
Litchi	Seeds, pericarp	High phenolic content	Food, pharmaceutical	Babbar et al. (2011)
Mango	Peel (ripe and raw)	Gallic acid, protocatechuic acid, gentisic acid, mangiferin, ellagic acid, gentisyl-protocatechuic acid, syringic acid, gallic acid and quercetin; raw mango peels (glycosylated iriflophenone, derivatives of maclurin); β-carotene violaxanthin and lutein	Food, pharmaceutical	Maisuthisakul and Gordon (2009)
Olive	Olive mill waste	Biophenols (antioxidant, antimicrobial activity)	Pharmaceutical, food	Obied et al. (2005)
Pomegranate	Pith and carpellary membrane	Puncalagin	Nutraceutical, pharmaceutical	Kulkarni et al. (2004)
Rambutan	Rind	Geraniin (an elaggitannin); Anti-hyperglycemic	Nutraceutical, pharmaceutical, cosmetic	Palaniswamy et al. (2011), Perera et al. (2012)

Conclusions

Adequate utilisation of bio-waste is essential in order to maximize the use of available resources in Nature. Besides this, if we consider management of bio-waste, then we can analyze bio-waste disposal too. This is a potential challenge that must be overcome by utilizing bio-waste and by extracting bioactive compounds present in the bio-waste. Fruits and vegetables and their by-products (seeds, kernel, peels, etc.) are very rich in bioactive compounds and are good for health. Therefore, we need a comprehensive strategy for sustainable utilisation of the bio-waste left during industrial processing of fruits and vegetables. The extracted bioactive compounds can be utilized in food, nutraceutical and pharmaceutical sectors with the aim of developing products that have benefits for human health and well-being.

References

Agarwal, M., A. Kumar, R. Gupta and S. Upadhyaya. 2012. Extraction of polyphenol, flavanoid from *Emblica officinalis*, *Citrus limon*, *Cucumis sativus* and evaluation of their antioxidant activity. Oriental. J. Chem. 28: 993–998.

Ajila, C. M., S. G. Bhat and U. J. S. P. Rao. 2007. Valuable components of raw and ripe peels from two Indian mango varieties. Food Chem. 102(4): 1006–1011.

Ajila, C. M., K. Leelavathi and U. J. S. P. Rao. 2008. Improvement of dietary fiber content and antioxidant properties in soft dough biscuits with the incorporation of mango peel powder. J. Cer. Sci. 48(2): 319–326.

Ajila, C. M., M. Aalami and U. J. S. P. Rao. 2009. Mango peel powder: A potential source of antioxidant and dietary fiber in macaroni preparations. Innov. Food Sci. Emerg. Technol. 11(1): 219–224.

Ajila, C. M., M. Aalami, K. Leelavathi and U. J. S. P. Rao. 2010. Mango peel powder: A potential source of antioxidant and dietary fiber in macaroni preparations. Innov. Food Sci. Emerg. Technol. 11(1): 219–224.

Al-Farsi, M. A. and C. Y. Lee. 2008. Optimization of phenolics and dietary fiber extraction from date seeds. Food Chem. 108(3): 977–985.

Almana, H. A. and R. M. Mahmoud. 1994. Palm date seeds as an alternative source of dietary fiber in Saudi bread. Ecol. Food Nutr. 32: 261–270.

Almeida, J. M., V. A. Lima, P. C. Giloni-Lima and A. Knob. 2015. Passion fruit peel as novel substrate for enhanced β-glucosidases production by *Penicillium verruculosum*: Potential of the crude extract for biomass hydrolysis. Biomass Bioenerg. 72: 216–226.

Arjona, H. E., F. B. Matta and J. O. Garner. 1991. Growth and composition of passion fruit (*Passiflora edulis*) and maypop (*P. incarnata*). Hort. Sci. 26: 921–923.

Arogba, S. S. 2000. Mango (*Mangifera indica*) kernel: Chromatographic analysis of the tannin, and stability study of the associated polyphenol oxidase activity. J. Food Compos. Anal. 13: 149–156.

Asther, M., L. Lesage-Meessen, M. Haon, C. Stentelaire, M. Delattre, S. V. Hulle, G. Hennebert, J. F. Thibault and S. Van-Hulle. 1996. Fungal biotransformation of European agricultural by-products to natural vanillin: A two-step process. pp. 123–125. *In*: Food Ingredients Europe-Conference Proceedings, Paris, France, 12–14 November.

Ayala-Zavala, J. F., C. Rosas-Dominquez, V. Vega-Vega and G. A. Gonzalez-Aguilar. 2010. Antioxidant enrichment and antimicrobial protection of fresh-cut fruits using their own by-products: Looking for integral exploitation. J. Food Sci. 75: R175–R181.

Babbar, N., H. S. Oberoi, D. S. Uppal and R. T. Patil. 2011. Total phenolic content and antioxidant capacity of extracts obtained from six important fruit residues. Food Res. Int. 44(1): 391–396.

Bagchi, D., M. Bagchi, S. J. Stohs, S. D. Ray, C. K. Sen and H. G. Preuss. 2002. Cellular protection with proanthocyanidins derived from grape seeds. Ann. New York Acad. Sci. 957: 260–270.

Bakir, U., S. Yavascaoglu, F. Guvenc and A. Ersayin. 2001. An endo-beta-1,4-xylanase from *Rhizopus oryzae*: Production, partial purification and biochemical characterization. Enz. Micro. Technol. 29(6-7): 328–334.

Balasundram, N., K. Sundram and S. Samman. 2006. Phenolic compounds in plants and agri-industrial by-products: antioxidant activity, occurrence, and potential uses. Food Chemistry 99(1): 191–203.

Bavaresco, L. 2003. Role of viticultural factors on stilbene concentrations of grapes and wine. Drugs Exp. Clin. Res. 29: 181–187.

Bocco, A., M. -E. Cuvelier, H. Richard and C. Berset. 1998. Antioxidant activity and phenolic composition of citrus peel and seed extracts. J. Agri. Food Chem. 46: 2123–2129.

Bonnin, E., L. Lesage-Meessen, M. Asther and J. F. Thibault. 1999. Enhanced bioconversion of vanillic acid into vanillin by the use of 'natural' cellobiose. J. Sci. Food Agric. 79(3): 484–486.

Botella, C., I. De ory, C. Webb, D. Cantero and A. Blandino. 2005. Hydrolytic enzyme production by *Aspergillus awamori* on grape pomace. Biochem. Eng. J. 26(2-3): 100–106.

Botella, C., A. Diaz, I. De ory, C. Webb and A. Blandino. 2007. Xylanase and pectinase production by *Aspergillus awamori* on grape pomace in solid state fermentation. Process Biochem. 42(1): 98–101.

C.S.F. 2013. Cranberry Seed Flour. Fruit essentials, P.O. Box 282, New Richmond, WI 54017, US Patent 6,391,345. http://www.fruitessentials.com/documents/bulkcsf-specs.pdf (Accessed on 8th June 2016).

Chang, S., C. Tan, E. N. Frankel and D. M. Barrett. 2000. Low density lipoprotein antioxidant activity of phenolic compounds and polyphenol oxidase activity in selected clingstone peach cultivars. J. Agric. Food Chem. 48: 147–151.

Chantaro, P., S. Devahastin and N. Chiewchan. 2008. Production of antioxidant high dietary fiber powder from carrot peels. LWT—Food Sci. Technol. 41(10): 1987–1994.

Chau, C. F. and Y. L. Huang. 2003. Comparison of the chemical composition and physicochemical properties of different fibers prepared from the peel of *Citrus sinensis* L. cv. Liucheng. J. Agric. Food Chem. 51(9): 2615–2618.

Chau, C. -F., C. -H. Chen and M. -H. Lee. 2004. Comparison of the characteristics, functional properties, and *in vitro* hypoglycemic effects of various carrot insoluble fiber-rich fractions. LWT—Food Sci. Technol. 37(2): 155–160.

Chen, Y., B. Huang, M. Huang and B. Cai. 2011. On the preparation and characterization of activated carbon from mangosteen shell. J. Taiwan Inst. Chem. Eng. 42: 837–842.

Cheok, C. Y., N. M. Adzahan, R. A. Rahman, N. H. Z. Abedin, N. Hussain, R. Sulaiman and G. H. Chong. 2016. Current trends of tropical fruit waste utilization. Crit. Rev. Food Sci. Nutr. DOI:1 0.1080/10408398.2016.1176009.

Choonut, A., M. Saejong and K. Sangkharak. 2014. The production of ethanol and hydrogen from pineapple peel by *Saccharomyces cerevisiae* and *Enterobacter aerogenes*. Energy Procedia 52: 242–249.

Coll, M. D., L. Coll, J. Laencine and F. A. Tomas-Barberan. 1998. Recovery of flavanons from wastes of industrially processed lemons. Z. Lebensm Unters F. A. 206: 404–407.

Da Silva, L. M. R., E. A. T. de Figueiredo, N. M. P. S. Ricardo, I. G. P. Vieira, R. W. de Figueiredo, I. M. Brasil and C. L. Gomes. 2014. Quantification of bioactive compounds in pulps and by-products of tropical fruits from Brazil. Food Chem. 143: 398–404.

De Castro, M. L. and L. E. García-Ayuso. 1998. Soxhlet extraction of solid materials: An outdated technique with a promising innovative future. Analyt. Chim. Acta 369(1): 1–10.

Del Valle, M., M. Camara and M. E. Torija. 2006. Chemical characterization of tomato pomace. J. Sci. Food Agric. 86(8): 1232–1236.

Delsart, C., R. Ghidossi, C. Poupot, C. Cholet, N. Grimi, E. Vorobiev, V. Milisic and M. M. Peuchot. 2012. Enhanced extraction of valuable compounds from merlot grapes by pulsed electric field. American J. Enol. Viticul. 63(2): 205–211.

Dilas, S., J. Čanadanović-Brunet and G. Ćetković. 2009. By-products of fruits processing as a source of phytochemicals. Chem. Ind. Eng. Quart. 15(4): 191–202.

Dorta, E., M. G. Lobo and M. González. 2012. Reutilization of mango byproducts: Study of the effect of extraction solvent and temperature on their antioxidant properties. J. Food Sci. 77(1): C80–C88.

Dorta, E., M. G. Lobo and M. González. 2013a. Optimization of factors affecting extraction of antioxidants from mango seed. Food Bioprocess Tech. 6(4): 1067–1081.

Dorta, E., M. G. Lobo and M. González. 2013b. Improving the efficiency of antioxidant extraction from mango peel by using microwave-assisted extraction. Plant Foods Human Nutr. 68(2): 190–199.

Elleuch, M., S. Besbes, O. Roiseux, C. Blecker, C. Deroanne, N. E. Drira and H. Attia. 2008. Date flesh: Chemical composition and characteristics of the dietary fiber. Food Chem. 111: 676–682.

Esquivel, P., F. C. Stintzing and R. Carle. 2007. Comparison of morphological and chemical fruit traits from different pitaya genotypes (*Hylocereus* sp.) grown in Costa Rica. J. Appl. Bot. Food Qual. 81: 7–14.

Farha, S., E. Chatterjee, S. G. Manuel, S. A. Reddy and R. D. Kale. 2012. Isolation and characterization of bioactive compounds from fruit wastes. Biotech. Mol. Bio. 6: 92–94.

Femenia, A., A. C. Lefebvre, J. -Y. Thebaudin, J. A. Roberston and C. -M. Bourgeois. 1997. Physical and sensory properties of model foods supplemented with cauliflower fiber. J. Food Sci. 62(4): 635–639.

Femenia, A., J. A. Robertson, K. W. Waldron and R. R. Selvendran. 1998. Cauliflower (*Brassica oleracea* L.), globe artichoke (*Cynara scolymus*) and chicory witloof (*Cichorium intybus*) processing by-products as sources of dietary fiber. J. Sci. Food Agric. 77(4): 511–518.

Foo, L. Y. and Y. Lu. 1999. Isolation and identification of procyanidins in apple pomace. Food Chem. 64: 511–518.

Friedmen, M. 1997. Chemistry, biochemistry and dietary role of potato polyphenols. J. Agric. Food Chem. 45: 1523–1540.

Fuleki, T. and J. M. Ricardo da Silva. 1997. Catechin and procyanidin composition of seeds from grape cultivars grown in Ontario. J. Agri. Food Chem. 45: 1156–1160.

Galanakis, C. M. 2012. Recovery of high added-value components from food wastes: Conventional, emerging technologies and commercialized applications. Trends Food Sci. Technol. 26(2): 68–87.

García Herrera, P., M. C. Sánchez-Mata and M. Cámara. 2010. Nutritional characterization of tomato fiber as a useful ingredient for food industry. Innov. Food Sci. Emerg. Technol. 11(4): 707–711.

González-Centeno, M. R., C. Rosselló, S. Simal, M. C. Garau, F. Lopez and A. Femenia. 2010. Physico-chemical properties of cell wall materials obtained from ten grape varieties and their byproducts: Grape pomaces and stems. LWT—Food Sci. Technol. 43(10): 1580–1586.

González-Montelongo, R., M. G. Lobo and M. González. 2010. Antioxidant activity in banana peel extracts: Testing extraction conditions and related bioactive compounds. Food Chem. 119(3): 1030–1039.

Gordon, M. H. 1996. Dietary antioxidants in disease prevention. Nat. Prod. Rep. 265: 273.

Gorecka, D., B. Pacholek, K. Dziedzic and M. Gorecka. 2010. Raspberry pomace as a potential fiber source for cookies enrichment. Acta Sci. Pol. Technol. Alim. 9(4): 451–462.

Gorinstein, S., O. Martin-Belloso, Y. S. Park, R. Haruenkit, A. Lojek, M. Ciz and A. Caspi. 2001. Comparison of some biochemical characteristics of different citrus fruits. Food Chem. 74(3): 309–315.

Gorinstein, S., Z. Zachwieja, M. Folta, H. Barton, J. Piotrowicz, M. Zemser et al. 2001. Comparative contents of dietary fiber, total phenolics, and minerals in persimmons and apples. J. Agric. Food Chem. 49(2): 952–957.

Guderjan, M., S. Töpfl, A. Angersbach and D. Knorr. 2005. Impact of pulsed electric field treatment on the recovery and quality of plant oils. J. Food Engin. 67(3): 281–287.

Gupta, K. and V. K. Joshi. 2000. Fermentative utilization of waste from food processing industry. pp. 1171–1193. *In*: L. R. Verma and V. K. Joshi (eds.). Postharvest Technology of Fruits and Vegetables: Handling, Processing, Fermentation and Waste Management. New Delhi: Indus Pub Co.

Harborne, J. B., H. Baxter and G. P. Moss. 1999. Phytochemical Dictionary: Handbook of Bioactive Compounds from Plants (2nd Ed.). London: Taylor and Francis.

Hernández, Y., M. G. Lobo and M. González. 2009. Factors affecting sample extraction in the liquid chromatographic determination of organic acids in papaya and pineapple. Food Chem. 114(2): 734–741.

Heim, K. E., A. R. Tagliaferro and D. J. Bobilya. 2002. Flavonoid antioxidants: Chemistry, metabolism and structure–activity relationships. J. Nutr. Biochem. 13: 572–584.

Hollman, P. H. and M. B. Katan. 1999. Dietary flavonoids: Intake, health effects and bioavailability. Food Chem. Toxicol. 37(9): 937–942.

Ignat, I., I. Volf and V. I. Popa. 2011. A critical review of methods for characterisation of polyphenolic compounds in fruits and vegetables. Food Chem. 126(4): 1821–1835.

Issara, U., U. Zzaman and T. A. Yang. 2014. Rambutan seed fat as a potential source of cocoa butter substitute in confectionary product. Int. Food Res. J. 21: 25–31.

Jaime, L., E. Molla, A. Fernández, M. A. Martín-Cabrejas, F. J. López-Andréu and R. M. Esteban. 2002. Structural carbohydrate differences and potential source of dietary fiber of onion (*Allium cepa* L.) tissues. J. Agric. Food Chem. 50(1): 122–128.

Jayaprakasha, G. K., R. P. Singh and K. K. Sakariah. 2001. Antioxidant activity of grape seed (*Vitis vinifera*) extracts on peroxidation models *in vitro*. Food Chem. 73: 285–290.

Joshi, V. K., A. Kumar and V. Kumar. 2012. Antimicrobial, antioxidant and phyto-chemicals from fruit and vegetable wastes: A review. Int. J. Food Ferment. Technol. 2(2): 123–136.

Kallithraka, S., C. Garcia-Viguera, P. Bridle and J. Bakker. 1995. Survey of solvents for the extraction of grape seed phenolics. Phytochem. Anal. 6: 265–267.

Kammerer, D., A. Claus, A. Schieber and R. Carle. 2005. A novel process for the recovery of polyphenols from grape (*Vitis vinifera* L.) pomace. J. Food Sci. 70: 157–163.

Ketnawa, S., P. Chaiwut and S. Rawdkuen. 2011. Aqueous two-phase extraction of bromelain from pineapple peels cv. Phu Lae and its biochemical properties. Food Sci. Biotech. 20: 1219–1226.

Ketsa, S., R. E. Paull and M. E. Saltveit. 2011. Mangosteen (*Garcinia mangostana* L.). pp. 1–30. *In*: E. M. Yahia (ed.). Postharvest Biology and Technology of Tropical and Subtropical Fruits. Volume 4: Mangosteen to White sapote. Oxford: Woodhead Publishers.

Khoddami, A., M. A. Wilkes and T. H. Roberts. 2013. Techniques for analysis of plant phenolic compounds. Molecules 18(2): 2328–2375.

Krishna, C. and M. Chandrasekaran. 1995. Economic utilization of cabbage wastes through solid-state fermentation by native microflora. J. Food Sci. Technol. 32(3): 199–201.

Kujala, T., J. Loponen and K. Pihlaja. 2001. Betalains and phenolics in red beetroot (*Beta vulgaris*) peel extracts: Extraction and characterisation. Zeitschrift fu"r Naturforschung C 56: 343–348.

Kulkarni, A. P., S. M. Aradhya and S. Divakar. 2004. Isolation and identification of a radical scavenging antioxidant-punicalagin from pith and carpellary membrane of pomegranate fruit. Food Chem. 87(4): 551–557.

Kunamneni, A., K. Permaul and S. Singh. 2005. Amylase production in solid state fermentation by the thermophilic fungus *Thermomyces lanuginosus*. J. Biosci. Bioeng. 100(2): 168–171.

Kunzek, H., S. Müller, S. Vetter and R. Godeck. 2002. The significance of physico chemical properties of plant cell wall materials for the development of innovative food products. European Food Res. Tech. 214(5): 361–376.

Kurz, C., R. Carle and A. Schieber. 2008. Characterisation of cell wall polysaccharide profiles of apricots (*Prunus armeniaca* L.), peaches (*Prunus persica* L.), and pumpkins (*Cucurbita* sp.) for the evaluation of fruit product authenticity. Food Chem. 106: 421–430.

Lanza, E., K. H. Ko and J. K. Palmer. 1976. Aroma production by cultures of *Ceratocystis moniliformis*. J. Agric. Food Chem. 24(6): 1247–1250.

Lapornik, B., M. Prošek and A. G. Wondra. 2005. Comparison of extracts prepared from plant by-products using different solvents and extraction time. J. Food Eng. 71(2): 214–222.

Larrauri, J. A. 1999. New approaches in the preparation of high dietary fibre powders from fruit by-products. Trends Food Sci. Technol. 10(1): 3–8.

Laufenberg, G., N. Schulze and K. Waldron. 2009. A modular strategy for processing of fruit and vegetable wastes into value-added products. pp. 286–353. *In*: K. Waldron (ed.). Handbook of Waste Management and Co-product Recovery in Food Processing, Vol. 2. Oxford: Woodhead Publishing.

Lee, W. -J., M. -H. Lee and N. -W. Su. 2011. Characteristics of papaya seed oils obtained by extrusion-expelling process. J. Sci. Food Agric. 91(13): 2348–2354.

Lesage-Meessen, L., C. Stentelaire, A. Lomascolo, D. Couteau, M. Asther, S. Moukha, E. Record, J. C. Sigoillot and M. Asther. 1999. Fungal transformation of ferulic acid from sugar beet pulp to natural vanillin. J. Sci. Food Agric. 79(3): 487–490.

Li, Y., C. Guo, J. Yang, J. Wei, J. Xu and S. Cheng. 2006. Evaluation of antioxidant properties of pomegranate peel extract in comparison with pomegranate pulp extract. Food Chem. 96(2): 254–260.

Liu, Q., R. Tarn, D. Lynch and N. M. Skjodt. 2007. Physicochemical properties of dry matter and starch from potatoes grown in Canada. Food Chem. 105(3): 897–907.

Llobera, A. and J. Cañellas. 2007. Dietary fibre content and antioxidant activity of 'Manto Negro' red grape (*Vitis vinifera*): Pomace and stem. Food Chem. 101(2): 659–666.

Llobera, A. and J. Canellas. 2008. Antioxidant activity and dietary fibre of 'Prensal Blanc' white grape (*Vitis vinifera*) by-products. Int. J. Food Sci. Technol. 43(11): 1953–1959.

Lommen, A., M. Godejohann, D. P. Venema, P. C. H. Hollman and M. Spraul. 2000. Application of directly coupled HPLC-NMR-MS to the identification and confirmation of quercetin glycosides and phloretin glycosides in apple peel. Anal. Chem. 72: 1793–1797.

Lu, Y. and L. Y. Foo. 1997. Identification and quantification of major polyphenols in apple pomace. Food Chem. 59(2): 187–194.

Lu, Y. and L. Y. Foo. 1998. Constitution of some chemical components of apple seed. Food Chem. 61: 29–33.

Lutfullah, G., H. Tila, A. Hussain and A. Ali Khan. 2014. Evaluation of plants extracts for proximate chemical composition, antimicrobial and antifungal activities. American-Eurasian J. Agric. Environ. Sci. 14: 964–970.

Maisuthisakul, P. and M. H. Gordon. 2009. Antioxidant and tyrosinase inhibitory activity of mango seed kernel by product. Food Chem. 117(2): 332–341.

Matsuda, N. 1997. Green chile peels and seeds as compared to oat bran and rice bran as a dietary fiber source for human food. MS Thesis, Department of Home Economics, New Mexico State University, Las Cruces, NM.

McKee, L. H. and T. A. Latner. 2000. Underutilized sources of dietary fiber: A review. Plant Foods Human Nutr. 55: 285–304.

Miljkovic, D. and G. S. Bignami. 2002. Nutraceuticals and Methods of Obtaining Nutraceuticals from Tropical Crops. USA Patent Application number: 10/992.502.

Mitra, S. K., P. K. Pathak, H. L. Devi and I. Chakraborty. 2013. Utilization of seed and peel of mango. Acta Hort. 992: 593–596.

Naczk, M. and F. Shahidi. 2006. Phenolics in cereals, fruits and vegetables: Occurrence, extraction and analysis. J. Pharma. Biomed. Anal. 41: 1523–1542.

Obied, H. K., M. S. Allen, D. R. Bedgood, P. D. Prenzler, K. Robards and R. Stockmann. 2005. Bioactivity and analysis of biophenols recovered from olive mill waste. J. Agric. Food Chem. 53(4): 823–837.

Onyeneho, S. N. and N. S. Hettiarachchy. 1993. Antioxidant activity, fatty acids and phenolic acids composition of potato peels. J. Sci. Food Agric. 62: 345–350.

Palanisamy, U. D., L. T. Ling, T. Manaharan and D. Appleton. 2011. Rapid isolation of geraniin from *Nephelium lappaceum* rind waste and its anti-hyperglycemic activity. Food Chem. 127(1): 21–27.

Pan, X., G. Niu and H. Liu. 2003. Microwave-assisted extraction of tea polyphenols and tea caffeine from green tea leaves. Chemical Engineering and Processing: Proc. Intensific. 42(2): 129–133.

Pandey, A., C. R. Soccol and D. Mitchell. 2000. New developments in solid state fermentations. I. Bioprocesses and products. Process Biochem. 35: 1153–1169.

Panouillé, M., M. C. Ralet, E. Bonnin and J. F. Thibault. 2007. Recovery and reuse of trimmings and pulps from fruit and vegetable processing. pp. 417–447. *In*: K. Waldron (ed.). Handbook of Waste Management and Co-product Recovery in Food Processing. Cambridge: Woodhead Publishing Limited.

Parni, B. and Y. Verma. 2014. Biochemical properties in peel, pulp and seeds of *Caricapapaya*. Plant Arch. 14: 565–568.

Pazmino-Duran, E. A., M. M. Giusti, R. E. Wrolstad and M. B. A. Gloria. 2001. Anthocyanins from banana bracts (*Musa paradisiaca*) as potential food colorant. Food Chem. 73: 327–332.

Perera, A., D. Appleton, L. H. Ying, S. Elendran and U. D. Palanisamy. 2012. Large-scale purification of geraniin from *Nepheliumlappaceum* rind waste using reverse-phase chromatography. Sep. Purify Tech. 98: 145–149.

Popa, V. I., M. Dumitru, I. Volf and N. Anghel. 2008. Lignin and polyphenols as allelochemicals. Indus. Crops Prod. 27: 144–149.

Porter, L. J. 1989. Tannins. pp. 389–419. *In*: J. B. Harborne (ed.). Methods in Plant Biochemistry. Vol. 1. Plant Phenolics. London: Academic Press.

Price, K. R. and M. J. C. Rhodes. 1997. Analysis of the major flavonol glycosides present in four varieties of onion (*Allium cepa*) and changes in composition resulting from autolysis. J. Sci. Food Agric. 74: 331–339.

Ramachandran, S., A. K. Patel, K. M. Nampoothiri, F. Francis, V. Nagy, G. Szakacs and A. Pandey. 2004. Coconut oil cake—A potential raw material for the production of alpha-amylase. Bioresource Technol. 93(2): 169–174.

Reddy, L. V., O. V. S. Reddy and Y. J. Wee. 2011. Production of ethanol from mango (*Mangifera indica* L.) peel by *Saccharomyces cerevisiae* CFTRI101. Af. J. Biotech. 10(20): 4183–4189.

Robbins, R. J. 2003. Phenolic acids in foods: An overview of analytical methodology. J. Agric. Food Chem. 51: 2866–2887.

Rodríguez Couto, S. 2008. Exploitation of biological wastes for the production of value-added products under solid-state fermentation conditions. Biotech. J. 3(7): 859–870.

Rodriguez de Sotillo, D., M. Hadley and E. T. Holm. 1994. Phenolics in aqueous potato peel extract: Extraction, identification and degradation. J. Food Sci. 59: 649–651.

Rodriguez, R., A. Jimenez, R. Guillen, A. Heredia and J. Fernandez-Bolanos. 1997. Postharvest changes in white asparagus during refrigerated storage. J. Agric. Food Chem. 47: 3551–3557.

Rodriguez, R., A. Jimenez, J. Fernandez-Bolanos, R. Guillen and A. Heredia. 2006. Dietary fiber from vegetable products as source of functional ingredients. Trends Food Sci. Technol. 17: 3–15.

Rosales, E., S. R. Couto and M. A. Sanroman. 2005. Reutilisation of food processing wastes for production of relevant metabolites: application to laccase production by *Trametes hirsuta*. J. Food Eng. 66(4): 419–423.

Rosales, E., S. R. Couto and M. A. Sanroman. 2007. Increased laccase production by *Trametes hirsuta* grown on ground orange peelings. Enz. Micro. Technol. 40(5): 1286–1290.

Rudra, S. G., J. Nishad, N. Jakhar and C. Kaur. 2015. Food industry waste: Mine of nutraceuticals. Int. J. Sci. Environ. Technol. 4(1): 205–229.

Sabu, A., A. Pandey, M. J. Daud and G. Szakacs. 2005. Tamarind seed powder and palm kernel cake: two novel agro residues for the production of tannase under solid state fermentation by *Aspergillus niger* ATCC 16620. Bioresource Technol. 96(11): 1223–1228.

Saito, M., H. Hosoyama, T. Ariga, S. Kataoka and N. Yamaji. 1998. Antiulcer activity of grape seed extract and procyanidins. J. Agric. Food Chem. 46: 1460–1464.

Saxena, A., A. S. Bawa and P. S. Raju. 2011. Jackfruit (*Artocarpus heterophyllus* Lam.). pp. 275–298. *In*: E. M. Yahia (ed.). Postharvest Biology and Technology of Tropical and Subtropical Fruits, Vol. 3. Woodhead Publishing Limited. Cambridge, UK.

Schieber, A., W. Ullrich and R. Carle. 2000. Characterization of polyphenols in mango puree concentrate by HPLC with diode array and mass spectrometric detection. Innov. Food Sci. Emerg. Technol. 1: 161–166.

Schieber, A., F. C. Stintzing and R. Carle. 2001. By-products of plant food processing as a source of functional compounds—Recent developments. Trends Food Sci. Technol. 12(11): 401–413.

Seker, I. T., O. Ozboy-Ozbas, I. Gokbulut, S. Ozturk and H. Koksel. 2010. Utilization of apricot kernel flour as fat replacer in cookies. J. Food Process Preserv. 34(1): 15–26.

Seyis, I. A. N. 2005. Xylanase production from *Trichoderma harzianum* 1073 D3 with alternative carbon and nitrogen sources. Food Technol. Biotechnol. 43(1): 37–40.

Shah, M. P., G. V. Reddy, R. Banerjee, P. R. Babu and I. L. Kothari. 2005. Microbial degradation of banana waste under solid state bioprocessing using two lignocellulolytic fungi (*Phylosticta* spp. MPS-001 and *Aspergillus* spp. MPS-002). Process Biochem. 40(1): 445–451.

Sharma, S. K. and M. L. Maguer. 1996. Lycopene in tomatoes and tomato pulp fractions. Italian J. Food Sci. 2: 107–113.

Sharma, S. K., S. Bansal, M. Mangal, A. K. Dixit, R. K. Gupta and A. K. Mangal. 2016. Utilization of food processing by-products as dietary, functional and novel fiber: A review. Crit. Rev. Food Sci. Nutr. 56(10): 1647–1661.

Shrikhande, A. J. 2000. Wine by-products with health benefits. Food Res. Int. 33: 469–474.

Shu, Y. Y., M. Y. Ko and Y. S. Chang. 2003. Microwave-assisted extraction of ginsenosides from ginseng root. Microchem. J. 74(2): 131–139.

Siriphanich, J. 2011. Durian (*Duriozibethinus* Merr.). pp. 80–114. *In*: E. M. Yahia (ed.). Postharvest Biology and Technology of Tropical and Subtropical Fruits, Oxford: Woodhead Publishing.

Sirisompong, W., W. Jirapakkul and U. Klinkesorn. 2011. Response surface optimization and characteristics of rambutan (*Nephelium lappaceum* L.) kernel fat by hexane extraction. LWT—Food Sci. Technol. 44: 1946–1951.

Someya, S., Y. Yoshiki and K. Okubo. 2002. Antioxidant compounds from bananas (*Musa cavendish*). Food Chem. 79: 351–354.

Sonia, N. S., C. Mini and P. R. Geethalekshmi. 2016. Vegetable peels as natural antioxidants for processed foods—A review. Agric. Rev. 37(1): 35–41.

Sood, A., P. Kaur and R. Gupta. 2012. Phytochemical screening and antimicrobial assay of various seeds extract of cucurbitaceae family. Int. J. App. Biol. Pharma. Tech. 3: 401–409.

Soong, Y. Y. and P. J. Barlow. 2004. Antioxidant activity and phenolic content of selected fruit seeds. Food Chem. 88(3): 411–417.

Souquet, J. -M., V. Cheynier, F. Brossaud and M. Moutounet. 1996. Polymeric proanthocyanidins from grape skins. Phytochem. 43: 509–512.

Soxhlet, F. 1879. Die gewichtsanalytische Bestimmung des Milchfettes. Dingler's Polytech. J. 232: 461–465.

Subagio, A., N. Morita and S. Sawada. 1996. Carotenoids and their fatty-acid esters in banana peel. J. Nutr. Sci. Vitamin 42: 553–566.

Thebaudin, J. Y., A. C. Lefebvre, M. Harrington and C. M. Bourgeois. 1997. Dietary fibers: Nutritional and technological interest. Trend Food Sci. Technol. 8: 42–48.

Thed, S. T. and R. D. Phillips. 1995. Changes of dietary fiber and starch composition of processed potato products during domestic cooking. Food Chem. 52(3): 301–304.

Tomás-Barberán, F. A., F. Ferreres and M. I. Gil. 2000. Antioxidant phenolic metabolites from fruit and vegetables and changes during postharvest storage and processing. Stud. Nat. Prod. Chem. 23: 739–795.

Torres, J. L. and R. Bobet. 2001. New flavanol derivatives from grape (*Vitis vinifera*) byproducts: Antioxidant aminoethylthio-flavan-3-olconjugates from a polymeric waste fraction used as a source of flavanols. J. Agric. Food Chem. 49: 4627–4634.

Turksoy, S. and B. Ozkaya. 2011. Pumpkin and carrot pomace powders as a source of dietary fiber and their effects on the mixing properties of wheat flour dough and cookie quality. Food Sci. Technol. Res. 17(6): 545–553.

Valenzuela, A., S. Nieto, B. K. Cassels and H. Speisky. 1992. Inhibitory effect of boldine on fish oil oxidation. J. American Oil Chem. Soc. 68(12): 935–937.

Vankar, P. S. 2004. Essential oils and fragrances from natural sources. Resonance 9(4): 30–41.

Vergara-Valencia, N., E. Granados-Pérez, E. Agama-Acevedo, J. Tovar, J. Ruales and L. A. Bello-Pérez. 2007. Fibre concentrate from mango fruit: Characterization, associated antioxidant capacity and application as a bakery product ingredient. LWT—Food Sci. Technol. 40(4): 722–729.

Visioli, F., G. Bellomo and C. Galli. 1998. Free radical-scavenging properties of olive oil polyphenols. Biochem. Biophys. Res. Commun. 247: 60–64.

Visioli, F., A. Romani, N. Mulinacci, S. Zarini, D. Conte, F. F. Vincierei and C. Galli. 1999. Antioxidant and other biological activities of olive mill waste waters. J. Agric. Food Chem. 47: 3397–3401.

Wachirasiri, P., S. Julakarangka and S. Wanlapa. 2009. The effects of banana peel preparations on the properties of banana peel dietary fiber concentrate. Songklanakarin J. Sci. Technol. 31(6): 605–611.

Wijngaard, H. and N. Brunton. 2009. The optimization of extraction of antioxidants from apple pomace by pressurized liquids. J. Agric. Food Chem. 57(22): 10625–10631.

Wolfe, K. L. and R. H. Liu. 2003. Apple peels as a value-added food ingredient. J. Agric. Food Chem. 51: 1676–1683.

WSDE (Washington State Department of Ecology) Report. 1994. Pollution Prevention in Fruit and Vegetable Food Processing Industries. https://fortress.wa.gov/ecy/publications/publications/94056.pdf

Zeyeda, N. N., M. A. M. Zeitoun and O. M. Barbary. 2008. Utilisation of some vegetables and fruits waste as natural antioxidants. Alex J. Food Sci. Technol. 5: 1–11.

Extraction, Isolation and Utilisation of Bioactive Compounds from Fruit Juice Industry Waste

*Suwimol Chockchaisawasdee** and *Costas E. Stathopoulos*

Introduction

Juice manufacturing is an important segment of the food industry. Within the fruit and vegetable drink market, approximately 50 per cent, 30 per cent, and 20 per cent of the market share belong to juice drinks mixed with a pure juice (0–24 per cent juice content), pure juice (100 per cent), and nectars (25–99 per cent juice content), respectively. In 2011, the global consumption of commercial juices and nectars was approximately 39 billion litres, which was equivalent to approximately USD 107 billion in market value (AIJN 2014). The most popular juices are of orange and apple; others include juices from lemon, grape, grapefruit, peach, pomegranate, berries, and exotic fruits, such as pineapple, mango, mangosteen and passion fruit (McLellan and Padilla-Zakour 2004, AIJN 2014, Reyes-De-Corcuera et al. 2014). Nevertheless, with the success and high growth of the functional food products in the last decade, demand for juice products with health benefits continues to rise, as is the demand for products in a variety of packages with increased emphasis on functionality, new flavors or blends. With the drives of new production and packaging technology, together with the launch of new super-premium juices, the global market of the juice industry is still expecting a steady growth (AIJN 2014, López 2014, Leatherhead Food Research 2014).

With a high production volume, inevitably the juice industry generates a large quantity of waste as a consequence. Waste streams from fruit juice processing are produced both in solid and liquid forms (McLellan and Padilla-Zakour 2004). Liquid

School of Science, Engineering and Technology, University of Abertay, Dundee DD1 1HG, UK.
  E-mail: C.Stathopoulos@abertay.ac.uk
* Corresponding author: S.Chockchaisawasdee@abertay.ac.uk

waste streams are mainly the discharge of cleaning water and process water which has low-to-medium biological oxygen demand (BOD) values and can be treated by aerobic or anaerobic systems (Arvanitoyannis and Varzakas 2008). Solid waste, on the other hand, is highly polluted and more difficult to treat (Kosseva 2011). Conventionally these wastes are disposed by using as animal feed or fertilizer (Van Dyk et al. 2013). Although they are discarded from the process as they cannot be further utilized, fruit solid waste retains high concentrations of several bioactive compounds. The peels of several fruits (for example apple, peach, pomegranate) contain higher amounts of bioactive compounds than the edible parts (Gorinstein et al. 2001, Li et al. 2006). Substantial evidence points out that all parts of fruit solid waste are rich in health-benefit phytochemicals (Widmer and Montanari 1994, Balasundram et al. 2006, Ayala-Zavala et al. 2011, O'Shea et al. 2012, Dhillon et al. 2013, Mirabella et al. 2014). Rather than using them conventionally for feeds and fertilizers, alternative valorisation of these unwanted materials to create higher value-added products is a better option. This topic has attracted great interest among researchers and industry alike in the last few decades.

This chapter will focus on the recovery of bioactive compounds, particularly phenolic compounds and dietary fiber, from fruit juice industry solid waste. It aims to provide comprehensive information on the use of such waste as a source of high value-added component. Extraction, isolation and potential applications of phenolic compounds and dietary fiber recovered from fruit solid waste in the food industry will be discussed.

Waste from Fruit Juice Industry

The types of fruits for juice processing can be broadly classified into pome fruit (e.g., apple, pear), citrus (e.g., orange, lemon, lime, tangerine, grapefruit), stone fruits (e.g., peach, nectarines, cherry), berries (e.g., grapes, pomegranate, cranberry, blackcurrant), and exotic fruit (mango, pineapple, mangosteen, passion fruit). Manufacturing of fruit juices consists of a series of unit operations that vary, depending on the nature of the raw material and the characteristics of the desired final product (McLellan and Padilla-Zakour 2004). A general process includes pre-treatment steps, juice extraction, and post-extraction treatment steps. Figure 1 illustrates a general flow diagram of juice processing process according to the fruit type including operation steps where solid waste is generated. Detailed manufacturing, including the objectives of each step, of juices from different types of fruit can be found in the literature (McLellan and Padilla-Zakour 2004, Horváth-Kerkai and Stéger-Máté 2013, Reyes-De-Corcuera et al. 2014).

Solid waste in fruit juice manufacturing is generated throughout the processing line. They are parts of raw materials that cannot be utilised in the production of the intended products (Commission Regulations 442/1975/EEC, 689/1991/EEC), which include pomace, peels, seeds, stones and stems. Estimation of manufacturing fruit waste is not clear since there is no distinct universal definition of food waste (Monier et al. 2010, Buzby and Hyman 2012). According to a report published by the Food and Agriculture Organization, the 2007 production volume of fruits and vegetables worldwide was 1,650 million tonnes, of which approximately 12 per cent (or 198 million tonnes) was wasted at processing stage (Gustavsson et al. 2011). Geographically, high percentages

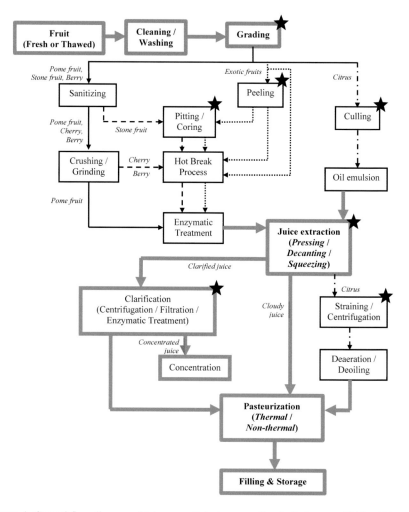

Figure 1. General flow diagram of juice manufacturing according to fruit types. Thick solid grey lines represent common unit operations in all fruit juice processing. Dash and solid thin black lines represent variations in processes between different fruit types. Unit operations with a star on the top right indicate the point where the solid wastes are generated.

of fruit and vegetable manufacturing wastes (20–25 per cent) were generated in Sub-Saharan Africa, North Africa, West and Central Asia, South and Southeast Asia and Latin America, while those percentages in Europe, North America and Oceania, and industrialized Asia were small (2 per cent) (Gustavsson et al. 2011). Raw materials important to juice industry on a global scale include citrus (orange), pome fruit (apple), stone fruits, berries, grape and exotic fruits (pineapple and mango). The nature and approximate percentages of these wastes from juice manufacturing are shown in Table 1.

Of all the fruits important to the international juice trade, citrus, particularly oranges, is the largest fruit crop and juice produced worldwide. Apart from oranges,

Table 1. Global production quantities of fruits and juices in 2013, and approximate percentages and nature of solid waste generated from fruit juice manufacturing.

Fruit	Total quantity of fruit produced (tonnes)	Total quantity of juice produced[1] (tonnes)	Solid waste	Approximate percentage of waste from raw material (w/w)
Apples	80,822,521	569,962 1,452,370 (c)	Pomace, skin, seeds, stem	25–30
Citrus			Pomace, peel, seeds	50
- Total	135,169,941			
- Oranges	71,305,973	2,133190 1,697,084 (c)		
- Lemons and limes	14,949,082	96,913 (lemon) 83,740 (lemon, c)		
- Grapefruits	8,255,486	233,177 115,157 (c)		
- Tangerines and mandarins	28,666,714	2,381		
Grapes	77,181,122	761,712	Pomace, skin, seed, stems	20
Berries			Pomace, skin, seed, stem	5
- Cranberries	540,259	N/A		
- Currants	706,910	N/A		
Stone Fruits			Pomace, skin, stones, stems	N/A
- Peaches and nectarines	21,638,953	N/A		
- Cherries (sweet and sour)	3,643,083	N/A		
- Plums	11,528,337	6 (c)		
Exotic Fruits			Skin, core	33–50
- Pineapples	24,778,262	941,177 331,575 (c)		
- Mangoes and mangosteens	42,663,770	255,162 (mango)	Peel, stone	35–60

[1]Numbers without (c) are quantities of single strength juices; numbers with (c) signify quantities of concentrated juices.

Sources: Widmer and Montanari (1994), Tran and Mitchell (1995), Larrauri et al. (1996), Arvanitoyannis and Varzakas (2008), Ajila et al. (2012), Dhillon et al. (2013), Kosseva (2013), FAOSTAT (2016)

other citrus fruits of importance include lemons, limes, grapefruits, tangerine and mandarins. With only approximately 50 per cent juice recovery from fresh weight, a considerably high quantity of citrus pomace (50 per cent peel comprises albedo and flavedo, 0.1–5 per cent seeds, pulp, carpellary membrane) are generated as waste (Rezzadori et al. 2012). On dry weight basis, citrus pomace contains high contents of sugars, protein, essential oil (peel and seeds), pectin (highest concentration in peel) and dietary fibers (Marín et al. 2007).

In the second place after citrus, the apple juice industry also generated several million tonnes of solid waste (Bhushan et al. 2008). Apple pomace accounts for 25–30 per cent of the total processed fruit weight and consists of peel, core, seed, calyx, stem and soft tissue (Foo and Lu 1999). Fresh pomace contains high moisture content (70–80 per cent) and is highly perishable as high amounts of carbohydrates (10–22 per cent, with up to 50 per cent fermentable sugars) is present (Gullón et al. 2007, Dhillon et al. 2013).

Grape juice is not as highly popular among consumers as orange and apple juices are (AIJN 2014, Reyes-De-Corcuera et al. 2014). Indeed, the majority (80 per cent) of fresh grape produced goes to wine making (Martí et al. 2014). Grape juice is not normally consumed in large amounts alone because it is either too sweet (about 200 g/L sugars) or too acidic (up to 10 g/L tartaric acid) and usually blended with other juices for a more balanced taste and flavor (Kashyap et al. 2001). Nonetheless, together with wine production, several million tonnes of grape residue are produced annually (Oreopoulou and Tzia 2007). After juice pressing (both in the wine or juice manufacturing) approximately 20 per cent of processed grape is discarded. The residue consists of 10–20 per cent grape pomace and 3–6 per cent stalks (Martí et al. 2014).

Berry juices are marketed as 'superfruit' juices and interest in consumption of food in this category has increased (López 2014). Different berries (blueberry, raspberry, strawberry, currants and pomegranate) are processed as juices. Generally, in berry juice manufacturing, solid wastes usually come from pre-treatment (washing and sorting) and juice pressing. The waste from pre-treatment stage consists of damaged fruits, stems and stalks, while that from pressing is pomace (Tomás Barberán 2007). Percentages of berry waste vary, depending on the nature of the fruits. For example, cranberry solid waste (pomace, stems) is 5 per cent of processed fruit weight (Arvanitoyannis and Varzakas 2008) while pomegranate solid waste (husk, membrane, seeds) is 50 per cent of processed fruit weight (Tomás Barberán 2007).

Although the world production volumes of stone fruit juices are not large in global scale (FAOSTAT 2016), plum, peach, apricot and cherry, are well used and popular for juice production particularly in Europe (AIJN 2014). Like in berry juice manufacturing, solid waste of stone fruits is generated during pre-treatment and juice-pressing steps. The waste includes damaged fruits, stems, stalks and pomace.

Exotic fruit juice manufacturing is another segment that generates a considerable quantity of waste. Pineapple, mango, and passion fruit are among the most important fruits for the juice industry (Schieber et al. 2001, Mirabella et al. 2014). Exotic fruits popular for juice manufacturing (e.g., pineapple, mango, passion fruit) have high percentages of inedible/unusable parts. Passion fruit waste could be as high as 75 per cent of raw material as it has thick rind which accounts for 90 per cent of the waste (Arvanitoyannis and Varzakas 2008). Although passion fruit seeds are edible, they are not a part of the final products and are removed as waste (Chau and Huang 2004).

Typically, disposal of fruit solid waste may be achieved by incineration or utilisation as animal feeds and fertilisers (Van Dyk et al. 2013). Only in some cases, fruit waste is used as raw material to produce secondary products on an industrial scale. For example, grape seeds have long been known for their oil-rich characteristics, with the first mention of grape-seed oil as a possible industry made approximately in 1780 (Rabak 1921). Apart from traditional uses as feeds and fertilizers, in some developing countries the waste may be simply discarded on the outskirts of the cities, causing major pollution to the environment, or disposed of in local landfills (McLellan and Padilla-Zakour 2004). Disposal of fruit waste causes a very high cost to the industry. In the USA alone, disposal of apple pomace is estimated to be higher than USD 10 million annually (Shalini and Gupta 2010). With regards to their use as animal feeds, not all fruit wastes are suitable for animal feeds as they may contain too low protein or too high lignin content (Van Dyk et al. 2013). Most fruit solid waste also contains high moisture content, which requires drying to prolong their shelf-life if they are going to be used further. Energy and transport costs together with low sale prices make return on investment unattractive and this has led to alternative valorisation concepts (Laufenberg et al. 2003).

Interest in the alternative use of waste streams to create high value products beyond disposal or fertilisation has increased drastically in the last few decades. With the high growth of functional food and waste utilisation concepts, fruit solid waste has been heavily researched as cheap sources for bioactive compounds. Substantial evidence has established that fruit solid waste retains a wide range of high-value functional compounds. With the continuity of new discoveries on extractions, isolation and characterization techniques, the use of such waste as cheap raw material to produce bioactive compounds on a commercial scale becomes tangible on a large scale.

Bioactive Compounds from Fruit Juice Industry Waste

Bioactive compounds from fruits (also known as phytochemicals) possess certain biological activities, namely antimicrobial, anticancer, anti-inflammatory, immuno-stimulatory and antioxidant activity which can exert physiological effects and may enhance the human health (Hollman and Katan 1999, Szajdek and Borowska 2008, González-Molina et al. 2010, Johnson 2013). There are many classes of bioactive compounds, which are categorized according to their molecular identity or biopolymer constituents (Campos-Vega and Oomah 2013). Figure 2 illustrates the classification of prominent functional compounds recovered from fruit solid waste that have been extensively investigated.

Phenolic compounds

Phenolic compounds are a broad group of chemical components that are structurally diverse (Naczk and Shahidi 2004). They are secondary metabolites found in plant species and more than 8,000 phenolic compounds have been identified (Croteau et al. 2000). Major classes of phenolic compounds found in fruit waste include flavonoids (flavonols, flavones, flavonones, flavanols, anthocyanins), phenolic acids (hydroxybenzoic acids, hydroxycinnamic acids), tannins, stilbenes and lignans (Balasundram et al. 2006, Ignat et al. 2011, Gnanavinthan 2013). Flavonoids are the

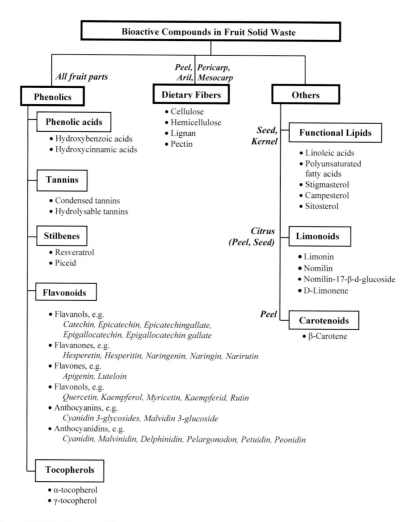

Figure 2. Major classes of bioactive compounds recovered from fruit juice industry wastes.

largest class of phenolic compounds with over 4,000 identified substances (Ignat et al. 2011). Molecular structure of phenolic compounds is found to be an important determinant in their scavenging capacity and oxidation potential (Shi et al. 2001). Several papers on bioavailability and metabolism of various phenolic compounds have been published (Hollman and Katan 1999, Scalbert and Williamson 2000, Felgines et al. 2003, Larrosa et al. 2009).

Isolation, quantification and characterisation of phenolic compounds in fruit solid waste have been studied as they are present in all types and parts of fruits waste (Table 2). Flavonoids and phenolic acids are the most common classes of phenolic compounds present. The profiles of these substances from different cultivars and fruit sources can widely differ, both in terms of components and concentrations. For instance, Wolfe et al. (2003) demonstrated that phenolic compounds were most

Table 2. Selected reports on major bioactive compounds in solid waste from fruit juice industry.

Solid fruit waste	Phenolic compounds	Dietary fiber	Others bioactive components	References
Apple pomace/ Apple skin	catechin, epicatechin, caffeic acid, chlorogenic acid, p-coumaric acid, rutin derivatives, 3-hydroxyphloridzin, phlorerin 2'-xyloglucoside, phloridzin, quercertin-glycosides, cyanidin-glycosides	60–90 per cent TDF; Cellulose, hemicellulose, lignin, pectin (12 per cent)	Terpenes (ursolic acid)	Lu and Foo (1997), Schieber et al. (2003), Nawirska and Kwaśniewska (2005), Bai et al. (2010), Çam and Aaby (2010), Pingret et al. (2012), Reis et al. (2012), Grigoras et al. (2013), Sun-Waterhouse et al. (2013)
Apple seed	Phloridzin, Chlorogenic acid, p-coumatylquinmic acid, quercertin-glycosides, 3-hydroxyphloridzin, phlorerin 2'-xyloglucoside, tocopherols (α, β, γ, δ)		20–24 per cent oil (linoleic acid, oleic acid)	Lu and Foo (1998), Schieber et al. (2003), Tian et al. (2010), Górnaś (2015)
Citrus pomace/ Citrus peel	Hesperidin, eriocitrin, narirutin, naringin, nobiletin, tangeretin, neohesperidin, neoeriocitrin, caffeic acid, p-coumaric acid, ferulic acid, sinapinic acid	30–70 per cent TDF; Pectins (16 per cent in pomace, 25 per cent in peel), cellulose, hemicellulose, lignin	Terpenes (limonin, nomilin, d-limonene), Carotenoids	Bracke et al. (1991), Ohta et al. (1993), Larrauri et al. (1996), Grigelmo-Miguel and Martin-Belloso (1998), Marín et al. (2007), Pourbafrani et al. (2010), Prabasari et al. (2011), Mamma and Christakopoulos (2014), Esparza-Martínez et al. (2016)
Citrus seed	Eriocitrin, hesperidin, naringin, narirutin, neohesperidin, neoeriocitrin, caffeic acid, p-coumaric acid, ferulic acid, sinapinic acid		Terpenes (limonin, nomilin, nomilin-17-β-d-glucoside)	Ozaki et al. (1991), Ohta et al. (1993), Bocco et al. (1998), Yu et al. (2005)
Grape pomace/ Grape skin	Anthocyanins, catechins, epicatechin, gallic acid, rutin, quercetin and kaempferol, epicatechin gallate, epigallocatechin	65–80 per cent TDF; Cellulose, pectin, hemicellulose, lignin		Souquet et al. (2000), Kammerer et al. (2005), Pinelo et al. (2005), Llobera and Cañellas (2007), Ruberto et al. (2007), Rockenbach et al. (2011)
Grape seed	Catechin, epicatechin, gallic acid, epicatechin gallate, epigallocatechir gallate, epigallocatechin gallate, epigallocat:chin, procyanidins, resveratrol	Cellulose, pectin, hemicellulose, lignin	7–19 per cent Oil (linoleic acid, oleic acid)	Molero Gómez et al. (1996), Guendez et al. (2005), Bozan et al. (2008), Köhler et al. (2008), Spranger et al. (2008), Prado et al. (2012), Da Porto et al. (2013)

Table 2 contd. ...

...Table 2 contd.

Solid fruit waste	Phenolic compounds	Dietary fiber	Others bioactive components	References
Grape stem	Quercetin 3-glucuronide, catechin, caffeoyltartaric acid, dihydroquercetin 3-rhamnoside (astilbin), tannins, resveratrol, viniferin	Cellulose (30.3 per cent), hemicelluloses (21.0 per cent), lignin (17.4 per cent)		Souquet et al. (2000), Rayne et al. (2008), Ping et al. (2011), Prozil et al. (2012)
Pomegranate peel/ Pomegranate mescarp	anthocyanins, ellagitannins (ellagic acid, gallic acid and punicalagin), gallotannins gallagyl esters, hydroxybenzoic acids, hydroxycinnamic acids and dihydroflavonol	30–60 per cent TDF (cellulose, Klason lignin, uronic acid, pectin)		Cerdá et al. (2003), Lansky and Newman (2007), Fischer et al. (2011), Johanningsmeier and Harris (2011), Fawole et al. (2012), Ismail et al. (2012), Hasnaoui et al. (2014)
Pomegranate seed	Gallic acid, ellagic acid γ-tocopherol	Lignins, lignin derivatives	Sterols (daucosterol, campesterol, stigmasterol, β-sitosterol)	Dalimov et al. (2003), Wang et al. (2004), Lansky and Newman (2007)
Mango peel	Anthocyanins, quercetin-glycosides, kaempferol-glycoside, xanthone-glycosides, cyanidin 3-O-galactoside anthocyanidin hexoside, γ-tocopherol, Quercertin, mangiferin pentodise Syringic, ellagic, gallic Condensed tannins	30–70 per cent TDF; cellulose, hemicellulose, lignin, pectin (12–20 per cent)	β-carotene	Larrauri et al. (1996), Berardini et al. (2005), Berardini et al. (2005), Ajila et al. (2007), Vergara-Valencia et al. (2007), Martínez et al. (2012)
Mango seed kernels	Tannins, gallic acid, coumarin, caffeic acid, vanillin, mangiferin, ferulic acid, cinnamic acid, ellagic acid, gallocatechin, acylated cyaniding, β-Sitosterol, δ-Avenasterol Campesterol, Stigmasterol α-Tocopherol, γ-Tocopherol		12 per cent Oil (oleic acid, linoleic acid)	Arogba (2000), Puravankara et al. (2000), Abdalla et al. (2007), Barreto et al. (2008), Maisuthisakul and Gordon (2009)

Source	Bioactive compounds	Fibre/carbohydrate content	Oil/fatty acid content	References
Mangosteen rind	Tannins, xanthones (α-mangostin, β-mangostin, 3-isomangostin, 9-hydroxycalabaxanthone, gartanin, and 8-desoxygartanin), anthocyanins, proanthocyanidins, catechin			Jung et al. (2006), Fu et al. (2007), Ji et al. (2007), Zadernowski et al. (2009), Wittenauer et al. (2012)
Mangosteen seed			21 per cent Unsaturated fatty acids (stearic acid, oleic acid, linoleic acid, gadoleic acid, and eicosadienoic acid)	Hawkins and Kridl (1998), Ajayi et al. (2007)
Passion fruit peel	Phenolic acids, flavonoids	70–80 per cent TDF (Cellulose, hemicellulose, pectic substances)		Silva et al. (2008), Kliemann et al. (2009), Martinez et al. (2012), López-Vargas et al. (2013)
Passion fruit seed	Tocopherols	50 per cent TDF (cellulose, pectic substances, hemicellulose)	30 per cent oil (linoleic acid, oleic acid)	Chau and Huang (2004), Malacrida and Jorge (2012), López-Vargas et al. (2013)
Blackcurrant pomace	Delphinidin-3-O-glucoside, delphinidin-3-O-rutinoside, cyanidin-3-O-glucoside, cyanidin-3-O-rutinoside	Cellulose, hemicellulose, pectin (2.7 per cent), lignin		Kapasakalidis et al. (2006), Sójka et al. (2009), Holtung et al. (2011)
Sour cherry pomace	Neochlorogenic acid, 3-p-coumaroylquinic acid, chlorogenic acid, quercetin glucoside and rutinoside, kaempferol-rutinoside, isorhamnetin-rutinoside, quercetin, kaempferol, isorhamnetin, anthocyanins	Cellulose, hemicellulose, pectin (1.5 per cent), lignin		Nawirska and Kwaśniewska (2005), Kołodziejczyk et al. (2013)

localised in apple peel. Apple seeds contain a smaller range of phenolic compounds than the skin with phloridzin as major component (80–90 per cent) (Lu and Foo 1998, Fromm et al. 2012). Variation in phenolic profiles is also evident in citrus. Lemon seed mainly contains high amounts of eriocitrin and hesperidin, while the peel is rich in neoeriocitrin, naringin and neohesperidin (Bocco et al. 1998). Hesperidin is the most abundant flavonoid in Valencia, Navel, Temple and Ambersweet orange peels (Manthey and Grohmann 1996) and naringin is the most abundant flavonoid in grapefruit peel (Wu et al. 2007). Grape pomace is rich in anthocyanins, catechins, procyanidins, flavonol glycosides, phenolic acids (Rodríguez Montealegre et al. 2006). The phenolic compounds in grape seeds are essentially all flavonoids, particularly, flavan-3-ol. Grape skin is rich in resveratrol. Pomegranate peel contains higher phenolic compounds, especially phenolic acids, than the pulp (Li et al. 2006). Mangosteen peels are rich in tannins and anthocyanins (Wittenauer et al. 2012). Detail of selected reports regarding phenolic compounds in specific parts and sources of fruit juice waste is included in Table 2.

Dietary fiber

The relationship between dietary fiber and health has long been established (Buttriss and Stokes 2008). The beneficial physiological effects in humans include decrease in intestinal transit time and increase in faecal bulk fermentable by colonic microflora, reduction in cholesterol levels in the blood and reduction in insulin responses (Laurentin et al. 2003). Dietary fiber can also impart some functional properties which can improve food characteristics, such as increase water-holding capacity, oil-holding capacity, emulsification and gel formation (Belitz et al. 1999). Dietary fiber is a class of complex carbohydrates and can be divided into soluble and insoluble fibers. Fruit solid waste is an excellent source of soluble dietary fiber, such as pectin and gums, as well as insoluble dietary fibers, such as cellulose, hemicellulose and lignin. Dietary fiber is associated with plant cell walls and tissues. Therefore, it is mostly located in peels, skins, pericarps and stalks. High percentages of dietary fiber can be recovered from the pomace of apples, grapes, citrus, pear, cherry, berries, passion fruit and mangos (Larrauri et al. 1996, Larrauri et al. 1996, Nawirska and Kwaśniewska 2005, Garau et al. 2007, Elleuch et al. 2011, Martínez et al. 2012, Amaya-Cruz et al. 2015).

Dietary fiber and pectin from citrus and apple peels have been produced on commercial scale (Rezzadori et al. 2012, Martí et al. 2014). Pectin yield depends on both technological factors and fruit physiology. In citrus, apart from extraction methods, intrinsic factors, such as the type of citrus and the portion of waste considerably affect pectin yield (Widmer and Montanari 1994, Marín et al. 2007, Martí et al. 2014). Pectin recovered from apple pomace has superior gelling properties but is inferior in colour to citrus pectin. Removal of oxidised phenolic compounds improves the colour of apple pomace pectin without compromising its gelling properties (Schieber et al. 2003).

Extraction of Bioactive Compounds from Fruit Juice Industry Waste

Many factors need to be considered in order to achieve the best results in the extraction of phenolic compounds and dietary fiber from fruit waste. Understanding the nature of target compounds and of raw materials, as well as waste matrices is crucial to the success of the operation. Apart from the aforementioned factors, process types and operating parameters used in the recovery process are also important determinants in the yield and quality of the recovered compounds.

In general, recovery of target compounds from fruit waste comprises (1) pre-treatment, (2) extraction, (3) isolation and purification, and (4) product formulation (Fig. 3). Details of the overall general recovery process of bioactive compounds from food waste has been previously described (Oreopoulou and Tzia 2007, Galanakis 2012).

Phenolic compounds

Important factors affecting the efficiency of the extraction of phenolics include their chemical nature, sample preparation (drying method, particle size, storage time

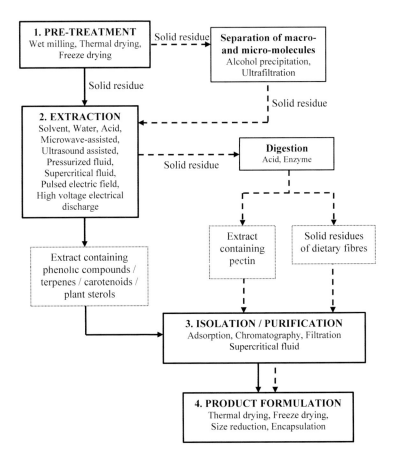

Figure 3. A general flow diagram of recovery stages of bioactive compounds from fruit wastes. Solid lines represent the recovery stages of phenolic compounds, terpenes, carotenoids and phytosterols; dash lines represent the recovery stages of dietary fiber and pectin.

and conditions), the extraction method employed (mode of extraction, extracting medium, solvent-to-solid ratio, contact time, temperature), and presence of interfering substances (Pinelo et al. 2005, Naczk and Shahidi 2006, Valls et al. 2009, Çam and Aaby 2010, Candrawinata et al. 2014). The solubility of phenolic compounds is greatly affected by the polarity of solvent, extracting conditions, degree of polymerization of phenolic compounds, as well as interaction of phenolic compounds with other food constituents and formation of insoluble complexes (Naczk and Shahidi 2004). Phenolic extracts of fruit waste are therefore always a mixture of different classes of phenolic compounds that are soluble in the extraction system applied. Conventional extractions result in dilute extracts; therefore, concentration of the extracts may be required (Galanakis 2012).

The method chosen to dry fruit waste prior to extraction can considerably affect the yield. Comparison studies of the effects of different drying methods on the yield of phenolic compounds extracted from grape skins (oven drying and freeze-drying) (de Torres et al. 2010), apple pomace (vacuum-drying, oven drying and freeze-drying) (Lavelli and Corti 2011), mango peel and kernel (freeze-drying, vacuum-drying, oven-drying and infra-red-drying) (Sogi et al. 2013) have been reported. In all the studies, freeze-drying was less aggressive than thermal drying methods, especially for anthocyanins and anthocyanidins, and it was possible to maintain high antioxidant activity in the fruit matrices. It should be noted that, although freeze-drying is found to be gentler than thermal drying methods, losses of phenolic compounds are still observed (Paes et al. 2014). In the sample preparation step, if desired, unwanted phenolic and non-phenolic substances such as wax, fat, terpene and chlorophyll can be removed by washing off with nonpolar solvent (such as hexane) before extraction of target compounds (Huber and Rupasinghe 2009).

There is no standard or completely satisfactory method to extract phenolic compounds in fruit waste, as the method suitable for one material may not be suitable for others. Effective solvents generally used for extracting phenolic compounds from fruit waste are ethyl acetate, ethyl ether, ethanol, acetone, methanol, including their aqueous mixtures (Ignat et al. 2011, O'Shea et al. 2012). Nonpolar solvents (hexane, petroleum ether) are suitable for extraction of tocopherols and certain phenolic terpenes (Oreopoulou and Tzia 2007). In solvent extraction, interference compounds, such as sugars and organic acids are generally co-eluted into the extract. Removal of these compounds can be achieved by passing the crude extract through C18 solid phase extraction prior to separation of phenolic components or antioxidant activity determination (Li et al. 2006, Reis et al. 2012). Methanol is found to be a highly effective solvent, especially for anthocyanin extraction (Kapasakalidis et al. 2006). Wijngaard et al. (2009) and Wijngaard and Brunton (2010) also reported that methanol was the most effective solvent in comparison to ethanol and acetone on apple pomace phenolic yield and antioxidant activity.

Though some solvents are found to be highly effective, some have become the less preferred option with the growing concerns on safety issues arising from their toxicity. In addition, some solvents potentially pose health and safety challenges for production and impart harmful impurities into the phenolic extracts, especially if the extracts are intended for food applications. Due to this, investigation of 'green' extractions using water, solely or in high proportion, as extracting medium has been explored. Experimental results suggest that water, including diluted acidic solutions and buffers, are not as effective as organic solvents, but can be an acceptable extracting medium. Pinelo et al. (2005) compared the extraction efficiencies of solvents (ethanol, methanol, water), together with other operating parameters (time, temperature and solvent-to-solid ratio) in grape pomace. They reported that, regardless of the solvent used, the highest yield of phenolic was obtained from the conditions with the highest temperature (50°C), extracting time (90 min) and solvent-to-solid ratio (5 mL/g), in the range studied. Water was able to recover approximately 60 per cent of phenolics relatively to the amount obtained from methanol. More recent studies showed that the efficiency of water as an extracting medium can be further improved by increasing the extracting temperature together with optimized solvent-to-solid ratio and extraction time in grape pomace (Çam and Aaby 2010), and apple pomace (Candrawinata et al. 2014). In addition to using water, solvents and their aqueous solutions, acidified extracting media can also improve yields, especially for anthocyanins. Weak organic acids (formic acid, acetic acid, citric acid and tartaric acid) and low concentrations of strong acids (trifluoroacetic acid, hydrochloric acid), are beneficial for extracting anthocyanins (Revilla et al. 1998, Ju and Howard 2003).

Besides the conventional solvent extraction, other techniques have been introduced in an attempt to improve the extraction process to obtain extracts with higher yield and functional activities. Over the last decade, application of compressed fluid extraction (pressurized liquids, supercritical fluids), microwave, sonication (as pre-treatments or sole extraction methods) have become strong candidates of choice. These technologies have proven beneficial for improving yield/biological activities of target compounds from fruit waste, more economical to run, and highly acceptable as green processes when applied with a carefully chosen extracting media (Galanakis 2012, Galanakis 2013).

Compressed fluid extraction must operate under medium-to-high pressure. Extraction methods using this approach include pressurized liquid extraction (PLE), subcritical water extraction (SWE) and supercritical fluid extraction (SFE). Operating principle of PLE is to use liquids (extraction media) at temperatures above their normal boiling point and under enough pressure to keep the extracting fluid in the liquid state. When applied PLE with water as extracting medium, the process is called subcritical water extraction (SWE). PLE enables rapid extraction (3–20 min) of analytes in a closed sample cartridge filled with an extraction fluid, under high pressures (3.3–20.3 MPa) and temperatures (40–200°C) (Richter et al. 1996). The most important operating parameter in PLE applications is temperature. In general, recovery of higher bioactive yields in higher amounts at higher temperatures has been observed but simultaneously, too high a temperature may be detrimental to biological activities of the extracts. Šťavíková et al. (2011) used pressurized water (15 MPa) to extract anthocyanins from grape skins and found that the recovery of anthocyanins as well as radical-scavenging abilities of the extracts were dependent on extraction temperature (up to 80°C). This trend, however, is not observed when using methanol or ethanol pressurized under the same pressure as optimal temperature was found to be 40°C (Polovka et al. 2010). Therefore, this parameter should be studied and selected for each type of matrix or bioactive being extracted. Other parameters (e.g., pressure and time) are also important but have a less critical effect (Herrero et al. 2013).

Supercritical fluid extraction (SFE) is another application of compressed fluid extraction. SFE operates at temperature and pressure close to the critical point of the solvent used. Based on this operating principle, the most utilized critical fluid has been carbon dioxide (CO_2) because of its low critical temperature and pressure (31.1°C, and 7.4 MPa, respectively) (Hawthorne 1990). Low operating temperature is beneficial to extraction of phenolic compounds which are thermolabile. Carbon dioxide has low toxicity and is safe for food application and SC-CO_2 is considered safe and green (Reverchon and De Marco 2006). In SC-CO_2 system, other solvents are not generally necessary, although the presence of co-solvents (such as methanol, ethanol, water) may be beneficial, especially in the case of anthocyanin extraction (Bleve et al. 2008, Wijngaard et al. 2012). This is because CO_2 has low polarity and small quantities of co-solvents (generally lower than 15 per cent) are commonly required to modify the effectiveness of CO_2 in extracting more polar compounds. Key operating parameters that need to be optimized in SFE applications include sample particle size, temperature, pressure, time, co-solvents and solvent-to-solid ratio (da Silva et al. 2016).

Performance of conventional solvent extraction, PLE, and SFE has been compared (Paes et al. 2014). Anthocyanin extraction from blueberry pomace using conventional solvent extraction (methanol, ethanol, acetone), PLE (acidified water, ethanol, 50 per cent v/v aqueous ethanol, 50 per cent v/v ethanol in acidified water, acetone) and SC-CO_2 have been investigated. Among the methods and conditions tested, the authors reported that PLE and SFE were effective on the extraction of phenolics, antioxidants and anthocyanins from blueberry waste, particularly PLE with water and/or ethanol, and SC-CO_2 with 5 per cent water and 5 per cent ethanol as co-solvents. Interestingly, Garcia-Mendoza et al. (2015) combined SC-CO_2 and PLE (ethanol) into sequential extraction steps to extract phenolic compounds from mango peel. The results showed that the extraction yield improved as non-polar phenolic compounds were recovered by SC-CO_2 in the first stage, while polar phenolic components were extracted by pressurized ethanol during the second stage.

Microwave-assisted extraction was reported to accelerate extraction time and improve phenolic yields (Inoue et al. 2010). Microwaves are non-ionizing radiations with frequencies between 300–300 GHz. Microwaves can interact with polar solvent (such as ethanol, methanol, water) and heat the solvent rapidly, causing moisture loss in the cells. The steam generated then swells and penetrates the sample matrix, resulting in cell walls disruption and fast migration of phenolics into the solvent (Wang and Weller 2006). Important operating parameters include type of solvent, solvent-to-solid ratio, microwave energy, extracting time and temperature (Hayat et al. 2010, Zhang et al. 2011, Rezaei et al. 2013). MAE process conditions for phenolic recovery have been investigated in a number of fruit solid waste (Table 3).

Ultrasound-assisted extraction is one of the emerging extraction techniques with many advantages, such as being rapid, reproducible, economical and clean. Ultrasound with frequencies higher than 20 Hz creates transient cavitation (bubbles) in the sample matrix, leading to cell-wall disruption and diffusion of phenolics into the solvent without significantly increasing the temperature (Soria and Villamiel 2010). Indeed, optimization of operating parameters (temperature, solvent system, sonication power, sonication time, solvent-to-solid ratio, particle size) needs to be carried out to achieve yield improvement. Optimization of operating factors, such as particle size, the extraction solvent, solid/solvent ratio, temperature, extraction time, the electrical acoustic intensity, liquid height and duty cycle have been studied in various types of fruit waste. The effect of ultrasound on phenolic extraction has been tested in various

Table 3. Extraction, separation and characterization of phenolic compounds in selected fruit waste.

Fruit waste	Sample preparation and extraction conditions	Separation/Characterization method	Major compounds identified	References
Solvent extraction				
Peach peel	Frozen; Solvent: 80 per cent aqueous methanol	LC-DAD; C18; Mobile phase: 50 mM ammonium phosphate, pH 2.6 (A), 80 per cent acetonitrile and 20 per cent buffer A (B), 200 mM orthophosphoric acid, pH 1.5 (C); Detection: 316 nm (hydroxycinnamates), 520 nm (anthocyanins), 280 nm (flavan-3-ols), 365 nm (flavanols)	Chlorogenic acid, procyanidin, catechin, isoquercetin B1, neochlorogenic acid, malvin, rutin	Chang et al. (2000)
Peach peel (yellow-, white-fleshed) Nectarine Peel (yellow-, white-fleshed), Plum peel	Frozen; Solvent: water/methanol 2:8 containing 2 mM NaF (5 g) was Solid-to-solvent ratio: 5:10 g/mL	HPLC-DAD/ESI-MS; C18; Mobile phase: 95 per cent water + 5 per cent methanol (A), 88 per cent water + 12 per cent methanol (B), 20 per cent water + 80 per cent methanol (C), methanol (D); Detection: 280, 340, 510 nm	Chlorogenic acid, catechin epicatechin, neochlorogenic acid, procyandin B1, rutin, cyanidin 3-rutinoside	Tomás-Barberán et al. (2001)
Mango peel, Mango kernel	Freeze dried, defatted; Solvent: methanol Temperature: RT	LC-DAD/ESI-MS; C18; Mobile phase: 2 per cent acetic acid in water (A), methanol (B) Detection: 278, 340 nm	Peel: Penta-O-galloyl-glucoside, methyl gallate, mangiferin, tetra-O-galloyl-glucoside, maclurin di-O-galloyl-glucoside, isoquercitrin Kernel: penta-O-galloylglucoside, methyl gallate, mangiferin, tetra-O-galloyl glucoside, gallic acid	Barreto et al. (2008)

Table 3 contd. ...

...Table 3 contd.

Fruit waste	Sample preparation and extraction conditions	Separation/Characterization method	Major compounds identified	References
Grape cane	Dried, ground; Solvent: aqueous ethanol (36–80 per cent v/v), Temperature: 30–70°C, Solvent-to-solid ratio: 50–90:1 mL/g	LC-DAD; C18; Mobile phase: 50 mM phosphoric acid (A), methanol (B); Detection: 280 nm	trans-resveratrol equivalent compounds	Karacabey and Mazza (2010)
Pomegranate peel and mesocarp	Lyophilized; Methanol	LC-DAD/ESI-MS; C18 column; Anthocyanins—Mobile phase: 5 per cent v/v formic acid in water (A), water, formic acid and methanol (10/10/80, v/v/v, B); Detection: 520 nm; Other phenolics—Mobile phase: 2 per cent v/v acetic acid in water (A), 0.5 per cent acetic acid in water and methanol (10/90, v/v; B); Detection: 280, 320 nm	Peel: anthocyanins (cyadinin-3,5-diglucoside, pelargonidin-3,5-giglucoside), ellagitannins (granatin B, castalagin der, galloyl-HHDP-hex, bis-HHDP-hex), gallic acid. Mesocarp: ellagitannins (galloyl-HHDP-gluconic acid, granatin B ellagic acid der, digalloyl-HHDP-gluconic acid)	Fischer et al. (2011)
Pomegranate peel	Air drying, ground; 80 per cent methanol	MALDI-TOF MS; 337 nm pulsed nitrogen laser, polarity-positive (alternatively negative); flight path-linear, 20 kV acceleration voltages, 100–150 pulses per spectrum	flavonoid tetramers, pentagalloyl glucose, hydrolyzable tannins, ellagitannins	Saad et al. (2012)
Apple seeds	Lyophilized, ground, defatted; Aqueous acetone (30:70 v/v) followed by liquid–liquid extraction with ethyl acetate	LC-DAD/ESI-MS; C18; Mobile phase: 2 per cent acetic acid in water (A), 0.5 per cent acetic acid in water and methanol (30:70, B); Detection: 280 nm (dihydrochalcones, flavanols), 320 nm (hydroxycinnamic acids), 370 nm (flavonols)	Phloridzin, epicatechin, catechin	Fromm et al. (2012)

Source	Extraction	Analysis	Compounds	Reference
Sour cherry pomace	Water (70°C), followed by extract purification on Amberlite XAD-7HP column	LC-DAD/ESI-MS; C18 column; Mobile phase: 10 per cent v/v formic acid in water (A), 50:40:10 v/v/v acetonitrile:water:formic acid (B); Detection: 320 nm (hydroxycinnamic acids), 360 nm (quercetin, kaempferol, isorhamnetin glycosides and their aglycones), 520 nm (anthocyanins) LC-UV; C18 column; 280 nm (flavanols)	Cyanidin-glucoside-rutinoside, chlorogenic acid, neochlorogenic acid, p-coumaroylquinic acid, quercetin, kaempferol, isorhamnetin glycosides	Kolodziejczyk et al. (2013)
Supercritical Fluid Extraction				
Apple pomace	Freeze drying, ground; PLE: 25–75 per cent aqueous ethanol Temperature: 160–193°C, Pressure: 10.3 MP Extraction time: 5 min	LC-DAD; C18; Mobile phase: acetic acid in 2 mM sodium acetate (pH 2.55, v/v, A), 100 per cent acetonitrile (B); Detection: 280 nm (hydroxybenzoic acids, dihydrochalcones, procyanidins, flavanols), 320 nm (hydroxycinnamic acid derivatives), 360 nm (flavonols)	chlorogenic acid, flavonols, and phloretin glucosides	Wijngaard et al. (2009)
Pomegranate peels	Sun drying, ground; PLE: water Pressure: 102.1 atm Extraction time: 5 min	LC-DAD; C18; Mobile phase: water/acetic acid (98:2, v/v, A), methanol (B); Detection: 260, 280, 320, 360, and 378 nm. Spectrophotometry method for tannins	Punicalagin B, punicalagin A, ellagic acid derivatives, gallic acid Condensed tannins, hydrolysed tannins	Çam and Hışıl (2010)
Grape seeds	Air drying, ground; SC-CO2 with 5–20 per cent ethanol Temperature: 30, 50°C Pressure: 25–30 MPa Extraction time: 60 min	LC-DAD; C18; Mobile phase: 2 per cent acetic acid in water (A),100 per cent acetonitrile (B); Detection: 280 nm	Gallic acid, epigallocatechin, epigallocatechingallate, catechin, epicatechin, epicathechingallate	Yilmaz et al. (2011)

Table 3 contd. ...

...Table 3 contd.

Fruit waste	Sample preparation and extraction conditions	Separation/Characterization method	Major compounds identified	References
Pomegranate seeds	Ground, defatted by SC-CO2 (37.9 MPa, 47.0°C); PLE: water Pressure: 6.0 MPa. Temperature: 80–280°C Extraction time: 15–120 min	LC–ABTS•+; C18; Mobile phase: 0.2 per cent v/v formic acid (A), methanol (B); Detection: 280 nm	Caffeic acid derivative, catechin, kaempferol 3-O-rutinoside	He et al. (2012)
Microwave-Assisted (MAE) and Ultrasound-Assisted Extraction (UAE)				
Apple pomace	Lyophilized. Ground SC-CO2 + 25 per cent ethanol (25 MPa, 50°C)	LC-DAD/ESI/ACPI-MS; C18; Mobile phase: 0.1 per cent formic acid (B), acetonitrile (B); Detection: 280 nm	Quercetin-3-O-xyloside, Quercetin-3-O-rhamnoside Quercetin-3-O-arabinoside Quercetin-3-O-glucoside Epicatechin, Quercetin-3-O-galactoside Phloridzin, Catechin, Chlorogenic acid	Garcia-Mendoza et al. (2015)
Blackcurrant pomace Blackcurrant seeds	Oven dried, ground; UAE Solvent: methanol:water:formic acid (50:48:2 v/v/v). Solid:solvent ratio: 0.5:5 g/mL	LC-DAD/ESI-MS; C18 column; Mobile phase: 10 per cent v/v formic acid in water (A), 50:40:10 v/v/v acetonitrile:water:formic acid (B); Detection: 320 nm (hydroxycinnamic acids), 360 nm (quercetin and myricetin), 520 nm (anthocyanins)	Delphinidin-3-rutinoside, Delphinidin-3-glucoside, Cyanidin-3-rutinoside, Cyanidin-3-glucoside, Myricetin, Quercetin, Kaempferol	Sójka et al. (2009)
Apple pomace	Oven-dried (60°C); Reflux, MAE, UAE Solvent: Ethanol	LC-UV; C8; Mobile phase: 0.1 per cent acetic acid (A), 10 per cent acetonitrile (B); Detection: 280 nm	Procyanidins, cinnamic acid, chlorogenic acid, caffeic acid, syringic acid	Bai et al. (2010)

Citrus peel	Ground; Conventional extraction: 0–100 per cent ethanol, and DMSO:methanol,1:1, v/v MAE: 70 per cent ethanol	LC–DAD/ESI-MS, C^{13} NMR; C18; Mobile phase: 40 per cent methanol (A), 100 per cent methanol (B); Detection: 284 and 332 nm (flavonoids)	Hesperidin, narirutin, nobiletin	Inoue et al. (2010)
Orange peel	UAE; 20–80 per cent v/v Ethanol	LC–DAD; C18; Mobile phase: 0.5 per cent acetic acid (A), 100 per cent acetonitrile (B); Detection: 280 nm	Naringin, hesperidin	Khan et al. (2010)
Citrus peels	Fresh and dried; UAE at 60 kHz, Peel moisture content: 0 per cent, 75 per cent Time: 30, 90 min solid/water ratio: 1/10 g/mL	LC–DAD/ESI-MS; C18; Mobile phase: 0.1 per cent formic acid (A), acenonitrile (B); Detection: 280 nm	Hesperidin, neohesperidin, diosmin, nobiletin, tangderitin	Londoño-Londoño et al. (2010)
Grape skins	MAE Solvent: 50–80 per cent MeOH Temperature: 50–100°C, Time: 5–20 min Microwave power: 100–500 W) Solid:solvent ratio: 1;12.5–1:25 g/mL	LC–DAD; C18; Mobile phase: 5 per cent formic acid (A), methanol (B); Detection: 520 nm	Malvidin 3-glucoside, peonidin 3-glucoside, malvidin 3-acetylglucoside, petunidin 3-glucoside, malvidin 3-p-coumaroylglucoside, delphinidin 3-glucoside, malvidin 3-caffeoylglucoside	Liazid et al. (2011)
Grape seeds	Air drying, ground; MA aqueous two-phase extraction Solvents: 24–34 per cent (w/w) acetone and 14–22 per cent (w/w) ammonium citrate Time: 2 min	LC-UV; C18; Mobile phase: 0.3 per cent phosphoric acid (A), methanol (B); Detection: 280 nm	Catechin, gallic acid, epicatechin, trans-resveratrol, quercetin	Dang et al. (2014)

Table 3 contd. ...

...Table 3 contd.

Fruit waste	Sample preparation and extraction conditions	Separation/Characterization method	Major compounds identified	References
Lime pomace	Freeze-dried, tray-tried (60, 90 and 120°C); UAE: 80 per cent Methanol at RT in ultrasonic bath followed by methanol/H_2SO_4 hydrolysis for non-extractable phenolics	LC-DAD; C18; Mobile phase: 6 per cent acetic acid in 2 mM sodium acetate buffer (pH 2.55, v/v, A), acetonitrile (B); Detection: 260 nm (hydroxybenzoic acids, quercetin, rutin), 280 nm (flavans and flavanones), 320 nm (hydroxycinnamic acids, stilbenes), 360 nm (miricetin and kaempherol)	Hesperidin Eriocitrin Naringin Naringenin p-Coumaric Benzoic Ellagic Catechin	Esparza-Martinez et al. (2016)

studies and enhanced extraction observed (Khan et al. 2010, Virot et al. 2010, Pingret et al. 2012, Dahmoune et al. 2013).

Combination of the aforementioned extraction techniques in order to achieve better results has also been investigated. Applying ultrasound during SC-CO$_2$ with water as a co-solvent was found to remarkably increase extraction rate and yields of phenolics, anthocyanins, as well as the antioxidant activity of the extracts obtained from blackberry bagasse. Using ethanol as a co-solvent also exerted positive influence on the extraction of anthocyanins, but the effect was much less pronounced than water (Pasquel Reátegui et al. 2014).

Comparison of extraction performance of several extraction methods (conventional solvent extraction (methanol and ethanol), UAE, MAE and high pressure and temperature extraction (HPTE; water) to obtain phenolic compounds from grape seeds and skins has been reported (Casazza et al. 2010). The authors reported that HPTE provided the highest content of total phenolics both for seeds and skins, while MAE retained the highest antiradical power. Prolonged extraction time (over 30 min) was not necessarily beneficial because, though the amount of total polyphenols increased, the amount of flavonoids and the antiradical power decreased.

Emerging extraction techniques have recently been implemented for recovery of phenolic compounds from fruit waste. Application of electrotechnology, such as pulsed electric field (PEF), and high voltage electrical discharge (HVED), has acquired increased interest. Both techniques are non-thermal processes, which are highly beneficial for recovery of heat-sensitive compounds. PEF and HVED have been shown to be promising for intracellular extraction from by-products (Luengo et al. 2013, Boussetta and Vorobiev 2014).

PEF uses strong electric field to provoke pre-formation on the cell structure. This electroporation (or electropermeabilization) facilitates the release of target compounds from the fruit matrices (Wijngaard et al. 2012). PEF-assisted extraction generally involves direct electric pulses of high voltage are applied (upto 40 kV) for short duration (less than 10 min) at a repeated pulse (frequency), resulting in high electric field strength (1–10 kV/cm). Efficiency of PEF-assisted extraction is dependent on the PEF system configuration and extraction parameters. Similar to other methods, extraction temperature, sample particle size, solvent system and concentration, are important factors determining the extraction performance. Enhanced PEF extraction yields of phenolic compounds from orange peel (Luengo et al. 2013), anthocyanins from blueberry pomace (Zhou et al. 2015) and phenolic compounds and anthocyanins from raspberry pomace (Lamanauskas et al. 2016) have been reported. In another study by Medina-Meza and Barbosa-Cánovas (2015), PEF offered enhanced anthocyanin yield from grape peel but the yield was not impressive when the same PEF conditions were applied to plum peels.

HVED works based on chemical reactions and physical processes. HVED have electrical and mechanical effects on the product caused by shock waves. This technique introduces energy directly into an aqueous solution through a plasma channel formed by a high-current/high-voltage electrical discharge between two submerged electrodes. The large range of current (10^3–10^4 A), voltage (10^3–10^4 V) and frequency (10^{-2}–10^{-3} Hz) are typically applied (Boussetta and Vorobiev 2014). Extraction parameters affecting the extraction yield include solvent system, inter-electrode space, energy input, liquid-to-solid ratio and temperature. HVED has been satisfactorily used to extract phenolic compounds from grape pomace (Boussetta et al. 2009, Boussetta et al. 2009, Boussetta et al. 2011) and grape seeds (Liu et al. 2011).

HVED has been reported to be more efficient than PEF in the extraction of phenolic compounds from grape skin (Boussetta et al. 2009), grape pomace (Barba et al. 2015), and mango peel (Parniakov et al. 2016). It is rather interesting that PEF efficacy can be markedly improved when the treatment is performed at 50°C in the presence of ethanol (Boussetta et al. 2012) or with a supplementary aqueous extraction after PEF treatment (Parniakov et al. 2016).

Dietary fiber

Fruit waste is a rich source of dietary fiber (DF). Cellulose, hemicelluloses, pectin and lignin are typical fiber components found. The constituents are divided into soluble dietary fiber (SDF, i.e., pectin) and insoluble dietary fiber (IDF, i.e., cellulose, most hemicelluloses, lignin). They provide various functional effects beneficial to human health, as well as functional properties in food processing and food formulation without offering nutritional value. Upon hydration, soluble fibers are able to form a gel or a network, while insoluble fibers absorb large amounts of water (up to 20 times their weight in water) and expand into bulky materials (Thebaudin et al. 1997, Figuerola et al. 2005, Nawirska and Kwaśniewska 2005, O'Shea et al. 2012).

High dietary fiber concentrate/powder

High dietary fiber products can be prepared directly from fruit waste or, if desired, after the recovery of other bioactive compounds (Fig. 2). The simplest preparation method is merely grinding the dried fruit waste into fine particles. Conventional production of dietary fiber powder from fruit waste involves a few mechanical steps, i.e., wet milling, washing, drying and lastly, dry milling (Oreopoulou and Tzia 2007). All the steps, although relatively simple, need to be optimized as they affect the yield and characteristics of the obtained fiber (Larrauri 1999). An appropriate mean particle size from wet milling ensures an adequate wash without holding too large an amount of water which will make subsequent drying more difficult. In the washing step, washing time, water temperature, water-to-solid ratio are important parameters for maximizing the removal of undesirable components (i.e., sugars), which will improve functionality and color of the final product and retain desirable water-soluble components (i.e., soluble dietary fiber) (Larrauri et al. 1996, Lario et al. 2004). Operating drying parameters, such as temperature, time and drying rate, affect the degradation and thus the yield of target compounds (phenolic compounds, dietary fibers) (Garau et al. 2007). Lastly, appropriate particle size from dry milling also needs to be determined as it affects the characteristics of the final product, such as water- and oil-holding capacity and suspension in water (Oreopoulou and Tzia 2007). Selected reports on extraction conditions of dietary fiber products from fruit waste is shown in Table 4. Fruit waste is a good source for dietary fiber recovery and include pomaces of citrus, apple, pear, peach, passion fruit, mango and pomegranate (Grigelmo-Miguel and Martín-Belloso 1998, Grigelmo-Miguel et al. 1999, Grigelmo-Miguel and Martín-Belloso 1999, Larrauri 1999, Lario et al. 2004, Figuerola et al. 2005, Viuda-Martos et al. 2012, Ajila and Prasada Rao 2013, López-Vargas et al. 2013).

Table 4. Selected studies on preparation of dietary fiber products from fruit waste.

Dietary fiber product	Fruit waste	Extraction conditions/analysis method	References
Fiber concentrate	Passion fruit seeds	Cleaned, finely ground to 0.5 mm size, defatted; Enzymatic-gravimetric method: AOAC method 991.43	Chau and Huang (2004)
Fiber concentrate	Apple pomace, Citrus peels (grapefruit, lemon, orange)	Washing: water, 30°C Drying: Air tunnel drier, 60°C, 30 min Dry milling: 500–600 μm; Enzymatic-gravimetric method: Lee et al. (1992)	Figuerola et al. (2005)
Customized functional fiber	Citrus—whole, peel, pulp (sour range, satsuma, grapefruit, sweet orange)	Scalded in a water bath Drying: Oven at 50 ± 5°C, 24 hour Dry mill: 0.2 mm; Enzymatic-gravimetric method: Prosky et al. (1988)	Marín et al. (2007)
High dietary fiber	Apple—parenchyma tissues, pomace	Frozen, ground, then precipitate either in 72% ethanol or HEPES buffer; Enzymatic-chemical method: uronic acid content	Sun-Waterhouse et al. (2008)
High dietary fiber powder	Lime pomace	Washing: water, 95°C, 5 min Soaking: ethanol (95 per cent v/v) Drying: Oven at 60°C Dry mill: 38–63, 63–150, 150–250, 250–300 and 300–450 μm; Enzymatic-gravimetric method: AOAC method 991.43	Peerajit et al. (2012)
IDF and SDF	Mango peel	Enzymatic extraction: α-amylase, pepsin, pancreatin; Separation of IDF: filtration; Enzymatic-gravimetric method: Asp et al. (1983)	Ajila and Prasada Rao (2013)
Dietary fiber powder	Yellow passion fruit – pomace, albedo	Washing: water, 45°C, 8 min Drying: Oven 60°C, 24 hours Dry milling: less than 0.417 mm; Enzymatic-gravimetric method: AOAC method 991.43	López-Vargas et al. (2013)

Without any extraction step prior to the fiber preparation, dietary fibers obtained directly from fruit waste contain high amounts of bioactive compounds, such as phenolic compounds, terpenes, carotenoids, depending on the fruit sources (Saura-Calixto 2010). Lime peel dietary fiber powder is found to have much stronger antioxidant activity than orange peel dietary fiber powder as it contains a broader range of phenolic components (caffeic acid ferulic acids, naringin, hesperidin, myricetin, ellagic acid, quercetin, kaempferol (Larrauri et al. 1996). Presence of phenolic compounds can cause discoloration of the final product. Applications of alkaline solution/ozone ultrasonic assisted extraction has been patented as a decoloration method to improve the product's color (Chen and Li 2013).

Dietary fiber with lower IDF/SDF ratio is of better quality and is more desirable as a food ingredient. The composition of polysaccharide constituents in dietary fibers depends on the source of fruit waste. Fiber from cherry and blackcurrant pomaces contain low amounts of pectin and amounts of lignin, thus have much higher IDF/SDF ratio than fiber from apple pomace (Nawirska and Kwaśniewska 2005).

Pectin

Pectin is a family of complex polysaccharides of α-d-(1→4) galacturonic acid present in the primary cell wall and middle lamella of the plant tissues. All pectins are characterized by a high content of galacturonic acid (GalA), and, according to the regulation of FAO and EU, pectin must contain at least 65 per cent GalA (Rolin 2002). Conventionally, pectin from fruit waste can be extracted by the use of mineral acids, usually hydrochloric or nitric acid. The extract is separated from the solid residue and pectin is precipitated by the addition of ethanol. The precipitated pectin is then purified by washing with acidified, alkaline and finally neutral alcohol. Lastly, the product is dried to a desirable moisture content. Citrus peel and apple pomace have been used to produce pectin on an industrial scale (Oreopoulou and Tzia 2007, Martí et al. 2011, O'Shea et al. 2012). However, other fruit wastes are found to yield high amount of pectin, such as peach pomace (Pagan and Ibarz 1999, Pagan et al. 1999), passion fruit peels (Silva et al. 2008, Kliemann et al. 2009, Kulkarni and Vijayanand 2010), and mango peels (Rehman et al. 2004, Berardini et al. 2005).

Several studies have shown that, apart from the source and type of fruit waste used as raw material, the yield and quality of the obtained pectin is greatly affected by the extraction conditions (acid type and concentration, pH, extraction time) (Virk and Sogi 2004, Faravash and Ashtiani 2007, Kliemann et al. 2009). In general, yield is improved by low pH and high temperature or a long extraction time. However, these extraction criteria adversely affect gelling quality of pectin (Aravantinos-Zafiris and Oreopoulou 1992, Pagan et al. 1999). Phenolic compounds should be removed before pectin extraction as they cause undesirable light brown coloring in the pectin, especially when drying under temperatures higher than 60°C. Removal of phenolic compounds can be achieved by conventional and non-conventional extraction methods described in the previous section. Alternatively, implementation of resin absorption can successfully separate phenolic compounds, which can be subsequently recoverable from the raw materials (Schieber et al. 2003, Berardini et al. 2005).

Applications of MAE and UAE in pectin recovery from fruit waste demonstrate high potential because these techniques can shorten the extraction time, reduce the solvent consumption and improve the extraction yield and functional properties of the obtained pectins (Table 5). Bagherian et al. (2011) did a comparison study on pectin extraction from grapefruit peel using MAE, UAE and conventional methods. They reported that MAE provided highest pectin yield with the best characteristics within the shortest extraction time. The extraction yield was also further improved when UAE was applied as a pretreatment for MAE. Another more recent comparison study investigated the efficacies of four different methods (MAE, UAE, conventional extraction and enzymatic extraction) on pectin extraction from apple pomace (Li et

Table 5. Selected studies on preparation of pectin from fruit waste.

Fruit waste	Extraction conditions	References
Orange albedo	MAE under pressure, pH 1–2 Temperature (max.): 195°C Pressure (max.): 50 ± 2 psi Solid-to-solvent ratio: 1:25, 5:25 g/mL Microwave power: 630 W at 2450 MHz	Fishman et al. (1999)
Apple pomace	MAE, pH 1.22–1.78 Time: 10.6–17.4 min Solid:liquid ratio (w/v): 0.0333–0.0571 Microwave power: 320, 450, 580 W	Wang et al. (2007)
Berry pomaces (red currant, black currant, raspberry, elderberry)	MAE Frequency: 2.45 GHz. Solvent: water Solid:solvent ratio: 1:10 Time of 30 min	Bélafi-Bakó et al. (2012)
Orange peel	MAE, pH 1–2 Time: 60–180 s Solid-to-solvent ratio: 1:10–1:30 g/mL Microwave power: 160–480 W	Maran et al. (2013)
Passion fruit peel	MAE, pH 2 Acid: acetic, tartaric, nitric Time: 3–9 min Solid-to-solvent ratio: 1:25 g/mL	Seixas et al. (2014)
Pomegranate peel	UAE, pH 1–2 Temperature: 50–70°C Time: 12–25 min Solid-to-solvent ratio: 1:10–1:20 g/mL	Moorthy et al. (2015)
Grape fruit peel	UAE, pH 1.5 Power intensity: 10/18–14.26 W/cm^2 Sonication time: 20–40 min Temperature 60–80°C	Wang et al. (2015)

al. 2014). The results showed enzymatic extraction was the best extraction method in terms of improving the yield and functionality of extracted pectin. Pectin yield obtained from UAE was slightly higher than that from MAE; however, MAE drastically reduced the extraction time. In comparison to conventional extraction, all non-conventional methods studied gave pectins of higher yields and improved functionality at a shorter extraction time.

Isolation of Bioactive Compounds from Fruit Juice Industry Waste

Due to the complex nature of both fruit materials and bioactive compounds recovered from them, many analysis techniques have been explored and developed to isolate, quantify and characterize these bioactive compounds. Each technique has its own advantages and limitations. Common methods for isolation/quantification/characterization of phenolic compounds and dietary fibers are discussed below.

Phenolic compounds

Isolation of phenolic compounds can be achieved by various methods. Spectrophotometric methods—such as Folin–Ciocalteu, DPPH, ABTS, TEAC, FRAP assay—have been widely used for determination of phenolic compounds extracted from fruit waste. These assays are relatively simple to perform with low running cost (Ignat et al. 2011). Nevertheless, they offer little information in terms of what polyphenols are in the sample. They are non-selective; therefore, overestimation from interference presence in the samples is one common drawback. Comparison of experimental data is generally difficult as they are not standardised.

Liquid chromatography is a better choice for separation and quantification of phenolic compounds in fruit waste as it is more sensitive and compound-specific. In most cases in fruit waste phenolic studies, separation is achieved by reversed-phase C18 column with gradient elution. In general, a binary solvent system composed of acidified water (dilute formic acid or acetic acid) and a less polar organic solvent (ethanol, methanol or acetonitrile) is used, but tertiary or quaternary solvent systems are also reported (Chang et al. 2000, Tomás-Barberán et al. 2001). The acidic additive in the mobile phase is necessary to suppress the ionisation of the phenolic hydroxyl groups to obtain sharper peaks and minimised peak tailing. UV-Vis photodiode array detector (DAD) is a suitable detection mode for monitoring and quantifying different classes of phenolic compounds. As mentioned previously, phenolic compounds in fruit waste are always a mixture of different phenolic classes, with different maximum absorption. In general, phenolic acids are detected at 220–280 nm, flavones and flavonols at 350–365 nm and anthocyanins at 460–560 nm (Valls et al. 2009). DAD is able to scan light spectra in the UV-Vis range, thus allowing easier monitoring of any separated phenolic fractions. Sakakibara et al. (2003) developed LC-DAD method and made a library comprising HPLC retention times and spectra of aglycons for 100 standard chemicals and for simultaneously determining all phenolic compounds in a wide range of food samples (vegetables, fruits, and teas). LC-DAD system has been reported to successfully separate and quantify anthocyanins, procyadinins, flavonones, flavonols, flavan-3-ols, flavones and phenolic acids in various types of fruit waste (Table 4).

Although LC-DAD is able to satisfactorily separate and quantify phenolic compounds in fruit waste, it also presents limitations. As phenolic compounds are present ubiquitously in fruit waste and their structure can be extremely complex, standards of only certain known compounds are available, which is one major limitation of the use of LC systems. Mass spectrometry (MS) is an analytical technique that is used for elucidating the chemical structures of molecules and plays a very important role in the analysis of polyphenolic compounds. MS structural elucidation is based on ionisation of chemical compounds to generate charged molecules or molecule fragments and measuring their mass-to-charge ratios (Sparkman 2000). To date, LC coupled with MS (LC-MS) technique is the most powerful and effective method for separation and characterization of complex phenolic structures, such as procyanidins, proanthocyanidins, prodelphinidins and tannins, including elucidation of speculated or hypothesised structures (Flamini 2003). Among the methods used for the determination of phenolics in crude plant extracts, liquid chromatography coupled with electrospray ionization (ESI) has been widely used as it is a powerful tool due to the soft ionization, which facilitates the analysis of this polar, non-volatile and thermally labile class of compounds (Table 4). Matrix-assisted-laser-desorption-ionisation-time-of-flight (MALDI–TOF) techniques have also been used to characterize phenolic compounds in pomegranate peel (Saad et al. 2012). Sánchez-Rabaneda et al. (2004) employed LC/MS/MS and successfully identified 60 phenolic compounds from apple pomace, of which 23 components were described for the first time. The main advantages of MS/MS include exclusion of interferences and verification of the structures of the different compounds present in an extract.

Dietary fiber

Isolation and quantification of soluble (pectin), insoluble (lignin, cellulose, hemicellulose) and total dietary fiber (TDF) in dietary fiber products prepared from fruit waste can be achieved by various approaches. One of the easiest approaches used in fruit waste studies are non-enzymatic-gravimetric methods (Lario et al. 2004, Martí et al. 2011). Dietary fiber is characterized as crude fiber, acid detergent fiber (cellulose, lignin and acid insoluble hemicellulose), and neutral detergent fiber (neutral detergent insoluble hemicellulose, lignin and cellulose). This approach, however, does not measure soluble dietary fiber, leading to underestimation of dietary fiber in the samples (Southgate et al. 1978).

In many studies, dietary fiber in fruit wastes was determined using the AOAC Prosky method (AOAC method 985.29), which is enzymatic-gravimetric based (Table 4). The general procedure involves removal of starch and protein through treatment of enzymes (α-amylase, protease and amyloglucosidase), followed by alcohol precipitation, filtration and weighing of dietary fiber. Correction of protein and ash residue is also taken into account to prevent overestimation of dietary fiber (Prosky et al. 1984). Variation of the classical Prosky method was proposed and adopted as a standard method (AOAC method 991.43) (Lee et al. 1992).

Apart from the enzymatic-gravimetric method, enzymatic-chemical method is also used in the determination of dietary fiber in fruit wastes (Larrauri et al. 1996, Grigelmo-Miguel et al. 1999, Grigelmo-Miguel and Martín-Belloso 1999, Larrauri 1999, Nawirska and Kwaśniewska 2005). This procedure determines soluble dietary fiber and lignin. Similar to the enzymatic-gravimetric, starch and/or protein is hydrolyzed by enzymes. Isolation of soluble dietary fiber in the enzymatically hydrolyzed fraction can be achieved by alcohol precipitation or dialysis. Determination of sugars (either by spectrophotometry, gas-liquid chromatography or high-performance liquid chromatography) and uronic acids (colorimetry) can also be performed to obtain more information, if desired. The insoluble fraction collected from enzymatic treatment is further hydrolyzed by sulfuric acid to obtain acid non-hydrolysable residue quantified as Klason lignin (Englyst et al. 1994, Manas et al. 1994).

Potential Use of Bioactive Compounds from Fruit Juice Industry Waste

The potential use of phenolic compounds and dietary fiber products from fruit juice waste as novel functional food ingredients has a very high potential in the food industry. Over the last few decades, the demand for functional food has increased as consumers are more health-conscious and expect food to deliver health-promoting physiological effects besides providing nutrients and satiety. The global functional foods market was worth an estimated USD 43.27 billion in 2013. In comparison to the market values of year 2009, this figure has increased by 26.7 per cent, and continues to demonstrate annual growth in excess of the world food industry as a whole (Leatherhead Food Research 2014). Functional food ingredients derived from natural sources are highly sought-after in order to deliver products matching the consumers' demands for functional foods of natural ingredients. Due to this driving force, bioactive compounds recovered from fruit waste not only provide a solution to food manufacturers in terms of affordability and availability of the ingredients they are seeking, but also a more sustainable approach to using valuable resources which become more and more limited. Phenolic compounds and dietary fiber recovered from various fruit waste has been introduced into various types of food as functional additives, such as antioxidative, coloring, antimicrobial agents, as well as texture modifiers.

Kabuki et al. (2000) reported that mango seed kernel ethanol extract exhibits antimicrobial activities against a broad spectrum of bacteria, especially gram-positive. The antimicrobial activity of the mango seed kernel extract was stable against sterilization conditions, freezing conditions and a wide range of pHs which makes it suitable for use in food processing. Bergamot peel extract exhibited antimicrobial activity against gram-negative bacteria (Mandalari et al. 2007). Fattouch et al. (2008) compared polyphenolic profiles and antioxidant and antimicrobial activities of pome fruit peels (apple, pear and quince) and reported that apple and quince peel extracts were effective in inhibiting the growth of *Staphylococcus aureus*, *Pseudomonas aeruginosa* and *Bacillus cereus*. Extracts prepared from mangosteen pericarp exhibited strong pH-dependent bacteriostatic and bactericidal effects against *Listeria monocytogenes* and *Staphylococcus aureus* (Palakawong et al. 2013). Casquete et al. (2015) reported the citrus peel extracts (lemon, mandarin, sweet orange) demonstrated antimicrobial activity against a wide range of microorganisms and high pressure treatment did not alter those antimicrobial activities. Promising antimicrobial effects of raspberry pomace extract against *Escherichia coli*, *Salmonella* sp., *Listeria monocytogenes*, *Enterococcus faecium* has also been reported (Caillet et al. 2012). Pomegranate peel extract showed excellent antioxidant activity against *Staphylococcus aureus* and *Bacillus cereus* and helped in prolonging the shelf-life of chilled chicken products by 2–3 weeks (Kanatt et al. 2010).

With regards to antioxidant activity, phenolic compounds extracted from mango seed kernel powder were reported to prolong the shelf-life of buffalo ghee (Puravankara et al. 2000). Apple waste phenolic extracts was found to be as effective a natural antioxidant in stabilizing fish-oil (Sekhon-Loodu et al. 2013) and meat products (Yu et al. 2015). Flavanol oligomers obtained from grape pomace were reported as potent inhibitors of oxidation in emulsions and in frozen fish muscles (Pazos et al. 2005).

In many reports on the use of bioactive compounds from fruit waste in food products, antioxidant activity is reported to have a synergistic effect with the addition of dietary fiber (Saura-Calixto 2010). As described in the previous section, when dietary fiber is prepared directly from fruit waste without prior extraction step to remove other bioactive compounds, the resulting dietary fiber products generally contain high amounts of other bioactive components associated with the fruit source. Due to this, many reports on waste-derived dietary fiber as an antioxidant carrier can be found in the literature. Fruit-waste-derived dietary fiber products have low-caloric value and offer some functional properties, such as water-holding capacity, swelling capacity, increasing viscosity or gel formation which are essential in formulating certain food products. Addition of such dietary fibers into baked goods has been reported to improve functional properties of the dough as well as the finished product (Sudha et al. 2007, Vergara-Valencia et al. 2007, Ajila et al. 2008, Min et al. 2010, Sivam et al. 2011, Pečivová et al. 2014, Chareonthaikij et al. 2016). Functionality improvement (e.g., rheological improvement, SDF/IDF and dietary fiber level modifier, shelf-life extension and fat replacement) after the addition of dietary fibers into other food products, such as beverages (Sun-Waterhouse et al. 2010, Sun-Waterhouse et al. 2014), dairy (Sah et al. 2016), fish and meat (Cengiz and Gokoglu 2005, Sánchez-Alonso et al. 2007, Sáyago-Ayerdi et al. 2009), pasta (Ajila et al. 2010) and ready-to-eat snacks (Kayacier et al. 2014, O'Shea et al. 2014) have also been reported.

Apart from direct food product applications, another promising potential application of bioactive compounds recovered from fruit waste is in the development of active food packaging. The biological activities of phenolic compounds (particularly antimicrobial and antioxidative activity) and technological properties of dietary fibers (water permeability, viscosity, gelling and network formation) make it feasible to develop food packaging with enhance functionality (Appendini and Hotchkiss 2002, Lopez-Rubio et al. 2006, Janjarasskul and Krochta 2010, Arcan and Yemenicioğlu 2011, Martinez-Avila et al. 2014, Salgado et al. 2015).

Conclusion

The global market and production values of fruit juice have increased with the drives of production technology and functional food demands. Consequently, the fruit juice industry generates a huge quantity of waste. Alternative valorisation of fruit waste needs to be addressed, as conventional disposal methods are not the best way to utilize such material. Fruit solid waste from juice industry contains high levels of recoverable bioactive compounds associated with human health benefits and can be used as a cheap source for the production of high-value compounds. Extensive studies on extraction, separation and characterisation of phenolic compounds and dietary fibers from various fruit waste have been conducted. Nevertheless, more research is still needed throughout the recovery process, such as applications of 'green' extraction approaches and more powerful separation and characterization techniques in order to achieve higher yield and quality of bioactive extracts suitable for food application. In the food industry, the recovered bioactive compounds have tremendously high potential uses in the development of functional foods and active food packaging.

References

Abdalla, A. E. M., S. M. Darwish, E. H. E. Ayad and R. M. El-Hamahmy. 2007. Egyptian mango by-product 1. Compositional quality of mango seed kernel. Food Chem. 103: 1134–1140.
Adámez, J. D., E. G. Samino, E. V. Sánchez and D. González-Gómez. 2012. *In vitro* estimation of the antibacterial activity and antioxidant capacity of aqueous extracts from grape-seeds (*Vitis vinifera* L.). Food Control 24: 136–141.
[AIJN] European Fruit Juice Association. 2014. Liquid Fruit Market Report. AIJN, Brussels: 44 pp.

Ajayi, I. A., R. A. Oderinde, B. O. Ogunkoya, A. Egunyomi and V. O. Taiwo. 2007. Chemical analysis and preliminary toxicological evaluation of *Garcinia mangostana* seeds and seed oil. Food Chem. 101: 999–1004.

Ajila, C. M., K. A. Naidu, S. G. Bhat and U. J. S. Prasada Rao. 2007. Bioactive compounds and antioxidant potential of mango peel extract. Food Chem. 105: 982–988.

Ajila, C. M., K. Leelavathi and U. J. S. Prasada Rao. 2008. Improvement of dietary fiber content and antioxidant properties in soft dough biscuits with the incorporation of mango peel powder. J. Cereal Sci. 48: 319–326.

Ajila, C. M., M. Aalami, K. Leelavathi and U. J. S. Prasada Rao. 2010. Mango peel powder: A potential source of antioxidant and dietary fiber in macaroni preparations. Innov. Food Sci. & Emerg. Technol. 11: 219–224.

Ajila, C. M., S. K. Brar, M. Verma and U. J. S. Prasada Rao. 2012. Sustainable solutions for agro processing waste management: An overview. pp. 65–109. *In*: A. Malik and E. Grohmann (eds.). Environmental Protection Strategies for Sustainable Development, Springer, Dordrecht.

Ajila, C. M. and U. J. S. Prasada Rao. 2013. Mango peel dietary fiber: Composition and associated bound phenolics. J. Funct. Food 5: 444–450.

Amaya-Cruz, D. M., S. Rodríguez-González, I. F. Pérez-Ramírez, G. Loarca-Piña, S. Amaya-Llano, M. A. Gallegos-Corona and R. Reynoso-Camacho. 2015. Juice by-products as a source of dietary fiber and antioxidants and their effect on hepatic steatosis. J. Funct. Food 17: 93–102.

Anal, A. K. 2013. Food processing by-products. pp. 180–197. *In:* B. K. Tiwari, N. P. Brunton and C. S. Brennan (eds.). Handbook of Plant Food Phytochemicals. Wiley-Blackwell, Wast Sussex.

[AOAC] Association of Official Analytical Chemists Method 985.29. 1990. Total dietary fiber in food. Enzymatic-gravimetric method. Official Methods of Analysis of the Association of Official Analytical Chemists, 15th ed. AOAC International, Arlington, VA.

[AOAC] Association of Official Analytical Chemists Method 991.43. 1992. Total, soluble and insoluble dietary fiber in foods and food products, Enzymatic-gravimetric method. MES-TRIS buffer Official Methods of Analysis of the Association of Official Analytical Chemists, 15th ed., 3rd suppl. AOAC International, Arlington, VA.

Appendini, P. and J. H. Hotchkiss. 2002. Review of antimicrobial food packaging. Innov. Food Sci. Emerg. Technol. 3: 113–126.

Aravantinos-Zafiris, G. and V. Oreopoulou. 1992. The effect of nitric acid extraction variables on orange pectin. J. Sci. Food Agric. 60: 127–129.

Arcan, I. and A. Yemenicioğlu. 2011. Incorporating phenolic compounds opens a new perspective to use zein films as flexible bioactive packaging materials. Food Res. Int. 44: 550–556.

Arogba, S. S. 2000. Mango (*Mangifera indica*) kernel: Chromatographic analysis of tannin and stability study of the associated polyphenol oxidase activity. J. Food Comp. Anal. 13: 149–156.

Arvanitoyannis, I. S. and T. H. Varzakas. 2008. Fruit/fruit juice waste management: Treatment methods and potential uses of treated waste. pp. 569–628. *In*: L. S. Arvanitoyannis and T. H. Varzakas (eds.). Waste Management for the Food Industries. Academic Press, Amsterdam.

Ayala-Zavala, J. F., V. Vega-Vega, C. Rosas-Domínguez, H. Palafox-Carlos, J. A. Villa-Rodriguez, M. W. Siddiqui, J. E. Dávila-Aviña and G. A. González-Aguilar. 2011. Agro-industrial potential of exotic fruit by-products as a source of food additives. Food Res. Int. 44: 1866–1874.

Bagherian, H., F. Z. Ashtiani, A. Fouladitajar and M. Mohtashamy. 2011. Comparisons between conventional, microwave- and ultrasound-assisted methods for extraction of pectin from grapefruit. Chem. Eng. Process 50: 1237–1243.

Bai, X. -L., T. -L. Yue, Y. -H. Yuan and H. -W. Zhang. 2010. Optimization of microwave-assisted extraction of polyphenols from apple pomace using response surface methodology and HPLC analysis. J. Sep. Sci. 33: 3751–3758.

Balasundram, N., K. Sundram and S. Samman. 2006. Phenolic compounds in plants and agri-industrial by-products: Antioxidant activity, occurrence, and potential uses. Food Chem. 99: 191–203.

Barba, F. J., S. Brianceau, M. Turk, N. Boussetta and E. Vorobiev. 2015. Effect of alternative physical treatments (ultrasounds, pulsed electric fields, and high-voltage electrical discharges) on selective recovery of bio-compounds from fermented grape pomace. Food Bioprocess Technol. 8: 1139–1148.

Barreto, J. C., M. T. S. Trevisan, W. E. Hull, G. Erben, E. S. de Brito, B. Pfundstein, G. Würtele, B. Spiegelhalder and R. W. Owen. 2008. Characterization and quantitation of polyphenolic compounds in bark, kernel, leaves and peel of mango (*Mangifera indica* L.). J. Agric. Food Chem. 56: 5599–5610.

Bélafi-Bakó, K., P. Cserjési, S. Beszédes, Z. Csanádi and C. Hodúr. 2012. Berry pectins: Microwave-assisted extraction and rheological properties. Food Bioprocess Technol. 5: 1100–1105.

Belitz, H. and W. Grosch. 1999. Food Chemistry. Springer, Berlin.

Berardini, N., M. Knödler, A. Schieber and R. Carle. 2005. Utilization of mango peels as a source of pectin and polyphenolics. Innov. Food Sci. Emerg. Technol. 6: 442–452.

Berardini, N., R. Fezer, J. Conrad, U. Beifuss, R. Carle and A. Schieber. 2005. Screening of mango (*Mangifera indica* L.) cultivars for their contents of flavonol O- and xanthone C-glycosides, anthocyanins, and pectin. J. Agric. Food Chem. 53: 1563–1570.

Bhushan, S., K. Kalia, M. Sharma, B. Singh and P. S. Ahuja. 2008. Processing of apple pomace for bioactive molecules. Crit. Rev. Biotechnol. 28: 285–296.

Bleve, M., L. Ciurlia, E. Erroi, G. Lionetto, L. Longo, L. Rescio, T. Schettino and G. Vasapollo. 2008. An innovative method for the purification of anthocyanins from grape skin extracts by using liquid and sub-critical carbon dioxide. Sep. Purif. Technol. 64: 192–197.

Bocco, A., M. -E. Cuvelier, H. Richard and C. Berset. 1998. Antioxidant activity and phenolic composition of citrus peel and seed extracts. J. Agric. Food Chem. 46: 2123–2129.

Boussetta, N., J. -L. Lanoisellé, C. Bedel-Cloutour and E. Vorobiev. 2009. Extraction of soluble matter from grape pomace by high voltage electrical discharges for polyphenol recovery: Effect of sulphur dioxide and thermal treatments. J. Food Eng. 95: 192–198.

Boussetta, N., N. Lebovka, E. Vorobiev, H. Adenier, C. Bedel-Cloutour and J. -L. Lanoisellé. 2009. Electrically assisted extraction of soluble matter from chardonnay grape skins for polyphenol recovery. J. Agric. Food Chem. 57: 1491–1497.

Boussetta, N., E. Vorobiev, V. Deloison, F. Pochez, A. Falcimaigne-Cordin and J. L. Lanoisellé. 2011. Valorisation of grape pomace by the extraction of phenolic antioxidants: Application of high voltage electrical discharges. Food Chem. 128: 364–370.

Boussetta, N., E. Vorobiev, L. H. Le, A. Cordin-Falcimaigne and J. -L. Lanoisellé. 2012. Application of electrical treatments in alcoholic solvent for polyphenols extraction from grape seeds. LWT—Food Sci. Technol. 46: 127–134.

Boussetta, N. and E. Vorobiev. 2014. Extraction of valuable biocompounds assisted by high voltage electrical discharges: A review. Cr. Chim. 17: 197–203.

Bozan, B., G. Tosun and D. Özcan. 2008. Study of polyphenol content in the seeds of red grape (*Vitis vinifera* L.) varieties cultivated in Turkey and their antiradical activity. Food Chem. 109: 426–430.

Bracke, M., B. Vyncke, G. Opdenakker, J. -M. Foidart, G. De Pestel and M. Mareel. 1991. Effect of catechins and citrus flavonoids on invasion *in vitro*. Clin. Exp. Metastasis 9: 13–25.

Buttriss, J. L. and C. S. Stokes. 2008. Dietary fiber and health: An overview. Nutr. Bull. 33: 186–200.

Buzby, J. C. and J. Hyman. 2012. Total and per capita value of food loss in the United States. Food Policy 37: 561–570.

Caillet, S., J. Côté, J. -F. Sylvain and M. Lacroix. 2012. Antimicrobial effects of fractions from cranberry products on the growth of seven pathogenic bacteria. Food Control 23: 419–428.

Çam, M. and K. Aaby. 2010. Optimization of extraction of apple pomace phenolics with water by response surface methodology. J. Agric. Food Chem. 58: 9103–9111.

Çam, M. and Y. Hışıl. 2010. Pressurised water extraction of polyphenols from pomegranate peels. Food Chem. 123: 878–885.

Campos-Vega, R. and B. D. Oomah. 2013. Chemistry and classification of phytochemicals. pp. 5–48. *In*: B. K. Tiwari, N. P. Bruton and C. S. Brennan (eds.). Handbook of Plant Food Phytochemicals: Sources, Stability and Extraction. Wiley-Blackwell, West Sussex, UK.

Candrawinata, V. I., J. B. Golding, P. D. Roach and C. E. Stathopoulos. 2014. Total phenolic content and antioxidant activity of apple pomace aqueous extract: Effect of time, temperature and water to pomace ratio. Int. Food Res. J. 21: 2337–2344.

Casazza, A. A., B. Aliakbarian, S. Mantegna, G. Cravotto and P. Perego. 2010. Extraction of phenolics from *Vitis vinifera* wastes using non-conventional techniques. J. Food Eng. 100: 50–55.

Casquete, R., S. M. Castro, A. Martín, S. Ruiz-Moyano, J. A. Saraiva, M. G. Córdoba and P. Teixeira. 2015. Evaluation of the effect of high pressure on total phenolic content, antioxidant and antimicrobial activity of citrus peels. Innov. Food Sci. Emerg. Technol. 31: 37–44.

Cengiz, E. and N. Gokoglu. 2005. Changes in energy and cholesterol contents of frankfurter-type sausages with fat reduction and fat replacer addition. Food Chem. 91: 443–447.

Cerdá, B., R. Llorach, J. J. Cerón, J. C. Espín and F. A. Tomás-Barberán. 2003. Evaluation of the bioavailability and metabolism in the rat of punicalagin, an antioxidant polyphenol from pomegranate juice. Eur. J. Nutr. 42: 18–28.

Chang, S., C. Tan, E. N. Frankel and D. M. Barrett. 2000. Low-density lipoprotein antioxidant activity of phenolic compounds and polyphenol oxidase activity in selected clingstone peach cultivars. J. Agric. Food Chem. 48: 147–151.

Chareonthaikij, P., T. Uan-On and W. Prinyawiwatkul. 2016. Effects of pineapple pomace fiber on physicochemical properties of composite flour and dough, and consumer acceptance of fiber-enriched wheat bread. Int. J. Food Sci. Tech. 51: 1120–1129.

Chau, C. F. and Y. L. Huang. 2004. Characterization of passion fruit seed fibers—a potential fiber source. Food Chem. 85: 189–194.

Chen, X. and R. Li. 2013. Decolouring method of apple dietary fibers. State Intellectual Property Office of the People's Republic of China # CN103229993A.

Croteau, R., T. M. Kutchan and N. G. Lewis. 2000. Natural products (secondary metabolites). pp. 1250–1268. *In*: B. Buchanan, W. Gruissem and R. Jones (eds.). Biochemistry and Molecular Biology of Plants. American Society of Plant Physiologists, Rockville.

Da Porto, C., E. Porretto and D. A. Decorti. 2013. Comparison of ultrasound-assisted extraction with conventional extraction methods of oil and polyphenols from grape (*Vitis vinifera* L.) seeds. Ultrason. Sonochem. 20: 1076–1080.

da Silva, R. P. F. F., T. A. P. Rocha-Santos and A. C. Duarte. 2016. Supercritical fluid extraction of bioactive compounds. Trends Analyt. Chem. 76: 40–51.

Dahmoune, F., L. Boulekbache, K. Moussi, O. Aoun, G. Spigno and K. Madani. 2013. Valorization of Citrus limon residues for the recovery of antioxidants: Evaluation and optimization of microwave and ultrasound application to solvent extraction. Ind. Crops Prod. 50: 77–87.

Dalimov, D. N., G. N. Dalimova and M. Bhatt. 2003. Chemical composition and lignins of tomato and pomegranate seeds. Chem. Nat. Compd. 39: 37–40.

Dang, Y. -Y., H. Zhang and Z. -L. Xiu. 2014. Microwave-assisted aqueous two-phase extraction of phenolics from grape (*Vitis vinifera*) seed. J. Chem. Technol. Biot. 89: 1576–1581.

de Torres, C., M. C. Díaz-Maroto, I. Hermosín-Gutiérrez and M. S. Pérez-Coello. 2010. Effect of freeze-drying and oven-drying on volatiles and phenolics composition of grape skin. Anal. Chim. Acta. 660: 177–182.

Dhillon, G. S., S. Kaur and S. K. Brar. 2013. Perspective of apple processing wastes as low-cost substrates for bioproduction of high value products: A review. Renew. Sust. Energ. Rev. 27: 789–805.

Elleuch, M., D. Bedigian, O. Roiseux, S. Besbes, C. Blecker and H. Attia. 2011. Dietary fiber and fiber-rich by-products of food processing: Characterisation, technological functionality and commercial applications: A review. Food Chem. 124: 411–421.

Englyst, H. N., M. E. Quigley and G. J. Hudson. 1994. Determination of dietary fiber as non-starch polysaccharides with gas–liquid chromatographic, high-performance liquid chromatographic or spectrophotometric measurement of constituent sugars. Analyst 119: 1497–1509.

Esparza-Martínez, F. J., R. Miranda-López and S. H. Guzman-Maldonado. 2016. Effect of air-drying temperature on extractable and non-extractable phenolics and antioxidant capacity of lime wastes. Ind. Crops Prod. 84: 1–6.

[European Union] Commission Regulation (EEC) 442. 1975. Waste. Official Journal of European Community L194: 39–41.

[European Union] Commission Regulation (EEC) 689. 1991. Hazardous waste. Official Journal of European Community L377: 20–27.

[FAOSTAT] FAO Statistic Database. 2016. Food and Agriculture Organization of the United Nations. Retrieved 15 March 2016, from http://faostat3.fao.org/home/E.

Faravash, R. S. and F. Z. Ashtiani. 2007. The effect of pH, ethanol volume and acid washing time on the yield of pectin extraction from peach pomace. Int. J. Food Sci. Tech. 42: 1177–1187.

Fattouch, S., P. Caboni, V. Coroneo, C. Tuberoso, A. Angioni, S. Dessi, N. Marzouki and P. Cabras. 2008. Comparative analysis of polyphenolic profiles and antioxidant and antimicrobial activities of Tunisian pome fruit pulp and peel aqueous acetone extracts. J. Agric. Food Chem. 56: 1084–1090.

Fawole, O. A., U. L. Opara and K. I. Theron. 2012. Chemical and phytochemical properties and antioxidant activities of three pomegranate cultivars grown in South Africa. Food Bioprocess Tech. 5: 2934–2940.

Felgines, C., S. Talavéra, M. -P. Gonthier, O. Texier, A. Scalbert, J. -L. Lamaison and C. Rémésy. 2003. Strawberry anthocyanins are recovered in urine as glucuro-and sulfoconjugates in humans. J. Nutr. 133: 1296–1301.

Figuerola, F., M. L. Hurtado, A. M. Estévez, I. Chiffelle and F. Asenjo. 2005. Fiber concentrates from apple pomace and citrus peel as potential fiber sources for food enrichment. Food Chem. 91: 395–401.

Fischer, U. A., R. Carle and D. R. Kammerer. 2011. Identification and quantification of phenolic compounds from pomegranate (*Punica granatum* L.) peel, mesocarp, aril and differently produced juices by HPLC-DAD–ESI/MSn. Food Chem. 127: 807–821.

Fishman, M. L., H. K. Chau, P. Hoagland and K. Ayyad. 1999. Characterization of pectin, flash-extracted from orange albedo by microwave heating, under pressure. Carbohyd. Res. 323: 126–138.

Flamini, R. 2003. Mass spectrometry in grape and wine chemistry. Part I: Polyphenols. Mass Spectrom. Rev. 22: 218–250.

Foo, L. Y. and Y. Lu. 1999. Isolation and identification of procyanidins in apple pomace. Food Chem. 64: 511–518.

Fromm, M., S. Bayha, R. Carle and D. R. Kammerer. 2012. Characterization and quantitation of low and high molecular weight phenolic compounds in apple seeds. J. Agric. Food Chem. 60: 1232–1242.

Fu, C., A. E. K. Loo, F. P. P. Chia and D. Huang. 2007. Oligomeric proanthocyanidins from mangosteen pericarps. J. Agric. Food Chem. 55: 7689–7694.

Galanakis, C. M. 2012. Recovery of high added-value components from food wastes: Conventional, emerging technologies and commercialized applications. Trends Food Sci. Tech. 26: 68–87.

Galanakis, C. M. 2013. Emerging technologies for the production of nutraceuticals from agricultural by-products: A viewpoint of opportunities and challenges. Food Bioprod. Process 91: 575–579.

Garau, M. C., S. Simal, C. Rosselló and A. Femenia. 2007. Effect of air-drying temperature on physico-chemical properties of dietary fiber and antioxidant capacity of orange (*Citrus aurantium* v. *Canoneta*) by-products. Food Chem. 104: 1014–1024.

Garcia-Mendoza, M. P., J. T. Paula, L. C. Paviani, F. A. Cabral and H. A. Martinez-Correa. 2015. Extracts from mango peel by-product obtained by supercritical CO_2 and pressurized solvent processes. LWT—Food Sci. Technol. 62: 131–137.

Gnanavinthan, A. 2013. Introduction to the major classes of bioactives present in fruit. pp. 1–18. *In*: M. Skinner and D. Hunter (eds.). Bioactives in Fruit: Health Benefits and Functional Foods. John Wiley & Sons, West Sussex.

González-Molina, E., R. Domínguez-Perles, D. A. Moreno and C. García-Viguera. 2010. Natural bioactive compounds of *Citrus limon* for food and health. J. Pharmaceut. Biomed. 51: 327–345.

Gorinstein, S., O. Martín-Belloso, Y. -S. Park, R. Haruenkit, A. Lojek, M. Číž, A. Caspi, I. Libman and S. Trakhtenberg. 2001. Comparison of some biochemical characteristics of different citrus fruits. Food Chem. 74: 309–315.

Górnaś, P. 2015. Unique variability of tocopherol composition in various seed oils recovered from by-products of apple industry: Rapid and simple determination of all four homologues (α, β, γ and δ) by RP-HPLC/FLD. Food Chem. 172: 129–134.

Grigelmo-Miguel, N., S. Gorinstein and O. Martín-Belloso. 1999. Characterisation of peach dietary fiber concentrate as a food ingredient. Food Chem. 65: 175–181.

Grigelmo-Miguel, N. and O. Martín-Belloso. 1998. Characterization of dietary fiber from orange juice extraction. Food Res. Int. 31: 355–361.

Grigelmo-Miguel, N. and O. Martín-Belloso. 1999. Comparison of dietary fiber from by-products of processing fruits and greens and from cereals. LWT—Food Sci. Technol. 32: 503–508.

Grigoras, C. G., E. Destandau, L. Fougère and C. Elfakir. 2013. Evaluation of apple pomace extracts as a source of bioactive compounds. Ind. Crops Prod. 49: 794–804.

Guendez, R., S. Kallithraka, D. P. Makris and P. Kefalas. 2005. Determination of low molecular weight polyphenolic constituents in grape (*Vitis vinifera* sp.) seed extracts: Correlation with antiradical activity. Food Chem. 89: 1–9.

Gullón, B., E. Falqué, J. L. Alonso and J. C. Parajó. 2007. Evaluation of apple pomace as a raw material for alternative applications in food industries. Food Technol. Biotech. 45: 426–433.

Gustavsson, J., C. Cederberg, U. Sonesson, R. Van Otterdijk and A. Meybeck. 2011. Global Food Losses and Food Waste. Food and Agriculture Organization of the United Nations, Rome, 29 pp.

Hasnaoui, N., B. Wathelet and A. Jiménez-Araujo. 2014. Valorization of pomegranate peel from 12 cultivars: Dietary fiber composition, antioxidant capacity and functional properties. Food Chem. 160: 196–203.

Hawkins, D. J. and J. C. Kridl. 1998. Characterization of acyl-ACP thioesterases of mangosteen (*Garcinia mangostana*) seed and high levels of stearate production in transgenic canola. Plant J. 13: 743–752.

Hawthorne, S. B. 1990. Analytical-scale supercritical fluid extraction. Anal. Chem. 62: 633A–642A.

Hayat, K., X. Zhang, U. Farooq, S. Abbas, S. Xia, C. Jia, F. Zhong and J. Zhang. 2010. Effect of microwave treatment on phenolic content and antioxidant activity of citrus mandarin pomace. Food Chem. 123: 423–429.

He, L., X. Zhang, H. Xu, C. Xu, F. Yuan, Ž. Knez, Z. Novak and Y. Gao. 2012. Subcritical water extraction of phenolic compounds from pomegranate (*Punica granatum* L.) seed residues and investigation into their antioxidant activities with HPLC–ABTS+ assay. Food Bioprod. Process. 90: 215–223.

Herrero, M., M. Castro-Puyana, J. A. Mendiola and E. Ibañez. 2013. Compressed fluids for the extraction of bioactive compounds. Trends Analyt. Chem. 43: 67–83.

Hollman, P. C. H. and M. B. Katan. 1999. Dietary flavonoids: Intake, health effects and bioavailability. Food Chem. Toxicol. 37: 937–942.

Holtung, L., S. Grimmer and K. Aaby. 2011. Effect of processing of black currant press-residue on polyphenol composition and cell proliferation. J. Agric. Food Chem. 59: 3632–3640.

Horváth-Kerkai, E. and M. Stéger-Máté. 2013. Manufacturing fruit beverages and concentrates. pp. 213–228. *In*: N. K. Sinha, J. S. Sidhn, J. Barta, J. S. B. WM and M. P. Cano (eds.). Handbook of Fruits and Fruit Processing, 2nd Edition. John Wiley & Sons, West Sussex.

Huber, G. M. and H. P. V. Rupasinghe. 2009. Phenolic profiles and antioxidant properties of apple skin extracts. J. Food Sci. 74: C693–C700.

Ignat, I., I. Volf and V. I. Popa. 2011. A critical review of methods for characterisation of polyphenolic compounds in fruits and vegetables. Food Chem. 126: 1821–1835.

Inoue, T., S. Tsubaki, K. Ogawa, K. Onishi and J. Azuma. 2010. Isolation of hesperidin from peels of thinned Citrus unshiu fruits by microwave-assisted extraction. Food Chem. 123: 542–547.

Ismail, T., P. Sestili and S. Akhtar. 2012. Pomegranate peel and fruit extracts: A review of potential anti-inflammatory and anti-infective effects. J. Ethnopharmacol. 143: 397–405.

Janjarasskul, T. and J. M. Krochta. 2010. Edible packaging materials. Annu. Rev. Food Sci. Technol. 1: 415–448.

Ji, X., B. Avula and I. A. Khan. 2007. Quantitative and qualitative determination of six xanthones in *Garcinia mangostana* L. by LC–PDA and LC–ESI-MS. J. Pharmaceut. Biomed. 43: 1270–1276.

Johanningsmeier, S. D. and G. K. Harris. 2011. Pomegranate as a functional food and nutraceutical source. Annu. Rev. Food Sci. Technol. 2: 181–201.

Johnson, I. T. 2013. Phytochemicals and health. pp. 49–67. *In*: B. K. Tiwari, N. P. Brunton and C. S. Brennan (eds.). Handbook of Plant Food Phytochemicals: Sources, Stability and Extraction. Wiley-Blackwell, West Sussex, UK.

Ju, Z. Y. and L. R. Howard. 2003. Effects of solvent and temperature on pressurized liquid extraction of anthocyanins and total phenolics from dried red grape skin. J. Agric. Food Chem. 51: 5207–5213.

Jung, H. -A., B. -N. Su, W. J. Keller, R. G. Mehta and A. D. Kinghorn. 2006. Antioxidant xanthones from the Pericarp of *Garcinia mangostana* (Mangosteen). J. Agric. Food Chem. 54: 2077–2082.

Kabuki, T., H. Nakajima, M. Arai, S. Ueda, Y. Kuwabara and S. Dosako. 2000. Characterization of novel antimicrobial compounds from mango (*Mangifera indica* L.) kernel seeds. Food Chem. 71: 61–66.

Kammerer, D., A. Claus, A. Schieber and R. Carle. 2005. A novel process for the recovery of polyphenols from grape (*Vitis vinifera* L.) pomace. J. Food Sci. 70: C157–C163.

Kanatt, S. R., R. Chander and A. Sharma. 2010. Antioxidant and antimicrobial activity of pomegranate peel extract improves the shelf life of chicken products. Int. J. Food Sci. Tech. 45: 216–222.

Kapasakalidis, P. G., R. A. Rastall and M. H. Gordon. 2006. Extraction of polyphenols from processed black currant (*Ribes nigrum* L.) residues. J. Agric. Food Chem. 54: 4016–4021.

Karacabey, E. and G. Mazza. 2010. Optimisation of antioxidant activity of grape cane extracts using response surface methodology. Food Chem. 119: 343–348.

Kashyap, D. R., P. K. Vohra, S. Chopra and R. Tewari. 2001. Applications of pectinases in the commercial sector: a review. Bioresource Technol. 77: 215–227.

Kayacier, A., F. Yüksel and S. Karaman. 2014. Response surface methodology study for optimization of effects of fiber level, frying temperature, and frying time on some physicochemical, textural, and sensory properties of wheat chips enriched with apple fiber. Food Bioprocess Tech. 7: 133–147.

Khan, M. K., M. Abert-Vian, A. -S. Fabiano-Tixier, O. Dangles and F. Chemat. 2010. Ultrasound-assisted extraction of polyphenols (flavanone glycosides) from orange (*Citrus sinensis* L.) peel. Food Chem. 119: 851–858.

Kliemann, E., K. N. De Simas, E. R. Amante, E. S. Prudêncio, R. F. Teófilo, M. M. C. Ferreira and R. D. M. C. Amboni. 2009. Optimisation of pectin acid extraction from passion fruit peel (*Passiflora edulis* flavicarpa) using response surface methodology. Int. J. Food Sci. Tech. 44: 476–483.

Köhler, N., V. Wray and P. Winterhalter. 2008. Preparative isolation of procyanidins from grape seed extracts by high-speed counter-current chromatography. J. Chromatogr. A 1177: 114–125.

Kołodziejczyk, K., M. Sójka, M. Abadias, I. Viñas, S. Guyot and A. Baron. 2013. Polyphenol composition, antioxidant capacity, and antimicrobial activity of the extracts obtained from industrial sour cherry pomace. Ind. Crops Prod. 51: 279–288.

Kosseva, M. R. 2011. Management and processing of food wastes. pp. 557–593. *In*: M. -Y. Murray (ed.). Comprehensive Biotechnology, Volume 6 Environmental Biotechnology and Safety (2nd Edition). Academic Press, Burlington.

Kosseva, M. R. 2013. Sources, characterization, and composition of food industry wastes. pp. 37–60. *In*: M. R. Kosseva and C. Webb (eds.). Food Industry Wastes. Academic Press, San Diego.

Kulkarni, S. G. and P. Vijayanand. 2010. Effect of extraction conditions on the quality characteristics of pectin from passion fruit peel (*Passiflora edulis* f. *flavicarpa* L.). LWT—Food Sci. Technol. 43: 1026–1031.

Lamanauskas, N., G. Pataro, Č. Bobinas, S. Šatkauskas, P. Viskelis, R. Bobinaitė and G. Ferrari. 2016. Impact of pulsed electric field treatment on juice yield and recovery of bioactive compounds from raspberries and their by-products. Žemdirbystė 103: 83–90.

Lansky, E. P. and R. A. Newman. 2007. *Punica granatum* (pomegranate) and its potential for prevention and treatment of inflammation and cancer. J. Ethnopharmacol. 109: 177–206.

Lario, Y., E. Sendra, J. García-Pérez, C. Fuentes, E. Sayas-Barberá, J. Fernández-López and J. A. Pérez-Alvarez. 2004. Preparation of high dietary fiber powder from lemon juice by-products. Innov. Food Sci. Emerg. Technol. 5: 113–117.

Larrauri, J. A. 1999. New approaches in the preparation of high dietary fiber powders from fruit by-products. Trends Food Sci. Tech. 10: 3–8.

Larrauri, J. A., P. Rupérez, B. Borroto and F. Saura-Calixto. 1996. Mango peels as a new tropical fiber: Preparation and characterization. LWT—Food Sci. Technol. 29: 729–733.

Larrauri, J. A., P. Rupérez, L. Bravo and F. Saura-Calixto. 1996. High dietary fiber powders from orange and lime peels: Associated polyphenols and antioxidant capacity. Food Res. Int. 29: 757–762.

Larrosa, M., C. Luceri, E. Vivoli, C. Pagliuca, M. Lodovici, G. Moneti and P. Dolara. 2009. Polyphenol metabolites from colonic microbiota exert anti-inflammatory activity on different inflammation models. Mol. Nutr. Food Res. 53: 1044–1054.

Laufenberg, G., B. Kunz and M. Nystroem. 2003. Transformation of vegetable waste into value added products: (A) the upgrading concept; (B) practical implementations. Bioresource Technol. 87: 167–198.

Laurentin, A., D. Morrison and C. Edwards. 2003. Dietary fiber in health and disease. Nutr. Bull. 28: 69–72.

Lavelli, V. and S. Corti. 2011. Phloridzin and other phytochemicals in apple pomace: Stability evaluation upon dehydration and storage of dried product. Food Chem. 129: 1578–1583.

[Leatherhead Food Research] Future Directions for the Global Functional Foods Market: 2014 Market Report. 2014. Leatherhead Food Research, Surrey, 190 pp.

Lee, S. C., L. Prosky and J. W. De Vries. 1992. Determination of total, soluble, and insoluble dietary fiber in foods: Enzymatic-gravimetric method, MES-TRIS buffer: Collaborative study. J. AOAC 75: 395–416.

Li, B. B., B. Smith and M. M. Hossain. 2006. Extraction of phenolics from citrus peels: I. Solvent extraction method. Sep. Purif. Technol. 48: 182–188.

Li, X., X. He, Y. Lv and Q. He. 2014. Extraction and functional properties of water-soluble dietary fiber from apple pomace. J. Food Process Eng. 37: 293–298.

Li, Y., C. Guo, J. Yang, J. Wei, J. Xu and S. Cheng. 2006. Evaluation of antioxidant properties of pomegranate peel extract in comparison with pomegranate pulp extract. Food Chem. 96: 254–260.

Liazid, A., R. F. Guerrero, E. Cantos, M. Palma and C. G. Barroso. 2011. Microwave-assisted extraction of anthocyanins from grape skins. Food Chem. 124: 1238–1243.

Liu, D., E. Vorobiev, R. Savoire and J. -L. Lanoisellé. 2011. Intensification of polyphenols extraction from grape seeds by high voltage electrical discharges and extract concentration by dead-end ultrafiltration. Sep. Purif. Technol. 81: 134–140.

Llobera, A. and J. Cañellas. 2007. Dietary fiber content and antioxidant activity of Manto Negro red grape (*Vitis vinifera*): pomace and stem. Food Chem. 101: 659–666.

Londoño-Londoño, J., V. R. de Lima, O. Lara, A. Gil, T. B. C. Pasa, G. J. Arango and J. R. R. Pineda. 2010. Clean recovery of antioxidant flavonoids from citrus peel: Optimizing an aqueous ultrasound-assisted extraction method. Food Chem. 119: 81–87.

López, F. 2014. New trends in fruit juices. pp. 27–40. *In*: V. Falguera and A. Ibarz (eds.). Juice Processing: Quality, Safety and Value-added Opportunities. CRC Press, Boca Raton.

Lopez-Rubio, A., R. Gavara and J. M. Lagaron. 2006. Bioactive packaging: Turning foods into healthier foods through biomaterials. Trends Food Sci. Tech. 17: 567–575.

López-Vargas, J. H., J. Fernández-López, J. A. Pérez-Álvarez and M. Viuda-Martos. 2013. Chemical, physico-chemical, technological, antibacterial and antioxidant properties of dietary fiber powder obtained from yellow passion fruit (*Passiflora edulis* var. *flavicarpa*) co-products. Food Res. Int. 51: 756–763.

Lu, Y. and L. Y. Foo. 1997. Identification and quantification of major polyphenols in apple pomace. Food Chem. 59: 187–194.

Lu, Y. and L. Y. Foo. 1998. Constitution of some chemical components of apple seed. Food Chem. 61: 29–33.

Luengo, E., I. Álvarez and J. Raso. 2013. Improving the pressing extraction of polyphenols of orange peel by pulsed electric fields. Innov. Food Sci. Emerg. Technol. 17: 79–84.

Maisuthisakul, P. and M. H. Gordon. 2009. Antioxidant and tyrosinase inhibitory activity of mango seed kernel by product. Food Chem. 117: 332–341.

Malacrida, C. R. and N. Jorge. 2012. Yellow passion fruit seed oil (*Passiflora edulis* f. *flavicarpa*): Physical and chemical characteristics. Braz. Arch. Biol. Technol. 55: 127–134.

Mamma, D. and P. Christakopoulos. 2014. Biotransformation of citrus by-products into value added products. Waste & Biomass Valorization 5: 529.

Manas, E., L. Bravo and F. Saura-Calixto. 1994. Sources of error in dietary fiber analysis. Food Chem. 50: 331–342.

Mandalari, G., R. N. Bennett, G. Bisignano, D. Trombetta, A. Saija, C. B. Faulds, M. J. Gasson and A. Narbad. 2007. Antimicrobial activity of flavonoids extracted from bergamot (*Citrus bergamia* Risso) peel, a byproduct of the essential oil industry. J. Appl. Microbiol. 103: 2056–2064.

Manthey, J. A. and K. Grohmann. 1996. Concentrations of hesperidin and other orange peel flavonoids in citrus processing byproducts. J. Agric. Food Chem. 44: 811–814.

Maran, J. P., V. Sivakumar, K. Thirugnanasambandham and R. Sridhar. 2013. Optimization of microwave assisted extraction of pectin from orange peel. Carbohyd. Polym. 97: 703–709.

Marín, F. R., C. Soler-Rivas, O. Benavente-García, J. Castillo and J. A. Pérez-Alvarez. 2007. By-products from different citrus processes as a source of customized functional fibers. Food Chem. 100: 736–741.

Martí, N., D. Saura, E. Fuentes, V. Lizama, E. García, M. J. Mico-Ballester and J. Lorente. 2011. Fiber from tangerine juice industry. Ind. Crops Prod. 33: 94–98.

Martí, N., J. Lorente, M. Valero, A. Ibarz and D. Saura. 2014. Recovery and use of by-products from fruit juice production. pp. 41–74. *In*: V. Falguera and A. Ibarz (eds.). Juice Processing: Quality, Safety, and Value-Added Opportunities. CRC Press, Boca Raton.

Martinez-Avila, G. C. G., A. F. Aguilera, S. Saucedo, R. Rojas, R. Rodriguez and C. N. Aguilar. 2014. Fruit wastes fermentation for phenolic antioxidants production and their application in manufacture of edible coatings and films. Crit. Rev. Food Sci. Nutr. 54: 303–311.

Martínez, R., P. Torres, M. A. Meneses, J. G. Figueroa, J. A. Pérez-Álvarez and M. Viuda-Martos. 2012. Chemical, technological and *in vitro* antioxidant properties of mango, guava, pineapple and passion fruit dietary fiber concentrate. Food Chem. 135: 1520–1526.

McLellan, M. R. and O. I. Padilla-Zakour. 2004. Juice processing. pp. 73–96. *In*: D. M. Barrett, L. Somogyi and H. Ramaswamy (eds.). Processing Fruits: Science and Technology. CRC Press, Boca Raton.

Medina-Meza, I. G. and G. V. Barbosa-Cánovas. 2015. Assisted extraction of bioactive compounds from plum and grape peels by ultrasonics and pulsed electric fields. J. Food Eng. 166: 268–275.

Min, B., I. Y. Bae, H. G. Lee, S. -H. Yoo and S. Lee. 2010. Utilization of pectin-enriched materials from apple pomace as a fat replacer in a model food system. Bioresource Technol. 101: 5414–5418.

Mirabella, N., V. Castellani and S. Sala. 2014. Current options for the valorization of food manufacturing waste: A review. Journal of Cleaner Production 65: 28–41.

Molero Gómez, A., C. Pereyra López and E. Martinez de la Ossa. 1996. Recovery of grape seed oil by liquid and supercritical carbon dioxide extraction: a comparison with conventional solvent extraction. The Chemical Engineering Journal and the Biochemical Engineering Journal 61: 227–231.

Monier, V., M. Shailendra, V. Escalon, C. O'Connor, T. Gibon, G. Anderson, M. Hortense and H. Reisinger. 2010. Preparatory Study on Food Waste Across EU 27. European Commission (DG ENV) Directorate C-Industry. 2010. Final Report, Paris, 210 pp.

Moorthy, I. G., J. P. Maran, S. M. Surya, S. Naganyashree and C. S. Shivamathi. 2015. Response surface optimization of ultrasound assisted extraction of pectin from pomegranate peel. International Journal of Biological Macromolecules 72: 1323–1328.

Naczk, M. and F. Shahidi. 2004. Extraction and analysis of phenolics in food. J. Chromatogr. A 1054: 95–111.

Naczk, M. and F. Shahidi. 2006. Phenolics in cereals, fruits and vegetables: Occurrence, extraction and analysis. J. Pharm. Biomed. Anal. 41.

Nawirska, A. and M. Kwaśniewska. 2005. Dietary fiber fractions from fruit and vegetable processing waste. Food Chem. 91: 221–225.

O'Shea, N., E. Arendt and E. Gallagher. 2014. Enhancing an extruded puffed snack by optimising die head temperature, screw speed and apple pomace inclusion. Food Bioprocess Tech. 7: 1767–1782.

Ohta, H., C. H. Fong, M. Berhow and S. Hasegawa. 1993. Thin-layer and high-performance liquid chromatographic analyses of limonoids and limonoid glucosides in Citrus seeds. J. Chromatogr. A 639: 295–302.

Oreopoulou, V. and C. Tzia. 2007. Utilization of plant by-products for the recovery of proteins, dietary fibers, antioxidants, and colorants. pp. 209–232. *In*: V. Oreopoulou and W. Russ (eds.). Utilization of By-products and Treatment of Waste in the Food Industry. Springer, Boston.

O'Shea, N., E. K. Arendt and E. Gallagher. 2012. Dietary fiber and phytochemical characteristics of fruit and vegetable by-products and their recent applications as novel ingredients in food products. Innov. Food Sci. Emerg. Technol. 16: 1–10.

Ozaki, Y., C. H. Fong, Z. Herman, H. Maeda, M. Miyake, Y. Ifuku and S. Hasegawa. 1991. Limonoid glucosides in citrus seeds. Agricultural and Biological Chemistry 55: 137–141.

Paes, J., R. Dotta, G. F. Barbero and J. Martínez. 2014. Extraction of phenolic compounds and anthocyanins from blueberry (*Vaccinium myrtillus* L.) residues using supercritical CO$_2$ and pressurized liquids. J. Supercrit. Fluid 95: 8–16.

Pagan, J. and A. Ibarz. 1999. Extraction and rheological properties of pectin from fresh peach pomace. J. Food Eng. 39: 193–201.

Pagan, J., A. Ibarz, M. Llorca and L. Coll. 1999. Quality of industrial pectin extracted from peach pomace at different pH and temperatures. J. Sci. Food Agric. 79: 1038–1042.

Palakawong, C., P. Sophanodora, P. Toivonen and P. Delaquis. 2013. Optimized extraction and characterization of antimicrobial phenolic compounds from mangosteen (*Garcinia mangostana* L.) cultivation and processing waste. J. Sci. Food Agric. 93: 3792–3800.

Parniakov, O., F. J. Barba, N. Grimi, N. Lebovka and E. Vorobiev. 2016. Extraction assisted by pulsed electric energy as a potential tool for green and sustainable recovery of nutritionally valuable compounds from mango peels. Food Chem. 192: 842–848.

Pasquel, R., J. Luis, A. P. da Fonseca Machado, G. F. Barbero, C. A. Rezende and J. Martínez. 2014. Extraction of antioxidant compounds from blackberry (*Rubus* sp.) bagasse using supercritical CO_2 assisted by ultrasound. J. Supercrit. Fluid 94: 223–233.

Pazos, M., J. M. Gallardo, J. L. Torres and I. Medina. 2005. Activity of grape polyphenols as inhibitors of the oxidation of fish lipids and frozen fish muscle. Food Chem. 92: 547–557.

Pečivová, P., K. Juříková, I. Burešová, M. Černá and J. Hrabě. 2014. The effect of pectin from apple and arabic gum from acacia tree on quality of wheat flour dough. Acta Universitatis Agriculturae et Silviculturae Mendelianae Brunensis 59: 255–264.

Peerajit, P., N. Chiewchan and S. Devahastin. 2012. Effects of pretreatment methods on health-related functional properties of high dietary fiber powder from lime residues. Food Chem. 132: 1891–1898.

Pinelo, M., M. Rubilar, M. Jerez, J. Sineiro and M. J. Núñez. 2005. Effect of solvent, temperature, and solvent-to-solid ratio on the total phenolic content and antiradical activity of extracts from different components of grape pomace. J. Agric. Food Chem. 53: 2111–2117.

Ping, L., N. Brosse, P. Sannigrahi and A. Ragauskas. 2011. Evaluation of grape stalks as a bioresource. Ind. Crops Prod. 33: 200–204.

Pingret, D., A. -S. Fabiano-Tixier, C. Le Bourvellec, C. M. G. C. Renard and F. Chemat. 2012. Lab and pilot-scale ultrasound-assisted water extraction of polyphenols from apple pomace. J. Food Eng. 111: 73–81.

Polovka, M., L. Šťavíková, B. Hohnová, P. Karásek and M. Roth. 2010. Offline combination of pressurized fluid extraction and electron paramagnetic resonance spectroscopy for antioxidant activity of grape skin extracts assessment. J. Chromatogr. A 1217: 7990–8000.

Pourbafrani, M., G. Forgács, I. S. Horváth, C. Niklasson and M. J. Taherzadeh. 2010. Production of biofuels, limonene and pectin from citrus wastes. Bioresource Technol. 101: 4246–4250.

Prabasari, I., F. Pettolino, M. -L. Liao and A. Bacic. 2011. Pectic polysaccharides from mature orange (*Citrus sinensis*) fruit albedo cell walls: Sequential extraction and chemical characterization. Carbohyd. Polym. 84: 484–494.

Prado, J. M., I. Dalmolin, N. D. D. Carareto, R. C. Basso, A. J. A. Meirelles, J. V. Oliveira, E. A. C. Batista and M. A. A. Meireles. 2012. Supercritical fluid extraction of grape seed: Process scale-up, extract chemical composition and economic evaluation. J. Food Eng. 109: 249–257.

Prosky, L., N. -G. Asp, I. Furda, J. W. DeVries, T. F. Schweizer and B. F. Harland. 1984. Determination of total dietary fiber in foods and food products: Collaborative study. J. AOAC 68: 677–679.

Prozil, S. O., D. V. Evtuguin and L. P. Cruz Lopes. 2012. Chemical composition of grape stalks of *Vitis vinifera* L. from red grape pomaces. Ind. Crops Prod. 35: 178–184.

Puravankara, D., V. Boghra and R. S. Sharma. 2000. Effect of antioxidant principles isolated from mango (*Mangifera indica* L.) seed kernels on oxidative stability of buffalo ghee (butter-fat). J. Sci. Food Agric. 80: 522–526.

Rabak, F. 1921. Grape-seed oil. Journal of Industrial & Engineering Chemistry 13: 919–921.

Rayne, S., E. Karacabey and G. Mazza. 2008. Grape cane waste as a source of trans-resveratrol and trans-viniferin: High-value phytochemicals with medicinal and anti-phytopathogenic applications. Ind. Crops Prod. 27: 335–340.

Rehman, Z. U., A. M. Salariya, F. Habib and W. H. Shah. 2004. Utilization of mango peels as a source of pectin. J. Chem. Soc. Pak. 26: 73–76.

Reis, S. F., D. K. Rai and N. Abu-Ghannam. 2012. Water at room temperature as a solvent for the extraction of apple pomace phenolic compounds. Food Chem. 135: 1991–1998.

Reverchon, E. and I. De Marco. 2006. Supercritical fluid extraction and fractionation of natural matter. J. Supercrit. Fluid 38: 146–166.

Revilla, E., J. -M. Ryan and G. Martín-Ortega. 1998. Comparison of several procedures used for the extraction of anthocyanins from red grapes. J. Agric. Food Chem. 46: 4592–4597.

Reyes-De-Corcuera, J. I., R. M. Goodrich-Schneider, S. A. Barringer and M. A. Landeros-Urbina. 2014. Processing of fruit and vegetable beverages. pp. 339–362. *In*: S. Clark, S. Jung and B. Lamsal (eds.). Food Processing. John Wiley & Sons, West Sussex.

Rezaei, S., K. Rezaei, M. Haghighi and M. Labbafi. 2013. Solvent and solvent to sample ratio as main parameters in the microwave-assisted extraction of polyphenolic compounds from apple pomace. Food Sci. Biotechnol. 22: 1.

Rezzadori, K., S. Benedetti and E. R. Amante. 2012. Proposals for the residues recovery: Orange waste as raw material for new products. Food Bioprod. Process. 90: 606–614.

Richter, B. E., B. A. Jones, J. L. Ezzell, N. L. Porter, N. Avdalovic and C. Pohl. 1996. Accelerated solvent extraction: A technique for sample preparation. Anal. Chem. 68: 1033–1039.

Rockenbach, I. I., E. Rodrigues, L. V. Gonzaga, V. Caliari, M. I. Genovese, A. E. de Souza Schmidt Gonçalves and R. Fett. 2011. Phenolic compounds content and antioxidant activity in pomace from selected red grapes (*Vitis vinifera* L. and *Vitis labrusca* L.) widely produced in Brazil. Food Chem. 127: 174–179.

Rodríguez Montealegre, R., R. Romero Peces, J. L. Chacón Vozmediano, J. Martínez Gascueña and E. García Romero. 2006. Phenolic compounds in skins and seeds of ten grape *Vitis vinifera* varieties grown in a warm climate. J. Food Comp. Anal. 19: 687–693.

Rolin, C. 2002. Commercial pectin preparations. pp. 222–241. *In*: G. B. Seymour and J. P. Knox (eds.). Pectins and their Manipulation. Blackwell Publishing, Oxford.

Ruberto, G., A. Renda, C. Daquino, V. Amico, C. Spatafora, C. Tringali and N. De Tommasi. 2007. Polyphenol constituents and antioxidant activity of grape pomace extracts from five Sicilian red grape cultivars. Food Chem. 100: 203–210.

Saad, H., F. Charrier-El Bouhtoury, A. Pizzi, K. Rode, B. Charrier and N. Ayed. 2012. Characterization of pomegranate peels tannin extractives. Ind. Crops Prod. 40: 239–246.

Sah, B. N. P., T. Vasiljevic, S. McKechnie and O. N. Donkor. 2016. Physicochemical, textural and rheological properties of probiotic yogurt fortified with fiber-rich pineapple peel powder during refrigerated storage. LWT—Food Sci. Technol. 65: 978–986.

Sakakibara, H., Y. Honda, S. Nakagawa, H. Ashida and K. Kanazawa. 2003. Simultaneous determination of all polyphenols in vegetables, fruits, and teas. J. Agric. Food Chem. 51: 571–581.

Salgado, P. R., C. M. Ortiz, Y. S. Musso, L. Di Giorgio and A. N. Mauri. 2015. Edible films and coatings containing bioactives. Curr. Opin. Food Sci. 5: 86–92.

Sánchez-Alonso, I., A. Jiménez-Escrig, F. Saura-Calixto and A. J. Borderías. 2007. Effect of grape antioxidant dietary fiber on the prevention of lipid oxidation in minced fish: Evaluation by different methodologies. Food Chem. 101: 372–378.

Sánchez-Rabaneda, F., O. Jauregui, R. M. Lamuela-Raventós, F. Viladomat, J. Bastida and C. Codina. 2004. Qualitative analysis of phenolic compounds in apple pomace using liquid chromatography coupled to mass spectrometry in tandem mode. Rapid Commun. Mass Sp. 18: 553–563.

Saura-Calixto, F. 2010. Dietary fiber as a carrier of dietary antioxidants: An essential physiological function. J. Agric. Food Chem. 59: 43–49.

Sáyago-Ayerdi, S. G., A. Brenes and I. Goñi. 2009. Effect of grape antioxidant dietary fiber on the lipid oxidation of raw and cooked chicken hamburgers. LWT—Food Sci. Technol. 42: 971–976.

Scalbert, A. and G. Williamson. 2000. Dietary intake and bioavailability of polyphenols. J. Nutr. 130: 2073S–2085S.

Schieber, A., F. C. Stintzing and R. Carle. 2001. By-products of plant food processing as a source of functional compounds—recent developments. Trends Food Sci. Tech. 12: 401–413.

Schieber, A., P. Hilt, P. Streker, H. -U. Endreß, C. Rentschler and R. Carle. 2003. A new process for the combined recovery of pectin and phenolic compounds from apple pomace. Innov. Food Sci. Emerg. Technol. 4: 99–107.

Seixas, F. L., D. L. Fukuda, F. R. B. Turbiani, P. S. Garcia, L. de O Carmen, S. Jagadevan and M. L. Gimenes. 2014. Extraction of pectin from passion fruit peel (*Passiflora edulis* f. *flavicarpa*) by microwave-induced heating. Food Hydrocolloid. 38: 186–192.

Sekhon-Loodu, S., S. N. Warnakulasuriya, H. P. V. Rupasinghe and F. Shahidi. 2013. Antioxidant ability of fractionated apple peel phenolics to inhibit fish oil oxidation. Food Chem. 140: 189–196.

Shalini, R. and D. K. Gupta. 2010. Utilization of pomace from apple processing industries: A review. J. Food Sci. Tech. 47: 365–371.

Shi, Y. -Q., T. Fukai, H. Sakagami, W. -J. Chang, P. -Q. Yang, F. -P. Wang and T. Nomura. 2001. Cytotoxic flavonoids with isoprenoid groups from *Morus mongolica* 1. J. Nat. Prod. 64: 181–188.

Silva, I. M. D. A., L. V. Gonzaga, E. R. Amante, R. F. Teófilo, M. M. C. Ferreira and R. D. M. C. Amboni. 2008. Optimization of extraction of high-ester pectin from passion fruit peel (*Passiflora edulis* flavicarpa) with citric acid by using response surface methodology. Bioresource Technol. 99: 5561–5566.

Sivam, A. S., D. Sun-Waterhouse, G. I. N. Waterhouse, S. Y. Quek and C. O. Perera. 2011. Physicochemical properties of bread dough and finished bread with added pectin fiber and phenolic antioxidants. J. Food Sci. 76: H97–H107.

Sogi, D. S., M. Siddiq, I. Greiby and K. D. Dolan. 2013. Total phenolics, antioxidant activity and functional properties of 'Tommy Atkins' mango peel and kernel as affected by drying methods. Food Chem. 141: 2649–2655.

Sójka, M., S. Guyot, K. Kołodziejczyk, B. Król and A. Baron. 2009. Composition and properties of purified phenolics preparations obtained from an extract of industrial blackcurrant (*Ribes nigrum* L.) pomace. J. Hortic Sci. Biotech. 84: 100–106.

Soria, A. C. and M. Villamiel. 2010. Effect of ultrasound on the technological properties and bioactivity of food: A review. Trends Food Sci. Tech. 21: 323–331.

Souquet, J. -M., B. Labarbe, C. Le Guernevé, V. Cheynier and M. Moutounet. 2000. Phenolic composition of grape stems. J. Agric. Food Chem. 48: 1076–1080.

Southgate, D. A. T., S. Bingham and J. Robertson. 1978. Dietary fiber in the British diet. Nature 274: 51–52.

Sparkman, D. O. 2000. Mass Spectrometry Desk Reference. Pittsburgh, Global View Publishing.

Spranger, I., B. Sun, A. M. Mateus, V. de Freitas and J. M. Ricardo-da-Silva. 2008. Chemical characterization and antioxidant activities of oligomeric and polymeric procyanidin fractions from grape seeds. Food Chem. 108: 519–532.

Šťavíková, L., M. Polovka, B. Hohnová, P. Karásek and M. Roth. 2011. Antioxidant activity of grape skin aqueous extracts from pressurized hot water extraction combined with electron paramagnetic resonance spectroscopy. Talanta 85: 2233–2240.

Sudha, M. L., V. Baskaran and K. Leelavathi. 2007. Apple pomace as a source of dietary fiber and polyphenols and its effect on the rheological characteristics and cake making. Food Chem. 104: 686–692.

Sun-Waterhouse, D., J. Farr, R. Wibisono and Z. Saleh. 2008. Fruit-based functional foods I: Production of food-grade apple fiber ingredients. Int. J. Food Sci. Tech. 43: 2113–2122.

Sun-Waterhouse, D., S. Nair, R. Wibisono, S. S. Wadhwa, C. Massarotto, D. I. Hedderley, J. Zhou, S. R. Jaeger and V. Corrigan. 2010. Insights into smoothies with high levels of fiber and polyphenols: Factors influencing chemical, rheological and sensory properties. World Acad. Sci. Eng. Technol. 65: 276–285.

Sun-Waterhouse, D., C. Luberriaga, D. Jin, R. Wibisono, S. S. Wadhwa and G. I. N. Waterhouse. 2013. Juices, fibers and skin waste extracts from white, pink or red-fleshed apple genotypes as potential food ingredients. A Comparative Study. Food Bioprocess Tech. 6: 377–390.

Sun-Waterhouse, D., K. Bekkour, S. S. Wadhwa and G. I. N. Waterhouse. 2014. Rheological and chemical characterization of smoothie beverages containing high concentrations of fiber and polyphenols from apple. Food Bioprocess Tech. 7: 409–423.

Szajdek, A. and E. J. Borowska. 2008. Bioactive compounds and health-promoting properties of berry fruits: A review. Plant Food Hum. Nutr. 63: 147–156.

Thebaudin, J. Y., A. C. Lefebvre, M. Harrington and C. M. Bourgeois. 1997. Dietary fibers: Nutritional and technological interest. Trends Food Sci. Tech. 8: 41–48.

Tian, H. -L., P. Zhan and K. -X. Li. 2010. Analysis of components and study on antioxidant and antimicrobial activities of oil in apple seeds. International J. Food Sci. Nutr. 61: 395–403.

Tomás Barberán, F. A. 2007. High-value co-products from plant foods: Nutraceuticals, micronutrients and functional ingredients. pp. 448–489. *In*: K. Waldron (ed.). Handbook of Waste Management and Co-product Recovery in Food Processing. Woodhead Publishing, Cambridge.

Tomás-Barberán, F. A., M. I. Gil, P. Cremin, A. L. Waterhouse, B. Hess-Pierce and A. A. Kader. 2001. HPLC-DAD-ESIMS analysis of phenolic compounds in nectarines, peaches and plums. J. Agric. Food Chem. 49: 4748–4760.

Tran, C. T. and D. A. Mitchell. 1995. Pineapple waste-a novel substrate for citric acid production by solid-state fermentation. Biotechnology Letters 17: 1107–1110.

Valls, J., S. Millán, M. P. Martí, E. Borràs and L. Arola. 2009. Advanced separation methods of food anthocyanins, isoflavones and flavanols. J. Chromatogr. A 1216: 7143–7172.

Van Dyk, J. S., R. Gama, D. Morrison, S. Swart and B. I. Pletschke. 2013. Food processing waste: Problems, current management and prospects for utilisation of the lignocellulose component through enzyme synergistic degradation. Renew. Sust. Energ. Rev. 26: 521–531.

Vergara-Valencia, N., E. Granados-Pérez, E. Agama-Acevedo, J. Tovar, J. Ruales and L. A. Bello-Pérez. 2007. Fiber concentrate from mango fruit: Characterization, associated antioxidant capacity and application as a bakery product ingredient. LWT—Food Sci. Technol. 40: 722–729.

Virk, B. S. and D. S. Sogi. 2004. Extraction and characterization of pectin from apple (*Malus Pumila* Cv Amri) peel waste. Int. J. Food Prop. 7: 693–703.

Virot, M., V. Tomao, C. Le Bourvellec, C. M. C. G. Renard and F. Chemat. 2010. Towards the industrial production of antioxidants from food processing by-products with ultrasound-assisted extraction. Ultrason. Sonochem. 17: 1066–1074.

Viuda-Martos, M., Y. Ruiz-Navajas, A. Martin-Sánchez, E. Sánchez-Zapata, J. Fernández-López, E. Sendra, E. Sayas-Barberá, C. Navarro and J. A. Pérez-Álvarez. 2012. Chemical, physico-chemical and functional properties of pomegranate (*Punica granatum* L.) bagasses powder co-product. J. Food Eng. 110: 220–224.

Wang, L. and C. L. Weller. 2006. Recent advances in extraction of nutraceuticals from plants. Trends Food Sci. Tech. 17: 300–312.

Wang, R. -F., W. -D. Xie, Z. Zhang, D. -M. Xing, Y. Ding, W. Wang, C. Ma and L. -J. Du. 2004. Bioactive compounds from the seeds of *Punica granatum* (Pomegranate). J. Nat. Prod. 67: 2096–2098.

Wang, S., F. Chen, J. Wu, Z. Wang, X. Liao and X. Hu. 2007. Optimization of pectin extraction assisted by microwave from apple pomace using response surface methodology. J. Food Eng. 78: 693–700.

Wang, W., X. Ma, Y. Xu, Y. Cao, Z. Jiang, T. Ding, X. Ye and D. Liu. 2015. Ultrasound-assisted heating extraction of pectin from grapefruit peel: Optimization and comparison with the conventional method. Food Chem. 178: 106–114.

Widmer, W. and A. M. Montanari. 1994. Citrus Waste Streams as a Source of Phytochemicals. 107th Annual Meeting of the Florida State Horticultural Society. Orlando/Florida, USA. Vol. 107, pp. 284–288.

Wijngaard, H. H., C. Rößle and N. Brunton. 2009. A survey of Irish fruit and vegetable waste and by-products as a source of polyphenolic antioxidants. Food Chem. 116: 202–207.

Wijngaard, H. H. and N. Brunton. 2010. The optimisation of solid–liquid extraction of antioxidants from apple pomace by response surface methodology. J. Food Eng. 96: 134–140.

Wijngaard, H. H., M. B. Hossain, D. K. Rai and N. Brunton. 2012. Techniques to extract bioactive compounds from food by-products of plant origin. Food Res. Int. 46: 505–513.

Wittenauer, J., S. Falk, U. Schweiggert-Weisz and R. Carle. 2012. Characterisation and quantification of xanthones from the aril and pericarp of mangosteens (*Garcinia mangostana* L.) and a mangosteen containing functional beverage by HPLC–DAD–MSn. Food Chem. 134: 445–452.

Wolfe, K., X. Wu and R. H. Liu. 2003. Antioxidant activity of apple peels. J. Agric. Food Chem. 51: 609–614.

Wu, T., Y. Guan and J. Ye. 2007. Determination of flavonoids and ascorbic acid in grapefruit peel and juice by capillary electrophoresis with electrochemical detection. Food Chem. 100: 1573–1579.

Yilmaz, E. E., E. B. Özvural and H. Vural. 2011. Extraction and identification of proanthocyanidins from grape seed (*Vitis Vinifera*) using supercritical carbon dioxide. J. Supercrit. Fluid 55: 924–928.

Yu, H., C. Qin, P. Zhang, Q. Ge, M. Wu, J. Wu, M. Wang and Z. Wang. 2015. Antioxidant effect of apple phenolic on lipid peroxidation in Chinese-style sausage. J. Food Sci. Tech. 52: 1032–1039.

Yu, J., L. Wang, R. L. Walzem, E. G. Miller, L. M. Pike and B. S. Patil. 2005. Antioxidant activity of citrus limonoids, flavonoids, and coumarins. J. Agric. Food Chem. 53: 2009–2014.

Zadernowski, R., S. Czaplicki and M. Naczk. 2009. Phenolic acid profiles of mangosteen fruits (*Garcinia mangostana*). Food Chem. 112: 685–689.

Zhang, H. -F., X. -H. Yang and Y. Wang. 2011. Microwave assisted extraction of secondary metabolites from plants: Current status and future directions. Trends Food Sci. Tech. 22: 672–688.

Zhou, Y., X. Zhao and H. Huang. 2015. Effects of pulsed electric fields on anthocyanin extraction yield of blueberry processing by-products. J. Food Process Pres. 39: 1898–1904.

CHAPTER 12

Valorization of Waste and By-products from the Agrofood Industry using Fermentation Processes and Enzyme Treatments

Phuong Nguyen Nhat Minh,[1,2,3,4] *Thien Trung Le,*[1] *John Van Camp*[2] and *Katleen Raes*[3],*

Introduction

Modern technology in food processing industries has led to increased food production and as a consequence, also of food by-products and waste. Generation of by-products and waste which consists primarily of organic residues of processed raw materials during processing is unavoidable. These residues can cause pollution, management and economic problems. Instead of throwing and landfilling these by-products and waste, there is an increasing interest to turn these waste and by-products into useful products. Therefore, different methods to use food industrial residues as a source of high value-added products are under development. Current methods for further utilisation of product-specific by-products and waste have developed along traditional lines and been closely bound to develop the best useful management methods. Two general methods of traditional by-products and waste utilisation include the use of the waste and by-products either as animal feed or as fertilizer. Current valorization, a relatively

---

[1] Department of Food Engineering, Faculty of Food Science and Technology, Nong Lam University, Ho Chi Minh City, Vietnam.
[2] Department of Food Safety and Food Quality, Faculty of Bioscience Engineering, Ghent University, Ghent, Belgium.
[3] Department of Industrial Biological Sciences, Faculty of Bioscience Engineering, Ghent University – Campus Kortrijk, Kortrijk, Belgium.
[4] Department of Food Technology, College of Agriculture and Applied Biology, Can Tho University, Can Tho City, Vietnam.
* Corresponding author: katleen.raes@ugent.be

new concept in the field of industrial residue management promoting the principle of sustainable development, has been developed to produce industrial chemicals, micronutrients, enzymes, and precious metabolites that have industrial value. These products are mainly produced by chemical and biotechnological processes. In addition, part of these residues is used to produce bio-energy through a vast range of processes including chemical, thermochemical and biochemical treatments. As a result, a healthy environment and sustainable development are ensured. In this chapter, two main processes, i.e., enzyme treatment and fermentation for the valorization of food waste and by-products, are discussed. Especially, fundamental principles and highlighting findings of concerned studies are focused at.

Valorization of Waste and By-products from the Agrofood Industry

Types and availability of waste and by-products from the agrofood industry

Agrofood industries are those industries which depend on agricultural products as raw materials. In these industries, food processing results in a range of waste and by-products according to the type of raw material processed and the respective processing technologies employed. During food processing, the food source material and the processing aids that enter the food production process exist either as a desired product or as a product-specific waste (Russ and Schnappinger 2007). Although most of the desired components as soluble conjugates forms are already extracted from the source material, waste and by-products may contain other potentially useful components, such as cell wall-bound forms that merit consideration as a raw material in related industries. Mirabella et al. (2014) clearly define food waste concepts, that is, plant-based food waste, derived from vegetable and fruits and cereal products; animal-based food waste, related to dairy products and meat and derivatives; and miscellaneous. The main categories of food waste and by-products are summarized in Fig. 1.

Vegetables and fruits

Food industries produce huge amounts of vegetable and fruit waste and by-products which have high biodegradability, soluble constituents and methane emissions leading

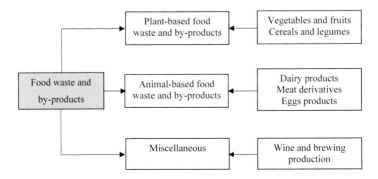

Figure 1. Summary of food waste and by-products from different categories.

to environmental problems (Misi and Forster 2002). Waste from fruit and vegetables processing generally contains large amounts of total suspended solids (TSS) and present high biochemical (BOD) and chemical oxygen demand (COD), which influence possible recovery solutions and treatment costs. Various manufacturing processes that generate food processing by-products/waste in terms of solid and liquid waste are cleaning, removal of leaves, skin and seeds, blanching, washing and cooling, packaging and clean up (Chandrasekaran et al. 2013). For example, large quantities of by-products including peels, pulp, rag (membranes and cores) and seeds are produced in the citrus juice industry, which may account for 50 per cent of the total fruit weight (Kao and Chen 2013).

Cereals and legumes

The most important cereals for human consumption are wheat, rice, corn, millet, oat, rye and barley. They are cultivated on approximately 80 per cent of the world's arable land and supply around 50 per cent of the world's population with food. Wheat and rye are ground to flour and grits in flourmills, oats are made into rolled oats in oat mills, and rice is hulled in rice mills. Corn is primarily processed to starch and oil, whereas almost all barley is turned into malt (Russ and Schnappinger 2007). The waste and by-products produced in this process are bran, middlings, broken grains, seeds, shells, husks, fine dust, chaff, straw, seedlings from malt, malt dust (Russ and Meyer-Pittroff 2004). Otherwise, legumes which are beans, lentils, soybeans among others, produce straw and pods as main by-products with a global production capacity of more than 460 million metric tons of residues every year (Santana-Méridas et al. 2012).

Dairy products

Dairy products are a major and important part of the food industry. The production process of dairy products results in a significant amount of by-products including skim milk, buttermilk, ghee residues and whey (Surajit et al. 2013). They are characterized by high organic load (e.g., whey proteins, non-protein nitrogen compounds and lactose), considerable variations in pH (4.2–9.4) as well as increased content of suspended solids (0.4–2 g/L) (Kosseva et al. 2003). Among these by-products, whey is generated as a by-product during the preparation of a number of dairy products, such as cheese, channa, caseins, paneer, shrikhand and so on (Surajit et al. 2013). The world whey production is about 120 million tons and some of this amount may remain unutilized (Mukhopadhyay et al. 2005). Although the exact amount of buttermilk production in the world is not known, it is estimated that about 3.2 million tons of buttermilk are produced annually in India alone (Surajit et al. 2013).

Meat and derivatives

In the world, the frequency of meat consumption has increased. However, although extensively consumed in the past, the demand for less valuable products such as blood, entrails and parts containing high amount of connective tissues (e.g., tail, tongue and ears) declined. For this reason, the meat industry discards a large quantity of slaughterhouse by-products, which mainly include skin, bones, entrails, fatty tissues, feet, skull, etc. (Mirabella et al. 2014). The by-products of meat constitute nearly 60–70 per cent of the slaughtered carcass, of which nearly 40 per cent forms edible and 20 per cent inedible by-products (Bhaskar et al. 2007). Therefore, the meat industry tries to find efficient recovery solutions for that waste that could pose a serious risk to both the environment and human health. However, unlike fruit and vegetable waste, the recovery of meat by-products is bound by severe hygiene and health limitations. The most dangerous disease is Bovine spongiform encephalopathy (BSE), for which the European Union promulgated legislative measures in order to prevent products containing BSE end up in the distribution chain (Regulation 999/2001 and 853/2004).

Egg products

In egg processing the waste consists almost only eggshells, regardless of which egg product is made, including deep-frozen, spray-dried or chemically preserved eggs, egg yolks, egg-derived lecithin or egg oil. The high variation in the level of accumulated waste can be attributed to the variation in the size of the eggs and the thickness of their shells (Russ and Meyer-Pittroff 2004). Normally, the egg shell represents about 11 per cent of the total weight of the egg and is predominantly composed of calcium carbonate (94 per cent), calcium phosphate (1 per cent), organic matter (4 per cent), and magnesium carbonate (1 per cent) (Rathinaraj and Sachindra 2013). On the other hand, inedible eggs produced during egg processing, defective products (e.g., scrap pastry), unhatched eggs and other types of eggs such as double yolked and eggs with blood spot and meat spot are good sources of phospholipids besides other nutraceuticals. In the United States, inedible eggs are estimated to account for about 2 per cent of the egg production (Shah et al. 2004).

Miscellaneous

Waste during the beer production process is obtained by filtration and separation. It consists of the filtration or separation residues or of sludge, which contains the filtration aid and residues, mostly kieselguhr and organic material (Russ and Meyer-Pittroff 2004). However, a technique using microfiltration does not need filtration aids. Water is used to extract the desired substances from coarsely ground malt. The liquid extract, called wort, is separated from the spent grains that stay behind as waste. Further, a protein–tannin complex, known as hot-break and cold-break material, is also separated from the wort. When the fermentation is complete, the yeast utilized during the fermentation process, also needs to be separated from the beer (Russ and Meyer-Pittroff 2004).

The waste from wine production can be divided into three categories: pomace, clarification sediment and yeast sediment (consisting of yeast cells and tartar). The generated amount of pomace varies according to what kind of squeezer system is used and is about 62 per cent (Shyam and Chandrasekaran 2013). During fermentation, the sediment must be removed to clarify the fermented liquid. The amount of sediment depends upon the type of processing methods used, the condition of the fruits at the time of harvest and the type of wine being produced (Russ and Schnappinger 2007). Normally, the total sediment (lees) is estimated to count for about 14 per cent (Shyam and Chandrasekaran 2013).

Valorization processes

The waste and by-products of agrofood industries are one of the major concerns for most food factories. Awareness and stringent legislations of environmental issues and technological development utilizing diverse products as raw material have significantly contributed to sustainable waste management practices and valorization of food industry waste throughout the world. Various methods including physical, thermal, chemical and biological methods are applied to manage the food industry waste. The main approaches to valorize waste and by-products adapted from Kasi et al. (2013) and Santana-Meridas et al. (2012) are summarized in Fig. 2.

Combustion is a common way to remove agricultural waste in the field, resulting in environmental and health problems (Saxena et al. 2009, Lim et al. 2012). Otherwise, removing them in excess of effective erosion control below the tolerable limit may result in soil deterioration and declining yields. Thus, it is important to return the amount of generated residues into the field as mulch or plant cover to improve soil quality mainly in erodible lands (Lal 2005). Pretreatments such as physical, chemical and biological ones are necessary to convert by-products and waste into energy or materials in order to breakdown the structure of the residues, minimize transportation cost and facilitate the subsequent processing (Naveen et al. 2014).

A part of agricultural residues is used to produce bio-energy due to the characteristics (non-edible, high energy potential, etc.) of the lignocellulosic materials (Ruane et al. 2010). Similarly, residual biomass from agricultural activities can be converted into a variety of industrial products such as animal feed (López et al. 2005, Kafilzadeh and Maleki 2012), organic acids, aroma compounds, microbial pigments (Sowbhagya and Chitra 2010, Liang et al. 2016). In addition, a wide range of bioactive natural products can be recovered through several biological and chemical processes (Bengoechea et al. 1997, Galanakis 2012, Dong et al. 2010).

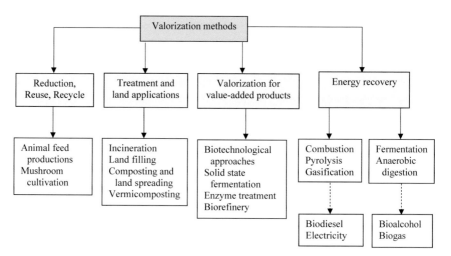

Figure 2. Flow diagram of valorization methods for agrofood industry waste and by-products adapted from Kasi et al. (2013) and Santana-Meridas et al. (2012).

Fermentation Technology

In the context of selection of suitable methods for valorization of agro-industrial waste, fermentation technology has received great attention, since it is able to produce the desired products with high quality and without any toxicity associated to the organic solvents (Martins et al. 2011). In addition, the agro-industrial waste is considered as ideal substrate for fermentation because it is a rich source of carbohydrate, crude fiber, vitamins and minerals. Moreover, the utilisation of these agro-industrial wastes helps in solving pollution problems (Pandey et al. 1999a).

Fundamental principles

The fermentation process could be divided into several paths (Bhavbhuti et al. 2012), depending on the end products obtained. For example, in the alcoholic fermentation process, enzymes produced by yeast convert carbohydrates into ethanol and carbon dioxide. In the fermentation process by homofermentative lactic acid bacteria (LBA), lactic acid is produced from glucose as a primary product whereas fermentation process by heterofermentative LBA, glucose is converted into lactic acid, ethanol and carbon dioxide.

Many factors affect the fermentation process. Among these are microorganisms, media, substrates, inoculation concentrations, fermentation process variables and waste/by-products management (Jayachandran et al. 2013). Selection of an appropriate organism for the fermentation process is a very important step. They are selected for their advantageous properties in terms of process performance and impact on the final product quality. In general, the organisms used for food products must comply with the criterion of being Generally Recognized as Safe (GRAS) (Charles 2005). The organism may be collected from culture collection centers or may be isolated from appropriate sources. A decisive component in the success of the fermentation process is the composition of the medium. The selection of suitable medium constituents and optimization of the concentrations of the various constituents are important for fermentation. The medium must be economically effective and nontoxic. The optimized medium may contain natural raw materials like carbon/nitrogen/energy sources. However, sometimes macronutrients as well as micronutrients are also added for better fermentation. Inoculum concentrations and the mode of inoculation are important factors influencing microbial growth and fermentation. A low concentration of inoculum leads to a long lag phase, whereas a high inoculum concentration results in dilution of the rate-limiting substrate in the medium, contributing to a lower specific growth rate (Jayachandran et al. 2013). Moreover, it is important to consider physicochemical parameters as well, such as temperatures, pH, water activity, oxygen, radiation, pressure and 'static' agents which have an impact on the growth of microorganisms during fermentation (Charles 2005).

Fermentation can be classified into various types based on mode of cultivation, water activity, oxygen requirement, nutrient metabolism and the number of inoculums. The type of fermentation utilized will depend upon the requirement of the food industry and the characteristics of the product under consideration (Jayachandran et al. 2013). The types of fermentation are listed in Fig. 3. Based on the mode of cultivating the microorganisms, fermentation can be classified as batch, continuous, or

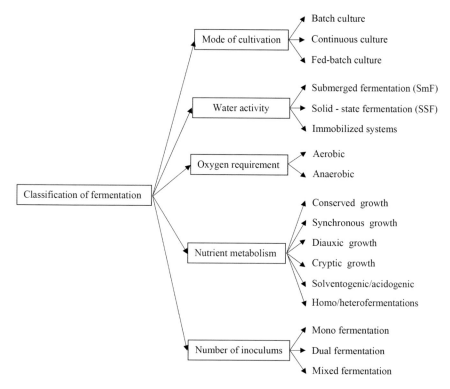

Figure 3. Classification of fermentation processing adapted from Jayachandran et al. (2013).

fed-batch. The process of fermentation may be classified on the basis of water activity requirement of the microorganisms as well as the mode of growth of microorganisms in the fermentation medium being either submerged, solid-state, slurry-state, or surface culture. Another way to classify fermentation processes is based on the oxygen requirements in which fermentation may be classified as aerobic or anaerobic. According to the nutrient metabolism and physiological status of the culture, the growth can be conserved, synchronous, diauxic, or cryptic. Fermentation may be solventogenic (products are solvents like acetone, butanol, and ethanol) or acidogenic (products are acids such as acetic acid, citric acid and lactic acid) based on the nature of the product formed. Final classification of fermentation is based on the number of inoculums. Mono fermentation, dual fermentation and mixed fermentation belong to this group. Mixed fermentations are those in which more than two organisms are involved. Most natural fermentations employed in fermented food production are mixed fermentations.

Recent studies: application of fermentation technology in valorization of waste and by-products from the plant-based agrofood industry

The selection criteria chosen to identify the relevant articles were related to the objectives of this chapter. Selected studies: (1) focusing on high value-added products using fermentation and enzyme technology in the production, (2) related to waste and by-products from the plant-based agrofood industry, (3) limiting to papers published from 2000 to 2016, and (4) providing suggestions for a possible recovery strategy.

Bioactive compounds

In the last decade, there has been an increasing trend towards utilisation of low-cost agricultural and agro-industrial residues as substrates to produce bioactive natural compounds. Techniques which have been recently proposed to obtain these compounds include the use of supercritical fluids, high pressure processes, microwave-assisted extraction and ultrasound-assisted extraction (Farías-Campomanes et al. 2013, Corrales et al. 2008, Bai et al. 2010, Allaf et al. 2013). However, the production of bioactive natural compounds by fermentation process is also an interesting alternative, since it is able to provide high quality and activity extracts while precluding any toxicity associated to the organic solvents (nee'Nigam 2009). The obtained products are determined by the used substrates (by-products) and fermentation microorganisms. Some of the common substrates used in solid state fermentation are wheat bran, rice and rice straw, hay, fruit and vegetable waste, paper pulp, bagasse and coconut coir (Pandey et al. 1999). Among them, by-products from the fruit and vegetable processing industry can be targeted for production of value-added phenolic ingredients. Bioconversion of by-products by a solid-state process using food-grade fungi provide a unique strategy to extract and convert phenolic compounds with improved nutraceutical, functional and sensory properties. Table 1 shows examples of bioactive secondary metabolites that are produced by solid-state fermentation (SSF) and submerged fermentation (SmF).

Pomegranate waste, the by-product of the pomegranate juice-processing industry, is a good source of ellagic acid. In the last few years, pomegranate peel was successfully used as support and nutrient source for the production of ellagic acid by SSF with *Aspergillus niger* GH1. Aguilar et al. (2008) showed that ellagic acid

Table 1. Examples of secondary metabolites produced/released by various microorganisms under SSF and SmF.

Product	Microorganism	Reference
Cephamycin C	*Streptomyces clavuligerus*	(Hölker et al. 2004)
Coconut aroma	*Trichoderma* sp.	(Hölker et al. 2004)
Ergot alkaloids	*Claviceps fusiformis*	(Hölker et al. 2004)
Giberellic acid	*Giberella fujikuroi*	(Hölker et al. 2004)
Iturin	*Bacillus subtillis*	(Hölker et al. 2004)
Lycopene	*Fusarium solani pisi*	(Azabou et al. 2016)
Ellagic acid	*Pediococcus pentosaceus*	(Puupponen-Pimiä et al. 2016)
Kaempferol and kaempferol-3-glucoside	*Aspergillus awamori*	(Lin et al. 2014)
Anthocyanin	*Aspergillus* sp. and *Rhizopus* sp.	(Lee et al. 2008)
Gallic acid	*Bacillus pumilus*	(Cho et al. 2009)
Vanillin	*Phanerochaete chrysosporium*	(dos Santos Barbosa et al. 2008)

accumulated during 96 hours of fermentation, reaching up to 0.9 per cent. In recent studies, the litchi pericarp and its aqueous-organic extracted residues were fermented by *A. awamori* in order to produce bioactive compounds (Lin et al. 2014). The study identified that rutin, which is present in litchi pericarp, could be deglycosylated to form quercetin and quercetin-3-glucoside which is also biotransformed into kaempferol and kaempferol-3-glucoside during the fermentation. In the work of Puupponen-Pimia et al. (2016), accumulation of ellagic acid and simultaneous degradation of ellagitannins was found in the fermentation of cloudberry, resulting in a decrease of ellagitannins from 221 to 198 mg/g and an increase from 221 to 528 mg/g in ellagic acid after 14 days of fermentation with *Pediococcus pentosaceus* VTT E-072742. Huynh et al. (2015) studied the utilisation of cauliflower outer leaves as a substrate in order to release phenolic compounds. Fermentation with *Aspergillus sojae* obtained the highest level of total phenolic compounds (321 mg rutin equivalents (RE)/100 g fresh weight (FW)) after 24 hours compared to 6 other fungal strains, which was three times higher compared to the unfermented sample (113 mg RE/100 g FW). The profile of phenolic compounds changed from mainly glycosidic froms of kaempferol before fermentation into kaempferol-3-O-diglucoside as the predominant kaempferol metabolite after fermenting the cauliflower outer leaves. Another study proved that crude extracts from SSF with *Fusarium solani pisi* of tomato processing by-products was a fruitful method to enhance the lycopene recovery (Azabou et al. 2016). Moreover, along with lycopene, other bioactive compounds were detected, such as, phenolic compounds and flavonoids, which were around 152 mg gallic acid equivalents (GAE)/g and 15 mg quercetin equivalents (QE)/g, respectively. These values were higher than those measured in fresh tomatoes (0.75 mg GAE/g and 0.4 mg QE/g (Azabou et al. 2016). However, a fermentation process does not always increase all phenolic compounds. In some cases, a decrease in some components is observed, as they are metabolized into other forms having a lower toxicity towards the fermenting microorganisms. Perez-Gregorio et al. (2011) reported that from fresh mulberry the largest amount of flavonoids was extracted, whereas from fermented mulberry lower levels were extracted. This may be explained by the decomposition of anthocyanins.

Industrial enzymes

Amylolytic enzymes, lignocellulolytic enzymes, together with other related enzymes like pectinases, xylanases and tannases, are among the most important group of enzymes in term of commercial value. These enzymes are employed in processing of starchy and ligno-cellulosic materials for the production of sugar, fuel and other value-added products. Therefore, the industrial demand for these enzymes is very high. In addition, waste and by-products from the plant-based agrofood industry meet the requirements to be used as substrates for the production of industrial enzymes, also by using fermentation technology as shown in Table 2.

Amylolytic enzymes

Amylolytic enzymes are divided into α-amylase (1,4-α-D-glucan glucanohydrolase, EC 3.2.1.1), β-amylase (EC 3.2.1.2), maltogenic amylase (EC 3.2.1.33), glucoamylases (γ-amylase, amyloglucosidase, EC 3.2.1.3), isoamylase (EC 3.2.1.68) and pullulanases

Table 2. The production of industrial enzymes from waste and by-products using fermentation technology.

Enzyme (s)	Substrate	Microorganism	Fermentation type	Activity*	Reference
Amylolytic enzymes					
α-amylase	Loquat kernels (*Eriobotrya japonica Lindley*)	*Penicillium expansum*	SSF	1012 U/g	(Erdal and Taskin 2010)
α-amylase	Banana peel	*Bacillus subtilis*	SSF	7 U/g	(Unakal et al. 2012)
α-amylase	Mango seed kernels	*Fusarium solani*	SmF	0.9 U/g	(Kumar et al. 2013)
α-amylase	Vegetable waste	*Anoxybacillus amylolyticus*	SSF	102 U/g	(Finore et al. 2014)
α-amylase	Orange waste	*Streptomyces* sp.	SmF	8.26 U/mL/min	(Mouna imen and Mahmoud 2015)
α, β and γ -amylase	Potato starchy waste	*Bacillus amyloliquefaciens*	SmF	2.4 U/mL/min	(Abd-Elhalem et al. 2015)
Lignocellulolytic enzymes					
Cellulase	Kinnow (*Citrus reticulata*) pulp	*Trichoderma reesei*	SSF	13.4 U/g (FPase)	(Oberoi et al. 2010)
Cellulase	Fruit waste	*Streptomyces* sp.	SSF	20 U/mL/min	(Rathan and Ambili 2011)
Cellulase	Palm kernel cake	*Bacillus* sp.	SSF	2.6 FPU/mL/hr	(Norsalwani and Norulaini 2012)
Cellulase	Banana waste	*Cellulomonas cartae, Pseudomonas fluorescence, Pseudomonas putida* and *Bacillus megaterium*	SSF	1.7 U/mL/hr (FPase)	(Dabhi et al. 2014)

Table 2 contd. ...

...Table 2 contd.

Enzyme (s)	Substrate	Microorganism	Fermentation type	Activity*	Reference
Xylanase	Grape pomace	*Aspergillus awamori*	SSF	40 U/g DW	(Botella et al. 2005)
Xylanase	Sugarcane bagasse	*Trichoderma harzianum*	SSF	120 U/L/ min	(Gelain et al. 2015)
Pectinases					
Pectin lyase (Pl), Polygalacturonase (Pg)	Orange bagasse, banana and mango peels, sugarcane bagasse	*Penicillium viridicatum*	SSF	3540 U/g DW (Pl) 55 U/g DW (Pg)	(Silva et al. 2002)
Pectinase	Grape waste	*Aspergillus foetidus*	SSF	0.35 U/ mg	(Venkatesh et al. 2009)
Pectinase	Orange peel and ground nut oil cake	*S. cerevisiae*	SmF	6285 U/ mL/min	(Poondla et al. 2016)
Pectinase	Citrus pulp and sugarcane bagasse	*Aspergillus oryzae*	SSF	33–41 U/g	(Biz et al. 2016)
Tannases					
Tannases	Cashew bagasse	*Aspergillus* spp.	SSF	12.3 U/g DW	(Liu et al. 2016)
Tannases	Grape peel	*P. chrysogenum* and *T. viridae*	SSF	84 U/g DW	(Paranthaman et al. 2009)
Tannases	Barbados cherry and mangaba fruit waste	*Aspergillus* and *Penicillium*	SSF	42 U/mL/ min	(Lima et al. 2014)
Tannases	Citrus residues	*Paecilomyces variotii*	SSF	16 U/g DW	(Madeira et al. 2015)

*Activity is highly relative and was expressed as unit (U)/g (mg) or U/mL (L)/min (hr), depending on enzyme assays

DW: dry weight of substrate

(EC 3.2.1.41). Amylolytic enzymes hydrolyze starch and similar oligo- and poly-saccharides into low molecular weight sugars, like glucose, fructose and maltose (Robert and Maarten 2010).

Seed kernels from fruits are seen as valuable waste with a high content of starch. Many studies have been performed using kernels as a substrate for the production of amylolytic enzymes. Typically, mango seed kernels have been used as substrate for the production of α-amylase by *Fusarium solani* NAIMCC-F-02956, using submerged fermentation. Within the optimum conditions of a substrate concentration of 5 per cent (w/v), pH 4.0 and temperature 30°C at the 9th day of incubation, α-amylase has been successfully produced obtaining 0.9 U/g (one unit of α-amylase (U) is the amount of enzyme that produced one μmole of maltose/min/mL under assay conditions and was expressed as U/g of dry substrate) (Kumar et al. 2013). Other wastes or by-products from fruits were also used for the production of amylolytic enzymes. Recently, Mouna imen and Mahmoud (2015) showed that by applying halophilic *Streptomyces* sp. 20r to orange waste powder under submerged fermentation, it was possible to perform a cost-effective production of α-amylase on this agro-residue. Besides *Bacillus subtilis* was also applied to banana peel for the production of α-amylase. The optimized parameters were incubation time, 24 hours, substrate content, 50 g; temperature, 35°C and pH, 7 under SSF for maximum yield of amylase, with less or more than 7 U/g (one unit of α-amylase (U) is the amount of enzyme that produced one μmole of maltose/min/mL under assay conditions and was expressed as U/g of dry substrate) (Unakal et al. 2012).

Lignocellulolytic enzymes

Lignocellulolytic enzymes constitute a large group of mainly enzymes, including ligninolytic enzymes (peroxidases and oxidases) and hydrolytic enzymes (cellulases and hemicellulases). Lignocellulolytic enzymes are biocatalysts that are responsible for degradation of cellulose, hemicellulose and lignin (Mtui 2012). Information is available on the production of *lignocellulolytic enzymes*, such as cellulases and hemicellulases from microorganisms. Attempts are being made to study SSF processes using waste and by-products from agrofood industry for the production of these enzymes at lab-scale or on semi-pilot-scale. Cellulase production was produced by using banana waste as substrate in SSF with *Cellulomonas cartae*, *Pseudomonas fluorescence*, *Pseudomonas putida* and *Bacillus megaterium*. This bacterial consortia exhibited high level of enzyme activities. Maximum specific activities of enzymes were obtained between 15 to 25 days of culture growth (Dabhi et al. 2014). A recent study showed that xylanase production (xylanase belongs to hemicellulase group and acts on hemicellulose, a polymer of pentose sugars) could be estimated using a mathematical model based on the growth of filamentous fungus *Trichoderma harzianum* P49P11 on sugarcane bagasse (Gelain et al. 2015).

Pectinases

Pectinases are a group of enzymes that hydrolyze pectins. Pectinases may be broadly classified into two groups, namely depolymerizing [(break α 1–4 linkages in principal pectin chain), e.g., galacturonase and pectin lyase] and demethoxylating enzymes (e.g., pectin esterase that esterifies pectin to pectic acid by removing methoxy residues) (Mrudula and Anitharaj 2011). Pectinase is used in the degradation of plant material and is particularly applied for low-temperature extraction of juice from many kind of fruits, for example, banana juice (Sagu et al. 2014). There are many studies on pectinase production using agrofood waste/by-products as substrate. In a recent study, cost-effective media formulation using orange peel and groundnut oil cake was obtained for pectinase production. With the optimized fermentation conditions of orange peel (5 per cent, w/v), groundnut oil cake (4 per cent, w/v), $MnSO_4$ (0.08 per cent, w/v), and incubation period (48 hours) under SmF with *S. cerevisiae* PVK4, enzyme activity showed a highest value of 6285 U/mL (one unit of pectinase activity (U) was defined as the amount of enzyme that liberates one µmole of D-galacturonic acid/min/mL under assay conditions and was expressed as U/mL), a nine-fold increase in comparison with the control basal medium (YEPD broth) (Poondla et al. 2016). Another study in pectinase production on a pilot scale was conducted by Biz et al. (2016). *Aspergillus oryzae* was cultivated on a pilot-scale packed-bed bioreactor, on 15 kg of a substrate containing 51.6 per cent citrus pulp and 48.4 per cent sugarcane bagasse (w/w, dry basis). Pectinase yields of 33–41 U/g (one unit of pectinase activity (U) was defined as the amount of enzyme that liberates one µmole of D-galacturonic acid/min/mL under assay conditions and was expressed as U/g of dry substrate) were obtained and they were able to release the amount of D-galacturonic acid similar to that obtained with a commercial pectinase.

Tannase

The enzyme tannin acyl hydrolase, commonly referred to as tannase (EC.3.1.1.20) is involved in biodegradation of tannins into gallic acid and glucose (Chávez-González et al. 2012). There are many studies on the production of tannase using fungi or bacterial species and agro-residues as a substrate. An interesting study was carried out by Lima et al. (2014) in which evaluation of the production of tannase by *Aspergillus* and *Penicillium* species through SSF, using leaves and agro-industrial waste, barbados cherry and mangaba fruit, as substrate was obtained. From this study, *P. montanense* was selected as the best producer of tannase with the highest activity of 42 U/mL (one unit of tannase activity (U) was defined as the amount of enzyme required to release one µmole of gallic acid/min/mL under assay conditions, and was expressed as U/mL) after 72 hours of fermentation residue, using barbados cherry with 3.5 per cent tannic acid and 70 per cent moisture. The tannase of *P. montanense* was stable over a wide pH range and temperatures and showed the highest activity at pH 9.0 and 50°C. A recent study demonstrated the benefits of producing tannase (from a low-cost process) for the food industry (Madeira et al. 2015). In this study, citrus residue as a substrate and the fungus strain of *Paecilomyces variotii* were used for the production of tannase by SSF. The tannase produced showed optimum activity at pH 5.0, with 70°C and 80 per cent stability between pH 4.0–6.5 and 20–60°C. The enzyme was then applied to orange juice and the results showed that tannase on orange juice decreased the levels of hesperidin, naringin and increased their aglycon form, hesperetin and naringenin.

Organic acids

Organic acids, which possess acidic properties, are water-soluble compounds containing one or more carboxyl groups and other non-amino functional groups (Piero 2008). Organic acids are regarded as building block chemicals which can be produced by microbial processing (Sauer et al. 2008). Organic acids can be used to manufacture a variety of products in many industries and processing units, such as food processing, nutrition and feed industry, pharmaceuticals, oil and gas stimulation units, etc. Microorganisms, namely bacterial (e.g., *Lactobacillus* sp.) and fungal species (e.g., *Aspergillus* sp., *Yarowia lipolytica*), are used commercially for the production of organic acids (Sauer et al. 2008). Table 3 lists the organic acids produced during SSF in several agro-industrial residues.

Citric acid is a carboxylic acid and popularly used in food and pharmaceutical industries. Citric acid may be produced by chemical synthesis but this way, it is not cheaper than fungal fermentation (Ghasem 2007). The biochemical pathways related to citric acid accumulation and the role of tricarboxylic acid cycle (TCA) in fungi has been well established. Citric acid accumulation can be divided into three processes: (1) The breakdown of hexoses to pyruvate and acetyl-CoA by glycolysis, (2) formation of oxaloacetate, and (3) condensation of acetyl-CoA and oxaloacetate to citric acid (Yigitoglu 1992).

In fungal fermentation, citric acid is produced mainly by SSF using *Aspergillus niger* from alternative carbon sources, such as agro-industrial residues (Table 3). The production of citric acid via the fermentation route has been widely studied in the last few decades. In view of the utilisation of agricultural by-products, banana peels were selected as a possible alternative substrate for the production of citric acid due to its high nutrient content (Karthikeyan and Sivakumar 2010). In this study, koji fermentation was conducted using the peels of banana (*Musa acuminata*) as substrate for the production of citric acid, using *Aspergillus niger.* At a moisture content of 70 per cent, 28°C temperature, an initial pH 3, 10^8 spores/ml as inoculum and 72 hour incubation was found to be suitable for maximum citric acid yield (approximately 180 g/kg DW). Using co-culture consortia of microorganisms during the fermentation, to produce citric acid, was investigated by Ali et al. (2016). In this study, the authors used agro-based waste materials (apple pomace, peanut shell and a mixture of apple pomace and peanut shell in a 50:50 ratio) as substrate for SSF with co-culture consortia of *Aspergillus ornatus* and *Alternaria alternata* to enhance the citric acid production. The results showed a maximum citric acid yield of 2.64 mg/mL in the presence of arginine as an additional nutritional ingredient at 30°C in an apple pomace-based medium at 50 per cent moisture content with pH of 5 and a substrate content (25 g) after 48 hours of fermentation. In comparison with the control sample (without arginine), the citric acid yield was about 0.5 mg/mL.

Acetic acid is a fermented product of both aerobic and anaerobic fermentation and is the main component in vinegar. It is the primary metabolite of acetic acid bacteria produced from the bioconversion of ethanol through two reactions catalyzed by the membrane-bound pyrroloquinoline quinone (PQQ)-dependent alcohol dehydrogenase (ADH) and aldehyde dehydrogenase (ALDH). The ethanol is oxidized to acetaldehyde by ADH and then ALDH converts acetaldehyde to acetic acid to release into the surrounding environment (Gullo et al. 2014). A number of successful studies have been conducted to establish acetic acid production from fruit waste (e.g., apple pomace, kiwi fruit peel, pineapple waste) (Hang and Woodams 1986, Hang et al. 1987, Raji et al. 2012, Singh and Singh 2007). Raji et al. (2012) demonstrated the production of acetic acid from pineapple peels through two steps. The peels were first fermented for 48 hours, using *S. cerevisiae* for conversion of sugar to ethanol. Subsequently, ethanol was oxidized to acetic acid by *Acetobacter aceti* for nine days of incubation. The maximum production of acetic acid was about 4.77 per cent at

Table 3. Organic acids produced during SSF using agro-industrial residues as substrates.

Organic acid	Substrate	Microorganism	Yield/ Concentration*	Reference
Citric acid				
	Apple pomace	*Aspergillus niger*	124 g/kg DW	(Shojaosadati and Babaeipour 2002)
	Pineapple waste	*Yarrowia lipolytica*	202 g/kg DW	(Imandi et al. 2008)
	Banana peels	*Aspergillus niger*	180 g/kg DW	(Karthikeyan and Sivakumar 2010)
	Pineapple pulp waste	*Aspergillus niger*	5.25 g/kg DW	(Bezalwar et al. 2013)
	Apple pomace, peanut shell	*Aspergillus ornatus, Alternaria alternata*	2.6 mg/mL	(Ali et al. 2016)
Acetic acid				
	Pineapple peel	*Saccharomyces cerevisiae,* and *acetobacter rancens*	2% (v/v)	(Singh and Singh 2007)
	Pineapple peels	*Saccharomyces cerevisiae* and *acetobacter aceti*	4.77% (v/v)	(Raji et al. 2012)
	Papaya peel	*Saccharomyces cerevisiae,* and *acetobacter aceti*	5.23% (v/v)	(Vikas and Mridul 2014)
Lactic acid	Mango peels Orange peels	*Lactobacillus casei, Lactobacillus delbrueckii*	63.33 g/L 54.54 g/L	(Mudaliyar et al. 2012)
	Sapota peels	*Rhizopus oryzae*	72 g/L	(Kumar and Shivakumar 2014)
	Wasted potato	*Lactobacillus rhamnosus*	46 g/L	(Djukić-Vuković et al. 2016)
Butyric acid	Sugarcane bagasse	*Clostridium tyrobutyricum*	21 g/L	(Wei et al. 2013)
	Jerusalem artichoke	*Clostridium tyrobutyricum*	27.5 g/L	(Huang et al. 2011)
	Cane molasses	*Clostridium tyrobutyricum*	35 g/L	(Jiang et al. 2009)

*DW: dry weight of substrate

optimal conditions. Comparing to standards, this value satisfied the specifications regarding the acetic acid concentration in vinegar. For example, in Canada, acetic acid concentration of vinegars remains between 4.1–12.3 per cent (Panda et al. 2016).

Lactic acid is a common organic acid that can be derived from renewable resources like mango peel, potato peel, cassava waste, etc. Lactic acid can be produced either by fermentation or by chemical synthesis. Ninety per cent of the world's production of lactic acid is through bacterial fermentation. Metabolic pathways for lactic acid production from various sugars by lactic acid bacteria are divided into pentose phosphate/Glycolytic pathway (homolactic acid metabolism) and phosphoketolase pathway (heterolactic acid metabolism) (Abdel-Rahman et al. 2013). Homo-fermentative lactic acid bacteria produce nearly pure lactic acid (90 per cent) (Vijayakumar et al. 2008). There are several studies outlining the use of fruit and vegetable waste as well as agricultural residue to produce lactic acid. Peels of potato, green peas, sweet corn, orange and mango were used for producing lactic acid through fermentation, using the strains of *Lactobacillus casei* and *Lactobacillus delbrueckii*. The highest lactic acid yield of 63.3 g/L was obtained for mango peels by *L. casei*, whereas it was 54.5 g/L for orange peels by *L. delbrueckii*. The amount of lactic acid from the other substrates was obtained at lower levels, ranging from 13.4 to 38.9 g/L (Mudaliyar et al. 2012). These value could be compared with the yield of lactic acid obtained from a normal carbon source, for example, the yield of lactic acid of 67 g/L for waste sugarcane bagasse (Adsul et al. 2007) and 28 g/L for defatted rice bran (Tanaka et al. 2006). In a recent study, three types of waste substrates—wasted bread and wasted potato stillage from bioethanol production and beer production—were studied as substrates for the production of L(+) lactic acid by *Lactobacillus rhamnosus* ATCC 7469 (Djukić-Vuković et al. 2016). A maximal lactic acid productivity of 1.28 g/L/hr, lactic acid concentration of 46 g/L and a highest lactic acid yield of 0.8 g/g were obtained on wasted potato stillage media after 36 hours of fermentation (Djukić-Vuković et al. 2016).

These results listed above support a general view of organic acid production by fermentation technology, using agrofood industrial residues in lab scale. Waste and by-products of fruits and vegetables could be completely used as alternative substrates with low economic value in fermentation for the production of value-added products on a pilot scale. However, when once the fermentation process is developed to obtain the desired products in a large scale, the requisite acquirements must be satisfied, such as (1) which kind of substrates and microorganisms are suitably selected; (2) biosynthetic pathways to the desired product; (3) microorganism's behavior in fermentation processes and substrates (e.g., wide substrate utilisation range and defined media requirements, leading to simpler product purification); and (4) strictly monitoring parameters of the fermentation process.

Bioalcohol

Bioalcohol (bioethanol, ethyl alcohol) is a liquid biofuel which can be produced from several different biomass resources. Most agricultural biomass containing sugars, starches, cellulose can be used as a potential substrate for ethanol fermentation by microbial processes (Lin and Tanaka 2006). Nearly all bioethanol fuel is produced by fermentation. Ethanol currently provides over 40 per cent of the fuel consumed by cars and light trucks (Balat et al. 2008). Approximately 4.5 billion gallons of bioethanol are mainly produced from sugarcane by fermentation every year in Brazil (Balat et al. 2008). Potato-starch residue stream produced during chips manufacture is used as an economical source for bioethanol production by *Saccharomyces cerevisiae*. After treatment with 1 per cent H_2SO_4 at 100°C for one hour, the potato residue was fermented in both aerobic and semi-anaerobic conditions. The results showed that the maximum yield of ethanol (5.5 g/L) was achieved under semi-anaerobic conditions at 35°C by *S. cerevisiae* y-1646 after 36 hours when $ZnCl_2$ (0.4 g/L) was added (Hashem and Darwish 2010). Recently, a trial for producing bioethanol using coculturing of *Trichoderma harzianum*, *Aspergillus sojae* and *Saccharomyces cerevisiae* was investigated by Evcan and Tari (2015). The optimum hydrolysis conditions of apple pomace were a temperature of 110°C, 40 min, 4 per cent phosphoric acid and 1:10 solid/liquid ratio (w/v). Besides, inoculation concentrations of 6 per cent (w/v) for *A. sojae* and *T. harzianum* and 4 per cent (w/v) for *S. cerevisiae* were determined as the optimum conditions for fermentation with the vented aeration method and agitation speed of 200 rpm that gave the highest bioethanol concentration and ethanol yield on total reducing sugar content (YP/S) as 8.7 g/L and 0.9 g/g, respectively.

Biogas

Methane-rich biogas (biomethane) is a versatile renewable energy source, which can be used for replacement of fossil fuels in power and heat production, and used as gaseous vehicle fuel. Methane (CH_4) is the primary fuel present in natural gas. It can be produced from biodegradation of organic materials of biological origin (biomass) in anaerobic conditions (Zhongtang et al. 2010). Studies for the production of biogas from agricultural waste and residues obtained successful results over the last few decades. Olive pomace from the manufacturing process of olive oil was used as substrate for biogas production in the experiment of Tekin and Dalgic (2000). The authors mixed finely ground olive pomace in water, using anaerobic digesters of 1 L working volume at 37°C. After 10 days, under culture sources supplied from waste waters from landfill areas, the maximum methane composition was observed to be around 80 per cent (v/v), corresponding to a hydraulic retention time of 20 days and 10 per cent total solid during semi-continuous digestion. In recent studies, improving and developing biogas production processes to obtain higher yield was investigated by many researchers. In particular, improvement of biogas production from orange peel waste by leaching of limonene was conducted by Wikandari et al. (2015). Limonene is present in orange peel waste and is known as an antimicrobial agent, which impedes biogas production when digesting the peels. The authors removed limonene in orange peel using hexane as solvent before digestion in batch reactors for 33 days. The biogas production showed more than threefold increase from 0.06 to 0.22 m^3 methane/kg volatile solids.

Enzyme Technology

General introduction

Enzymes are catalysts that are responsible for various biological functions. Enzymes are proteins composed by a number of amino acids which are covalently bound through the peptide bond. Enzymes can be produced from any living organism, either by extracting them from their cells (intracellular enzymes) or by recovering them from cell exudates (extracellular enzymes) (Andrés 2008). Enzymes catalyze specific biochemical reactions at different temperatures. A small quantity of active enzyme can catalyze a large quantity of substrate into the end product without being consumed itself (Sowbhagya and Chitra 2010). The catalytic activities of enzymes bring a number of advantages which are efficient at low concentrations, active under mild pH and temperature conditions, high substrate specificity, low toxicity and the ease of termination of activity. Therefore, they are applied in several industries, such as food, beverage, leather, detergent industry, etc. (Sarita and Sukumaran 2013).

Plant tissues and animal organs were the most important sources of enzymes at the onset of enzyme biotechnology. About 70 per cent of the enzymes were extracted from plant tissues or exudates and animal organs in 1960. After a few decades, this tendency had changed and most industrial enzymes are now produced from microbial sources because of the excellent cell systems of microorganisms (Andrés 2008). Microbial enzymes are produced mainly through submerged fermentation under tightly controlled environmental conditions. However, solid-state fermentation has also a good potential for the production of enzymes, especially those from filamentous organisms that are particularly suited for growth on the surface of solid substrates (Pandey et al. 1999).

Enzymes used in bioconversion of food by-products

Enzymes are used in a lot of industries, such as food, animal feed, detergents, textiles, leather, pulp and paper, diagnostics, and therapy (Sanchez and Demain 2010, Liguori et al. 2013). The major class of enzymes used in the valorization of food by-products is apparently carbohydrate-degrading enzymes (Sarita and Sukumaran 2013). Carbohydrate-degrading enzymes are used to hydrolyze starch (found in some by-products) in the production of sugar-rich syrups that can be fermented by microbes to produce a large number of industrially important compounds. Besides, enzymes are used as digestive aids, in the clarification of fruit juices and in food and feed processing. In the food industry, calf intestinal rennet is traditionally used for the preparation of cheese, while papain from papaya trees is used as a meat tenderizer. Most of the cereal processing industries generate lignocellulosic waste, which can be converted into fermentable sugars by the action of cellulases and hemicellulases. The use of amylases can liberate glucose from starchy waste, which then can be used as the substrate for producing a large number of industrially important compounds by microbial fermentation (Wolfgang 2007, Sarita and Sukumaran 2013).

Enzyme-mediated valorization of waste and by-products from the plant-based agrofood industry: Product categories and recent studies

Bioactive compounds

In the plant matrices some phenolic compounds are retained in the polysaccharide-lignin network by hydrogen or hydrophobic bonding, which are not accessible with a solvent in a normal extraction process. Pretreatment with single or multiple enzymes like cellulase, α-amylase and pectinase has been considered as an effective way to release bound compounds and increase overall yield because of breaking the cell wall and hydrolyzing the structural polysaccharides. This declaration has been proved by a number of scientists over the last few decades. In their most recent study, Mushtaq et al. (2015) used cocktail enzyme mixtures (pectinase, cellulase from *Aspergillus niger* and protease from *Streptomyces griseus*) for the pretreatment of pomegranate peel before applying supercritical fluid extraction of phenolic antioxidants. The results indicated that the pretreatment under optimized conditions involving a cocktail enzyme concentration (3.8 per cent) at 49°C and pH 6.7 for 85 min during supercritical fluid extraction offered the maximum total phenolics of 302 mg GAE/g of extracts and extract yield 65.9 per cent, whereas with control samples (no enzyme applied) this was 181 mg GAE/g of extracts and 36 per cent, respectively. In another study, an enzyme cocktail namely 'Kemzyme' Kemine, composed of pectinase, endo-1,3 (4)-β-glucanase, α-amylase, endo-1,4-β-xylanase and bacillolysine (protease) was applied to watermelon rind as a pretreatment before solvent extraction to recover phenolics. The results indicated that enzyme-assisted solvent extraction enhanced the liberation of antioxidant phenolics three-fold on fresh weight basis (FW) as compared to conventional solvent extraction (no enzyme applied) (Mushtaq et al. 2015). Another scientists' group used a mixture of pectinolytic and cellulolytic enzyme preparations (ratio 2:1) to treat grape pomace for enzyme-assisted extraction of polyphenols. The highest amount of phenolic compounds was found after 2 hours of treatment with a dosage of 4.5 g/kg (based on dry matter) at 40°C and pH 4.0 (Maier et al. 2008). Aqueous pre-extraction of the pomace followed by enzymatic treatment resulted in significantly improved extraction yields reaching 91.9, 92.4 and 63.6 per cent for phenolic acids, non-anthocyanin flavonoids and anthocyanins, respectively (Maier et al. 2008). In a recent study, tomato waste (containing both peels and seeds) was treated with a crude enzyme extract produced by solid-state fermentation of *Fusarium solani pisi* to obtain lycopene. Lycopene recovery was slightly higher when extracted by using the crude enzyme extract compared to either pectinases or cellulases preparations (Azabou et al. 2016).

Biofuels

Starches or lignocellulosic residues in waste or by-products from food processing can be hydrolyzed into soluble sugars. These are fermented for conversion into liquid biofuels. One of the major limitations to lignocellulosic biomass conversion is the enzymatic degradation of the complex matrix of lignocellulosic polymers that form plant-cell walls to release hexoses (glucoses) and pentoses (primarily xyloses) for fermentation (Margret et al. 2010). Pretreatment with cellulases or hemicellulases is the preferred method because of the higher conversion yields and less corrosive and toxic conditions compared to acid hydrolysis. Without any pretreatment, the conversion of native cellulose to sugar is extremely slow, since cellulose is well protected by the matrix of lignin and hemicellulose in macrofibrils. Therefore, pretreatment of these materials is necessary to increase the rate of hydrolysis of cellulose to fermentable sugars (Taherzadeh and Karimi 2007).

Several studies on bioethanol production showed increased ethanol yields when cellulase was supplemented with β-glucosidase in simultaneous saccharification and fermentation. Krishna et al. (2001) reported that the conversion of sugar in sugar cane to ethanol with supplementation of cellulase (40 Filter Paper Units (FPU) of cellulase/g substrate) and β-glucosidase (50 U/g substrate) in the saccharification process and with yeast, *Kluyveromyces fragilis* (2.8 per cent, w/v), and *Saccharomyces cerevisiae* (2.2 per cent, w/v) in the fermentation process showed an increase in ethanol yield after 96 hours. In another study, a simultaneous saccharification and fermentation process for ethanol production from various lignocellulosic woody (poplar and eucalyptus) and herbaceous (*Sorghum* sp. bagasse, wheat straw and *Brassica carinata* residue) materials was assayed, using the thermotolerant yeast strain *Kluyveromyces marxianus* CECT 10875. The experiments were performed in laboratory conditions at 42°C, 10 per cent (w/v) substrate concentration and enzymatic loading of 15 FPU/g substrate of commercial cellulase. The results indicated that it is possible to reach SSF yields in the range of 50–72 per cent of the maximum theoretical SSF yield, based on the glucose available in pretreated materials, in 72–82 hours. The ethanol contents obtained in fermentation media ranged from 16–19 g/L, depending on the material tested (Ballesteros et al. 2004).

The hydrolysis of cellulose and hemicellulose in anaerobic digestion of lignocellulosic biomass normally occurs at a very slow rate (Romano et al. 2009). Therefore, enzymes with hydrolytic activity including cellulase and hemicellulase were applied prior to or during the anaerobic digestion of biomass. In most cases, the effect of enzymes in enhancing biogas production was minimal and the cost of enzymes was high. Therefore, application of enzymatic pretreatment was limited. However, a study using mushroom compost extract with laccase and carboxymethylcellulose activity to pretreat pulp and paper sludge increased methane yield by 34.2 per cent (Lin et al. 2010). This bio-pretreatment was promising because of the short pretreatment time of 4 hours and low cost of used enzyme.

Food and feed

Food-processing waste can be easily converted to other food or feed for cattle and poultry by enzymatic processing. Starch generated as a waste during food processing (e.g., cassava starch, corn waste, potato-process waste, etc.) can serve as raw material for the production of various functional foods and sugars, such as glucose, maltose, maltodextrins, cyclodextrins, fructose, sugar alcohols, syrups (Akoh et al. 2008). Enzymes often used in bioconversion of starch are α-amylase, β-amylase, glucoamylase and gluco isomerase (Sarita and Sukumaran 2013). Crude amylases prepared from *Bacillus subtilis* ATCC 23350 and *Thermomyces lanuginosus* ATCC 58160 under solid state fermentation were used for enzymatic hydrolysis of maize starch to obtain higher glucose production (Kunamneni and Singh 2005). The optimum values for the tested variables to obtain maximum conversion efficiency were pre-cooking α-amylase dose 2.2 U/mg solids (one unit of α-amylase activity (U) was defined as the amount of enzyme that caused 10 per cent reduction in the starch-iodine colour, under the assay conditions), post-cooking α-amylase dose 3.4 U/mg solids, glucoamylase dose 0.073 U/mg solids (one unit of glucoamylase activity (U) was defined as the amount of enzyme that released one μmol of glucose/min, under the assay conditions.) at a saccharification temperature of 55°C. A maximum conversion efficiency of 96 per cent was achieved. These results provide an improvement in the basic enzymatic hydrolysis knowledge (Kunamneni and Singh 2005).

Akaracharanya et al. (2011) reported that glucose levels shot up by increasing glucoamylase levels from 0.4 U/g (11.8 g glucose/L) to 2.0 U/g and the maximum glucose yield of 22.6 g/L was achieved when cassava pulp underwent an enzymatic saccharification process, using both α-amylase and glucoamylase enzymes. Cellulosic waste can also be hydrolyzed to generate reducing sugars using cellulase. For example, purified cellulase produced from *Bacillus carboniphilus* CAS 3 using cellulosic waste as substrate was applied for enzymatic saccharification of rice straw. The yield of reducing sugars for further ethanol production was about 15.6 g/L after 96 hours (Annamalai et al. 2014). Besides, fruit and vegetable waste/by-products including seeds, peels, fibers are good sources for enzymatic processing to produce valuable products. All polysaccharides in these residues are easily hydrolyzed to monomeric sugars by mixtures of cellulolytic and pectinolytic enzymes.

Commodity chemicals

Several industrial chemicals including organic acids, surfactants, emulsifiers and a large number of other chemicals were produced by enzymatic methods or microbial fermentation using food processing waste or by-products as substrates. Food waste converted enzymatically into its component sugars is then fermented by appropriate microorganisms to produce the desired chemicals. Organic acids, such as citric and lactic acids, are very common examples. Normally, the products of enzymatic hydrolysis from corn or potato starch waste have been used as feedstock for the fermentative lactic acid production process. Recently, lignocellulosic biomass was tested as an alternative feedstock for lactic acid production. Although amylolytic lactobacillus strains and other lactic acid-producing organisms, such as *Rhizopus oryzae*, can directly metabolize starch to produce lactic acid, they do so with a very low fermentation rate giving a relatively low product yield and low product concentrations. Enzymatic hydrolysis of starch and fermentation of glucose to lactic acid are well-established industrial processes. In a recent study, enzyme complexes, including cellulase, β-glucosidase, xylanase and pectinase were used for enhancing enzymatic hydrolysis of lignocellulose in corncob (Zhang et al. 2010). The results showed that the conversion of glucan and xylan was promoted by increase in cellulase loading and reached a high level at a cellulase loading of 15 mg protein/g glucan (50 FPU/g glucan), at which the yield of glucose and xylose was 93 per cent and 82 per cent, respectively. The saturation limit for β-glucosidase was 2.9 mg protein/g glucan (60 cellobiase units/g glucan), at which the yield of glucose increased to 80 per cent after 24 hour hydrolysis. Compared with xylanase, pectinase was more effective to promote the hydrolysis of cellulose and hemicellulose. The supplementation of pectinase with 0.12 mg protein/g glucan could increase the yield of glucose and xylose by 7.5 per cent and 29.3 per cent, respectively (Zhang et al. 2010).

Conclusions and Prospects

Waste and by-products from the agrofood industry contain various nutrients that can be utilized to extract or convert into value-added products, for example, biofuels, bioactive compounds, organic acids, enzyme products. Although there are physical, chemical and biological operations in current practice, the biological strategies will receive more priority as basically the contents of the waste or by-products are more amenable to microorganisms and enzymes to be treated more efficiently. Biological processes that include the use of whole microbial cells as biocatalysts and enzymes for processing organic materials into desired products are considered environment-friendly. These do not include the use of inorganic compounds as catalysts for effecting bioconversions, which often end up as pollutants in the environment. The success of any valorization of waste and by-products from the agrofood industry depends on the reliable monitoring and control of the processes employed for this purpose. There is a need the development of product-specific methods as well as to target and maximize the yields obtained for valued products and necessarily a market for these products. In the future, new and improved methods based on modern developments and more sophistication in biochemical and biophysical instrumentation, such as in the case of spectrophotometry, chromatography, and *in situ* measurements based on microprocessor-embedded sensors (Wang et al. 2006) will enhance the efficiency of the existing methods of valorization through a reliable assessment and improvement of the bioprocesses employed.

References

Abd-Elhalem, B. T., M. El-Sawy, R. F. Gamal and K. A. Abou-Taleb. 2015. Production of amylases from *Bacillus amyloliquefaciens* under submerged fermentation using some agro-industrial by-products. Annals of Agricultural Sciences 60(2): 193–202.

Abdel-Rahman, M. Ali, Y. Tashiro and K. Sonomoto. 2013. Recent advances in lactic acid production by microbial fermentation processes. Biotechnology Advances 31(6): 877–902.

Adsul, M. G., A. J. Varma and D. V. Gokhale. 2007. Lactic acid production from waste sugarcane bagasse derived cellulose. Green Chemistry 9(1): 58–62.

Aguilar, C. N., A. Aguilera-Carbo, A. Robledo et al. 2008. Production of antioxidant nutraceuticals by solid-state cultures of pomegranate (*Punica granatum*) peel and creosote bush (*Larrea tridentata*) leaves. Food Technology and Biotechnology 46(2): 218–222.

Akaracharanya, A., J. Kesornsit, N. Leepipatpiboon, T. Srinorakutara, V. Kitpreechavanich and V. Tolieng. 2011. Evaluation of the waste from cassava starch production as a substrate for ethanol fermentation by *Saccharomyces cerevisiae*. Annals of Microbiology 61(3): 431–436.

Akoh, C. C., S. Chang, G. Lee and J. Shaw. 2008. Biocatalysis for the production of industrial products and functional foods from rice and other agricultural produce. Journal of Agricultural and Food Chemistry 56(22): 10445–10451.

Ali, S. R., Z. Anwar, M. Irshad, S. Mukhtar and N. T. Warraich. 2016. Bio-synthesis of citric acid from single and co-culture-based fermentation technology using agro-wastes. Journal of Radiation Research and Applied Sciences 9(1): 57–62.

Allaf, T., V. Tomao, K. Ruiz and F. Chemat. 2013. Instant controlled pressure drop technology and ultrasound assisted extraction for sequential extraction of essential oil and antioxidants. Ultrasonics Sonochemistry 20(1): 239–246.

Andrés, I. 2008. Enzyme Biocatalysis: Principles and Applications. Springer Science.

Annamalai, N., M. V. Rajeswari and T. Balasubramanian. 2014. Enzymatic saccharification of pretreated rice straw by cellulase produced from *Bacillus carboniphilus* CAS 3 utilizing lignocellulosic wastes through statistical optimization. Biomass and Bioenergy 68: 151–160.

Azabou, S., Y. Abid, H. Sebii, I. Felfoul, A. Gargouri and H. Attia. 2016. Potential of the solid-state fermentation of tomato by products by *Fusarium solani pisi* for enzymatic extraction of lycopene. LWT—Food Science and Technology 68: 280–287.

Bai, X., T. Yue, Y. Yuan and H. Zhang. 2010. Optimization of microwave-assisted extraction of polyphenols from apple pomace using response surface methodology and HPLC analysis. Journal of Separation Science 33(23-24): 3751–3758.

Balat, M., H. Balat and C. Öz. 2008. Progress in bioethanol processing. Progress in Energy and Combustion Science 34(5): 551–573.

Ballesteros, M., J. M. Oliva, M. J. Negro, P. Manzanares and I. Ballesteros. 2004. Ethanol from lignocellulosic materials by a simultaneous saccharification and fermentation process (SFS) with *Kluyveromyces marxianus* CECT 10875. Process Biochemistry 39(12): 1843–1848.

Bengoechea, M. L., A. I. Sancho, B. Bartolomé, I. Estrella, C. Gómez-Cordovés and M. T. Hernández. 1997. Phenolic composition of industrially manufactured purees and concentrates from peach and apple fruits. Journal of Agricultural and Food Chemistry 45(10): 4071–4075.

Bezalwar, P., A. V. Gomashe, H. M. Sanap and P. A. Gulhane. 2013. Production and optimization of citric acid by *Aspergillus niger* using fruit pulp waste. Int. J. Curr. Microbiol. App. Sci. 2(10): 347–352.

Bhaskar, N., V. K. Modi, K. Govindaraju, C. Radha and R. G. Lalitha. 2007. Utilization of meat industry by products: protein hydrolysate from sheep visceral mass. Bioresource Technology 98(2): 388–394.

Bhavbhuti, M. M., R. Z. Iwan ski and A. Kamal-Eldin. 2012. Introduction. pp. 1–6. *In*: Bhavbhuti M. Mehta, Afaf Kamal-Eldin and Robert Z. Iwanski (eds.). Fermentation Effects on Food Properties. CRC Press, Boca Raton, London, New York.

Biz, A., A. T. J. Finkler, L. O. Pitol, B. S. Medina, N. Krieger and D. A. Mitchell. 2016. Production of pectinases by solid-state fermentation of a mixture of citrus waste and sugarcane bagasse in a pilot-scale packed-bed bioreactor. Biochemical Engineering Journal 111: 54–62.

Botella, C., I. De Ory, C. Webb, D. Cantero and A. Blandino. 2005. Hydrolytic enzyme production by *Aspergillus awamori* on grape pomace. Biochemical Engineering Journal 26(2):100–106.

Chandrasekaran, M., S. M. Basheer, S. Chellappan, P. Karthikeyan and K. K. Elyas. 2013. Food processing industries: An overview. pp. 3–34. *In*: M. Chandrasekaran (ed.). Valorization of Food Processing By-products. CRC Press, Boca Raton, London, New York.

Charles, W. B. 2005. The science underpinning food fermentations food. pp. 1–38. *In*: Fermentation and Micro-organisms. Blackwell Science.

Chávez-González, M., L. V. Rodríguez-Durán, N. Balagurusamy et al. 2012. Biotechnological advances and challenges of tannase: An overview. Food and Bioprocess Technology 5(2): 445–459.

Cho, K. M., S. Y. Hong, R. K. Math et al. 2009. Biotransformation of phenolics (isoflavones, flavanols and phenolic acids) during the fermentation of cheonggukjang by Bacillus pumilus HY1. Food Chemistry 114(2): 413–419.

Corrales, M., S. Toepfl, P. Butz, D. Knorr and B. Tauscher. 2008. Extraction of anthocyanins from grape by-products assisted by ultrasonics, high hydrostatic pressure or pulsed electric fields: A comparison. Innovative Food Science & Emerging Technologies 9(1): 85–91.

Dabhi, B. K., R. V. Vyas and H. N. Shelat. 2014. Use of banana waste for the production of cellulolytic enzymes under solid substrate fermentation using bacterial consortium. Int. J. Curr. Microbiol. Appl. Sci. 3(1): 337–346.

Djukić-Vuković, A., D. Mladenović, M. Radosavljević, S. Kocić-Tanackov, J. Pejin and L. Mojović. 2016. Wastes from bioethanol and beer productions as substrates for l(+) lactic acid production–A comparative study. Waste Management 48: 478–482.

Dong, J., Y. Liu, Z. Liang and W. Wang. 2010. Investigation on ultrasound-assisted extraction of salvianolic acid B from *Salvia miltiorrhiza* root. Ultrasonics Sonochemistry 17(1): 61–65.

dos Santos Barbosa, Elisabete, Daniel Perrone, Ana Lúcia do Amaral Vendramini and Selma Gomes Ferreira Leite. 2008. Vanillin production by Phanerochaete chrysosporium grown on green coconut agro-industrial husk in solid state fermentation. BioResources 3(4): 1042–1050.

Erdal, S. and M. Taskin. 2010. Production of alpha-amylase by *Penicillium expansum* MT-1 in solid-state fermentation using waste Loquat (*Eriobotrya japonica* Lindley) kernels as substrate. Romanian Biotechnological Letters 15(3): 5342–5350.

Evcan, E. and C. Tari. 2015. Production of bioethanol from apple pomace by using cocultures: Conversion of agro-industrial waste to value added product. Energy 88: 775–782.

Farías-Campomanes, A. M., M. A. Rostagno and M. A. A. Meireles. 2013. Production of polyphenol extracts from grape bagasse using supercritical fluids: Yield, extract composition and economic evaluation. The Journal of Supercritical Fluids 77: 70–78.

Finore, I., P. Di Donato, A. Poli et al. 2014. Use of agro waste biomass for α-amylase production by *Anoxybacillus amylolyticus*: Purification and properties. Journal of Microbial & Biochemical Technology. 2014.

Galanakis, C. M. 2012. Recovery of high added-value components from food wastes: Conventional, emerging technologies and commercialized applications. Trends in Food Science & Technology 26(2): 68–87.

Gelain, L., J. G. da Cruz Pradella and A. C. da Costa. 2015. Mathematical modeling of enzyme production using *Trichoderma harzianum* P49P11 and sugarcane bagasse as carbon source. Bioresource Technology 198: 101–107.

Ghasem, D. N. 2007. Production of citric acid. pp. 280–286. *In*: Biochemical Engineering and Biotechnology. Elsevier.

Gullo, M., E. Verzelloni and M. Canonico. 2014. Aerobic submerged fermentation by acetic acid bacteria for vinegar production: Process and biotechnological aspects. Process Biochemistry 49(10): 1571–1579.

Hang, Y. D. and E. E. Woodams. 1986. Solid state fermentation of apple pomace for citric acid production. MIRCEN Journal of Applied Microbiology and Biotechnology 2(2): 283–287.

Hang, Y. D., B. S. Luh and E. E. Woodams. 1987. Microbial production of citric acid by solid state fermentation of kiwifruit peel. Journal of Food Science 52(1): 226–227.

Hashem, M. and S. M. I. Darwish. 2010. Production of bioethanol and associated by-products from potato starch residue stream by *Saccharomyces cerevisiae*. Biomass and Bioenergy 34(7): 953–959.

Hölker, U., M. Höfer and J. Lenz. 2004. Biotechnological advantages of laboratory-scale solid-state fermentation with fungi. Applied Microbiology and Biotechnology 64(2): 175–186.

Huang, J., J. Cai, J. Wang et al. 2011. Efficient production of butyric acid from Jerusalem artichoke by immobilized Clostridium tyrobutyricum in a fibrous-bed bioreactor. Bioresource Technology 102(4): 3923–3926.

Huynh, T. N., G. Smagghe, G. B. Gonzales, J. Van Camp and K. Raes. 2015. Extraction and bioconversion of kaempferol metabolites from cauliflower outer leaves through fungal fermentation. Biochemical Engineering Journal.

Imandi, S. B., V. V. R. Bandaru, S. R. Somalanka, S. R. Bandaru and H. R. Garapati. 2008. Application of statistical experimental designs for the optimization of medium constituents for the production of citric acid from pineapple waste. Bioresource Technology 99(10): 4445–4450.

Jayachandran, K., I. C. Nair, T. S. Swapna and A. Sabu. 2013. Fermentation of food processing by products. pp. 204–231. *In*: M. Chandrasekaran (ed.). Valorization of Food Processing By-products. CRC Press, Boca Raton, London, New York.

Jiang, L., J. Wang, S. Liang, X. Wang, P. Cen and Z. Xu. 2009. Butyric acid fermentation in a fibrous bed bioreactor with immobilized Clostridium tyrobutyricum from cane molasses. Bioresource Technology 100(13): 3403–3409.

Kafilzadeh, F. and E. Maleki. 2012. Chemical composition, *in vitro* digestibility and gas production of straws from different varieties and accessions of chickpea. Journal of Animal Physiology and Animal Nutrition 96(1): 111–118.

Kao, T. H. and B. H. Chen. 2013. Fruits and vegetables. pp. 517–557. *In*: M. Chandrasekaran (ed.). Valorization of Food Processing By-products. CRC Press, Boca Raton, London, New York.

Karthikeyan, A. and N. Sivakumar. 2010. Citric acid production by Koji fermentation using banana peel as a novel substrate. Bioresource Technology 101(14): 5552–5556.

Kasi, M., V. S. Chandrasekaran, P. Karthikeyan and S. Al-Sohaibani. 2013. Current state of the art of food processing by-products. pp. 35–62. *In*: M. Chandrasekaran (ed.). Valorization of Food Processing By-products. CRC Press, Boca Raton, London, New York.

Kosseva, M. R., C. A. Kent and D. R. Lloyd. 2003. Thermophilic bioremediation strategies for a dairy waste. Biochemical Engineering Journal 15(2): 125–130.

Krishna, S. H., T. J. Reddy and G. V. Chowdary. 2001. Simultaneous saccharification and fermentation of lignocellulosic wastes to ethanol using a thermotolerant yeast. Bioresource Technology 77(2): 193–196.

Kumar, D., K. K. Yadav, M. Muthukumar and N. Garg. 2013. Production and characterization of [alpha]-amylase from mango kernel by *Fusarium solani* NAIMCC-F-02956 using submerged fermentation. Journal of Environmental Biology 34(6): 1053.

Kumar, R. and S. Shivakumar. 2014. Production of L-Lactic acid from starch and food waste by amylolytic *Rhizopus oryzae* MTCC 8784. International Journal of Chemical Technology and Research 6: 527–537.

Kunamneni, A. and S. Singh. 2005. Response surface optimization of enzymatic hydrolysis of maize starch for higher glucose production. Biochemical Engineering Journal 27(2): 179–190.

Lal, R. 2005. World crop residues production and implications of its use as a biofuel. Environment International 31(4): 575–584.

Lee, I., Y. Hung and C. Chou. 2008. Solid-state fermentation with fungi to enhance the antioxidative activity, total phenolic and anthocyanin contents of black bean. International Journal of Food Microbiology 121(2): 150–156.

Liang, S., K. Gliniewicz, A. T. Gerritsen and A. G. McDonald. 2016. Analysis of microbial community variation during the mixed culture fermentation of agricultural peel wastes to produce lactic acid. Bioresource Technology 208: 7–12.

Liguori, R., A. Amore and V. Faraco. 2013. Waste valorization by biotechnological conversion into added value products. Applied Microbiology and Biotechnology 97(14): 6129–6147.

Lim, J. S., Z. Abdul Manan, S. R. Wan Alwi and H. Hashim. 2012. A review on utilisation of biomass from rice industry as a source of renewable energy. Renewable and Sustainable Energy Reviews 16(5): 3084–3094.

Lima, J. S. d., R. Cruz, J. C. Fonseca et al. 2014. Production, characterization of tannase from *Penicillium montanense* URM 6286 under SSF using agroindustrial wastes and application in the clarification of grape juice (*Vitis vinifera* L.). The Scientific World Journal. 2014.

Lin, S., Q. Zhu, L. Wen et al. 2014. Production of quercetin, kaempferol and their glycosidic derivatives from the aqueous-organic extracted residue of litchi pericarp with *Aspergillus awamori*. Food Chemistry 145: 220–227.

Lin, Y. and S. Tanaka. 2006. Ethanol fermentation from biomass resources: Current state and prospects. Applied Microbiology and Biotechnology 69(6): 627–642.

Lin, Y., D. Wang and S. Li. 2010. Biological pretreatment enhances biogas production in the anaerobic digestion of pulp and paper sludge. Waste Management & Research.

Liu, T. P. S. L., T. S. Porto, K. A. Moreira et al. 2016. Tannase production by *Aspergillus* spp. UCP1284 using cashew bagasse under solid state fermentation. African Journal of Microbiology Research 10(16): 565–571.

López, S., D. R. Davies, F. J. Giraldez, M. S. Dhanoa, J. Dijkstra and J. France. 2005. Assessment of nutritive value of cereal and legume straws based on chemical composition and *in vitro* digestibility. Journal of the Science of Food and Agriculture 85(9): 1550–1557.

Madeira, J. V., L. R. Ferreira, J. A. Macedo and G. A. Macedo. 2015. Efficient tannase production using Brazilian citrus residues and potential application for Orange juice valorization. Biocatalysis and Agricultural Biotechnology 4(1): 91–97.

Maier, T., A. Göppert, D. R. Kammerer, A. Schieber and R. Carle. 2008. Optimization of a process for enzyme-assisted pigment extraction from grape (*Vitis vinifera* L.) pomace. European Food Research and Technology 227(1): 267–275.

Margret, E. B. M., J. M. Brulc, E. A. Bayer, R. Lamed, H. J. Flint and B. A. White. 2010. Advanced technologies for biomass hydrolysis and saccharification using novel enzymes. pp. 199–212. *In*: Alain A. Vertes, Nasib Qureshi, Hans P. Blaschek and Hideaki Yukawa (eds.). Biomass to Biofuels: Strategies for Global Industries. Wiley, United Kingdom.

Martins, S., S. I. Mussatto, G. Martínez-Avila, J. Montañez-Saenz, C. N. Aguilar and J. A. Teixeira. 2011. Bioactive phenolic compounds: Production and extraction by solid-state fermentation. A review. Biotechnology Advances 29(3): 365–373.

Mirabella, N., V. Castellani and S. Sala. 2014. Current options for the valorization of food manufacturing waste: A review. Journal of Cleaner Production 65: 28–41.

Misi, S. N. and C. F. Forster. 2002. Semi-continuous anaerobic co-digestion of agro-wastes. Environmental Technology 23(4): 445–451.

Mouna imen, O. and K. Mahmoud. 2015. Statistical optimization of cultural conditions of an halophilic alpha-amylase production by halophilic *Streptomyces* sp. grown on orange waste powder. Biocatalysis and Agricultural Biotechnology 4(4): 685–693.

Mrudula, S. and R. Anitharaj. 2011. Pectinase production in solid state fermentation by *Aspergillus niger* using orange peel as substrate. Glob. J. Biotechnol. Biochem. 6(2): 64–71.

Mtui, G. Y. S. 2012. Lignocellulolytic enzymes from tropical fungi: Types, substrates and applications. Scientific Research and Essays 7(15): 1544–1555.

Mudaliyar, P., L. Sharma, and C. Kulkarni. 2012. Food waste management-lactic acid production by *Lactobacillus* species. Int. J. Adv. Biol. Res. 2(1): 34–38.

Mukhopadhyay, R., S. Chatterjee, B. P. Chatterjee, P. C. Banerjee and A. K. Guha. 2005. Production of gluconic acid from whey by free and immobilized Aspergillus niger. International Dairy Journal 15(3): 299–303.

Mushtaq, M., B. Sultana, F. Anwar, A. Adnan and S. S. H. Rizvi. 2015. Enzyme-assisted supercritical fluid extraction of phenolic antioxidants from pomegranate peel. The Journal of Supercritical Fluids 104: 122–131.

Mushtaq, M., B. Sultana, H. N. Bhatti and M. Asghar. 2015. RSM based optimized enzyme-assisted extraction of antioxidant phenolics from underutilized watermelon (*Citrullus lanatus* Thunb.) rind. Journal of Food Science and Technology 52(8): 5048–5056.

Naveen, K. M., R. Potumarthi, R. R. Baadhe and V. K. Gupta. 2014. Current bioenergy researches: Strengths and future challenges. pp. 1–21. *In*: Vijai K. Gupta, Maria G. Tuohy, Christian P. Kubicek, Jack Saddler, Feng Xu (eds.). Bioenergy Research: Advances and Applications. Elsevier.

nee'Nigam, P. S. 2009. Production of bioactive secondary metabolites. *In*: Biotechnology for Agro-Industrial Residues Utilisation: Springer.

Norsalwani, T. L. T. and N. A. N. Norulaini. 2012. Utilization of lignocellulosic wastes as a carbon source for the production of bacterial cellulases under solid state fermentation. International Journal of Environmental Science and Development 3(2): 136.

Oberoi, H. S., Y. Chavan, S. Bansal and G. S. Dhillon. 2010. Production of cellulases through solid state fermentation using kinnow pulp as a major substrate. Food and Bioprocess Technology 3(4): 528–536.

Panda, S. K., S. S. Mishra, E. Kayitesi and R. C. Ray. 2016. Microbial-processing of fruit and vegetable wastes for production of vital enzymes and organic acids: Biotechnology and scopes. Environmental Research 146: 161–172.

Pandey, A., P. Selvakumar, C. R. Soccol and P. Nigam. 1999. Solid state fermentation for the production of industrial enzymes. Current Science 77(1): 149–162.

Paranthaman, R., R. Vidyalakshmi, S. Murugesh and K. Singaravadivel. 2009. Effects of fungal co-culture for the biosynthesis of tannase and gallic acid from grape wastes under solid state fermentation. Glob. J. Biotechnol. Biochem. 4(1): 29–36.

Pérez Gregorio, M. R., J. Regueiro, F. Alonso-González, L. M. Pastrana-Castro and J. Simal-Gándara. 2011. Influence of alcoholic fermentation process on antioxidant activity and phenolic levels from mulberries (*Morus nigra* L.). LWT—Food Science and Technology 44(8): 1793–1801.

Piero, R. 2008. Organic acids. pp. 137–170. *In*: Nenad Blau, Marinus Duran and K. Michael Gibson (eds.). Laboratory Guide to the Methods in Biochemical Genetics. Springer.

Poondla, V., S. K. Yannam, S. N. Gummadi, R. Subramanyam and V. S. Reddy Obulam. 2016. Enhanced production of pectinase by *Saccharomyces cerevisiae* isolate using fruit and agro-industrial wastes: Its application in fruit and fiber processing. Biocatalysis and Agricultural Biotechnology 6: 40–50.

Puupponen-Pimiä, R., L. Nohynek, R. Juvonen et al. 2016. Fermentation and dry fractionation increase bioactivity of cloudberry (*Rubus chamaemorus*). Food Chemistry 197, Part A: 950–958.

Raji, Y. O., M. Jibril, I. M. Misau and B. Y. Danjuma. 2012. Production of vinegar from pineapple peel. International Journal of Advanced Scientific Research and Technology 3(2): 656–666.

Rathan, R. K. and M. Ambili. 2011. Cellulase enzyme production by *Streptomyces* sp. using fruit waste as substrate. Australian Journal of Basic and Applied Sciences 5(12): 1114–1118.

Rathinaraj, K. and N. M. Sachindra. 2013. Meats, poultry, and eggs. pp. 650–684. *In*: M. Chandrasekaran (ed.). Valorization of Food Processing By-products. CRC Press, Boca Raton, London, New York.

Robert, J. W. and V. O. Maarten. 2010. Enzymes in Food Technology. Wiley-Blackwell.

Romano, R. T., R. Zhang, S. Teter and J. A. McGarvey. 2009. The effect of enzyme addition on anaerobic digestion of JoseTall Wheat Grass. Bioresource Technology 100(20): 4564–4571.

Ruane, J., A. Sonnino and A. Agostini. 2010. Bioenergy and the potential contribution of agricultural biotechnologies in developing countries. Biomass and Bioenergy 34(10): 1427–1439.

Russ, W. and R. Meyer-Pittroff. 2004. Utilizing waste products from the food production and processing industries. Critical Reviews in Food Science and Nutrition 44(1): 57–62.

Russ, W. and M. Schnappinger. 2007. Waste related to the food industry: A challenge in material loops. pp. 1–14. *In*: Vasso Oreopoulou and Winfried Russ (eds.). Utilization of By-products and Treatment of Waste in the Food Industry. Springer, New York.

Sagu, S. T., E. J. Nso, S. Karmakar and S. De. 2014. Optimisation of low temperature extraction of banana juice using commercial pectinase. Food Chemistry 151: 182–190.

Sanchez, S. and A. L. Demain. 2010. Enzymes and bioconversions of industrial, pharmaceutical and biotechnological significance. Organic Process Research & Development 15(1): 224–230.

Santana-Méridas, O., A. González-Coloma and R. Sánchez-Vioque. 2012. Agricultural residues as a source of bioactive natural products. Phytochemistry Reviews 11(4): 447–466.

Sarita, G. B. and R. K. Sukumaran. 2013. Enzyme technologies for bioconversion of food processing by-products. pp. 233–259. *In*: M. Chandrasekaran (ed.). Valorization of Food Processing By-products. CRC Press, Boca Raton, London, New York.

Sauer, M., D. Porro, D. Mattanovich and P. Branduardi. 2008. Microbial production of organic acids: Expanding the markets. Trends in Biotechnology 26(2): 100–108.

Saxena, R. C., D. K. Adhikari and H. B. Goyal. 2009. Biomass-based energy fuel through biochemical routes: A review. Renewable and Sustainable Energy Reviews 13(1): 167–178.

Shah, A., C. C. Akoh, R. T. Toledo and M. Corredig. 2004. Isolation of a phospholipid fraction from inedible egg. The Journal of Supercritical Fluids 30(3): 303–313.

Shojaosadati, S. A. and V. Babaeipour. 2002. Citric acid production from apple pomace in multi-layer packed bed solid-state bioreactor. Process Biochemistry 37(8): 909–914.

Shyam, K. R. and M. Chandrasekaran. 2013. Beverages. pp. 590–614. *In*: M. Chandrasekaran (ed.). Valorization of Food Processing By-products. CRC Press, Boca Raton, London, New York.

Silva, D., E. d. S. Martins, R. d. Silva and E. Gomes. 2002. Pectinase production by *Penicillium viridicatum* RFC3 by solid state fermentation using agricultural wastes and agro-industrial by-products. Brazilian Journal of Microbiology 33(4): 318–324.

Singh, R. and S. Singh. 2007. Design and development of batch type acetifier for wine-vinegar production. Indian Journal of Microbiology 47(2): 153–159.

Sowbhagya, H. B. and V. N. Chitra. 2010. Enzyme-assisted extraction of flavorings and colorants from plant materials. Critical Reviews in Food Science and Nutrition 50(2): 146–161.

Surajit, M., M. Puniya, K. P. S. Sangu, S. S. Dagar, R. Singh and A. K. Puniya. 2013. Dairy by-products: Wastes or resources? pp. 617–648. *In*: M. Chandrasekaran (ed.). Valorization of Food Processing By-products. CRC Press, Boca Raton, London, New York.

Taherzadeh, M. J. and K. Karimi. 2007. Enzymatic-based hydrolysis processes for ethanol from lignocellulosic materials: A review. BioResources 2(4): 707–738.

Tanaka, T., M. Hoshina, S. Tanabe, K. Sakai, S. Ohtsubo and M. Taniguchi. 2006. Production of D-lactic acid from defatted rice bran by simultaneous saccharification and fermentation. Bioresource Technology 97(2): 211–217.

Tekin, A. R. and A. C. Dalgıç. 2000. Biogas production from olive pomace. Resources, Conservation and Recycling 30(4): 301–313.

Unakal, C., R. I. Kallur and B. B. Kaliwal. 2012. Production of α-amylase using banana waste by *Bacillus subtilis* under solid state fermentation. Eur. J. Exper. Biol. 2: 1044–1052.

Venkatesh, M., P. B. Pushpalatha, K. B. Sheela and D. Girija. 2009. Microbial pectinase from tropical fruit wastes. Journal of Tropical Agriculture 4: 67–69.

Vijayakumar, J., R. Aravindan and T. Viruthagiri. 2008. Recent trends in the production, purification and application of lactic acid. Chemical and Biochemical Engineering Quarterly 22(2): 245–264.

Vikas, O. V. and U. Mridul. 2014. Bioconversion of papaya peel waste in to vinegar using *Acetobacter aceti*. Int. J. Sci. Res. 3(11): 409–411.

Wang, N., N. Zhang and M. Wang. 2006. Wireless sensors in agriculture and food industry—Recent development and future perspective. Computers and Electronics in Agriculture 50(1): 1–14.

Wei, D., X. Liu and S. Yang. 2013. Butyric acid production from sugarcane bagasse hydrolysate by Clostridium tyrobutyricum immobilized in a fibrous-bed bioreactor. Bioresource Technology 129: 553–560.

Wikandari, R., H. Nguyen, R. Millati, C. Niklasson and M. J. Taherzadeh. 2015. Improvement of biogas production from orange peel waste by leaching of limonene. BioMed. Research International 2015.

Wolfgang, A. 2007. Enzymes in Industry. WILEY-VCH Verlag GmbH & Co. KGaA, Weinheim.

Yigitoglu, M. 1992. Production of citric acid by fungi. Journal of Islamic Academy of Sciences 5(2): 100–106.

Zhang, M., R. Su, W. Qi and Z. He. 2010. Enhanced enzymatic hydrolysis of lignocellulose by optimizing enzyme complexes. Applied Biochemistry and Biotechnology 160(5): 1407–1414.

Zhongtang, Y., M. Morrison and F. L. Schanbacher. 2010. Production and utilization of methane biogas as renewable fuel. pp. 403–434. *In*: Alain A. Vertes, Nasib Qureshi, Hans P. Blaschek and Hideaki Yukawa (eds.). Biomass to Biofuels: Strategies for Global Industries. Wiley, United Kingdom.

CHAPTER 13

Utilisation of Bioactive Compounds Derived from Waste in the Food Industry

Quan V. Vuong[1,]* and *Mirella A. Atherton*[2]

Introduction

Bioactive compounds are well known for their antioxidant quality, so it is no wonder that, they have been linked with various health benefits, such as prevention of cardiovascular disease, obesity, diabetes and cancer. Bioactive compounds are also associated with anti-microbial growth and prevention of lipid oxidation (Vuong et al. 2011). In food products, microbial activity and lipid oxidation are major problems. Microbial growth is well known to be a major cause of food deterioration. Microorganisms, such as bacteria, yeasts and molds can spoil the food during harvesting, handling, transportation, processing, storage and preservation. As microorganisms are found everywhere—the environment, human handling and processing equipment—they can contaminate and deteriorate the food as well as negatively affect the health of the consumer (Manay and Shadaksharaswamy 2001).

Lipid oxidation leads to the development of undesirable rancidity and potentially toxic reaction products (Vuong et al. 2011). Bioactive compounds have the ability to scavenge free radicals and this scavenging reaction allows them to trap superoxide anions or hydroxyl radicals, thus suppressing and terminating the free radical chain reaction that occurs during lipid peroxidation. In addition, bioactive compounds also have metal ion-chelating properties. Thus they can effectively chelate metal ions, which are catalysts, for the initiation and propagation of the lipid peroxidation chain reaction. Therefore, bioactive compounds as natural antioxidants have been found to prevent lipid peroxidation in food (Choe and Min 2009).

[1] School of Environmental and Life Sciences, University of Newcastle, Australia.
[2] University of Newcastle, Academic Division, Centre for English Language and Foundation Studies, 10 Chittaway Road, Ourimbah, NSW 2258 Australia.
* Corresponding author: vanquan.vuong@newcastle.edu.au

Bioactive compounds are known as natural pigments, which have been used in food. Natural pigments found in plants include chlorophyll, flavonoids/anthocyanins, betalains and carotenoids. Chlorophyll can absorb light in the red and blue regions to emit a green color. Chlorophyll has been used as a natural food coloring agent and has antioxidant as well as antimutagenic properties (Hosikian et al. 2010). Flavonoids or anthocyanins, a class of flavonoids, which are found in many flowers, leaves, fruits, seeds and other tissues, are known to be responsible for a wide range of colors, from orange-to-blue (Tanaka et al. 2008). Whereas, betalains, which are derived from tyrosine, are responsible for yellow-to-red colors, carotenoids are essential components of photosystems and impart yellow-to-red colors to flowers and fruits. In addition, flavonoids/anthocyanins are found to combine with carotenoids, resulting in a variety of colors in fruits and vegetable (Tanaka et al. 2008).

The changes in consumers' demands and legislation have given rise to preference for using natural coloring agents in food products, instead of using artificial colors (Hosikian et al. 2010). Bioactive compounds derived from plant materials have the potential to be used as natural coloring agents in food products. It should be noted that bioactive compounds are sensitive to heat, oxygen and light. It is challenging to retain these natural compounds when applied to food due to degradation. Therefore, proper technologies for minimizing the loss of bioactive compounds are needed. Despite the challenge of utilizing bioactive compounds in food, there are substantial opportunities for using bioactive compounds as potential antioxidant, antibacterial and antimutagen ingredients in food products.

As mentioned in the earlier chapters, food waste contains numerous bioactive compounds which can be isolated for further utilisation in food products. In Australia, food consumption has grown by 0.7 per cent per capita since 2015 (Australia Food and Drink Report 2016). It is important to understand that not only is the total population of a country a key variable in consumer demand, even the demographic profile is essential to understand issues, such as food waste. The long-term outlook is that intense consolidated activity is expected to continue across the global food and drink industry. Organic growth and cost-cutting have been identified as the key areas of focus since 2008. Manufacturers and retailers will look for synergies and greater market power in this competitive industry (Australia Food and Drink Report 2016). Isolation and reuse of bioactive compounds from food waste is one form of recycling and cost cutting that is expected to escalate.

Mechanisms of Action of Bioactive Compounds as Antioxidants in Oxidation of Lipids

Lipid oxidation is a major problem for oil-containing food products because their shelf-life and nutritional composition are significantly lessened during preservation, transportation, processing, storage and cooking (Uluata et al. 2015). Therefore, many attempts have been made to prevent or delay oxidation in food. Modification of oxygen from the closed food-packaging system or addition of edible oxygen scavengers into foods to remove oxygen within the food have been found effective to inhibit lipid oxidation in food (Uluata et al. 2015). Lipid oxidation occurs due to the reaction of lipids with atmospheric triplet oxygen or singlet oxygen. Triplet oxygen reacts with radicals and causes autoxidation. Singlet oxygen directly reacts with the double bonds of unsaturated fats and oils with high electron densities and causes oxidation (Eunok et al. 2005).

There are different chemical mechanisms, which are responsible for the lipid oxidation. Three mechanisms are commonly known for yielding different oxidation products including the free-radical mechanism, photo-oxidation and the process related to lipoxygenase activity (Wasowicz et al. 2004). The free-radical chain mechanism involves three steps: (1) the initiation step, in which homolytic hydrogen atom abstraction from a methylene group leads to alkyl radical formation; (2) the propagation step, in which the formation of peroxy radicals reacting with unsaturated fatty acids to form hydroperoxides; and (3) the termination step wherein the formation of non-radical products by interaction of the fatty acid radical, peroxy radical and terminal free radical chain occurs (Wasowicz et al. 2004).

The photo-oxidation mechanism involves the formation of hydroperoxides. Light accelerates lipid oxidation, especially in the presence of photosensitizers, which become excited upon absorption of light energy. The excited photosensitizers react with triplet oxygen and produce singlet oxygen through energy transfer. Singlet oxygen reacts with the double bond of unsaturated fatty acids to produce hydroperoxides without lipid radical formation (Choe and Min 2009, Wasowicz et al. 2004). The mechanism related to lipoxygenase activity is known as non-radical mechanism. Lipoxygenase is an iron-bound enzyme with Fe in its active center. Lipoxygenase oxidizes unsaturated fatty acids having a 1-*cis*, 4-*cis*-pentadiene system and resulting in oil deterioration (Choe and Min 2009). The active enzyme abstracts hydrogen atom from the methylene group of a polyunsaturated fatty acid with the iron being reduced to Fe (II). Antioxidants chelate the Fe (II) to inhibit lipoxygenase activity and thus slow down the oxidation of unsaturated fatty acids (Wasowicz et al. 2004).

In general, bioactive compounds can slow down lipid oxidation by a combination of scavenging free radicals, chelating prooxidative metals, quenching singlet oxygen and photosensitizers and inactivating lipoxygenase (Choe and Min 2009).

Bioactive Compounds as Antioxidants against Microbial Growth

The exact mechanisms of antimicrobial action of bioactive compounds have not been fully described; however, the anti-microbial activities can be explained through the effect of bioactive compounds on microbial cells by attacking the phospholipid bilayer of the cell membrane, disrupting enzyme systems, compromising the genetic material of bacteria and oxidizing unsaturated fatty acids to result in the formation of fatty acid hydroperoxides (Tajkarimi et al. 2010).

Numerous studies have reported the link between bioactive compounds derived from plants with inhibition of bacteria, molds and fungi. In terms of anti-bacterial properties, bioactive compounds have been found to inhibit the growth of a wide range of bacteria, such as *A. hydrophila, Bacillus* sp. *C. jejuni, Clostridium perfringens, Escherichia coli, L. monocytogenes, Pseudomonas* sp., *Salmonella* sp. and *Shigella* sp., and *Staphylococcus aureus* (Aziz and Karboune 2016, Tajkarimi et al. 2010). Bioactive compounds are reported to slow down the growth of molds, such as *Aspergillus flavus, Aspergillus niger* and *Aspergillus parasiticus* (Kumar and Pandey 2013, Tajkarimi et al. 2010). In addition, bioactive compounds have been known to inhibit growth of various fungi, such as *Aspergillus niger*, *A. flavus*, *A. parasiticus*, and *spergillus* and *Penicillium* species (Tajkarimi et al. 2010).

In general, bioactive compounds inhibit the growth and activities of bacteria, mold and fungi, which are major problems linked with food deterioration and human health. Therefore, bioactive compounds derived from waste can be utilized in food products to extend their self-life and minimize their loss during transportation, processing, storage and cooking.

Bioactive Compounds as Natural Coloring Agents

Many bioactive compounds derived from plants are responsible for the color and the three major groups including chlorophyll, flavonoids and carotenoids are outlined in this section. Color is an important characteristic of food products because it contributes to the food sensory quality. In recent times, customer demands have changed from artificial colors to natural colors because artificial colors have been linked with negative health issues (Aberoumand 2011). Numerous studies have attempted to extract and isolate natural colored compounds from plants and utilize them for coloring in food products. However, it should be noted that natural coloring pigments are more expensive and less stable than artificial coloring agents and future research is required to reduce the cost and improve the stability of the natural coloring pigments.

Chlorophyll

There are two main types of chlorophyll which can be degraded into different types when exposed to weak acids, oxygen or light. Chlorophyll is found abundantly in many plants and plays an important role in photosynthesis. As it selectively absorbs light in the red and blue regions of the visible spectrum, it emits a green color (Hosikian et al. 2010). Chlorophyll has been extracted and isolated using different solvents and methods for further utilisation of foods. However, chlorophyll has its limitations when applied as a coloring agent in food products. For example, it is not water soluble; its pigment content is not precisely known and it tends to be unstable under different pH conditions of foods to which it is added. Furthermore, it is more expensive than artificial colorings and less stable. Therefore, chlorophyll requires a chemical modification by replacing the magnesium centre with a copper ion in order to improve its stability as a food coloring agent (Hosikian et al. 2010). In addition, it can be made to become water-soluble by saponification of oleoresin, and this water-soluble component is called chlorophyllin (Mortensen 2006). Of note, the use of chlorophyll in food products varies from country to country. For example, it is not allowed to be used in the US, but is permitted for use in EU and elsewhere (Mortensen 2006).

Flavonoids

Flavonoids are the most common and widely distributed group of plant phenolic compounds and are a major coloring component of plants (Kumar and Pandey 2013). Flavonoid pigments comprise anthocyanins, anthoxanthins and betalains. Anthocyains include a total of 19 types, with the six major ones being: pelargonidin, cyanidin, peonidin, delphinidin, petunidin and malvidin. Anthocyanins exhibit the red, purple and blue colors. Their color greatly depends on the number of hydroxyl groups present on their B-ring; the larger the number of groups, the bluer the color (Tanaka et al. 2008). The color can be also affected by pH. Under acidic conditions, anthocyains intensify the red color, but might change to green color under alkaline conditions (Brown 2008).

Anthoxanthins comprise of flavones, flavonols and flavonones and they are responsible for the cream or white color of cauliflower, white potato and turnips. Under acidic conditions, anthoxanthins can be whiter; however, they might turn into an undesirable yellow color in alkaline water. They even change to blue-black or red-brown under excessive heating or in the presence of iron or copper (Brown 2008). Betalains comprise two groups of pigments: the red–purple betacyanidins and the yellow betaxanthins, which are water-soluble. Betacyanidins are conjugates of cyclo-DOPA and betalamic acid, whereas betaxanthins are conjugates of amino acids or amines and betalamic acid (Mortensen 2006). Under acidic conditions, Betalains can change the purple-red to a brighter red color, whereas the red color can change to yellow in an alkaline medium (Brown 2008).

Carotenoids

Carotenoids have a deep red, yellow or orange color and are lipid soluble (Mortensen 2006). Carotenoids include carotenes, lycopenes and xanthophylls. Carotenes show the reddish-orange color of carrots and winter squashes. Carotenes comprise of alpha-, beta- and gamma-carotenes (Brown 2008), of which, beta-carotene is popular and has been applied for coloring dairy products, which typically contain a high fat content. Thus, beta-carotene is often added to margarine and cheese. Lycopene is orange in color. However, it is hardly used as a colorant because it is an expensive pigment and is very prone to oxidative degradation (Mortensen 2006). Xanthophylls are responsible for the light yellow pigment. Carotenoids are sensitive to heat and light; thus the content of carotenoids can be affected when applied to food (Brown 2008, Mortensen 2006).

Utilisation of Bioactive Compounds in Meat and Meat Products

Meat and meat products have high levels of lipids, ranging from 4.5–11 per cent, and thus they are susceptible to lipid oxidation (Tang et al. 2001). Of note, levels of lipids vary according to the type of meat or different parts of the animals are used (Vuong et al. 2011). Bioactive compounds derived from plant materials as well as waste have been utilized in meat and meat products to inhibit microbial growth and slow down the enzyme activities, that is, to extend the shelf-life and/or improve quality of meat and meat products (Table 1).

During storage or processing, meat and meat products can get exposed to light and oxygen besides being affected by microorganisms. Addition of bioactive compounds with the potent antioxidant can minimize the lipid oxidation by chelating iron, which is the major active catalyst for oxidative rancidity in meat (Tang et al. 2001). Moreover, bioactive compounds can also trap superoxide, hydroxyl and peroxyl

Table 1. Utilisation of plant and waste extracts in meat and meat products.

Plant extracts/ bioactive compounds	Meat and meat products	Role of actions	References
Green tea, grape seed polyphenols and ascorbic acid	Dry-cured sausages during the ripening period	Decrease Thiobarbituric acid reactive substances (TBARS), residual nitrite and N-nitrosamines	(Li et al. 2013)
Cherry and blackcurrant leaf extracts containing pigallocatechin and glycosides of quercetin and kaempferol	Pork sausages chilled for 14 and 28 days	Lower malondialdehyde (MDA) generation, increased the shelf life of vacuum-packed sausages, and decreased microbial growth	(Nowak et al. 2016)
Tea polyphenol, grape seed extract, gingerol and α-tocopherol	Dry-cured bacons at the end of ripening	Decrease pH, thiobarbituric acid reactive substances content, and total volatile basic nitrogen. Decrease aerobic plate counts, Enterobacteriaceae, Micrococcaceae, yeast, and molds	(Wang et al. 2015)
Ampelopsis grossedentata extract and its major component dihydromyricetin	Cooked ground beef	Lower oxidation of lipids	(Ye et al. 2015)
Cysteinylcaffeic acids (which are coupling products of caffeic acid and cysteine)	Fresh meat	Retain the fresh meat color	(Miura et al. 2014)
Olive leaf extracts	Frozen pork patties	Improvement of sensory attributes by delaying oxidation of lipids	(Botsoglou et al. 2014)
Ethanol extract of Kitaibelia vitifolia	Fermented dry sausages without nitrite addition refrigerated for 20th days	Inhibition of growth of *Escherichia coli*	(Kurcubic et al. 2014)
Ginkgo leaf extracts	Pork meatballs refrigerated over 21 days	Stabilising effect on cholesterol and inhibition of the formation of oxidized derivatives	(Kobus-Cisowska et al. 2014)
Combination of Alpinia katsumadai seed extract and epigallocatechin gallate	Chicken-meat juice and sterile minced meat	Inhibition of food-borne pathogenic strains of *Listeria monocytogenes*, *Escherichia coli* and *Campylobacter jejuni*	(Klancnik et al. 2014)
Lemon balm (*Melissa officinalis* L.) extract	Hamburger Patties during Refrigerated Storage	Improvement of sensory quality and delay lipid peroxidation	(Lee et al. 2014)

Table 1 contd. ...

...Table 1 contd.

Plant extracts/ bioactive compounds	Meat and meat products	Role of actions	References
Lycopene addition	Cured turkey meat	Increase the content of essential hypocholesterolemic fatty acids and decrease the content of saturated hypercholesterolemic fatty acids	(Skiepko et al. 2016)
Nitraria retusa extracts	Beef patties refrigerated for 9 days	Inhibition of the formation of the TBARS and microbial growth	(Mariem et al. 2014)

radicals and suppress free radical chain reactions, thereby preventing and terminating the lipid oxidation in meat and meat products (Wasowicz et al. 2004). Furthermore, addition of bioactive compounds can minimize the microbial growth by attacking the phospholipid bilayer of the cell membrane, or disrupting the enzyme systems (Tajkarimi et al. 2010). In addition, bioactive compounds (catechins) have been found to improve and protect the color of meat products, such as minced beef patties packed under aerobic acid addition and modified atmosphere pakaging (MAP) conditions, with 80 per cent O_2 and 20 per cent CO_2, during refrigerated storage for seven days; or beef patties packed under aerobic conditions during refrigerated storage for nine days (Bañón et al. 2007, Tang et al. 2006).

Waste generated from agriculture and the food industry containing diversified bioactive compounds is a great potential source for utilisation in meat and meat products to improve their quality and extend their shelf-life. Numerous studies have applied extracts from waste for utilisation in meat and meat products. For example, extracts from grape seeds, cherry and blackcurrant leaves, Ginkgo and olive leaves have been effectively applied to improve quality, minimize lipid oxidation and inhibit microbial growth in sausages, bacon, meat patties and meat juice (Table 1), indicating that waste can be potentially used as by-products for improvement of meat and meat product quality.

Utilisation of Bioactive Compounds in Seafood and Seafood Products

Seafood and seafood products are very susceptible to lipid peroxidation because they contain a high content of polyunsaturated fatty acids. Polyunsaturated fatty acids are responsible for the initial development of oxidation, which is initiated by the enzymes—lipoxygenase and peroxidase (Vuong et al. 2011). In addition, other enzymes, such as serine proteinase, metalloproteinase and cysteine proteinase, which hydrolyze proteins in seafood have resulted in deterioration of flavor, texture and taste of seafood during preservation (Vuong et al. 2011). Therefore, lipid oxidation and enzymatic deterioration are the major causes responsible for degradation of seafood.

Table 2. Utilisation of plant and waste extracts in seafood and seafood products.

Plant extracts/ bioactive compounds	Seafood and seafood products	Role of actions	References
Red grape pomace extracts	Minced rainbow trout refrigerated for 6 days	Delay lipid oxidation and cadaverine formation	(Gai et al. 2015)
Pomegranate fruit peel extract	Fish (Bigeye Trevally, *Caranx sexfasciatis*)	Reduction of *L. monocytogenes*	(Al-Zoreky 2009)
Purple rice bran extract	Patties made from minced channel catfish (*Ictalurus punctatus*) belly flap meat	Maintain quality during cooking and inhibit microbial growth	(Min et al. 2009)
Quercetin, cinnamic acid and 4-hexylresorcinol	Pacific white shrimp (*Litopenaeus vannamei*) during cold storage for 12 days	Prevention of melanosis and protein degradation	(Qian et al. 2015)
Epigallocatechin gallate	Skin, muscle, and blood of rainbow trout	Inhibition of metalloproteinase activities	(Saito et al. 2002)
Pomegranate peel extract	Pacific white shrimp (*Litopenaeus vannamei*) refrigerated for 10 days	Retard the mesophilic, psychrophilic, lactic acid bacteria and enterobacteriaceae counts	(Basiri et al. 2015)
Green tea extract in combination with ascorbic acid	Pacific white shrimp (*Litopenaeus vannamei*)	Inhibition of polyphenoloxidase and melanosis as well as retardation of quality loss of shrimp	(Nirmal and Benjakul 2012a)
Catechin and ferulic acid	Pacific white shrimp (*Litopenaeus vannamei*)	Inhibition of melanosis with different modes of inhibition towards polyphenoloxidase	(Nirmal and Benjakul 2012b)
Tea polyphenol coating combined with ozone water washing	Black sea bream (*Sparus macrocephalus*)	Decrease nucleotide breakdown, lipid oxidation, protein decomposition, and microbial growth. Maintain better sensory characteristics through texture and color	(Feng et al. 2012)

Numerous studies have applied extracts from plants and waste to improve quality and extend the shelf-life of seafood and seafood products (Table 2). Red grape pomace, which is considered as waste, was found to effectively prevent lipid oxidation in the minced rainbow trout (Gai et al. 2015). Other waste, such as pomegranate fruit peels and rice bran, was reported to maintain the quality of patties made of catfish mince during cooking and inhibit microbial growth of the catfish minced patties and fish (Bigeye Trevally, *Caranx sexfasciatis*) (Al-Zoreky 2009, Min et al. 2009). These studies revealed that waste containing high levels of bioactive compounds can be utilized for improving and extending the shelf-life of seafood and seafood products.

Utilisation of Bioactive Compounds in Milk and Dairy Products

Milk and dairy products can deteriorate due to oxidation of lipids, proteins or sterols. Oxidation in dairy products leads to off-flavour formation and changes, which negatively affect their shelf-life and quality. Oxidation occurs in dairy products mainly due to their exposure to light (Mortensen et al. 2010). Microbial and enzyme activities are other major causes leading to deterioration of milk and milk products (Sørhaug and Stepaniak 1997). As bioactive compounds have antimicrobial properties and can inhibit some enzyme activities as well as minimize the lipid oxidation, bioactive compounds can be potentially fortified with milk and dairy products to extend their shelf-life and improve their quality.

Many studies have been conducted to utilise plant and waste extracts, which contain high levels of bioactive compounds, into the various dairy products to improve their quality and extend their shelf-life (Table 3). Extracts prepared from olive and grape

Table 3. Utilisation of plant and waste extracts in milk and dairy products.

Plant extracts/ bioactive compounds	Milk and dairy products	Role of actions	References
Olive and grape pomace extracts	Fermented milk	Improve antioxidant capacity	(Aliakbarian et al. 2015)
Grape polyphenols	Yoghurt	Improve antioxidant capacity	(Ye et al. 2014)
Grape (*Vitis vinifera*) seed extracts	Yoghurt	Exhibition of higher antiradical and antioxidant activity	(Chouchouli et al. 2013)
Grape pomace extracts	Semi-hard and hard cheeses	Improvement of polyphenol level and antioxidant capacity	(Marchiani et al. 2016)
Green tea extract	Cheddar-type cheese	Improvement of the antiradical activity of cheese	(Giroux et al. 2013)
Catechin	Low-fat hard cheese	Improvement of polyphenol level and antioxidant capacity	(Rashidinejad et al. 2015)
Malabar spinach pigment-rich extract	Ice-cream	Improvement of overall sensorial quality and more preference by the consumers	(Kumar et al. 2015)
Oleaster (*Elaeagnus angustifolia* L.) with rich phenolic compounds	Ice-cream	Improvement of nutritive and functional values	(Çakmakçı et al. 2015)

pomace waste have been applied into fermented milk, yogurt and cheese to increase antioxidant properties of the products and also improve their antiradical properties (Aliakbarian et al. 2015, Ye et al. 2014). Other products, like cheddar-type cheese and ice-cream, have been fortified with green tea extract or catechins to improve their antioxidant level (Giroux et al. 2013, Kumar et al. 2015) and they can also provide additional health benefits to the consumers. Of note, extracts or bioactive compounds derived from plants and waste do not only exhibit antimicrobial, antienzyme capacity and prevent lipid oxidation, they are also natural coloring agents. Further studies are needed to utilize bioactive compounds in milk and dairy products.

Utilisation of Bioactive Compounds in Plant Food Products

Extracts and bioactive compounds derived from plants and waste can potentially be used in a wide range of plant food products to prolong their shelf-life and to provide antioxidant supplementation to consumers. Different extracts and bioactive compounds derived from plants and waste have been applied into various plant food products to improve their quality (Table 4). For example, extracts from grape seeds with high levels of polyphenols have been applied to cookies and potato chips to inhibit the formation of acrylamide, which is known to be toxic and is usually formed in a variety of heat-treated commercial starchy foods (Xu et al. 2015, Zhu et al. 2011). In addition, extracts and bioactive compounds derived from plants and waste are utilised into other products, such as ice-cream and biscuits to add more antioxidant values to the products (de Camargo et al. 2014, Upadhyay et al. 2007), indicating that there is a great potential for fortification of bioactive compounds into plant food products for quality improvement.

It is interesting to note that extracts and bioactive compounds derived from plants and waste have been also used in post-harvest for fresh produce. They have been incorporated into edible coatings to add more antioxidant values, prevent browning reactions and more importantly, to inhibit microbial growth. The advantages include improving the quality of fresh produce and extending their shelf-life (Table 4). For example, apple pectin was incorporated into edible coating for fresh-cut 'Rojo Brillante' persimmon to prevent browning and to inhibit the growth of mould, yeast and psychrophilic aerobic bacteria (Sanchís et al. 2016). *Salvia fruticosa* mill extract was added into edible coating to effectively control the growth of *B. cinerea* fungus without deteriorating the quality and physico-chemical properties of grapes (Kanetis et al. 2016). In general, there is a great potential for utilisation of bioactive compounds derived from plants and plant waste in food products for improving both quality and extended shelf-life.

Perspectives of Utilisation of Extracts and Bioactive Compounds Derived from Waste in the Food Industry

Bioactive compounds have shown their great potential for utilisation in various food products. As they can inhibit lipid oxidation, slow down some enzyme activities, prevent microbial growth and can be natural coloring pigments, they can potentially be applied in many animal and plant food products to improve their quality and extend

Table 4. Utilisation of plant and waste extracts in plant food products.

Plant extracts/ bioactive compounds	Plant food products	Role of actions	References
Addition of 2 per cent proanthocyanidins from grape seeds	Cookie and potato starch-based models	Inhibition of acrylamide formation	(Zhu et al. 2011)
Muscadine grape polyphenols	Potato chip	Inhibition of acrylamide formation	(Xu et al. 2015)
Green and yellow tea leaf extracts	Cookies	Improvement of the nutritional value, antioxidative potential and stability of the lipid fraction	(Gramza-Michalowska et al. 2016)
Sour cherry pomace extract encapsulated in whey and soy proteins	Cookies	Improvement of functional characteristics of fortified cookies and their preservation	(Tumbas Saponjac et al. 2016)
Peanut skins	Cookies	Increase of soluble fiber, total phenolic content and the corresponding antioxidant capacities	(de Camargo et al. 2014)
Groundnut cake and soybean cake, by-product of agriculture industry	Biscuits	Decrease the breaking strength and increase the sensory parameters	(Behera et al. 2013)
Pomace of carrot	Soft served ice cream	Supplementation of carotenoids	(Upadhyay et al. 2007)
Pomace of carrot	Carrot condensed milk product	Supplementation of carotenoids	(Singh et al. 2006)
Extract of *Salvia fruticosa* Mill. with chitosan on edible coating	Table grapes	Effectively control *B. cinerea* without deteriorating quality and physico-chemical properties of grapes	(Kanetis et al. 2016)
Tea polyphenols incorporated into alginate-based edible coating	Fresh winter jujube	Significantly reduce red indices, total chlorophylls content, respiration rate, electrolyte leakage and malonaldehyde content while maintaining the ascorbic acid content, total phenol content and the activities of antioxidant enzymes	(Zhang et al. 2016)
Edible coatings based on apple pectin	Fresh-cut 'Rojo Brillante' persimmon	Inhibition of browning and growth of moulds, yeasts and psychrophilic aerobic bacteria	(Sanchís et al. 2016)
Cinnamic acid as antioxidant agent into xanthan gum based edible coating	Fresh-cut Asian pear (*Pyrus pyrifolia* L. cv. 'Nashpati') and European pear (*Pyrus communis* L. cv. 'Babughosha') stored at 4°C	Retardation of the oxidative browning, decline of ascorbic acid level, degradation of total phenolics content and reduction in antioxidant capacity	(Sharma and Rao 2015)

their shelf-life. The changes in market demands and legislation show promising trends towards the use of natural agents instead of artificial ones for food products. This has opened up more opportunities for bioactive compounds derived from plants and waste. However, several challenges remain for the utilisation of bioactive compounds in food products and these include:

- High cost of utilisation
- Unstable when exposure to high temperature, light and/or oxygen
- May become toxic when applying high level of bioactive compounds
- May change properties or interact with other food components, such as proteins, vitamins

The challenges can be overcome by application of new techniques. For instance, the cost of bioactive compounds can be reduced by application of cost-effective techniques to prepare the extracts or bioactive compounds. Stability of bioactive compounds can be improved by different techniques, such as encapsulation or application in optimal conditions to minimize the loss of bioactive compounds. As a large quantity of waste has been generated from agriculture and the food industry, this source of material is cheap and accessible for extraction and isolation of bioactive compounds for further utilisation. However, this waste needs to be treated and processed as recommended in Fig. 1. The loss of bioactive compounds from waste is high, which means that the waste needs to be treated as soon as it is disposed from agriculture or the food

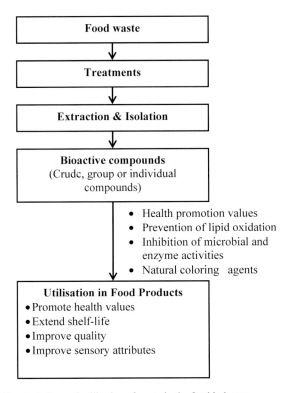

Figure 1. Extraction, isolation and utilisation of waste in the food industry.

factories. Dehydration or cold treatment is needed to minimize any loss; however, further studies are needed to identify the most suitable and economical method for treatment each type of waste. There are many factors affecting the efficient extraction of bioactive compounds from the materials but the most cost-effective methods to extract bioactive compounds from the waste have yet to be determined.

Bioactive compounds can exhibit various properties in groups or individually. In some cases, crude extract or group of bioactive compounds might exhibit stronger properties than individual compounds due to synergistic effects. Studies are also required to test the effective use of crude extract, groups of compounds or individual compounds in food products. Finally, application of bioactive compounds in food products require proper studies to indicate the right level, the right time and the right methods of addition of bioactive compounds into food for different purposes.

Conclusion

A large volume of waste, which is generated annually, is causing concern for the food industry as well as for the environment. Waste from different sources can contain high levels of bioactive compounds, which can potentially be recovered for further utilisation in the food industry. Bioactive compounds carry health benefits, are associated with anti-microbial growth and prevention of lipid oxidation. Bioactive compounds can be seen as exciting novel potential additives for the food industry. Numerous studies have applied bioactive compounds in various food products to extend their shelf-life and improve their quality. Research and development are needed to determine efficient and cost effective ways to isolate bioactive compounds and increase the range of potential uses of by-products from the food industry waste.

References

Aberoumand, A. 2011. A review article on edible pigments properties and sources as natural biocolorants in foodstuff and food industry. World Journal of Dairy & Food Sciences 6: 71–78.
Al-Zoreky, N. S. 2009. Antimicrobial activity of pomegranate (*Punica granatum* L.) fruit peels. International Journal of Food Microbiology 134(3): 244–248.
Aliakbarian, B., M. Casale, M. Paini, A. A. Casazza, S. Lanteri and P. Perego. 2015. Production of a novel fermented milk fortified with natural antioxidants and its analysis by NIR spectroscopy. LWT—Food Science and Technology 62(1, Part 2): 376–383.
Australian Food and Drink Report Retrieved on 20 October 2016 at http://store.bmiresearch.com/australia-food-drink-report.html.
Aziz, M. and S. Karboune. 2016. Natural antimicrobial/antioxidant agents in meat and poultry products as well as fruits and vegetables: A review. Critical Reviews in Food Science and Nutrition 00-00.
Bañón, S., P. Díaz, M. Rodríguez, M. D. Garrido and A. Price. 2007. Ascorbate, green tea and grape seed extracts increase the shelf life of low sulphite beef patties. Meat Sci. 77(4): 626–633.
Basiri, S., S. S. Shekarforoush, M. Aminlari and S. Akbari. 2015. The effect of pomegranate peel extract (PPE) on the polyphenol oxidase (PPO) and quality of Pacific white shrimp (*Litopenaeus vannamei*) during refrigerated storage. LWT—Food Science and Technology 60(2, Part 1): 1025–1033.
Behera, S., K. Indumathi, S. Mahadevamma and M. L. Sudha. 2013. Oil cakes—A by-product of agriculture industry as a fortificant in bakery products. Int. J. Food Sci. Nutr. 64(7): 806–814.
Botsoglou, E., A. Govaris, I. Ambrosiadis, D. Fletouris and N. Botsoglou. 2014. Effect of olive leaf (*Olea europea* L.) extracts on protein and lipid oxidation of long-term frozen n-3 fatty acids-enriched pork patties. Meat Sci. 98(2): 150–157.
Brown, A. 2008. Understanding Food: Principles and Preparation. Australia: Thomson Wadsworth.

Çakmakçı, S., E. F. Topdaş, P. Kalın, H. Han, P. P. Şekerci, L. Köse and İ. Gülçin. 2015. Antioxidant capacity and functionality of oleaster (*Elaeagnus angustifolia* L.) flour and crust in a new kind of fruity ice cream. International Journal of Food Science & Technology 50(2): 472–481.

Choe, E. and D. B. Min. 2009. Mechanisms of antioxidants in the oxidation of foods. Comprehensive Reviews in Food Science and Food Safety 8(4): 345–358.

Chouchouli, V., N. Kalogeropoulos, S. J. Konteles, E. Karvela, D. P. Makris and V. T. Karathanos. 2013. Fortification of yoghurts with grape (*Vitis vinifera*) seed extracts. LWT—Food Science and Technology 53(2): 522–529.

de Camargo, A. C., C. M. Vidal, S. G. Canniatti-Brazaca and F. Shahidi. 2014. Fortification of cookies with peanut skins: Effects on the composition, polyphenols, antioxidant properties and sensory quality. J. Agric. Food Chem. 62(46): 11228–11235.

Eunok Choe and D. B. Min. 2005. Chemistry and reactions of reactive oxygen species in foods. Journal of Food Science 70: R142–R159.

Feng, L., T. Jiang, Y. Wang and J. Li. 2012. Effects of tea polyphenol coating combined with ozone water washing on the storage quality of black sea bream (*Sparus macrocephalus*). Food Chem. 135(4): 2915–2921.

Gai, F., M. Ortoffi, V. Giancotti, C. Medana and P. G. Peiretti. 2015. Effect of red grape pomace extract on the shelf life of refrigerated rainbow trout (*Oncorhynchus mykiss*) minced muscle. Journal of Aquatic Food Product Technology 24(5): 468–480.

Giroux, H. J., G. De Grandpré, P. Fustier, C. P. Champagne, D. St-Gelais, M. Lacroix and M. Britten. 2013. Production and characterization of Cheddar-type cheese enriched with green tea extract. Dairy Science & Technology 93(3): 241–254.

Gramza-Michalowska, A., J. Kobus-Cisowska, D. Kmiecik, J. Korczak, B. Helak, K. Dziedzic and D. Gorecka. 2016. Antioxidative potential, nutritional value and sensory profiles of confectionery fortified with green and yellow tea leaves (*Camellia sinensis*). Food Chem. 211: 448–454.

Hosikian, A., S. Lim, R. Halim and M. K. Danquah. 2010. Chlorophyll extraction from microalgae: A review on the process engineering aspects. International Journal of Chemical Engineering 2010: 11.

Kanetis, L., V. Exarchou, Z. Charalambous and V. Goulas. 2016. Edible coating composed of chitosan and Salvia fruticosa Mill. extract for the control of grey mould of table grapes. J. Sci. Food Agric: n/a–n/a.

Klancnik, A., S. Piskernik, F. Bucar, D. Vuckovic, S. S. Mozina and B. Jersek. 2014. Reduction of microbiological risk in minced meat by a combination of natural antimicrobials. J. Sci. Food Agric. 94(13): 2758–2765.

Kobus-Cisowska, J., E. Flaczyk, M. Rudzinska and D. Kmiecik. 2014. Antioxidant properties of extracts from Ginkgo biloba leaves in meatballs. Meat Sci. 97(2): 174–180.

Kumar, S. and A. K. Pandey. 2013. Chemistry and biological activities of flavonoids: An overview. The Scientific World Journal 2013: 16.

Kumar, S. S., P. Manoj, N. P. Shetty, M. Prakash and P. Giridhar. 2015. Characterization of major betalain pigments -gomphrenin, betanin and isobetanin from *Basella rubra* L. fruit and evaluation of efficacy as a natural colourant in product (ice cream) development. Journal of Food Science and Technology 52(8): 4994–5002.

Kurcubic, V. S., P. Z. Maskovic, J. M. Vujic, D. V. Vranic, S. M. Veskovic-Moracanin, D. G. Okanovic and S. V. Lilic. 2014. Antioxidant and antimicrobial activity of Kitaibelia vitifolia extract as alternative to the added nitrite in fermented dry sausage. Meat Sci. 97(4): 459–467.

Lee, H. J., Y. J. Choi, Y. I. Choi and J. J. Lee. 2014. Effects of lemon balm on the oxidative stability and the quality properties of hamburger patties during refrigerated storage. Korean J. Food Sci. Anim. Resour. 34(4): 533–542.

Li, L., J. Shao, X. Zhu, G. Zhou and X. Xu. 2013. Effect of plant polyphenols and ascorbic acid on lipid oxidation, residual nitrite and N-nitrosamines formation in dry-cured sausage. International Journal of Food Science and Technology 48(6): 1157–1164.

Manay, N. S. and M. Shadaksharaswamy. 2001. Food: Facts and Principles. New Delhi: New Age International Limited.

Marchiani, R., M. Bertolino, D. Ghirardello, P. L. H. McSweeney and G. Zeppa. 2016. Physicochemical and nutritional qualities of grape pomace powder-fortified semi-hard cheeses. Journal of Food Science and Technology 53(3): 1585–1596.

Mariem, C., M. Sameh, S. Nadhem, Z. Soumaya, Z. Najiba and E. G. Raoudha. 2014. Antioxidant and antimicrobial properties of the extracts from *Nitraria retusa* fruits and their applications to meat product preservation. Industrial Crops and Products 55: 295–303.

Min, B., M. H. Chen and B. W. Green. 2009. Antioxidant activities of purple rice bran extract and its effect on the quality of low-NaCl, phosphate-free patties made from channel catfish (*Ictalurus punctatus*) belly flap meat. Journal of Food Science 74(3): C268–C277.

Miura, Y., M. Inai, S. Honda, A. Masuda and T. Masuda. 2014. Reducing effects of polyphenols on metmyoglobin and the *in vitro* regeneration of bright meat color by polyphenols in the presence of cysteine. J. Agric. Food Chem. 62(39): 9472–9478.

Mortensen, A. 2006. Carotenoids and other pigments as natural colorants. Pure Appl. Chem. 78: 1477–1491.

Mortensen, G., U. Andersen and J. H. Nielsen (eds.). 2010. Chemical Deterioration and Physical Instability of Dairy Products. Boca Raton: CRC Press.

Nirmal, N. P. and S. Benjakul. 2012a. Effect of green tea extract in combination with ascorbic acid on the retardation of melanosis and quality changes of pacific white shrimp during iced storage. Food and Bioprocess Technology 5(8): 2941–2951.

Nirmal, N. P. and S. Benjakul. 2012b. Inhibition kinetics of catechin and ferulic acid on polyphenoloxidase from cephalothorax of Pacific white shrimp (*Litopenaeus vannamei*). Food Chem. 131(2): 569–573.

Nowak, A., A. Czyzowska, M. Efenberger and L. Krala. 2016. Polyphenolic extracts of cherry (*Prunus cerasus* L.) and blackcurrant (*Ribes nigrum* L.) leaves as natural preservatives in meat products. Food Microbiol. 59: 142–149.

Qian, Y. -F., J. Xie, S. -P. Yang, S. Huang, W. -H. Wu and L. Li. 2015. Inhibitory effect of a quercetin-based soaking formulation and modified atmospheric packaging (MAP) on muscle degradation of Pacific white shrimp (*Litopenaeus vannamei*). LWT—Food Science and Technology 63(2): 1339–1346.

Rashidinejad, A., E. J. Birch, D. Sun-Waterhouse and D. W. Everett. 2015. Total phenolic content and antioxidant properties of hard low-fat cheese fortified with catechin as affected by *in vitro* gastrointestinal digestion. LWT—Food Science and Technology 62(1, Part 2): 393–399.

Saito, M., K. Saito, N. Kunisaki and S. Kimura. 2002. Green tea polyphenols inhibit metalloproteinase activities in the skin, muscle, and blood of rainbow trout. Journal of Agricultural and Food Chemistry 50(24): 7169–7174.

Sanchís, E., S. González, C. Ghidelli, C. C. Sheth, M. Mateos, L. Palou and M. B. Pérez-Gago. 2016. Browning inhibition and microbial control in fresh-cut persimmon (*Diospyros kaki* Thunb. cv. *Rojo Brillante*) by apple pectin-based edible coatings. Postharvest Biology and Technology 112: 186–193.

Sharma, S. and T. V. R. Rao. 2015. Xanthan gum based edible coating enriched with cinnamic acid prevents browning and extends the shelf-life of fresh-cut pears. LWT—Food Science and Technology 62(1, Part 2): 791–800.

Singh, B., P. S. Panesar and V. Nanda. 2006. Utilization of carrot pomace for the preparation of a value added product. World Journal of Dairy & Food Sciences 1: 22–27.

Skiepko, N., I. Chwastowska-Siwiecka, J. Kondratowicz and D. Mikulski. 2016. Fatty acid profile, total cholesterol, vitamin content, and TBARS value of turkey breast muscle cured with the addition of lycopene. Poultry Science 95(5): 1182–1190.

Sørhaug, T. and L. Stepaniak. 1997. Psychrotrophs and their enzymes in milk and dairy products: Quality aspects. Trends in Food Science & Technology 8(2): 35–41.

Tajkarimi, M. M., S. A. Ibrahim and D. O. Cliver. 2010. Antimicrobial herb and spice compounds in food. Food Control 21(9): 1199–1218.

Tanaka, Y., N. Sasaki and A. Ohmiya. 2008. Biosynthesis of plant pigments: Anthocyanins, betalains and carotenoids. The Plant Journal 54: 733–749.

Tang, S., D. Sheehan, D. J. Buckley, P. A. Morrissey and J. P. Kerry. 2001. Anti-oxidant activity of added tea catechins on lipid oxidation of raw minced red meat, poultry and fish muscle. International Journal of Food Science & Technology 36(6): 685–692.

Tang, S. Z., S. Y. Ou, X. S. Huang, W. Li, J. P. Kerry and D. J. Buckley. 2006. Effects of added tea catechins on color stability and lipid oxidation in minced beef patties held under aerobic and modified atmospheric packaging conditions. Journal of Food Engineering 77(2): 248–253.

Tumbas Saponjac, V., G. Cetkovic, J. Canadanovic-Brunet, B. Pajin, S. Djilas, J. Petrovic and J. Vulic. 2016. Sour cherry pomace extract encapsulated in whey and soy proteins: Incorporation in cookies. Food Chem. 207: 27–33.

Uluata, S., D. J. McClements and E. A. Decker. 2015. how the multiple antioxidant properties of ascorbic acid affect lipid oxidation in oil-in-water emulsions. Journal of Agricultural and Food Chemistry 63(6): 1819–1824.

Upadhyay, A., S. P. Mishra and H. K. Sharma. 2007. Utilization of a by-product from Carrot. Indian Journal of Agricultural Biochemistry 20(2): 97–99.

Vuong, Q. V., C. E. Stathopoulos, M. H. Nguyen, J. B. Golding and P. D. Roach. 2011. isolation of green tea catechins and their utilization in the food industry. Food Reviews International 27(3): 227–247.

Wang, Y., F. Li, H. Zhuang, L. Li, X. Chen and J. Zhang. 2015. Effects of plant polyphenols and alpha-tocopherol on lipid oxidation, microbiological characteristics, and biogenic amines formation in dry-cured bacons. J. Food Sci. 80(3): C547–555.

Wasowicz, E., A. Gramza, M. Hêoe, H. H. Jelen, J. Korczak, M. Maecka and R. Zawirska-Wojtasiak. 2004. Oxidation of lipids in food. Polish Journal of Food and Nutrition Sciences 13/54: 87–100.

Xu, C., Y. Yagiz, S. Marshall, Z. Li, A. Simonne, J. Lu and M. R. Marshall. 2015. Application of muscadine grape (*Vitis rotundifolia* Michx.) pomace extract to reduce carcinogenic acrylamide. Food Chem. 182: 200–208.

Ye, L., H. Wang, S. E. Duncan, W. N. Eigel and S. F. O'Keefe. 2015. Antioxidant activities of Vine Tea (*Ampelopsis grossedentata*) extract and its major component dihydromyricetin in soybean oil and cooked ground beef. Food Chem. 172: 416–422.

Ye, Y. E., L. Li, M. Ye, M. Zhang, L. Y. Jing and L. Yang. 2014. Effects of grape polyphenols on the quality and antioxidant activity of yoghurt. Journal of Pure and Applied Microbiology 8(SPEC. ISS. 1): 53–58.

Zhang, L., S. Li, Y. Dong, H. Zhi and W. Zong. 2016. Tea polyphenols incorporated into alginate-based edible coating for quality maintenance of Chinese winter jujube under ambient temperature. LWT—Food Science and Technology 70: 155–161.

Zhu, F., Y. Z. Cai, J. Ke and H. Corke. 2011. Dietary plant materials reduce acrylamide formation in cookie and starch-based model systems. J. Sci. Food Agric. 91(13): 2477–2483.

CHAPTER 14

Potential Application of Bioactive Compounds from Agroindustrial Waste in the Cosmetic Industry

Francisca Rodrigues,[1,*] *Ana F. Vinha,*[1,2] *M. Antónia Nunes*[1]
and *M. Beatriz P. P. Oliveira*[1]

Introduction

The demand for natural cosmetics is stronger than ever, being now widely considered a serious threat to the worldwide economy and society. This situation can deteriorate due to climate change, biodiversity loss, environmental problems, desertification and ecosystem degradation (Behrens et al. 2007). Agro-food industries produce large amounts of by-products per year, normally discarded as waste. Only a minor percentage of it is used as animal feed or for composting, improving the environmental threat. The challenge is to recover and valorize food by-products, particularly for the cosmetic and pharmaceutical industry after the identification of nutrients and bioactive compounds. Actually, few studies support the potential use of these agro-industrial waste (Martinez-Saez et al. 2014, Mussatto et al. 2012). Natural ingredients, phytonutrients, microbial metabolites, dairy derived actives, minerals and animal protein components have long been considered beneficial for healthy skin aging (Prakash and Majeed 2009). Indeed, from a sustainable point of view, this new application could provide, in the near future, a recycling way for food companies, developing cost-effective processing methods, decreasing the negative impacts of waste on the environment and providing economical advantages.

In response, sustainable raw material sourcing and greener formulations, as well as low overall resource consumption should be considered (McPhee and Jain 2015).

[1] LAQV@REQUIMTE, Department of Chemical Sciences, Faculty of Pharmacy, University of Porto, Rua de Jorge Viterbo Ferreira, 228, 4050-313 Porto, Portugal.
[2] FP-ENAS (UFP Energy, Environment and Health Research Unit), CEBIMED (Biomedical Research Center), University Fernando Pessoa, Praça 9 de Abril, 349, 4249-004 Porto, Portugal.
* Corresponding author: franciscapintolisboa@gmail.com

In fact, the cosmetic field could be a new way to reuse these by-products, but first of all it is extremely important to identify interesting molecules present on these wastes and their effects on skin.

Skin has historically been used for the topical delivery of compounds. It is much more than a static, impenetrable shield against external insults. Rather, skin is a dynamic, complex, integrated arrangement of cells, tissues and matrix elements that regulate the body heat and water loss, whilst preventing the invasion of toxic substances and microorganisms. The re-use of bioactive compounds from food waste as cosmetic ingredients needs the knowledge of basic skin structure and absorption mechanisms. Skin has three major regions: epidermis, dermis and hypodermis. The skin barrier function is realized by a cornified layer of protein-rich dead cells (corneocytes) embedded in a lipid matrix, namely *stratum corneum* (Mathes et al. 2014). The lipid-enriched component consists predominantly of ceramides, as well as cholesterol and fatty acids (Schurer and Elias 1991). Interspersed amongst the keratinocytes, in the viable epidermis, are the cells that are involved in melanin production (melanocytes), sensory perception (Merkel cells) and immunological function (Langerhans and other cells) (Prow et al. 2011). Deeper skin layers are the viable epidermis (50–100 μm) with the basal membrane (consisting of at least one member of the protein family of laminin, type IV collagen and nidogen, and the proteoglycan perlecan) and the dermis (1–2 mm), where appendices like sweat glands and hair follicle are located (Franzen and Windbergs 2015, Schäfer-Korting et al. 2007). The dermis has an upper papillary layer containing loosely-arranged collagen fibers and a reticular layer with dense collagen fibers arranged parallel to the surface of the skin. As well as collagen, the dermal matrix comprises a high amount of elastin that provides the elastic properties of the skin (Mathes et al. 2014). This matrix is produced by fibroblasts, the main cell type of the dermis. Dermis is pervaded by blood and lymph vessels. Beneath the dermis lies the subcutis, also known as the hypodermis. The subcutis functions are insulator, conserving the body's heat, and shock-absorber. Next to fibroblasts, adipocytes are the most prominent cell type in this compartment (Mathes et al. 2014). Figure 1 presents an overview of skin structure, referring to the local of action of active substances detailed in the next sections.

Indeed, the efficacy of cosmetic active ingredients is related to their diffusion through the skin barrier (Rawlings and Matts 2005). However, small soluble molecules with simultaneous lipophilic and hydrophilic properties have a greater ability to cross *the stratum corneum* than particles, polymers or highly lipophilic substances (Rawlings and Matts 2005).

The aim of this chapter is to describe some of the most promising food by-products for the cosmetic industry, particularly focusing on their composition and skin effects.

Bioactive Compounds from Food By-products with Cosmetic Interest

In the last years, valorization of food waste or underutilized food has been addressed. According to Gustavsson et al. 1.3 billion tons of edible material is wasted in the world and this represents one-third of the global food production (Gustavsson et al. 2011).

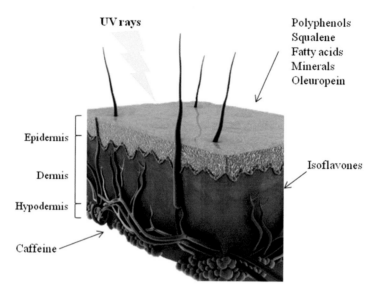

Figure 1. Schematic picture of the native skin sub-classified into three main compartments: epidermis, dermis and hypodermis.

Also, it is well known that food by-products are rich in vitamins, sugars, minerals and other bioactive compounds, such as polyphenols and flavonoids. Their appealing composition, together with the growing interest in natural ingredients and sustainable consumption, determine food by-products as economically attractive sources of high-value compounds for the cosmetic industry. Thus, it is extremely important to detail the composition of the most promising food by-product for this field.

Coffee by-products

Coffee is one of the most popular beverages in the world and the second largest traded commodity (Murthy and Naidu 2012). The processing of coffee beans aims to separate coffee beans, by removing the shell from the mucilaginous part. Depending upon the method of coffee processing, roasting and brewing, solid residues like pulp, husk, silverskin, parchment or mucilage are produced. Pulp and silverskin are of particular interest as potential cosmetic bioactive compounds. The organic components present in coffee pulp (dry weight) include tannins ($\approx$ 2–9 per cent), total pectic substances (6.5 per cent), reducing sugars (12.4 per cent), non-reducing sugars (2.0 per cent), caffeine (1.3 per cent), chlorogenic acid (2.6 per cent) and total caffeic acid (1.6 per cent) (Murthy and Naidu 2012). In turn, coffee silverskin, a thin closely skin that covers each coffee bean, is obtained after roasting. After complete removal of skin, the ground roasted coffee beans for coffee beverage production are released (Saenger et al. 2001). Regarding its macronutritional composition, it has a high content of dietary fiber (50–60 per cent, including 85 per cent and 15 per cent of soluble and insoluble fiber, respectively), protein (16.2–19 per cent) and minerals (4–5 per cent) (Borrelli et al. 2004, Napolitano et al. 2007, Pourfarzad et al. 2013). However, the exact mineral composition of coffee silverskin has not been clarified so far (Narita and Inouye 2014). Different authors have demonstrated the high antioxidant capacity, probably due to the concentration of phenolic compounds, particularly in chlorogenic acid (Borrelli et al. 2004, Bresciani et al. 2014, Costa et al. 2014, Rodrigues et al. 2015a).

Olive by-products

Olive oil is one of the most well-known natural ingredients in the world. Global production of olive oil has doubled in the last 20 years with a positive increasing socio-economic impact in Europe, especially in the Mediterranean countries (Meksi et al. 2012). At the same time, the amount of olive by-products discharged has increased (Rahmanian et al. 2014). Olive mill waste water (OMWW) and olive pomace are the two main olive by-products. Their chemical composition depends on the olive variety, culture conditions, olive origin and extraction process (Dermeche et al. 2013). Nevertheless, the final composition is relatively similar. Apart from water (83–92 per cent), the main components of OMWW are phenolic compounds, sugars and organic acids, besides valuable amounts of mineral nutrients, especially potassium (Dermeche et al. 2013). Olive pomace is mainly composed of cellulose (30 per cent), pectic polysaccharides (39 per cent) and hemicellulosic polymers rich in xylans and glucuronoxylans (14 per cent), xyloglucans (15 per cent) and mannans (2 per cent) (Lama-Muñoz et al. 2012). The mineral composition of olive cake comprises potassium, calcium and sodium (Dermeche et al. 2013). Besides, olive cake is rich in water (70–80 per cent), 2–3 per cent of residual olive oil, including squalene, and many phytochemicals (natural antioxidants, including tocopherols, flavonoids, quercetin, cinnamic acid, peptides and phenolic compounds) (Aziz et al. 1998, Kim et al. 1997, Masghouni and Hassairi 2000).

Castanea sativa by-products

Castanea sativa Mill. is a species of Fagaceae family that can be found in south Europe and Asia (China). It is composed by fruit (chestnut), pericarp (outer shell), integument (inner shell) and bur that surrounds the edible nuts (Braga et al. 2014). During processing, huge amounts of by-products are produced, namely, leaves, shells and bur. Several studies were conducted on chestnut by-products and were found to be a good source of phenolic compounds with marked biological activity, mainly antioxidant properties (Barreira et al. 2008, Barreira et al. 2012, Rodrigues et al. 2015d). As an example, Rodrigues et al. assessed the macronutrient composition and the amino acid and vitamin E profiles of *C. sativa* shell (Rodrigues et al. 2015d). The nutritional composition was characterized by a high moisture content and low fat amount. Arginine and leucine were the predominant essential amino acids. The shells contained high amounts of vitamin E, with γ-tocopherol being the predominant vitamer. The antimicrobial and antioxidant activities of *C. sativa* shell were determined. The highest antioxidant activity and total phenolic content were, respectively, an EC_{50} of 31.8 ± 1.3 μg mL^{-1} for DPPH and 796.8 mg gallic acid equivalents (GAE) per g db sample. No antimicrobial activity was observed.

Almeida et al. evaluated the hydroalcoholic extracts of *C. sativa* leaves (Almeida et al. 2010, Almeida et al. 2015). The composition of extracts was mainly composed of phenolic compounds, like chlorogenic and ellagic acid, rutin and quercetin. An *in vivo* patch test was performed on 20 volunteers and the results were very promising (Almeida et al. 2010). No differences were observed for Transepithelial water loss (TEWL) measurements in comparison to purified water 2 hours after patch removal (Almeida et al. 2015). The same extract was evaluated under stability studies in order to prove its efficacy overtime.

Grape by-products

Grape (*Vitis* spp.) is one of the oldest fruit crops to be being used in different processes such as wine, beverages or jelly production, representing one of the most economically important plants worldwide (Georgiev et al. 2014). It has been well known for centuries that grapes provide superior nutritional, antioxidant and overall health benefits to the body, hair and skin. Besides, to be considered beneficial to human health, they are widely marketed and used as food supplements.

The general composition of a grape is 0–5 per cent seeds, 2–6 per cent stems, 5–12 per cent skins, and 80–90 per cent juice. The highest concentrations of polyphenols are found in stems, skin and seeds (Dohadwala and Vita 2009). Nevertheless, grape products (such as wine and juice) are also known to contain high amounts of polyphenols, mostly phenolic acids, anthocyanins and simple and complex flavonoids as proanthocyanidins (Leifert and Abeywardena 2008).

During grape processing, huge amounts of by-products are generated, namely skin, seed and stem, that are normally blended back into the soil for its enrichment. The processing of wine-making residues into viable commercial products is a challenge for the cosmetic industry.

Grape pomace, a wine-making by-product, represents approximately 20 per cent of the total weight of processed grapes (Teixeira et al. 2014). It is constituted of grape seeds from 15–52 per cent, according to the technological and residue management process (Teixeira et al. 2014). Grape seeds are a potential source of bioactive compounds, containing fiber (40 per cent), essential oils (16 per cent), proteins (11 per cent), sugars and minerals (de Campos et al. 2008). Also, their extract is a rich source of phenolic compounds, mainly proanthocyanidins ($\approx$ 90 per cent) that can be found predominantly in red wine (Feringa et al. 2011). Grape seed oil has been extensively studied for it effect on health and wellbeing, namely due to it fatty acid profile and vitamin E content. Linoleic acid is the predominant fatty acid, followed by oleic and palmitic acids (Lachman et al. 2015).

In turn, grape skin represents almost 65 per cent of the total grape pomace and is a rich source of phenolic compounds, whose amounts depend on the vinification process (extraction process, solvent, temperature, etc.) (Teixeira et al. 2014). Anthocyanins, flavan-3-ols and hydroxycinnamates are also present in grape skin. As flavan-3-ols, simple monomers (+)-catechin and its isomer (–)-epicatechin, oligomeric and polymeric procyanidins, also known as condensed tannins, are described. Procyanidins are exclusively composed of epicatechin units and are abundantly found in plants. On the other hand, prodelphinidin is composed of epigallocatechin subunits, although they are less common (Tsang et al. 2005). The grape skin proanthocyanidins differ from those detected in seeds with the presence of prodelphinidins. Grape seed and skin extracts obtained from grape pomace contain significant amounts of flavan-3-ols and anthocyanins (Ky et al. 2014). Even though polyphenols can exist as monomers, they are found conjugated with sugars in Nature. These molecules can also be linked to other phenol groups, organic acids, proteins and lipids (Parada and Aguilera 2007).

Resveratrol is also commonly found in grape skins and seeds varying in content according to the grape variety and the vinification process. Also, it can be detected in the grape pomace (Casas et al. 2010).

Grape clusters or grape stems vary between 1–7 per cent of the total processed raw material. Due to its astringent properties, grape stems are removed before the vinification process (Teixeira et al. 2014). This by-product can be a valuable source of dietary fiber (60–90 per cent), protein (up to 7 per cent), soluble sugars (up to 2 per cent) and bioactive compounds (Barros et al. 2015). The major phenolic compounds present in grape stem extracts are rutin, quercetin 3-O-glucuronide, trans-resveratrol and astilbin (Xia et al. 2010). So far, this winery by-product is underexploited. Further studies are needed to increase the grape stem potential as a source of bioactive compounds.

In 2014, the Cosmetic Ingredient Review Expert Panel assessed the safety of 24 *Vitis vinifera* (grape)-derived ingredients, such as flower, leave, skin or root extracts (Fiume et al. 2014). The panel found that the function of these ingredients in cosmetics is mostly as a skin-conditioning agent, with antioxidant, flavoring and/or colorant properties (Fiume et al. 2014).

Medicago by-products

Legumes are important agricultural and commercial crops consumed in large quantities by both humans and animals. Waste and by-products of vegetable origin are currently rising, driven by both a net increase in plant product consumption and the changing consumer trend in ready-to-use products (Rodrigues et al. 2014a).

The genus *Medicago* is part of the botanical family of *Leguminosae* and includes about 56 different species mainly distributed in Mediterranean climatic areas (Farag et al. 2007). Alfalfa (*Medicago sativa*) is the main *Medicago* species widely grown throughout the world, predominantly as high quality forage for livestock, renewable energy production, phytoremediation, and as a source of phytochemicals (Nunes et al. 2008, Silva et al. 2013). It is also used as a food ingredient, consumed as sprouts in salads, in sandwiches or soups (Facciola 1990), as leaf protein concentrate (Koschuh et al. 2004), or as food supplement (Takyi et al. 1992). Despite this use, alfalfa has pharmacological value related to its isoflavone content, for use in certain human health dysfunctions, such as anaemia, diabetes, endometriosis, stomach ulcers, osteoporosis, menopausal symptoms, breast and prostate cancers and low bone density (Mortensen et al. 2009, Rostagno et al. 2009). Also, the main applications of *Medicago* species are as grazing feed for cattle and other livestock. However, like other plants of the same family, high levels of isoflavones were reported. These compounds are desirable as lamb feed, for quick weight gain, as well as for milk production. Reproductive disorders, namely low fertility, have also been reported in sheep and cattle fed on forages with high isoflavone content (Sabudak and Guler 2009, Sivesind and Seguin 2005). Thus, depending on the final objective for cattle, *Medicago* could not be desired.

Medicago species are important sources of phytochemicals, including carotenoids, saponins or phytoestrogens, which are known as antimicrobial agents, phytoanticipins, phytoalexins, structural barriers, modulators of pathogenicity, plant defence gene activators or fungitoxic agents (Barreira et al. 2015, Silva et al. 2013, Visnevschi-Necrasov et al. 2015, Von Baer et al. 2006). Over this panoply of compounds, phytoestrogens are the most suitable for cosmetic products. Phytoestrogen is a general term used to define a wide variety of compounds that are non-steroidal and are either of plant origin or derived from the *in vivo* metabolism of several plants used as food (Brzezinski and Debi 1999). Isoflavones are phenolic compounds with antioxidant activity and structural similarity to estradiol molecule (Bacaloni et al. 2005, Tham et al. 1998), being primarily found in plants of the *Fabaceae* family (Fletcher 2003). A number of epidemiological studies associate the consumption of isoflavone-rich foods with low incidence of the major hormone-dependent cancers (Adlercreutz 1995), cardiovascular diseases (Clarkson et al. 1995), osteoporosis (Tham et al. 1998) and climacteric complaints (Adlercreutz et al. 1992). Because of their structural similarity to ß-estradiol, health benefits of isoflavones have been evaluated in age-related and hormone-dependent diseases (Moraes et al. 2009). These compounds are largely described as in red clover, other Leguminosae plant (Delmonte et al. 2006, Klejdus et al. 2001, Mazur et al. 1998, Wu et al. 2003). Biochanin A (5,7-dihydroxy-4'-methoxyisoflavone) and formononetin (7-hydroxy-4'-methoxyisoflavone) are methylated precursors of genistein and daidzein, respectively, present in high amounts in red clover (Krenn et al. 2002, Wu et al. 2003). Rodrigues et al. evaluated extracts from seven species of *Medicago* (*M. minima, M. tornata, M. truncatula, M. rigidula, M. scutelata, M. segitalis* and *M. sativa*) and demonstrated their antioxidant and antimicrobial activities with low cytotoxic potential for skin cells (Rodrigues et al. 2013). Also, the isoflavones content was reported (Rodrigues et al. 2014a).

Therefore, estrogens can slow down the aging process of the skin in postmenopausal women. For this reason, isoflavones are interesting compounds for cosmetic formulations. When incorporated in creams, natural estrogens could result in morphological modifications to the aged skin, characterized mainly by an increased number of fibroblasts and glycosaminoglycans, and, consequently, collagen and hyaluronic acid.

Tomato by-products

Tomato processing is one of the most important food processes that generates valuable products to be exported all over the world. On a global scale, the annual production is close to 152.9 million tons (FAO 2009). Tomatoes are universally recognized as a health-promoting food, with already reported health benefits. These industries use manufacturing technologies to transform fresh tomato into tomato pulp, juice and canned tomato, resulting in large amounts of by-products, namely peel, pulp and seeds. Together, peel and seeds (already known as pomace) constitute the major agro industrial by-products obtained from tomato fruit processing (~ 5 per cent of the total weight of fruit) (Barbulova et al. 2015, Ilahy et al. 2016). Also, tomato pomace is an inexpensive and abundant source of natural antioxidants for nutraceutical, cosmetic and pharmaceutical industries.

Most of the available literature reports on the micronutrient contents and phytochemical characterization of tomatoes and tomato-based products, including lipids (25 per cent), proteins (35 per cent) and fiber (30 per cent) (Omer and Abdel-Magid 2015, Schieber et al. 2001), among other important phytochemicals (Pinela et al. 2012, Vinha et al. 2014a). In tomato skin and seeds were found the essential amino acids and higher amounts of minerals in seeds (Fe, Mn, Zn and Cu). The main interest of tomato peel is in the presence of bioactive compounds and antioxidant activity, in particular the high levels of lycopene, up to five times higher than pulp (Al-Wandawi et al. 1985, Ilahy et al. 2016, Vinha et al. 2014a). Lycopene content in tomato varies from 30 to 200 mg/kg in fresh fruit and from 430 to 2950 mg/kg on a dry basis, representing more than 85 per cent of the total carotenoids present (Vasapollo et al. 2004). Toor et al. reported that tomato peels and seeds of three cultivars ('Excell', 'Tradiro' and 'Flavorine') contributed about 53 per cent to the total phenolics, 52 per cent to the total flavonoids, 48 per cent to the total lycopene, 43 per cent to the total ascorbic acid and 52 per cent to the total antioxidant activity of the total fruit (Toor and Savage 2005). Likewise, Chandra et al. confirmed that peel and seed fractions of different tomato varieties grown in India presented high contents of lycopene, ascorbic acid and total phenolics (Chandra and Ramalingam 2011, Chandra et al. 2012). The same authors also reported a higher antioxidant activity in seed and peel fractions. More recently, Vinha et al. appraised the effects of peel and seed removal of four Portuguese tomato cultivars ('Redondo', 'Rama', 'Chucha' and 'Cereja') (Vinha et al. 2014a). The final results revealed that peeling decrease on an average by 71 per cent in lycopene, 50 per cent in β-carotene, 32 per cent in total phenolics and 14 per cent in ascorbic acid contents, as well as ~ 10 per cent in antioxidant activity. On another hand, seed removal also decreases the content of bioactive compounds (11 per cent of carotenoids and 24 per cent of total phenolics) as well as antioxidant activity (5 per cent).

Despite the diversity of bioactive compounds in tomato, including neochlorogenic acid, naringenin, kaempferol, quercitrin, rutin (Vinha et al. 2014b), lycopene is in high demand in the cosmetic industry. Due to the high degree of double-bond conjugation the compound is a strong antioxidant, exhibiting a higher singlet oxygen-quenching ability compared to β-carotene or α-tocopherol (Palozza et al. 2010). Thus, alternative sources for production of natural lycopene are warranted.

Quercus by-products

Quercus spp. (Fagaceae) represents an important group of evergreen or deciduous trees from temperate and tropical zones, comprising around 450 species worldwide (Sánchez-Burgos et al. 2013, Tejerina et al. 2011). Oak constituted significant components of European prehistoric landscape with its potential as a natural resource underestimated and relatively under-exploited (Šálková et al. 2011). Acorns and their by-products have diverse beneficial effects, mainly due to the presence of specific functional groups of phytochemicals. Actually, acorns are typically acknowledged for their high importance in the rural economy as an animal feed, particularly for pigs.

However, their nutritional value and high phytochemical contents have aroused the interest of many researchers on the lookout for undervalued natural sources to be used as new active ingredients in the cosmetic industry. The acorn is a one-seeded nut (occasionally two seeds), which has no endosperm and an achlorophyllous embryo, enclosed in a hard, fibrous shell, and borne in a cup-shaped cupule (Nikolić and Orlović 2002). The fruit is a source of starch (up to 50 per cent), containing relatively low amounts of protein (~ 2 per cent) and fat (~ 2 per cent), similar organoleptic features to those presented by chestnuts. Acorns are a natural source of oleic acid (Estévez et al. 2004) and contain high amounts of α-linolenic acid (Petrovic et al. 2004). This fatty acid is active in eicosanoid synthesis, promoting benefic effects on the decrease of seric triglycerides and in the increase of HDL-cholesterol levels. Vitamin E, mainly α- and γ-tocopherols, was also present. In general, γ-tocopherol is the most abundant vitamer, reaching levels almost 9 times higher than those detected for α-tocopherol (Tejerina et al. 2011). These contents are important to the cosmetic industry since aging is normally associated with a decrease in the plasma levels of γ-tocopherol.

Still, different acorn components have strong antioxidant activity, which might be related to other biological functions, such as antimutagenicity, anticarcinogenicity, antiaging effects, and the reduction of risks or symptoms of cardiovascular diseases, diabetes, microbial infection and inflammatory diseases (Karimi and Moradi 2015, Rakić et al. 2006, Silva et al. 2008). This activity is attributed to the presence of high amounts of phenolic compounds in the acorn extracts (Rakić et al. 2006, Tejerina et al. 2011). Despite the phylogenetic variability, phenolic acids (particularly, gallic and ellagic acids derivative compounds), flavonoids (particularly flavan-3-ols) and tannins are somehow ubiquitous in all *Quercus* species. Nevertheless, there are obvious differences between species. Kim et al. for instance, in a study conducted on five different *Quercus* species (*Q. acuta*, *Q. glauca*, *Q. myrsinaefolia*, *Q. phylliraeoides* and *Q. salicina*), reported particularly high levels (especially in *Q. salicina*) of gentisic and chlorogenic acids, as well as in flavonoids like naringin and rutin; none of these compounds were detected in any other *Quercus* species (Kim et al. 2012). The shell (regardless of species) is characterized by its richness in tannins, important gustatory components responsible for the astringency of many fruits and vegetables with recognized pharmacological activities. In addition, the reaction between tannins and proteins constitutes the basis for the insecticidal, antifungal and antibacterial properties. Sung et al. observed a strong antibacterial activity against vancomycin-resistant *Staphylococcus aureus* and *Bacillus coagulans* in the aqueous extract of acorn shell (Sung et al. 2012).

As it can be easily concluded from Table 1, the phenolic profiles vary greatly among food by-products. For example, *Medicago* is rich in isoflavones while grapes contain high amounts of catequin, epicathequin and epigallocatechin.

Table 2 summarizes the ingredients type obtained from the food by-products mentioned above and their effect on skin.

Bioactive Compounds of Food By-products and their Effects on Skin

Food by-products are particularly rich in antioxidants. The antioxidant activity is an excellent example of a functional benefit that extracts from food by-products can

Table 1. Individual compounds with cosmetic interest reported in different food by-products.

Compound	Chemical structure	By-product	Reference
Caffeic acid		Coffee	(Benoit et al. 2006, Bresciani et al. 2014)
Resveratrol		Grape	(Casas et al. 2010)
Formononetin		*Medicago*	(Krenn et al. 2002, Wu et al. 2003)
Biochanin A		*Medicago*	(Krenn et al. 2002, Wu et al. 2003)
Catechin		Grape	(Ky et al. 2014, Parada and Aguilera 2007, Tsang et al. 2005)
Epicatechin		Grape	(Ky et al. 2014, Parada and Aguilera 2007, Tsang et al. 2005)
Ellagic acid		*C. sativa Quercus*	(Almeida et al. 2010)

Table 1 contd. ...

...Table 1 contd.

Compound	Chemical structure	By-product	Reference
Gallic acid		*Quercus*	(Kim et al. 2012, Rakić et al. 2006, Tejerina et al. 2011)
Quercetin		*C. sativa*	(Almeida et al. 2010)
Epigallocatechin		Grape	(Tsang et al. 2005)
Chlorogenic acid		Coffee	(Bresciani et al. 2014)
Rutin		*C. sativa* Tomato	(Almeida et al. 2010, Vinha et al. 2014b)
Caffeine		Coffee	(Bresciani et al. 2014, Rodrigues et al. 2015b)
Lycopene		Tomato	(Ilahy et al. 2016, Palozza et al. 2010, Vinha et al. 2014a, Vinha et al. 2014b)

Table 2. Examples of functional and/or biologically active plant ingredients.

Ingredient type/function	Examples	Food by-product
Antioxidants	Catequin	Coffee
	Epicathequin	*Medicago*
	Chlorogenic acid	*C. sativa*
	Rutin	Tomato
	Quercetin	*Quercus*
		Grape
		Olive
Anti-ageing	Isoflavones	*Medicago*
	Oleuropein	Olive
	Hydroxytyrosol	
Photo-protection	Polyphenols	*C. sativa*
Hydration/Moisturising	Polysaccharides	Olive
	Hemicellulosic polymers	Tomato
	Minerals	
	Squalene	
Natural colourants	Lycopene	Tomato

deliver to the skin. Cosmetic treatment of the current generation against wrinkles relies on the antioxidant properties of some ingredients. Many of these compounds and extracts are emerging candidates for moderating the effects of the aging process on skin by limiting the biochemical consequences of oxidation (Angerhofer et al. 2009).

Free radicals are highly reactive molecules with unpaired electrons whose formation is widely accepted as the pivotal mechanism that leads to skin aging. Oxidative stress happens when the balance between production and elimination of reactive oxygen species (ROS), reactive nitrogen species (RNS) and reactive sulfur species (RSS) is compromised, leading to overproduction of oxidative species (Craft et al. 2012). The main targets of pro-oxidants, like ROS, RNS and RSS are proteins, DNA and RNA molecules, sugars and lipids (Craft et al. 2012). The damaging effects of these reactive species are induced internally (during normal metabolism) and externally, through various oxidative stresses, such as ultraviolet (UV) light. Indeed, the production of free radicals increases with age, while the endogenous defense mechanisms decrease, leading to accelerated aging. One of the skin's defense mechanisms is antioxidant defense where enzymes and other antioxidant substances react directly with the radical species, preventing them from reaching their biological target. Antioxidant substances have the ability to bind free radicals and prevent the therapy of various skin diseases, as well as slowing the skin's aging process. The topical application of antioxidants, such as vitamin C or E, coenzyme Q10 and polyphenolic compounds may strengthen the skin's endogenous protection system, protecting it from the harmful effects of ROS and oxidative damage to the skin. Phenolic compounds are involved in the defense process against deleterious oxidative damage, thus protecting against oxidative stress-related diseases. This can result in visible signs of healthy, more vibrant skin. Three of the better-known bioflavonoids—quercetin, hesperidin, and rutin—have intrigued researchers with their abilities to prevent and reverse wrinkles, reduce the appearance of age spots and fight spider veins (Zhu and Gao 2008).

ROS and UV irradiation are also involved in skin diseases like erythema, cancer, psoriasis, acne, cutaneous vasculitis, allergic, irritant contact dermatitis and photoaging (Matsumura and Ananthaswamy 2004). Dermatitis is a polymorphic inflammation of skin characterized by redness, itching, swelling and blistering, with formation of crusts and scales during the healing process (Rios et al. 2005). Phenolics and terpenoids are considered to be the most effective inhibitors of contact dermatitis (Rios et al. 2005). Many of these compounds act by a non-specific mechanism like antioxidants, but also may act via specific mechanisms, such as the inhibition of the mediators implicated in the immune response, showing anti-inflammatory properties and being able to reduce symptoms of skin inflammation and contact hypersensitivity (Nichols and Katiyar 2010).

In recent years, resveratrol has been subjected of intense research due to a range of unique anti-aging properties. Resveratrol is one of the most commonly known polyphenols to be considered as a powerful antioxidant that improves the skin's texture while diminishing the appearance of wrinkles and strengthening the skin. In fact, this polyphenol acts on cellular signaling mechanisms related to UV-mediated photoaging, including MAP kinases, nuclear factor kappa B (NF-κB), and matrix metalloproteinases (Baxter 2008). Bastianetto et al. demonstrates that resveratrol possesses an *in vitro* protective action against cell death after exposure of HaCaT cells to the nitric oxide free radical donor sodium nitroprusside (Bastianetto et al. 2010). Also, according to Giardina et al. *in vitro* skin fibroblasts treated with resveratrol have a dose-related increase in the rate of cell proliferation and in inhibition of collagenase activity (Giardina et al. 2010).

The procyanidins are biopolymers comprising catechin or epicatechin monomer units in varying chain lengths. Proanthocyanidins have antioxidant activity and aid in the stabilization of collagen and maintenance of elastin—the two critical proteins in connective tissue (Han and Nimni 2012).

Regarding fatty acids, olive by-products appear to be most interesting due to the high content of polyunsaturated ones, being particularly the linoleic acid. According to Schure et al. the lipid-enriched component of skin consists predominantly of ceramides (about 50 per cent by weight) as well as cholesterol and free fatty acids (Schurer and Elias 1991). These compounds have particular impact as components of the lipid film in the skin surface, maintaining hydration, softness and elasticity, as well as the protective barrier (Rodrigues et al. 2015c). In fact, ROS modifies the phospholipids in the membrane by peroxidation of their polyunsaturated fatty acids, with cell destruction and eventual cell death (Viola and Viola 2009). In this way, fatty acids have multiple functions in the epidermis, such as the permeability barrier; blocking the lipids in sebum produced by sebaceous glands; signaling keratinocytes to regulate epidermal homeostasis; and promoting the acidification of the *stratum corneum*. Also, during life, the fatty acids content decreases, leading to a less moist and emollient skin where, importance lies on the use of enriched skin care products (Rodrigues et al. 2015c).

Cellulose and derivatives detected in olive by-products are of huge importance for cosmetics. Due to its chemical structure, these compounds may act as functional ingredients in cosmetic formulations, by improving the physical and structural properties of hydration, oil-holding capacity, emulsion and oxidative stability, viscosity, texture, sensory characteristics and shelf-life of products (Elleuch et al. 2011, Rodríguez-Gutiérrez et al. 2014). On another hand, squalene shows several advantages for skin tissues, such as antioxidant properties at the cutaneous level against solar rays, acting as a biological filter of singlet oxygen (Micol et al. 2005). Besides, squalene may also act as a sink for highly lipophilic xenobiotics, promoting their elimination from the organism (Ghanbari et al. 2012). Concerning other interesting characteristics of squalene, emollient properties already described support their use as ingredients in dermo-protective creams and other cosmetic formulations for moisturizing or as emollients (Stavroulias and Panayiotou 2005).

Another interesting compound for cosmetic products is caffeine, found in huge amounts in coffee by-products. It is increasingly used in cosmetics due to its high biological activity and ability to penetrate the skin barrier (Herman and Herman 2013). According to Bresciani et al. the amount of caffeine in coffee silverskin is very high and similar to that of coffee beans (Bresciani et al. 2014). This alkaloid stimulates the degradation of fats during lipolysis through inhibition of the phosphodiesterase activity and has potent antioxidant properties (Herman and Herman 2013). Indeed, some claims, such as anti-cellulite, are based on the implicit assumption that this bioactive substance is effectively released from the formulation into the epidermis and probably through the epidermis into the dermal and subcutaneous tissues, preventing the excessive accumulation of fat in cells and providing a slimming effect (Bolzinger et al. 2008, Herman and Herman 2013). Another favorable effect of caffeine is its anti-aging properties. According to different authors, extracts from the two most common species of coffee directly combat UV damage to improve strength, resilience and elasticity of facial skin (Velazquez et al. 2009). *Coffea arabica* seed oil significantly improves collagen and elastin production while *Coffea robusta* has a high concentration of chlorogenic acid (which reduces redness associated with excessive sunlight exposure) and caffeine (which limits photodamage, decreases skin roughness, wrinkle formation and appearance of crow's feet) (Velazquez et al. 2009).

Minerals are also interesting ingredients detected in food by-products, as previously mentioned. As it is well known, minerals are an important component of natural moisturizing factor (NMF), which is responsible for the osmotic force to attract water in *stratum corneum*, particularly, corneocytes (Roberts et al. 2007). According to Rodrigues et al. NMF is responsible for hydration, stiffness and pH of *stratum corneum* (Rodrigues et al. 2015c), being essentially composed of aminoacids ($\approx$ 40 per cent), mineral ions ($\approx$ 18 per cent), lactate ($\approx$ 12 per cent), sugars ($\approx$ 8 per cent), urea ($\approx$ 7 per cent) and water soluble ions ($\approx$ 8.5 per cent) (Rawlings and Matts 2005).

Changes in skin aging and function occur at variable rates due to hormonal factors, unique to each individual. Over time, the skin experiences a progressive increase in extensibility and a reduction in elasticity, thereby becoming more frail and susceptible to trauma. This, in turn leads to an increased risk of skin injury (e.g., lacerations, tears, ulcerations, bruising), and an impairment of wound healing (Thompson and Maibach 2010). Since the average woman in the developed nation spends about one-third of her life after the onset of menopause, the benefits and risks of estrogen replacement therapy (ERT) have become major areas for research focus (Thompson and Maibach 2010). ERT increases skin collagen content and preserves thickness, thereby reducing wrinkling. Skin moisture content improves with ERT, as it increases the skin's hyaluronic acid, acid mucopolysaccharides, sebum levels, and possibly maintains *stratum corneum* barrier function (Thompson and Maibach 2010). Collagen becomes progressively sparse, disordered, and atrophied as skin ages—one of the chief reasons for the skin transformation resulting due to aging. Different studies establish that menopause leads to estrogen deficiency and research over the past 60 years demonstrates skin thickness, estrogen content and skin collagen are closely correlated (Thompson and Maibach 2010). Topical estrogen administration could preserve the dermal collagen with significant improvement in skin appearance (Stevenson and Thornton 2007). Estrogen has significant effects on skin physiology and modulates epidermal keratinocytes, dermal fibroblasts and melanocytes, in addition to skin appendages, the hair follicle and the sebaceous gland (Stevenson and Thornton 2007). Many of the effects of estrogen on human skin are based on the changes that are seen in post-menopausal women, with a number of studies documenting the differences seen following menopause. It may be noted that there is also a variation in skin thickness during the menstrual cycle, with skin thickness lowest at the start of the menstrual cycle, when estrogen and progesterone levels are low; then increases with the rising levels of estrogen (Stevenson and Thornton 2007). In skin tissue, estrogen increases vascularization and shows effects at various levels of dermal tissue (Schmidt et al. 1996). Estrogen compounds were shown to be effective in the treatment of skin aging symptoms of perimenopausal women, increasing epidermal hydration, skin elasticity, skin thickness and also reducing skin wrinkles while improving the content and quality of collagen and the level of vascularization (Phillips et al. 2001, Sator et al. 2001, Schmidt et al. 1996).

Based on its double conjugated bonds, lycopene absorbs radiation at long wavelength of the visible spectrum, allowing to state that lycopene may reduce the damaging effects that UV light can have on skin and against cumulative effects of sun exposure (cancer) (Tanaka et al. 2012). Moreover, previous *in vitro* and *in vivo* studies showed that lycopene demonstrates a beneficial role in chronic diseases, such as cardiovascular disease (Mordente et al. 2011), atherosclerosis (Lorenz et al. 2012), cancer (Tanaka et al. 2012) and neurodegenerative disorders (Prema et al. 2015). The knowledge that free radical formation causes damage to DNA, collagen and elastin, which are essential for elasticity and the renewal process of the skin, upcoming topical formulations can be developed throught incorporation of tomato by-products.

Natural Bioactive Compounds from Food By-products in Final Cosmetic Products

Although food by-products are rich in interesting compounds for the cosmetic industry, few topical formulations have been developed with these active ingredients. Nevertheless, coffee and *C. sativa* by-products are used in topical formulations with good results. A particular example is coffee silverskin. The high antioxidant capacity of coffee silverskin, especially due to chlorogenic acid and other phenolic compounds, has been detailed (Borrelli et al. 2004, Bresciani et al. 2014, Narita and Inouye 2012, Rodrigues et al. 2014, Sato et al. 2011). Also, the caffeine content is high and similar to that of coffee beans (Bresciani et al. 2014, Narita and Inouye 2012). These compounds are believed to provide *in vivo* protection against free radical damage. Like coffee beans, coffee silverskin contains several classes of health compounds, such as phenolics, diterpenes, xanthines and vitamin precursors (Alves et al. 2009). Rodrigues et al. evaluated the cell viability and cytotoxicity of coffee silverskin in different monolayer's skin cell lines (Rodrigues et al. 2015a). No cytotoxicity was observed. Also, a number of *in vitro* three-dimensional models tests were developed to assess the potential skin or eye irritants by the cosmetic industry, such as the reconstructed human epidermis test (EpiSkin™) or the Human Corneal Epithelial Model (SkinEthic™ HCE), respectively. The same research group performed these assays in three different extracts of coffee silverskin. The histology of the models after application of extracts was also analyzed. The *in vitro* results demonstrated that extracts were not classified as irritants and the histological analyses proved that they did not affect both models' structure. The contents of caffeine, 5-hydroxymethyl furfural and chlorogenic acid were quantified after the epidermal assay, proving that caffeine was in similar amounts of coffee beans. This mean that this new extract could clearly be used as an anti-cellulitis agent, for example. The *in vivo* test carried out with the most promising extract (hydroalcoholic) showed that, with respect to irritant effects, they can be regarded as safe for topical application, as no skin irritation was observed. Rodrigues et al. also incorporated this coffee waste on different cosmetic formulations, such as a body cream and a hand cream (Rodrigues et al. 2016a, Rodrigues et al. 2016c). *In vitro* results were completely satisfactory, since no toxicity was observed in neither keratinocytes or fibroblasts. Also, the *in vivo* assays demonstrated that both formulations were safe. The total satisfaction with both products was revealed by sensorial analyses performed in volunteers. Recently, the same research group developed a facial formulation with coffee silverskin and the same cream enriched with hyaluronic acid (Rodrigues et al. 2016b). Facial formulations were applied twice a day by volunteers (n = 20 for each formulation) during 28 days. The influence on skin hydration and viscoelastic properties was investigated with validated devices (Corneometer® and Cutometer®). Wrinkles' depth, roughness, volume of cavities and Visioface® images were analysed at time 0 and after 28 days. Volunteers were asked about efficacy perception. Results revealed that penetration of coffee silverskin extract in pig ear skin after 8 hours is about 20 per cent. No cytotoxicity was observed in both the formulations. Significant changes in skin hydration and viscoelastic parameters were detected for both the formulations, without any differences between them. However, no differences were observed regarding the depth of wrinkles, roughness and volume of cavities for both formulations. In this way, coffee silverskin represents an effective ingredient for cosmetic creams that are intended to increase the skin hydration and firmness.

Almeida et al. evaluated the topical applications of ethanol/water extracts from *C. sativa* leaves incorporated in a skin care formulation (Almeida et al. 2010, Almeida et al. 2015). A strong absorption at 280 nm could forecast a possible effectiveness of chestnut leaf extract through topical administration to prevent UV radiation-induced skin damage (Almeida et al. 2010). Finally, as previously reported, oligomeric proanthocyanidins are phytonutrients that belong to the polyphenol family of antioxidants. Hughes-Formella et al. demonstrated the anti-inflammatory and skin hydration properties of oligomeric proanthocyanidins from grapes in the skin of human volunteers (extracted from *Vitis vinifera* seeds), after incorporation in specific topical creams or lotion formulations (Hughes-Formella et al. 2007).

Trends for Utilisation of Natural Bioactive Compounds in Cosmetics

Over the last few years the amount of scientific literature about bioactive compounds of food by-products, particularly phenolics, has increased. This fact demonstrates the growing interest of academia in the exploitation of wastes as a potential source of active ingredients for cosmetics. However, it is essential to have interdisciplinary interest, not only from the industry but also from the society, in topical formulations based on food by-products. It should be highlighted that although exciting opportunities for re-use in the cosmetic field exist, the methods for complete utilisation on a large scale and at affordable prices should be developed. The development of biotechnological processes to convert food by-products into different isolated molecules is of particular interest. An interesting field could be the isolation of specific compounds from food by-products using green techniques, such as supercritical fluids. These compounds would be added to cosmetics or even used to create food supplements. Also, for most of these food by-products, the bioavailability and toxicity need to be carefully assessed by *in vitro* and *in vivo* assays. In addition the stability of antioxidants has to be improved.

Also, nanotechnology could be a new perspective. The cosmetic industry was among the first to implement nanotechnological principles in product development (Müller et al. 2002). Future works may be extended to encapsulation of extracts and *in vivo* efficacy studies, so as to compare them with the present one.

Table 3 summarizes the strengths and weaknesses of active ingredients use from food by-products.

As observed in Table 3, the major strength point is the high amount of bioactive compounds, as previously described. In our opinion, taken into account the industry perspective, the regular supply of food by-products by the food industry could be the weakness. In fact, the establishment of cooperation between cosmetic and food industry is crucial and an essential point in all the processes.

Table 3. Strengths and weakness points of food by-products re-use in cosmetics.

(+) Strengths Points	(–) Weakness Points
High amount of bioactive compounds	Irregular supplier
Multiplicity of compounds	
Economical advantages	
Interest of society for sustainable ingredients	
Green cosmetics	

Conclusion

Food waste is nowadays considered a matter of treatment, minimization and prevention due to the environmental effects induced by their disposal. According to the Food and Agriculture Organization, one-third of the edible part of food produced for human consumption is lost or wasted globally. The perspectives to develop technologies, allowing the recovery, recycling and sustainable production of high-added value ingredients, are challenging. In particular, the field of cosmetics may benefit from these remaining materials, as the bioactive compounds can fulfil a real function and activity. Due to its versatile composition food by-products become an opportunity to obtain new active ingredients for cosmetics, benefiting from the advantages of huge amounts discarded daily at interesting prices, and providing new potential applications of the waste.

In fact, the food by-products previously detailed are still undervalued, even though their chemical composition reveals their high potential as a source of natural compounds, including antioxidants, fatty acids, minerals or caffeine, which can be used as functional and technological ingredients. Even though, a growing amount of scientific literature deals with clinical anti-aging studies, there is still a need for further investigation to demonstrate the efficacy of these different ingredients in topical formulations.

At this point, the performed work may be a good contribution to the food by-products and cosmetic fields, but much more remains to be done, particularly in what concerns industries' interests. In this way, strong and depth cooperation between academia and industry has to be established.

Acknowledgment

The authors are grateful for the financial support from the project Operação NORTE-01-0145-FEDER-00001, titled Qualidade e Segurança Alimentar—uma abordagem (nano)tecnológica. This work was also supported by the project UID/QUI/50006/2013–POCI/01/0145/FEDER/007265 with financial support from FCT/MEC through national funds and cofinanced by FEDER. Francisca Rodrigues and M. Antónia Nunes are grateful for the research grants from project UID/QUI/50006/2013.

References

Adlercreutz, H. 1995. Phytoestrogens: Epidemiology and a possible role in cancer protection. Environmental Health Perspectives 103: 103–112.
Adlercreutz, H., E. Hämäläinen, S. Gorbach and B. Goldin. 1992. Dietary phyto-oestrogens and the menopause in Japan. The Lancet 339: 1233.
Al-Wandawi, H., M. H. Abdul Rehman and K. A. Al Shaikhly. 1985. Tomato processing wastes as essential raw materials source. J. Agr. Food Chem. 33: 804–807.
Almeida, I. F., P. C. Costa and M. F. Bahia. 2010. Evaluation of functional stability and batch-to-batch reproducibility of a *Castanea sativa* leaf extract with antioxidant activity. AAPS Pharm. Sci. Tech. 11: 120–125.
Almeida, I. F., J. Maleckova, R. Saffi, H. Monteiro, F. Góios, M. H. Amaral, P. C. Costa, J. Garrido, P. Silva, N. Pestana and M. F. Bahia. 2015. Characterization of an antioxidant surfactant-free topical formulation containing *Castanea sativa* leaf extract. Drug Dev. Ind. Pharm. 41: 148–155.

Alves, R. C., S. Casal, M. R. Alves and M. B. Oliveira. 2009. Discrimination between arabica and robusta coffee species on the basis of their tocopherol profiles. Food Chem. 114: 295–299.

Angerhofer, C. K., D. Maes and P. U. Giacomoni. 2009. The use of natural compounds and botanicals in the development of anti-aging skin care products. pp. 205–263. *In*: D. Nava (ed.). Skin Aging Handbook. William Andrew Publishing, New York.

Aziz, N. H., S. E. Farag, L. A. Mousa and M. A. Abo-Zaid. 1998. Comparative antibacterial and antifungal effects of some phenolic compounds. Microbios. 93: 43–54.

Bacaloni, A., C. Cavaliere, A. Faberi, P. Foglia, R. Samperi and A. Laganà. 2005. Determination of isoflavones and coumestrol in river water and domestic wastewater sewage treatment plants. Anal. Chim. Acta. 531: 229–237.

Barbulova, A., G. Colucci and F. Apone. 2015. New trends in cosmetics: By-products of plant origin and their potential use as cosmetic active ingredients. Cosmetics 2: 82.

Barreira, J. C. M., S. Casal, I. C. F. R. Ferreira, A. M. Peres, J. A. Pereira and M. B. P. P. Oliveira. 2012. Chemical characterization of chestnut cultivars from three consecutive years: Chemometrics and contribution for authentication. Food Chem. Toxicol. 50: 2311–2317.

Barreira, J. C. M., I. C. F. R. Ferreira, M. B. P. P. Oliveira and J. A. Pereira. 2008. Antioxidant activities of the extracts from chestnut flower, leaf, skins and fruit. Food Chem. 107: 1106–1113.

Barreira, J. C. M., T. Visnevschi-Necrasov, E. Nunes, S. C. Cunha, G. Pereira and M. B. P. P. Oliveira. 2015. *Medicago* spp. as potential sources of bioactive isoflavones: Characterization according to phylogenetic and phenologic factors. Phytochem. 116: 230–238.

Barros, A., A. Gironés-Vilaplana, A. Texeira, N. Baenas and R. Domínguez-Perles. 2015. Grape stems as a source of bioactive compounds: Application towards added-value commodities and significance for human health. Phytochem. Rev. 14: 921–931.

Bastianetto, S., Y. Dumont, A. Duranton, F. Vercauteren, L. Breton and R. Quirion. 2010. Protective action of resveratrol in human skin: Possible involvement of specific receptor binding sites. PLoS One 5: 12935.

Baxter, R. A. 2008. Anti-aging properties of resveratrol: Review and report of a potent new antioxidant skin care formulation. J. Cosmet. Dermatol. 7: 2–7.

Behrens, A., S. Giljum, J. Kovanda and S. Niza. 2007. The material basis of the global economy: Worldwide patterns of natural resource extraction and their implications for sustainable resource use policies. Ecol. Econom. 64: 444–453.

Benoit, I., D. Navarro, N. Marnet, N. Rakotomanomana, L. Lesage-Meessen, J. C. Sigoillot, M. Asther and M. Asther. 2006. Feruloyl esterases as a tool for the release of phenolic compounds from agro-industrial by-products. Carbohydr. Res. 341: 1820–1827.

Bolzinger, M. A., S. Briançon, J. Pelletier, H. Fessi and Y. Chevalier. 2008. Percutaneous release of caffeine from microemulsion, emulsion and gel dosage forms. Eur. J. Pharm. Biopharm. 68: 446–451.

Borrelli, R. C., F. Esposito, A. Napolitano, A. Ritieni and V. Fogliano. 2004. Characterization of a new potential functional ingredient: Coffee silverskin. J. Agr. Food Chem. 52: 1338–1343.

Braga, N., F. Rodrigues and M. B. P. P. Oliveira. 2014. *Castanea sativa* by-products: A review on added value and sustainable application. Nat. Prod. Res. 29: 1–18.

Bresciani, L., L. Calani, R. Bruni, F. Brighenti and D. Del Rio. 2014. Phenolic composition, caffeine content and antioxidant capacity of coffee silverskin. Food Res. Int. 61: 196–201.

Brzezinski, A. and A. Debi. 1999. Phytoestrogens: The natural selective estrogen receptor modulators? Eur. J. Obstet. Gynecol. Reprod. Biol. 85: 47–51.

Casas, L., C. Mantell, M. Rodríguez, E. J. Ossa, A. Roldán, I. D. Ory, I. Caro and A. Blandino. 2010. Extraction of resveratrol from the pomace of Palomino fino grapes by supercritical carbon dioxide. J. Food Eng. 96: 304–308.

Chandra, H. M. and S. Ramalingam. 2011. Antioxidant potentials of skin, pulp, and seed fractions of commercially important tomato cultivars. Food Sci. Biotechnol. 20: 15–21.

Chandra, H. M., B. M. Shanmugaraj, B. Srinivasan and S. Ramalingam. 2012. Influence of genotypic variations on antioxidant properties in different fractions of tomato. J. Food Sci. 77: 1174–1178.

Clarkson, T. B., M. S. Anthony and C. L. Jr Hughes. 1995. Estrogenic soybean isoflavones and chronic disease: Risks and benefits. Trends Endocrinol. Metab. 6: 11–16.

Costa, A. S. G., R. C. Alves, A. F. Vinha, S. V. P. Barreira, M. A. Nunes, L. M. Cunha and M. B. P. P. Oliveira. 2014. Optimization of antioxidants extraction from coffee silverskin, a roasting by-product, having in view a sustainable process. Ind. Crop Prod. 53: 350–357.

Craft, B. D., A. L. Kerrihard, R. Amarowicz and R. B. Pegg. 2012. Phenol-based antioxidants and the *in vitro* methods used for their assessment. Comp. Rev. Food Sci. Food Saf. 11: 148–173.

de Campos, L. M., F. V. Leimann, R. C. Pedrosa and S. R. Ferreira. 2008. Free radical scavenging of grape pomace extracts from Cabernet sauvingnon (*Vitis vinifera*). Bioresour. Technol. 99: 8413–8420.

Delmonte, P., J. Perry and J. I. Rader. 2006. Determination of isoflavones in dietary supplements containing soy, red clover and kudzu: Extraction followed by basic or acid hydrolysis. J. Chromatogr. A 1107: 59–69.

Dermeche, S., M. Nadour, C. Larroche, F. Moulti-Mati and P. Michaud. 2013. Olive mill wastes: Biochemical characterizations and valorization strategies. Process Biochem. 48: 1532–1552.

Dohadwala, M. M. and J. A.Vita. 2009. Grapes and cardiovascular disease. J. Nutr. 139: 1788–1793.

Elleuch, M., D. Bedigian, O. Roiseux, S. Besbes, C. Blecker and H. Attia. 2011. Dietary fiber and fiber-rich by-products of food processing: Characterisation, technological functionality and commercial applications: A review. Food Chem. 124: 411–421.

Estévez, M., D. Morcuende, R. Ramírez, J. Ventanas and R. Cava. 2004. Extensively reared Iberian pigs versus intensively reared white pigs for the manufacture of liver pâté. Meat Sci. 67: 453–461.

Facciola, S. 1990. Cornucopia—A Source Book of Edible Plants. Kampong Publications, Vista, California.

FAO. 2009. Food and Agriculture Organization Corporate Statistical Database.

Farag, M. A., D. V. Huhman, Z. Lei and L. W. Sumner. 2007. Metabolic profiling and systematic identification of flavonoids and isoflavonoids in roots and cell suspension cultures of *Medicago truncatula* using HPLC–UV–ESI–MS and GC–MS. Phytochem. 68: 342–354.

Feringa, H. H., D. A. Laskey, J. E. Dickson and C. I. Coleman. 2011. The effect of grape seed extract on cardiovascular risk markers: A meta-analysis of randomized controlled trials. J. Am. Diet. Assoc. 111: 1173–1181.

Fiume, M. M., W. F. Bergfeld, D. V. Belsito, R. A. Hill, C. D. Klaassen, D. C. Liebler, J. G. Marks, R. C. Shank, T. J. Slaga, P. W. Snyder and F. A. Andersen. 2014. Safety assessment of *Vitis vinifera* (grape)-derived ingredients as used in cosmetics. Int. J. Toxicol. 33: 48S–83S.

Fletcher, R. J. 2003. Food sources of phyto-oestrogens and their precursors in Europe. Br. J. Nutr. 89: S39–S43.

Franzen, L. and M. Windbergs. 2015. Applications of raman spectroscopy in skin research—From skin physiology and diagnosis up to risk assessment and dermal drug delivery. Adv. Drug Deliv. Rev. 89: 91–104.

Georgiev, V., A. Ananga and V. Tsolova. 2014. Recent advances and uses of grape flavonoids as nutraceuticals. Nutrient 6: 391–415.

Ghanbari, R., F. Anwar, K. M. Alkharfy, A. -H. Gilani and N. Saari. 2012. Valuable nutrients and functional bioactives in different parts of olive (*Olea europaea* L.)—A review. Int. J. Mol. Sci. 13: 3291–3340.

Giardina, S., A. Michelotti, G. Zavattini, S. Finzi, C. Ghisalberti and F. Marzatico. 2010. Efficacy study *in vitro*: Assessment of the properties of resveratrol and resveratrol + N-acetyl-cysteine on proliferation and inhibition of collagen activity. Minerva Ginecol. 62: 195–201.

Gustavsson, J., C. Cederberg, R. Sonesson, R. van Otterdijk and A. Meybeck. 2011. Global food losses and food wastes: Extent, cause and prevention. *In*: Office of Knowledge Exchange, Research and Extension (ed.). Food and Agriculture Organization (FAO), Rome.

Han, B. and M. E. 2012. Composite Containing Collagen and Elastin as a Dermal Expander and Tissue Filler. Google Patents. Acessed on https://www.google.com/patents/US20120010146.

Herman, A. and A. P. Herman. 2013. Caffeine's mechanisms of action and its cosmetic use. Skin Pharmacol. Physiol. 26: 8–14.

Hughes-Formella, B., O. Wunderlich and R. Williams. 2007. Anti-inflammatory and skin-hydrating properties of a dietary supplement and topical formulations containing oligomeric proanthocyanidins. Skin Pharmacol. Physiol. 20: 43–49.

Ilahy, R., G. Piro, I. Tlili, A. Riahi, R. Sihem, I. Ouerghi, C. Hdider and M. S. Lenucci. 2016. Fractionate analysis of the phytochemical composition and antioxidant activities in advanced breeding lines of high-lycopene tomatoes. Food Funct. 7: 574–583.

Karimi, A. and M. T. Moradi. 2015. Total phenolic compounds and *in vitro* antioxidant potential of crude methanol extract and the correspond fractions of *Quercus brantii* L. acorn. J. Herb. Med. Pharmacol. 4: 35–39.

Kim, B. J., J. H. Kim, H. P. Kim and M. Y. Heo. 1997. Biological screening of 100 plant extracts for cosmetic use (II): Anti-oxidative activity and free radical scavenging activity. Int. J. Cosmet. Sci. 19: 299–307.

Kim, J. J., B. K. Ghimire, H. C. Shin, K. J. Lee, K. S. Song, Y. S. Chung, T. S. Yoon, Y. J. Lee, E. H. Kim and I. M. Chung. 2012. Comparison of phenolic compounds content in indeciduous *Quercus* species. J. Med. Plant. Res. 6: 5528–5239.

Klejdus, B., D. Vitamvásová-Sterbová and V. Kubán. 2001. Identification of isoflavone conjugates in red clover (*Trifolium pratense*) by liquid chromatography-mass spectrometry after two-dimensional solid-phase extraction. Anal. Chim. Acta. 450: 81–97.

Koschuh, W., G. Povoden, V. H. Thang, S. Kromus, K. D. Kulbe, S. Novalin and C. Krotscheck. 2004. Production of leaf protein concentrate from ryegrass (Lolium *perenne* x multiflorum) and alfalfa (*Medicago sauva* subsp. *sativa*). Comparison between heat coagulation/centrifugation and ultrafiltration. Desalination 163: 253–259.

Krenn, L., I. Unterrieder and R. Ruprechter. 2002. Quantification of isoflavones in red clover by high-performance liquid chromatography. J. Chromatogr. B 777: 123–128.

Ky, I., B. Lorrain, N. Kolbas, A. Crozier, P. -L. Teissedre. 2014. Wine by-products: Phenolic characterization and antioxidant activity evaluation of grapes and grape pomaces from six different french grape varieties. Molecules 19: 482.

Lachman, J., A. Hejtmánková, J. Táborský, Z. Kotíková, V. Pivec, R. Střalková, A.Vollmannová, T. Bojňanská and M. Dědina. 2015. Evaluation of oil content and fatty acid composition in the seed of grapevine varieties. LWT—Food Sci. Technol. 63: 620–625.

Lama-Muñoz, A., G. Rodríguez-Gutiérrez, F. Rubio-Senent and J. Fernández-Bolaños. 2012. Production, characterization and isolation of neutral and pectic oligosaccharides with low molecular weights from olive by-products thermally treated. Food Hydrocolloid 28: 92–104.

Leifert, W. R. and M. Y. Abeywardena. 2008. Cardioprotective actions of grape polyphenols. Nutr. Res. 28: 729–737.

Lorenz, M., M. Fechner, J. Kalkowski, K. Fröhlich, A. Trautmann, V. Böhm, G. Liebisch, S. Lehneis, G. Schmitz, A. Ludwig, G. Baumann, K. Stangl and V. Stangl. 2012. Effects of lycopene on the initial state of atherosclerosis in New Zealand white (NZW) rabbits. PLoS One 7: e30808.

Martinez-Saez, N., M. Ullate, M. A. Martín-Cabrejas, P. Martorell, S. Genovés, D. Ramón and M. D. Castillo. 2014. A novel antioxidant beverage for body weight control based on coffee silverskin. Food Chem. 150: 227–234.

Masghouni, M. and M. Hassairi. 2000. Energy applications of olive-oil industry by-products: - I. The exhaust foot cake. Biomass Bioenerg. 18: 257–262.

Mathes, S. H., H. Ruffner and U. Graf-Hausner. 2014. The use of skin models in drug development. Adv. Drug Deliv. Rev. 69-70: 81–102.

Matsumura, Y. and H. N. Ananthaswamy. 2004. Toxic effects of ultraviolet radiation on the skin. Toxicol. Appl. Pharmacol. 195: 298–308.

Mazur, W. M., J. A. Duke, K. Wahala, S. Rasku and H. Adlercreutz. 1998. Isoflavonoids and lignans in legumes: Nutritional and health aspects in humans. J. Nutr. Biochem. 9: 193–200.

McPhee, D. and R. Jain. 2015. Price, performance, supply and sustainability all-in-one—The Amyris case. H&PC Today 10: 71–72.

Meksi, N., W. Haddar, S. Hammami and M. F. Mhenni. 2012. Olive mill wastewater: A potential source of natural dyes for textile dyeing. Ind. Crop Prod. 40: 103–109.

Micol, V., N. Caturla, L. Pérez-Fons, V. Más, L. Pérez and A. Estepa. 2005. The olive leaf extract exhibits antiviral activity against viral haemorrhagic septicaemia rhabdovirus (VHSV). Antiviral Res. 66: 129–136.

Moraes, A. B., M. A. Haidar, J. M. Soares Júnior, M. J. Simões, E. C. Baracat and M. T. Patriarca. 2009. The effects of topical isoflavones on postmenopausal skin: Double-blind and randomized clinical trial of efficacy. Eur. J. Obstet. Gynecol. Reprod. Biol. 146: 188–192.

Mordente, A., B. Guantario, E. Meucci, A. Silvestrini, E. Lombardi, G. E. Martorana, B. Giardina and V. Bohm. 2011. Lycopene and cardiovascular diseases: An update. Current Med. Chem. 18: 1146–1163.

Mortensen, A., S. E. Kulling, H. Schwartz, I. Rowland, C. E. Ruefer, G. Rimbach, A. Cassidy, P. Magee, J. Millar, W. L. Hall, F. Kramer Birkved, I. K. Sorensen and G. Sontag. 2009. Analytical and compositional aspects of isoflavones in food and their biological effects. Mol. Nutr. Food Res. 53: S266–S309.

Müller, R. H., M. Radtke and S. A. Wissing. 2002. Solid lipid nanoparticles (SLN) and nanostructured lipid carriers (NLC) in cosmetic and dermatological preparations. Adv. Drug Deliv. Rev. 54: S131–S155.

Murthy, P. S. and M. M. Naidu. 2012. Sustainable management of coffee industry by-products and value addition—A review. Resour. Conserv. Recy. 66: 45–58.

Mussatto, S. I., E. M. S. Machado, L. M. Carneiro and J. A. Teixeira. 2012. Sugars metabolism and ethanol production by different yeast strains from coffee industry wastes hydrolysates. Appl. Energy 92: 763–768.

Napolitano, A., V. Fogliano, A. Tafuri and A. Ritieni. 2007. Natural occurrence of ochratoxin A and antioxidant activities of green and roasted coffees and corresponding byproducts. J. Agr. Food Chem. 55: 10499–10504.

Narita, Y. and K. Inouye. 2012. High antioxidant activity of coffee silverskin extracts obtained by the treatment of coffee silverskin with subcritical water. Food Chem. 135: 943–949.

Narita, Y. and K. Inouye. 2014. Review on utilization and composition of coffee silverskin. Food Res. Int. 61: 16–22.

Nichols, J. A. and S. K. Katiyar. 2010. Skin photoprotection by natural polyphenols: Anti-inflammatory, antioxidant and DNA repair mechanisms. Arch. Dermatol. Res. 302: 71–83.

Nikolić, N. P. and S. Orlović. 2002. Genotypic variability of morphological characteristics of English oak (*Quercus robur* L.) acorn. Proceedings for Natural Sciences, Matica Srpska Novi Sad. 102: 53–58.

Nunes, C., S. de Sousa Araújo, J. M. da Silva, M. P. S. Fevereiro and A. B. da Silva. 2008. Physiological responses of the legume model *Medicago truncatula* cv. Jemalong to water deficit. Environ. Exp. Bot. 63: 289–296.

Omer, H. A. A. and S. S. Abdel-Magid. 2015. Incorporation of dried tomato pomace in growing sheep rations. Glob. Vet. 14: 1–16.

Palozza, P., M. Colangelo, R. Simone, A. Catalano, A. Boninsegna, P. Lanza, G. Monego and F. O. Ranelletti. 2010. Lycopene induces cell growth inhibition by altering mevalonate pathway and Ras signaling in cancer cell lines. Carcinog 31: 1813–1821.

Parada, J. and J. M. Aguilera. 2007. Food microstructure affects the bioavailability of several nutrients. J. Food Sci. 72: 21–32.

Petrovic, S., S. Sobajic, S. Rakic, A. Tomic and J. Kukic. 2004. Investigation of kernel oils of *Quercus robur* and *Quercus cerris*. Chem. Nat. Comp. 40: 420–422.

Phillips, T. J., Z. Demircay and M. Sahu. 2001. Hormonal effects on skin aging. Clin. Geriatr. Med. 17: 661–672.

Pincla, J., L. Barros, A. M. Carvalho and I. C. F. R. Ferreira. 2012. Nutritional composition and antioxidant activity of four tomato (*Lycopersicon esculentum* L.) farmer' varieties in Northeastern Portugal homegardens. Food Chem. Toxicol. 50: 829–834.

Pourfarzad, A., H. Mahdavian-Mehr and N. Sedaghat. 2013. Coffee silverskin as a source of dietary fiber in bread-making: Optimization of chemical treatment using response surface methodology. LWT—Food Sci. Technol. 50: 599–606.

Prakash, L. and M. Majeed. 2009. Natural ingredients for anti-ageing skin care. H&PC Today 2: 44–46.

Prema, A., U. Janakiraman, T. Manivasagam and A. Justin Thenmozhi. 2015. Neuroprotective effect of lycopene against MPTP induced experimental Parkinson's disease in mice. Neurosci. Lett. 599: 12–19.

Prow, T. W., J. E. Grice, L. L. Lin, R. Faye, M. Butler, W. Becker, E. M. T. Wurm, C. Yoong, T. A. Robertson, H. P. Soyer and M. S. Roberts. 2011. Nanoparticles and microparticles for skin drug delivery. Adv. Drug Deliv. Rev. 63: 470–491.

Rahmanian, N., S. Jafari and C. Galanakis. 2014. Recovery and removal of phenolic compounds from olive mill wastewater. J. Am. Oil Chem. Soc. 91: 1–18.

Rakić, S., D. Povrenović, V. Tešević, M. Simić and R. Maletić. 2006. Oak acorn, polyphenols and antioxidant activity in functional food. J. Food Eng. 74: 416–423.

Rawlings, A. V. and P. J. Matts. 2005. *Stratum corneum* moisturization at the molecular level: An update in relation to the dry skin cycle. J. Investig. Dermatol. 124: 1099–1110.

Rios, J. L., E. Bas and M. C. Recio. 2005. Effects of natural products on contact dermatitis. Curr. Med. Chem. 4: 65–80.

Roberts, M. S., J. Bouwstra, F. Pirot and F. Falson. 2007. Skin hydration—A key determinant in topical absorption. pp. 115–128. *In*: K. A. Walters and M. S. Roberts (eds.). Dermatologic, Cosmeceutic, and Cosmetic Development -Therapeutic and Novel Approaches. Informa Healthcare, New York.

Rodrigues, F., A. Palmeira-de-Oliveira, J. das Neves, B. Sarmento, M. H. Amaral and M. B. P. P. Oliveira. 2013. *Medicago* spp. extracts as promising ingredients for skin care products. Ind. Crop Prod. 49: 634–644.

Rodrigues, F., I. Almeida, B. Sarmento, M. H. Amaral and M. B. P. P. Oliveira. 2014. Study of the isoflavone content of different extracts of *Medicago* spp. as potential active ingredient. Ind. Crop Prod. 57: 110–115.

Rodrigues, F., A. Palmeira-de-Oliveira, J. das Neves, B. Sarmento, M. H. Amaral and M. B. P. P. Oliveira. 2015a. Coffee silverskin: a possible valuable cosmetic ingredient. Pharm. Biol. 53: 386–394.

Rodrigues, F., C. Pereira, F. B. Pimentel, R. C. Alves, M. Ferreira, B. Sarmento, M. H. Amaral and M. B. P. P. Oliveira. 2015b. Are coffee silverskin extracts safe for topical use? An *in vitro* and *in vivo* approach. Ind. Crop Prod. 63: 167–174.

Rodrigues, F., F. B. Pimentel and M. B. P. P. Oliveira. 2015c. Olive by-products: Challenge application in cosmetic industry. Ind. Crop Prod. 70: 116–124.

Rodrigues, F., J. Santos, F. B. Pimentel, N. Braga, A. Palmeira-de-Oliveira and M. B. P. P. Oliveira. 2015d. Promising new applications of *Castanea sativa* shell: Nutritional composition, antioxidant activity, amino acids and vitamin E profile. Food Funct. 6: 2854–2860.

Rodrigues, F., C. Gaspar, A. Palmeira-de-Oliveira, B. Sarmento, M. H. Amaral and M. B. P. P. Oliveira. 2016a. Application of coffee silverskin in cosmetic formulations: Physical/antioxidant stability studies and cytotoxicity effects. Drug Dev. Ind. Pharm. 42: 99–106.

Rodrigues, F., R. Matias, M. Ferreira, M. H. Amaral and M. B. P. P. Oliveira. 2016b. *In vitro* and *in vivo* comparative study of cosmetic ingredients Coffee silverskin and hyaluronic acid. Exp. Dermatol. In press.

Rodrigues, F., B. Sarmento, M. H. Amaral and M. B. P. P. Oliveira. 2016c. Exploring the antioxidant potentiality of two food by-products into a topical cream: Stability, *in vitro* and *in vivo* evaluation. Drug Dev. Ind. Phar. 42: 880–889.

Rodríguez-Gutiérrez, G., F. Rubio-Senent, A. Lama-Muñoz, A. García and J. Fernández-Bolaños. 2014. Properties of lignin, cellulose, and hemicelluloses isolated from olive cake and olive stones: Binding of water, oil, bile acids, and glucose. J. Agr. Food Chem. 62: 8973–8981.

Rostagno, M. A., A. Villares, E. Guillamón, A. García-Lafuente and J. A. Martínez. 2009. Sample preparation for the analysis of isoflavones from soybeans and soy foods. J. Chromatogr. A 1216: 2–29.

Sabudak, T. and N. Guler. 2009. *Trifolium* L.—A review on its phytochemical and pharmacological profile. Phytother. Res. 23: 439–446.

Saenger, M., E. U. Hartge, J. Werther, T. Ogada and Z. Siagi. 2001. Combustion of coffee husks. Renew. Energ. 23: 103–121.

Šálková, T., M. Divišová, Š. Kadochová, J. Beneš, K. Delawská, E. Kadlčková, L. Němečková, K. Pokorná, V. Voska and A. Žemličková. 2011. Acorns as a food resource. An experiment with acorn preparation and taste. Interdiscip. Archaeol. 2: 139–147.

Sánchez-Burgos, J. A., M. V. Ramírez-Mares, M. M. Larrosa, J. A. Gallegos-Infante, R. F. González-Laredo, L. Medina-Torres and N. E. Rocha-Guzmán. 2013. Antioxidant, antimicrobial, antitopoisomerase and gastroprotective effect of herbal infusions from four *Quercus* species. Ind. Crop Prod. 42: 57–62.

Sato, Y., S. Itagaki, T. Kurokawa, J. Ogura, M. Kobayashi, T. Hirano, M. Sugawara and K. Iseki. 2011. *In vitro* and *in vivo* antioxidant properties of chlorogenic acid and caffeic acid. Int. J. Pharm. 403: 136–138.

Sator, P. G., J. B. Schmidt, M. O. Sator, J. C. Huber and H. Honigsmann. 2001. The influence of hormone replacement therapy on skin ageing: A pilot study. Maturitas 39: 43–55.

Schäfer-Korting, M., W. Mehnert and H. -C. Korting. 2007. Lipid nanoparticles for improved topical application of drugs for skin diseases. Adv. Drug Deliv. Rev. 59: 427–443.

Schieber, A., F. C. Stintzing and R. Carle. 2001. By-products of plant food processing as a source of functional compounds—Recent developments. Trends Food Sci. Tech. 12: 401–413.

Schmidt, J. B., M. Binder, G. Demschik, C. Bieglmayer and A. Reiner. 1996. Treatment of skin aging with topical estrogens. Int. J. Dermatol. 35: 669–674.

Schurer, N. Y. and P. M. Elias. 1991. The biochemistry and function of *stratum corneum* lipids. Adv. Lipid Res. 24: 27–56.

Silva, J. P., A. C. Gomes and O. P. Coutinho. 2008. Oxidative DNA damage protection and repair by polyphenolic compounds in PC12 cells. Eur. J. Pharmacol. 601: 50–60.

Silva, L. R., M. J. Pereira, J. Azevedo, R. F. Gonçalves, P. Valentão, P. G. Pinho and P. B. Andrade. 2013. *Glycine max* (L.) Merr., *Vigna radiata* L. and *Medicago sativa* L. sprouts: A natural source of bioactive compounds. Food Res. Int. 50: 167–175.

Sivesind, E. and P. Seguin. 2005. Effects of the environment, cultivar, maturity, and preservation method on red clover isoflavone concentration. J. Agr. Food Chem. 53: 6397–6402.

Stavroulias, S. and C. Panayiotou. 2005. Determination of optimum conditions for the extraction of squalene from olive pomace with supercritical CO_2. Chem. Biochem. Eng. Q 19: 371–381.

Stevenson, S. and J. Thornton. 2007. Effect of estrogens on skin aging and the potential role of SERMs. Clin. Interv. Aging 2: 283–297.

Sung, S. H., K. H. Kim, B. T. Jeon, S. H. Cheong, J. H. Park, D. H. Kim, H. J. Kweon and S. H. Moon. 2012. Antibacterial and antioxidant activities of tannins extracted from agricultural by-products. J. Med. Plant. Res. 6: 3072–3079.

Takyi, E. E. K., Y. Kido, T. Rikimaru and D. O. Kennedy. 1992. Possible use of alfalfa (*Medicago sativa*) as supplement in infant nutrition: Comparison of weight gained by rats fed on alfalfa and a popular weaning diet. J. Sci. Food Agr. 59: 109–115.

Tanaka, T., M. Shnimizu and H. Moriwaki. 2012. Cancer chemoprevention by carotenoids. Molecules 17: 3202.

Teixeira, A., N. Baenas, R. Dominguez-Perles, A. Barros, E. Rosa, D. Moreno and C. Garcia-Viguera. 2014. Natural bioactive compounds from winery by-products as health promoters: A review. Int. J. Mol. Sci. 15: 15638.

Tejerina, D., S. García-Torres, M. C. Vaca, F. M. Vásquez and R. Cava. 2011. Acorns (*Quercus rotundifolia* Lam.) and grass as natural sources of antioxidants and fatty acids in the "montanera" feeding of Iberian pig: intra- and inter-annual variations. Food Chem. 124: 997–1004.

Tham, D. M., C. D. Gardner and W. L. Haskell. 1998. Potential health benefits of dietary phytoestrogens: A review of the clinical, epidemiological, and mechanistic evidence. J. Clin. Endocrinol. Metab. 83: 2223–2235.

Thompson, Z. and H. I. Maibach. 2010. Biological effects of estrogen on skin. pp. 361–367. *In:* Miranda A. Farage and Howard I. Maibach (eds.). Textbook of Aging Skin. Springer, Berlin.

Toor, R. K. and G. P. Savage. 2005. Antioxidant activity in different fractions of tomatoes. Food Res. Int. 38: 487–494.

Tsang, C., C. Auger, W. Mullen, A. Bornet, J. M. Rouanet, A. Crozier and P. L. Teissedre. 2005. The absorption, metabolism and excretion of flavan-3-ols and procyanidins following the ingestion of a grape seed extract by rats. Br. J. Nutr. 94: 170–181.

Vasapollo, G., L. Longo, L. Rescio and L. Ciurlia. 2004. Innovative supercritical CO_2 extraction of lycopene from tomato in the presence of vegetable oil as co-solvent. J. Supercrit. Fluids. 29: 87–96.

Velazquez, C., C. Dieamant, S. Eberlin, C. Nogueira, D. Colombi, L. C. Di Stasi and M. L. de Souza Queiroz. 2009. Effect of green *Coffea arabica* L. seed oil on extracellular matrix components and water-channel expression in *in vitro* and *ex vivo* human skin models. J. Cosmet. Dermatol. 8: 56–62.

Vinha, A. F., R. C. Alves, S. V. P. Barreira, A. Castro, A. S. G. Costa and M. B. P. P. Oliveira. 2014a. Effect of peel and seed removal on the nutritional value and antioxidant activity of tomato (*Lycopersicon esculentum* L.) fruits. LWT—Food Sci. Technol. 55: 197–202.

Vinha, A. F., S. V. P. Barreira, A. S. G. Costa, R. C. Alves and M. B. P. P. Oliveira. 2014b. Pre-meal tomato (*Lycopersicon esculentum*) intake can have anti-obesity effects in young women? Int. J. Food. Sci. Nutr. 65: 1019–1026.

Viola, P. and M. Viola. 2009. Virgin olive oil as a fundamental nutritional component and skin protector. Clin. Dermatol. 27: 159–165.

Visnevschi-Necrasov, T., J. C. M. Barreira, S. C. Cunha, G. Pereira, E. Nunes and M. B. P. P. Oliveira. 2015. Phylogenetic insights on the isoflavone profile variations in Fabaceae spp.: Assessment through PCA and LDA. Food Res. Int. 76: 51–57.

Von Baer, D., R. Saelzer, M. Vega, P. Ibieta, L. Molina, E. Von Baer, R. Ibáñez and U. Hashagen. 2006. Isoflavones in *Lupinus albus* and *Lupinus angustifolius*: Quantitative determination by capillary zone electrophoresis, evolution of their concentration during plant development and effect on anthracnose causing fungus *Colletotrichum lupini*. J. Chil. Chem. Soc. 51: 1025–1029.

Wu, Q., M. Wang and J. E. Simon. 2003. Determination of isoflavones in red clover and related species by high-performance liquid chromatography combined with ultraviolet and mass spectrometric detection. J. Chromatogr. A 1016: 195–209.

Xia, E. -Q., G. -F. Deng, Y. -J. Guo and H. -B. Li. 2010. Biological activities of polyphenols from grapes. Int. J. Mol. Sci. 11: 622–646.

Zhu, W. and J. Gao. 2008. The use of botanical extracts as topical skin-lightening agents for the improvement of skin pigmentation disorders. J. Investig. Dermatol. Symp. Proc. 13: 20–24.

CHAPTER 15

Potential Use of Bioactive Compounds from Waste in the Pharmaceutical Industry

Antonietta Baiano

Introduction on Bioactive Compounds

The molecules found in Nature may be regarded as solutions to challenges overcome during molecular evolution (Caporale 1995). From a medical point of view, the terms 'bioactive' or 'biologically active' compounds define a substance having a direct effect on living tissues depending on the type of substance, amount and bioavailability (The American Heritage Medical Dictionary 2007, Bernhoft 2010). Nevertheless, the demonstration of these activities alone is not sufficient to define a compound as 'bioactive', since it must provide also beneficial effects on health (Guaadaoui et al. 2014). This specification is due to the scientific evidence that natural compounds can have several effects, ranging from being deadly to being healthy or curative. These molecules are also referred to as 'natural' products since they are produced by biological sources. Nevertheless, in recent times, the term 'natural' has been extended to include synthesis in a laboratory, biosynthesis *in vitro*, isolation from engineered organisms whereby the resultant molecules are chemically equivalent to the original natural ones (All natural 2007). It is useful to remember that the ability to synthesize bioactive molecules began in the early twentieth century, despite the development of pharmaceutical chemistry (Paulsen 2010). Furthermore, the term 'natural products' is referred to compounds produced as secondary metabolites, i.e., compounds that don't exert 'primary' functions in normal growth, development or reproduction of an organism and are produced as a consequence of the organism adapting to the surrounding environment or in order to act as a possible defense mechanism against predators (Maplestone et al. 1992). The biosynthesis of secondary metabolites use the fundamental processes of photosynthesis, glycolysis and the Krebs cycle. The most important building blocks involved in the biosynthesis of secondary metabolites are

Dipartimento di Scienze Agrarie degli Alimenti e dell'Ambiente – University of Foggia, Via Napoli 25 – 71122 Foggia, Italy. E-mail: antonietta.baiano@unifg.it those derived from the intermediates acetyl coenzyme A, shikimic acid, mevalonic acid and 1-deoxyxylulose-5-phosphate. They act through pathways involving different mechanisms and reactions, such as alkylation, decarboxylation, aldol, Claisen and Schiff base formation (Dewick 2002).

Natural products are usually molecules with a molecular weight below 3,000 Daltons and exhibit considerable structural diversity (Kinghorn et al. 2009). Studies concerning identification, isolation and characterization of natural products are included in the scientific field of pharmacognosy.

Natural products are mainly derived from plant (Joseph and Priya 2010) tissues. Nevertheless, fungi, bacteria and other living terrestrial and marine organisms (including groups of animals) can be used to produce compounds usable in a variety of application, from curing diseases to feeding humans. In particular, endophytes are interesting microorganisms to be used in preparation of bioactive products. 'Endophyte' are microorganisms that grow intra and intercellular in the tissues of higher plants without affecting the plants in which they live. According to the so-called 'plantendophyte co-evolution hypothesis' (Ji et al. 2009) endophytes assist the plant in chemical defense just by producing bioactive secondary metabolites.

Endophytes are recognized as potential sources of novel natural products to be used for pharmaceutical, agricultural and food application (Kumar and Sagar 2007). Their potential can be easily understood considering that each of the nearly 300,000 existing plant species can host one or more endophytes to reach the number of one million endophyte species (Strobel and Daisy 2003).

The product categories in which natural compounds can be found as active ingredients include drugs (pharmaceuticals), cosmetic ingredients (cosmeceuticals) and dietary supplements and natural health product ingredients (nutriceuticals) (Chernyak 2012). Bioactive compounds can be classified according to the following criteria: clinical function (intended as their pharmacological or toxicological effects); biological effects; botanical categorization; biochemical pathways and chemical classes (Bernhoft 2010).

Brief Historical and Legal *Excursus* of the Use of Bioactive Compounds in Pharmaceutical Products

Historically, natural products have been used for treatment of many diseases and illnesses (Dias et al. 2012), though many of these bioactive natural products are still being unidentified. In the past, the main source of knowledge on use of natural products was medicinal plants on which man experimented by trial and error for hundreds of centuries through palatability trials or untimely deaths (Kinghorn et al. 2011).

The therapeutic potential of plant products can be traced back to approximately 6,000 years (Paulsen 2010). Clay tablets from Mesopotamia, dating back to 2600 B.C., documented oils from *Cupressus sempervirens* (cypress) and *Commiphora* species (myrrh), which are still used in treatment of coughs, colds and inflammations (Cragg and Newman 2005, Dias et al. 2012). The Egyptian Ebers Papyrus (2900 B.C.) included over 700 plant-based drugs to be used for gargles, pills, infusions and ointments. In China, the *Materia Medica* (1100 B.C.), the *Shennong Herbal* (~ 100 B.C.) and the *Tang Herbal* (659 A.D.) included 52 prescriptions, 365 and 850 drugs, respectively. In ancient Greece, the physician Dioscorides (100 A.D.) recorded collection, storage and uses of medicinal herbs, whilst the philosopher and natural scientist Theophrastus (~ 300 B.C.) dealt with medicinal herbs (Cragg and Newman 2005, Dias et al. 2012).

In the 18th century, the effects of cardiotonic digitalis extracts was discovered for treatment of heart disease and of the use of the bark of the willow tree for treating fever. In 1804, morphine for analgesic and sedative effects was isolated from opium obtained from the seed pods of the poppy plant. In the same century, purified bioactive natural products were extracted from the Peruvian bark cinchoa (quinine) and cocoa (cocaine) (Beutler 2009). By 1829, scientists observed that the compound salicin, from willow trees, provided pain relief. In 1838, salicylic acid was isolated, but researchers found it harsh on the stomach. In the second half of the 19th century, a less-irritating compound, the acetylsalicylic acid, was synthesized (Jeffreys 2004).

Before the 20th century, the only medications available were crude and semi-pure extracts of plants, animals, microorganisms and minerals. In the 20th century, scientists understood that the effect of any drug in the human body was mediated by specific interactions of specific molecules introduced from outside with biological macromolecules. They then concluded that individual chemical compounds in extracts (instead of the whole extract) was required for the biological activity of the drug (Lahlou 2013).

After the Second World War, encouraged by the discovery of penicillin, intensive screening of microorganisms for new antibiotics was carried out (Beutler 2009), leading to the discovery of cephalosporins, tetracyclines, aminoglycosides, rifamycins, chloramphenicol, lipopeptides (Butler 2005, Mishra and Tiwari 2011) and to the isolation from Caribbean marine sponges of two nucleosides useful for the synthesis of vidarabine and cytarabine, to be used in viral diseases and cancer, respectively (Molinski et al. 2009). Another example of a bioactive compound is the anticancer taxol derived from the Yew tree, discovered in the 1970s and available in market in 1992 (Goodman and Walsh 2001). At present, half of the marketed agents in the drug arsenal are derived from biological sources, in particular from terrestrial plants, microorganisms and animals (Montaser and Luesch 2011).

As this brief history demonstrates, pharmaceutical organic chemists were for long interested in these bioactive compounds that were investigated for their chemical properties extensively out of academic interest, but due to their potential utility as dyes, polymers, fibers, glues, oils, waxes, flavoring agents, perfumes, drugs, antibiotics, insecticides and herbicides (Jimenez-Garcia et al. 2013). Although it has been largely demonstrated that natural products are an important source of compounds for drug production, their use decreased in the last two decades largely due to the technical barriers to their screening in high-throughput assays against molecular targets (Harvey et al. 2015). Furthermore, the difficulties associated with obtaining sufficient amounts of natural products pure enough for discovery and development activities had to be taken into account. In the traditional approach, the crude extracts have to be tested for biological activity and eventually fractionated until the active compounds are isolated and identified. This method is time-expensive and doesn't provide guarantees that the final results can be chemically workable or patentable (Rouhi 2003). The use of genomic and metabolomic approaches to the study of natural products could further contribute to a revival of interest in such products for drug discovery. In fact, the increasing commercial importance of these compounds has resulted in the investigation into the possibility of altering the production of bioactive plant metabolites by means of tissue culture technology and metabolomics (Jimenez-Garcia et al. 2013). In the case of compounds derived from wild plants, the difficulty in obtaining sufficient amounts of natural products is due to issues such as deforestation, environmental pollution in remote areas and global warming. To meet the demand for raw material without threat to the wild resources and to reduce the biological variability in the active ingredient content in medicinal plants from different collection areas, it is useful to perform controlled cultivation programs (Bhutani and Gohil 2010). In the case of marine organisms, troubles are related to extraction because of their high water and salt content, and to their unsuitableness (the compounds of interest must be extracted from specimens collected in the wild) (Molinski et al. 2009, Leal et al. 2012).

The pharmaceutical industry's interest in natural products diminished with the advent of promising new technologies, like combinatorial chemistry (CC, parallel synthesis techniques to create libraries with hundreds of thousands of compounds) and high throughput screening (HTS, rapid testing of a great numbers of compounds) (Ortholand and Ganesan 2004). It is useful to remember that, at present, only less than 10 per cent of the world's biodiversity has been evaluated for its biological activity. Thus a number of natural compounds await to be discovered (Cragg and Newman 2005).

Another fundamental issue is represented by the regulatory requirements of pharmaceutical products containing natural substances. The Center for Drug Evaluation and Research in the U.S.A. and the Health Canada in Canada evaluate new drugs before marketing them in order to ensure their safety (at least that their health benefits are greater than known risks) and effectiveness (Ciociola et al. 2014). In the European Union, drugs and medicinal products are regulated by the European Medicines Agency. The 2004/24/EC Directive on traditional herbal medicinal products, which is a subdirective of the 2001/83/EC Human Medicinal Products Directive, established a unique set of information on a herbal substance or herbal preparation for all EU Member States. According to the European medicines legislation, medicinal products containing herbal substances can be classified within one of the following three categories: products categorized as traditional medicinal use provisions ('traditional use') accepted on the basis of sufficient safety data and plausible efficacy; products categorized as well-established medicinal use provisions ('well-established use'). This is demonstrated with the provision of scientific literature establishing that the active substances of the medicinal products are in well-established medicinal use within the Union for atleast ten years, with recognized efficacy and an acceptable level of safety; a product that can be authorized after evaluation of the marketing authorization application consisting of only safety and efficacy data from the company's own development ('stand alone') or a combination of own studies and bibliographic data ('mixed application') (http://www.ema.europa.eu/ema/index.jsp?curl=pages/regulation/general/general_content_000208.jsp).

Bioactive Compounds of Interest for Pharmaceuticals

It is uncommon for natural products to show molecular structures with cyclic semirigid scaffolds, several chiral centers, more than 5 H-bond donors, more than 10 H-bond acceptors, more than 5 rotatable C-C bonds, a large polar surface area and a molecular weight higher than 500 (Lahlou 2013). The secondary metabolites of interest for pharmaceuticals can be classified into groups based on their chemical components, functions and biosynthesis. Terpenes, phenols, flavanoids, alkaloids, sterols, waxes, fats, tannins, sugars, gums, suberins, resin acids and carotenoids are the many classes of secondary metabolites (Gottlieb 1990). Secondary metabolites comprise 25,000 terpenoids, 12,000 alkaloids and 8,000 phenolic compounds (Genovese et al. 2009). The pathways for synthesis of secondary metabolites are the following: shikimic acid pathway, malonic acid pathway, mevalonic acid pathway and Non-mevalonate (MEP) pathway (Tiaz and Zeiger 2006). For example, phenolic compounds are produced by shikimic acid pathway (phenylpropanoids are formed) and acetic acid pathway (the main products include simple phenol) but the combination of both pathways leads to the formation of flavonoids, which are the most plentiful group of phenolic compounds in Nature. Terpenes are produced through mevalonic acid pathway and MEP pathway and alkaloids are synthesized from aromatic amino acids (shikimic acid pathway) and aliphatic amino acids (tricarboxylic acid cycle).

It is hypothesized that secondary metabolism utilizes amino acids and the acetate and shikimate pathways to produce intermediates that have adopted an alternate biosynthetic route. Modifications in the biosynthetic pathways may be due to natural causes (viruses, environmental changes) or unnatural causes (chemicals or radiations) (Sarker et al. 2006).

The following pages include a list and description of the main types of bioactive compounds:

Phenolic compounds

Phenolics are bioactive compounds found mostly in plants, where they play an important role in growth and reproduction while providing protection against pathogens or predators (Bravo 1998). These compounds have an aromatic ring bearing one or more hydroxyl substituents, and range from simple phenolic molecules to polymerised compounds (Bravo 1998). According to their carbon chain, phenolic compounds can be divided into 16 major classes: simple phenols, benzoquinones, phenolic acids, acetophenones, phenylacetic acids, hydroxycinnamic acids, phenylpropenes, coumarins and isocoumarins, chromones, naphthoquinones, xanthones, stilbenes, anthraquinones, flavonoids, lignans and neolignans, and lignins (Harborne 1989).

With reference to their distribution in Nature, phenolics can be divided into three classes: 'shortly distributed' (simple phenols, pyrocatechol, hydroquinone, resorcinol, aldehydes derived from benzoic acids); 'widely distributed' (flavonoids and their derivatives, coumarins and phenolic acids and their derivatives); 'polymers' (tannin and lignin) (Bravo 1998). Phenolic acids consist of two subgroups: hydroxybenzoic and hydroxycinnamic acids (Balasundram et al. 2006). Flavonoids account for over half of the 8,000 naturally-occurring phenolic compounds (Harborne et al. 1999). Variations in substitution patterns of ring C result in the following classes: flavonols, flavones, flavanones, flavanols (or catechins), isoflavones, flavanonols and anthocyanidins (Hollman and Katan 1999). Tannins are relatively high-molecular-weight compounds, subdivided into hydrolysable and condensed tannins and constitute the third important group of phenolics (Porter 1989).

Concerning their location in the plant (free in the soluble fraction of cell or bound to compounds of cell wall), phenolic compounds may also be classified as 'soluble' (simple phenol, flavonoids and tannins of low and medium molecular weight not bound to membranes compounds) and 'insoluble' (constituted by condensed tannins, phenolic acids and other phenolic compounds of low molecular weight bound to cell wall polysaccharides or proteins) (Sánchez-Moreno 2002).

Phenolics are involved in several physiological activities including stimulation of phagocytic cells and host-mediated tumor activity. Phenolics were traditionally used for protection of inflamed surfaces of the mouth and treatment of wounds, hemorrhoids and diarrhoea (Ogunleye and Ibitoye 2003) and for their antimicrobial activity (Nohynek et al. 2006, Veldhuizen et al. 2006). Phenolic compounds are also of considerable interest due to their antioxidant properties, which depend on the structure and in particular, on the number and positions of the hydroxyl groups and the nature of substitutions on the aromatic rings (Balasundram et al. 2006). Their antioxidant activity is due to different mechanisms of action—ability to scavenge free radicals; donation of hydrogen atoms or electron; chelation of metal cations.

Flavonols (quercetin, kaempferol and myricetin, for example) are the most widespread of the flavonoids found in vegetables, like onions, broccoli, apples, cherries and berries, tea and red wine. These compounds have anti-mutagenic, anti-carcinogenic and anti-hypertensive effects. Isoflavones (genistein, daidzein and coumestrol, for example) are found almost exclusively in leguminous plants such as soybean. They can exert pro- and anti-oestrogenic effects. Flavan-3-ols (include catechins and the larger proanthocyanidins) are found in tea, apples, apricots, cherries, red wines and dark chocolate. Flavones (apigenin and luteolin) are found in parsley, thyme and celery, for example. Anthocyanidins are mainly present in plants like anthocyanins, that are attached to sugar molecules. They are responsible for the characteristic red, blue and purple colored flowers, berry fruits and red wine. Flavanones are contained in citrus fruits (hesperetin in citrus peel; naringenin in grapefruit peel, imparting a bitter flavor) (Denny and Buttriss 2007).

Alkaloids

They include compounds such as cocaine, atropine, tnicotine noradrenaline, adrenaline (epinephrine), papaverine, morphine, melatonin and serotonin, quinine and capthotecin, lysergic acid diethylamide (LSD), theophylline and theobromine. Alkaloids contain basic (in some cases neutral or quaternary) nitrogen derived from an amino acid (or a purine, pyrimidine or other sources), in a heterocycle (or aliphatic in the case of protoalkaloids). They are generally classified according to the nitrogen-containing ring system (pyrrolidine, piperidine, etc.) and their biosynthetic origin, amino acids, amines, alcamides, cyanogenic glycosides and glucosinolates (Jimenez-Garcia et al. 2013). Furthermore, they can be classified according to their common molecular precursors, such as: pyridine (coniine derived from *Conium maculatum* and nicotine, an essential alkaloid present in *Nicotiana tabacum* and in many other plants belonging to the families Lycopodiaceae, Crassulaceae, Leguminosae, Chepopodiaceae and Compositeae); tropane group (atropine, cocaine); isoquinoline (morphine, codeine); purine (caffeine); and steroids (solanine) (Angelova et al. 2010, Khadem and Marles 2012). Alkaloids such as caffeine, nicotine, morphine, atropine and quinine have strong physiological and medicinal properties. Several alkaloids are toxic and, therefore, are involved in chemical defense against herbivores and microorganisms but also as protectants against damage by UV radiation. Concerning the usage of alkaloids in pharmaceuticals, rauwolscine, derived from *Rauwolfia canescens*, is a central nervous stimulator while coralyne is a DNA-binding agent, anti-cancer, antioxidant, anti-inflammatory agent (Paiva et al. 2010, Parsaeimehr et al. 2011).

Terpenes

Their structure includes 1 or more 5-carbon isoprene units. Terpenoids are classified according to the number of isoprene units they contain. Isoprene, which is synthesized and released by plants, comprises one unit and is classified as a hemiterpene —monoterpenes, sesquiterpenes, diterpenes, sesterpenes, triterpenes and tetraterpenes (Paiva et al. 2010). Terpenoids include compounds acting as flavors, fragrances, insect attractants and antibiotics. Terpenes institute biological activities which are anti-cancers, anti-microbials and anti-inflammatory (Liu et al. 2000). Terpenes include the hormones—gibberellins and abscisic acid; the pigments phytol and carotenoids (600 known naturally-occurred ones), resins and compounds such as pyrethrum (Jimenez-Garcie et al. 2013). Among the carotenoids, lycopene in red tomatoes, lutein in yellow peppers and α- and ß-carotene in orange carrots are to be cited. Important sources of carotenoids are carrots, tomatoes, peas, spinach and citrus fruits (Denny and Buttriss 2007).

Sterols

Terpenoids are able to reduce total and LDL-cholesterol since sterol and stanol esters have structures similar to dietary cholesterol, but are not absorbed by the human gastrointestinal tract, thus inhibiting the absorption of cholesterol from the diet and facilitate its elimination from the body (Denny and Buttriss 2007).

Glucosinolates

They are contained in Cruciferous vegetables, such as sprouts, broccoli, cabbage and watercress. When cruciferous vegetables are submitted to mechanical action, the breaking down of the walls of the plant cells is responsible for the contact between the glucosinolates contained within the cells and the enzyme 'myrosinase', which is contained within a different area of the plant cell (Mithen 2006). The action of myrosinase causes conversion of glucosinolates to isothiocyanates, which have effects on the human health. At least part of the potential health-promoting effect of isothiocyanates on the body seems to be due to their metabolism in the colon by gut bacteria (Denny and Buttriss 2007). The most promising evidence about the protective effect of plant bioactives on human health relates to lucosinolates. High intakes of brassicas show good correlation with the reduced risk of cancer at a number of sites (van Poppel et al. 1999).

Sulphur-containing compounds

Compounds, such as allicin and ajoene, are found in all the varieties of onion and garlic (Denny and Buttriss 2007).

Bioactive compounds can be classified according to their main source. In the following paragraphs, some compounds used in the treatment of human pathologies are listed within each considered class.

Bioactive compounds from plants

According to the World Health Organization, a great number of people still rely on plant-based traditional medicines for primary health care. The most widely used breast cancer drug is paclitaxel (Taxol®), isolated from the bark of *Taxus brevifolia*. Since the bark from about three mature 100-year old trees is required to provide 2 grams of the drug for a course of treatment, paclitaxel is now produced synthetically (Purvis 2000, Dewick 2002). Baccatin III is present in much higher quantities in the needles of *T. brevifolia*. Another antitumor molecule is ingenol 3-*O*-angelate, which is derived from a compound isolated from the sap of *Euphorbia peplus*. Calanolides from the *Calonphyllum* species gave results in the treatments of AIDS. Artemisinin, isolated from the plant *Artemisia annua*, is an approved antimalarial drug (Colegate and Molyneux 2008). Grandisines A and B are two indole alkaloids isolated from the leaves of the Australian rainforest tree, *Elaeocarpus grandis*, for use as analgesic agents (Carroll et al. 2005). Apomorphine is a derivative of morphine which is isolated from the poppy (*Papaver somniferum*) and is used to treat Parkinson's disease (Deleu et al. 2004).

Bioactive compounds from fungi

In this category, the most famous natural compound is penicillin (Kinghorn et al. 2011) produced by *Penicillium notatum* and discovered by Fleming in 1929. This discovery opened the way to production of synthetic penicillins. Fungi are a great source of pharmacologically active substances, showing antimicrobial (penicillin, norcardicin, imipenem, aztreonam, vancomycin, and erythromycin), antiviral (betulinic acid, bevirimat and ganoderic acid), and anticancer activities (amrubicin hydrochloride, doxorubicin, torreyanic acid) (Dias et al. 2012).

Bioactive compounds from marine sources

Plitidepsin isolated from the Mediterranean tunicate *Aplidium albicans* is effective in treating melanoma, small cell and non-small cell lung, bladder, non-Hodgkin lymphoma and acute lymphoblastic leukemia (Abraham et al. 1941). The brown alga, *Dictyota dichotoma*, produces compounds like 4-acetoxydictylolactone, dictyolides A and B and nordictyolide, which show antitumor activities (Faulkner 1988). The brown alga, *Dilophus ligatus*, produces the compound called crenuladial which has antimicrobial activity against *Staphylcoccus aureus* and other bacteria (Tringali et al. 1988). Natural insecticides isolated from the red alga, *Plocamium cartilagineum*, include cyclic polyhalogenated monoterpenes which are active against the Aster leafhopper (San-Martin et al. 1991). The C-nucleosides spongouridine and spongothymidine, isolated from the Caribbean sponge *Cryptotheca crypta*, possess antiviral activity (McConnell et al. 1994). Bryostatin 1, an antineoplastic compound isolated from the bryozoan *Bulgula neritina* (Chin et al. 2006), is considered as an anti-Alzheimer's drug (Alejandro et al. 2010).

A great number of known natural products calls for their inclusion in dedicated catalogues. Accurate research in the available database is a step to complete before embarking on time-consuming activities (isolation, purification, determination of the structure) so that no duplication of a suspected novel compound occurs. The main available database that can be used for virtual screening campaigns, include the following: Super Natural II; Universal Natural Product Database; Chinese Natural Product Database; Drug Discovery Portal; iSMART; database from historical medicinal plants; AfroDb and NuBBE (Harvey et al. 2015).

Future Trend in Utilisation of Bioactive Compounds Extracted from Agricultural and Food Waste in the Pharmaceutical Industry

Up to 42 per cent of food waste is produced by households, 38 per cent is created during food processing and 20 per cent is distributed along the whole chain (European Commission 2010). Thus, the processing of agricultural and food products results in the formation of high amounts of waste material (Martin et al. 2012) whose accumulation has strong environmental impact. In fact, every ton of food waste is equivalent to 4.5 tons of CO_2 emissions. The percentage of food waste and by-products generated during different processes are shown in Table 1.

Food waste is derived, in a decreasing order, from: vegetables and fruits; milk; meat; fish, and wine. The residues include straw, stem, leaves, husk, hulls and fiber, brans cobs, germ, gluten, steep liquor (cereals and cereal products); rotten fruits, stem, peel, pulp, pomace, lees, fibers (fruits and vegetables); shells of cocoa and coffee beans, spent liquors, molasses, solid waste from mashing (beverages such as coffee, tea, wines, beers); press solids and oil cakes, oil waste emulsions, shells of oil seeds (edible oils); fins, shells, bones, gut, fish oil (processed fish, crustaceans, and molluscs).

Table 1. Percentages of some food wastes and by-products generated during different processes.

Process	Waste and by-products percentages
Vegetable oil production	40–70
Wheat starch production	50
Corn starch production	41–43
Potato starch production	80
White wine production	20–30
Red wine production	20–30
Fruit and vegetable processing and preservation	5–30
Crustaceans processing	50–60
Fish canning	30–65
Fish filleting, curing, salting and smoking	50–75
Mollusk processing	20–50

The EU (European Union) has established that landfilling is no longer sustainable (Baiano 2014). As a consequence, the Landfill Directive was aimed at reducing biodegradable waste (food waste included) disposed to landfill (35 per cent of the 1995 level by 2016) (Directive 1999/31/EC). The Waste Framework Directive established the waste management hierarchy, which includes in the descending order of priority, the following actions: waste prevention; re-use; recycling; other recovery (e.g., energy recovery); and, lastly, disposal (Directive 2008/98/EC). According to this list, co-product exploitation is encouraged in order to prevent food waste being regarded as waste (Baiano 2014). Furthermore, exploitation activities are helped by evidence that waste and by-products are rich sources of bioactive compounds and this issue created a growing interest in recycling agro and food waste biomass, which could be a potential therapeutic agent (Kaneria et al. 2009, Aref et al. 2010).

The potentially marketable components present in food wastes and co-products need to be separated from the matrix through biochemical, chemical and physical approaches for selective extraction and modification of the targeted components. These operations must be performed while ensuring that the final products comply with the existing regulations. Next is the strategy for the exploitation of the bioactive compounds from agro waste (Spatafora and Tringali 2012). They include the identification and isolation of bioactive natural products as possible sources of lead compounds and the chemical and enzymatic modification of lead compound available from agro waste to obtain optimized analogues.

In Europe, large-scale facilities valorizing different food waste streams for biomolecules recovery have been established. For example, the Bio-based Industries Joint Undertaking (BBI JU) is a public-private partnership between the European Commission and the Bio-based Industries Consortium (BIC), whose aim is to bring together all relevant stakeholders (ranging from primary production to end-users) to establish innovative bio-based industries. The Joint Undertaking work plan has a total budget of €3.705 billion (Baiano 2014).

Excursus of possible pharmaceutical applications of bioactive compounds extracted from agricultural and food waste

The food and agricultural product processing generates enormous amounts of phenolics-rich by-products, which could be considered effective sources of phenolic antioxidants (Balasundram et al. 2006).

Raspberry solid wastes are an important source of bioactive compounds. In fact, they have high crude fiber content (comprising a heterogeneous mixture of non-starch polysaccharides, including cellulose, hemicellulose, pectin, β-glucans, gums and lignin) (Figuerola et al. 2005) with whom polyphenols are associated. The phenolic compounds can be released from the vegetal matrix, thus conferring antioxidant activity (Sáyago-Ayerdi et al. 2007). Furthermore, raspberry waste showed low protein content and this helped in the recovery of antioxidant because polyphenols can precipitate with proteins through mechanisms, such as hydrophobic and ionic interactions, and hydrogen and covalent bindings. Lastly, raspberry residues present low ash content, which positively affects the antioxidant capacity of raspberry extract because ash contains ions that can act as pro-oxidants (Maisuthisakul et al. 2006, Maisuthisakul et al. 2008). Enzymatic-assisted extraction with an hydro-ethanolic mixture (75:25, v/v) during 18 hours at 50°C is effective increasing phenolic content and antioxidant capacity of the extract (Laroze et al. 2010).

The peel of citrus fruits is a source of flavones and polymethoxylated flavones having potential antimicrobial activity with MIC, Minimum Inhibitory Concentration, in the range of 130 µg–50 mg/ml (Javed et al. 2011, Jwanny et al. 2012).

Spent coffee grounds (SCG) are a by-product generated during Espresso beverages or soluble coffee production and can be considered as a raw material of interest for the food and pharmaceutical industries. The crude aqueous extracts of SCG include the following phytochemical compounds: phenolics > flavonoids > carotenoids (mg/g dry waste), respectively. Caffeine found in SCG was ~ 0.82 g/100 g of dry waste. Furthermore, SCG extracts showed inhibition of *Staphylococcus aureus* and *Escherichia coli* growth for concentrations of 1.0 mg/mL and inhibition of *Candida albicans*, *C. krusei* and *C. parapsilosis* growth using concentration of 0.5 mg/mL (Sousa et al. 2015). These results suggest that spent coffee ground extracts are potentially useful in cosmetic formulations and soaps to be applied in superficial infections caused by microorganisms, as in acne or superficial candidosis, in order to improve the results of antimicrobial therapy, to solve problems of antimicrobial resistance and to lower the cost of therapy.

Nutraceutical molecules, such as tocopherols, tocotrienols, sterols and squalene are concentrated in the palm fatty acid distillate, collected from the physical refining of palm oil (Gapor 2000, Tan et al. 2007). They can be extracted through a process that includes treatment with alkyl alcohol and sodium methoxide to convert free fatty acids and glycerides into alkyl esters; the removal of the alkyl esters by distillation under reduced pressure, leaving tocopherols, tocotrienols, sterols and squalene in the residue; a cooling step to crystallize and separate sterols from tocopherols and tocotrienols; the passage of the filtrate containing tocopherols, tocotrienols and squalene through an ion-exchange column with anionic exchange resin to remove the squalene and produce a concentrated tocopherol and tocotrienol fraction in the solvent; the removal of the solvent and molecular distillation (Kawada et al. 1993, Gapor 1995). Since the extraction of palm oil requires the addition of water at certain stages, the resulting by-product is an aqueous solution containing phenolic compounds. These potent antioxidants can be extracted through a solvent-free process. The system includes a 3-phase decanter system which pellets the suspended solids and floats the oil; a membrane that removes residual oil; an ion-exchange membrane that removes ionic contaminants such as iron; a molecular weight cut-off membrane that removes the high-molecular-weight components. The resulting filtrate is rich in flavonoids, polyphenols and phenolic acids (Tg et al. 2013).

The by-products (peel, pulp's leftovers and seed) of pineapple, acerola, monbin, cashew apple, guava, soursop, papaya, mango, passion fruit, surinam cherry, sapodilla and tamarind have been studied for their content in bioactive compounds. The study highlighted the presence of resveratrol in guava and surinam cherry by-products and the presence of coumarin in passion fruit, guava and Surinam cherry by-products, thus demonstrating that they can be considered an important source of bioactive compounds having anti-inflammatory and antioxidant agents (Silva et al. 2014, Jahurul et al. 2015).

Elderberry pomace includes fruit skins and seeds and is a residue of the juice-processing industry. It is rich in anthocyanins and other phenolic compounds that can be obtained by applying supercritical carbon dioxide as a first step and pressurized solvent mixtures of carbon dioxide, ethanol and water in the second step (Seabra et al. 2010a, 2010b).

At present, about 210 million tons of grapes are produced annually and 15 per cent of them are addressed to the wine-making industry thus generating solid waste amounting to 30 per cent of the material used (Teixeira et al. 2014). The bioactive compounds extracted from winery by-products can be used as medicine due to their antioxidant, antimicrobial, anti-inflammatory and anticancer activities (Furiga et al. 2009, Apostolou et al. 2013, Rhodes et al. 2006, Balu et al. 2006). Another application of grape pomace is the production of pullulan, a non-ionic exopolysaccharide produced by *Aureobasidium* spp. Pullulan is biodegradable, impermeable to oxygen, non-hygroscopic, easily soluble in water and has high film-forming capabilities (Rekha and Chandra 2007). Pullulan can be used in pharmaceutical applications including denture adhesives and capsules for supplements (Farris et al. 2012).

Heat-stable proteins extracted from wastes of *Musa* sp., *Citrus limetta*, *Citrullus lanatus*, *Solanum lycopersicum* and *Psidium* sp. were tested for possible antimicrobial properties. The heat-stable proteins from *Solanum lycopersicum* wastes are able to suppress the growth of *Escherichia coli*, whereas those of *Musa* sp. and *Citrus limetta* inhibit the growth of *Pseudomonas* sp. (the latter also inhibits the growth of *Fusarium oxysporum*) (Farha et al. 2012).

One of the steps in the brewing process is filtration of beer through polyvinylpolypyrrolidone (PVPP) resin in order to stabilize the products by removing a substantial portion of the haze active and non-haze active polyphenols. These phenolic compounds can subsequently be recovered from the alkaline residual stream generated after cleaning the PVPP, by extraction with solvents (solid-phase extraction, SPE) or supercritical fluid extraction (SFE). The obtained extracts can be used as natural antioxidants for use in the pharmaceutical industry (Barbosa-Pereira et al. 2013).

Macadamia has edible kernel but skins and husks constitute waste with limited uses as by-products. Nevertheless, the skins can be submitted to extraction with a solvent (especially when organic solvents such as methanol, ethanol, acetonitrile and acetone are combined with water (50 per cent, v/v)) in order to obtain flavonoids and proanthocyanidins having high antioxidant properties (Dailey and Vuong 2015). Chestnut (*Castanea sativa* Mill.) and hazelnut (*Corylus avellana* L.) shells also represent important sources of gallic, chlorogenic, ferulic, *p*-coumaric, ellagic acids and of (+)-catechin, (−)-epicatechin and rutin when submitted to extraction with solvents, such as water, methanol/water 70/30(v/v), and absolute methanol (Nazzaro et al. 2012).

Potato peels represent a good source of the aglycone glycoalkaloid, solanidine, which have a high potential to synthesize novel anticancer and apoptotic drugs while the potato peel peptides show anti-inflammatory, anti-hypertensive and modest anti-oxidant activities (Kenny et al. 2013, Milner et al. 2011).

Agro-industrial wastes from the industrial processing of the olive oil, wine, citrus fruits, wheat and rapeseed have been treated through chemical and biotechnological methods for selective recovery of phenolic compounds. The recovered phenols are converted into new bioactive compounds by green chemistry methodologies. For example, flavonoids have been converted into new compounds having anti-tumoral activity (Bernini et al. 2003, Bernini et al. 2005); simple alkylated phenols into *p*-benzoquinones showing selective fungicide activities (Bernini et al. 2006) and cinnamic acids into *p*-vinyl phenols with the molecules exhibiting antioxidant properties (Bernini et al. 2007).

Cocoa (*Theobroma cacao* L.) is a major cash crop of the tropical world cultivated to obtain cocoa beans. Nevertheless, cocoa beans constitute only 10 per cent of the fresh weight of the cocoa fruit while the remaining 90 per cent by weight (mainly cocoa pulp and cocoa pod husk) is discarded as cocoa waste. Cocoa pod gum is extracted from cocoa pod husks by alcohol precipitation (Figueira et al. 1994, Samuel 2006). Cocoa pod gum can be used as binder in the pharmaceutical industry for binding pills, etc.

Peels of potato, green peas, sweet corn, orange and mango can be used as substrate for the production of lactic acid by means of bacterial strains, like *Lactobacillus casei* and *Lactobacillus delbrueckii*. The highest lactic acid production was obtained for mango peels by *L. casei*, while the lowest was obtained for corn by using the strain *L. delbrueckii* (Mudaliyar et al. 2012).

Date is the most popular fruit in the Middle East and its by-products can be used as substrate for producing value-added products in the pharmaceutical industry. Abou-Zeid et al. (1993) utilized date constituents as a nutrient for the production of oxytetracycline, an antibiotic active on Gram-positive and Gram-negative bacteria produced by different species of *Streptomyces*. Bleomycin is a family of glycopeptide-derived antibiotics having strong antitumor activity and employed for the treatment of several malignancies, including non-Hodgkin's lymphoma, squamous cell carcinoma and testicular tumors (Aras and Dilsizian 2008, Evens et al. 2008). Radwan et al. (2010) used a date syrup for the production of bleomycin through *Streptomyces mobaraensis*.

By-products from the preparation of propolis extracts can be used for the production of pharmaceutical films which are used as coating agents, drug-delivery platforms. They constitute important materials for taste-masking, moisture-resistance barriers and drug release control (Carvalho et al. 2010, Steele et al. 2011, Nascimento et al. 2012, Santos et al. 2014).

Waste material and their extracts from meat, poultry and fish processing industries can be used as medicine. Cholesterol, used for the synthesis of Vitamin D_3 can be extracted from the brain, the nervous system and the spinal cord. Melatonin, used for treatment of mental diseases, is extracted from the pineal gland. Bile is extracted from the gall bladder for use in the treatment of indigestion, constipation and bile tract disorders. Its components—prednisone and cortisone—can be used separately as medicines. Heparin can be extracted from the liver. Progesterone and estrogen can be extracted from pig ovaries (Jayathilakan et al. 2012). The waste streams from marine products are approximately 24 million tons and include viscera, heads, skins, fins, bones, trimmings and shellfish. Extraction helps to obtain structurally diverse molecules that possess bioactivities such as antioxidant, anticoagulant, anti-thrombotic, anti-cancer and immune-stimulatory activities (Suleria et al. 2016). Astaxanthin extracted from shrimp shells can used as regulator of the plasma HDL-cholesterol level. Shrimp and crab shells can be used to produce chitosan, a 'fat-binder' used for weight management. $CaCO_3$ from oyster shells is used as a calcium supplement. Extracts containing peptides (piscidins) with specific amino acid sequences are obtained from gills, belly flap muscle and skin of salmon (*Salmo salar*) and show good radical scavenging activity while no ACE (proteases angiotensin

I-converting enzyme) or DPP-4 (dipeptidyl peptidase 4) inhibiting activity could be detected (Falkenberg et al. 2014).

Skins, scales and bones are a good source of collagen, which can be extracted and enzymatically hydrolyzed to liberate physiologically-active peptides with antioxidant, antihypertensive and antimicrobial activities against different strains of bacteria. They have a protective effect on the cartilage, the capacity to stimulate bone formation and exercise, calciotropic and opioid effects (Alemán and Martínez-Alvarez 2013). Gelatin obtained through controlled hydrolysis of a water-insoluble collagen is used as an ingredient in a protective ointment, such as zinc gelatin for the treatment of ulcerated varicose veins; to produce a sterile sponge (by whipping it into foam, treating it with formaldehyde and drying it) used in surgery and also to implant a drug or antibiotic directly into a specific area and used as a plasma expander for blood in cases of severe shock and injury (Jayathilakan et al. 2012).

Conclusion

Waste streams or vegetal by-products obtained from the food and agriculture could be regarded as bio-renewable sources of valuable components for use in the pharmaceutical industry. The reuse of components of by-products and agricultural and food waste can prove a useful tool in increasing the eco-sustainability of food processing industries and to exercise a positive economic impact.

Identification of bioactive compounds in plants, microrganisms and animals is a growing field and processing of by-products facilitates commercial applications. Nevertheless, a limited number of natural sources (especially marine sources) were evaluated. In addition, efficient techniques must be explored for separation and isolation of bioactive substances of natural origin so that their value as natural products is not lost. An integrated biorefinery approach to waste biomass from agro-industry must be put into action. Finally, researchers need to develop methods to apply these bioactive compounds in human health promotion.

References

Abou-Zeid, A. A., N. A. Baeshin and A. O. Baghlaf. 1993. Utilization of date products in production of oxytetracycline by streptomyces rimosus. J. Chem. Technol. Biotechnol. 58: 77–79.

Abraham, E. P., E. Chain and C. M. Fletcher. 1941. Further observations on penicillin. Lancet 16: 177–189.

Alejandro, M., K. B. Glaser, C. Cuevas, R. S. Jacobs, W. Kem, R. D. Little, J. M. McIntosh, D. J. Newman, B. C. Potts and D. E. Shuster. 2010. The odyssey of marine pharmaceuticals: A current pipeline perspective. Trends Pharm. Sci. 31: 255–265.

Alemán, A. and O. Martínez-Alvarez. 2013. Marine collagen as a source of bioactive molecules: A review. Nat. Prod. J. 3: 105–114.

All natural. Nat. Chem. Biol. 2007; 3: 351. Available from: http://dx.doi.org/10.1038/nchembio0707-351

Angelova, S., M. Buchheim, D. Frowitter, A. Schierhorn and W. Roos. 2010. Overproduction of alkaloid phytoalexins in California poppy cells is associated with the co-expression of biosynthetic and stress-protective enzymes. Mol. Plant. 3: 927–939.

Apostolou, A., D. Stagos, E. Galitsiou, A. Spyrou, S. Haroutounian, N. Portesis, I. Trizoglou, A. W. Hayes, A. M. Tsatsakis and D. Kouretas. 2013. Assessment of polyphenolic content, antioxidant activity, protection against ROS-induced DNA damage and anticancer activity of *Vitis vinifera* stem extracts. Food Chem. Toxicol. 61: 60–68.

Aras, O. and V. Dilsizian. 2008. Targeting tissue angiotensinconverting enzyme for imaging cardiopulmonary fibrosis. Curr. Cardiol. Rep. 10: 128–134.

Aref, H. L., K. B. H. Salah, J. P. Chaumont, A. W. Fekih, M. Aouni and K. Said. 2010. *In vitro* antimicrobial activity of four Ficus carica latex fractions against resistant human pathogens. Pak. J. Pharm. Sci. 23: 53–58.

Baiano, A. 2014. Recovery of biomolecules from food wastes—A review. Molecules 19: 14821–14842.

Balasundram, N., K. Sundram and S. Samman. 2006. Phenolic compounds in plants and agri-industrial by-products: Antioxidant activity, occurrence and potential uses. Food Chem. 99: 191–203.

Balu, M., P. Sangeetha, G. Murali and C. Panneerselvam. 2006. Modulatory role of grape seed extract on age-related oxidative DNA damage in central nervous system of rats. Brain Res. Bull. 68: 469–473.

Barbosa-Pereira, L., A. Pocheville, I. Angulo, P. Paseiro-Losada and J. M. Cruz. 2013. Fractionation and purification of bioactive compounds obtained from a brewery waste stream. Biomed Res. Int. 2013: 1–11.

Bernhoft, A. 2010. A brief review on bioactive compounds in plants. pp. 11–17. *In*: A. Bernhoft (ed.). Bioactive Compounds in Plants—Benefits and Risks for Man and Animals. The Norwegian Academy of Science and Letters, Oslo.

Bernini, R., E. Mincione, M. Cortese, R. Saladino, G. Gualandi and M. C. Belfiore. 2003. Conversion of naringenin and hesperetin by heterogeneous Baeyer-Villiger reaction into lactones exhibiting apoptotic activity. Tetrahedron Lett. 44: 4823–4825.

Bernini, R., E. Mincione, P. Provenzano and G. Fabrizi. 2005. Catalytic oxidations of catechins to pbenzoquinones with hydrogen peroxide/methyltrioxorhenium. Tetrahedron Lett. 46: 2993–2996.

Bernini, R., E. Mincione, M. Barontini, G. Fabrizi, M. Pasqualetti and S. Tempesta. 2006. Convenient oxidation of alkylated phenols and methoxytoluenes to antifungal 1,4-benzoquinones with hydrogen peroxide/methyltrioxorhenium catalytic system in neutral ionic liquid. Tetrahedron 62: 7733–7737.

Bernini, R., E. Mincione, M. Barontini, G. Provenzano and L. Setti. 2007. Obtaining 4-vinylphenols by decarboxylation of natural 4-hydroxycinnamic acids under microwave irradiation. Tetrahedron 63: 9663–9667.

Beutler, J. A. 2009. Natural products as a foundation for drug discovery. Current Protoc. Pharmacol. 46:9 11 1–9 21.

Bhutani, K. K. and V. M. Gohil. 2010. Natural products drug discovery research in India: Status and appraisal. Indian J. Exp. Biol. 48: 199–207.

Bravo, L. 1998. Polyphenols: Chemistry, dietary sources, metabolism and nutritional significance. Nutr. Rev. 56: 317–333.

Butler, M. S. 2005. Natural products to drugs: Natural product derived compounds in clinical trials. Nat. Prod. Rep. 22: 162–95.

Caporale, L. H. 1995. Chemical ecology: A view from the pharmaceutical industry. Proc. Natl. Acad. Sci., USA 92: 75–82.

Carroll, A. R., G. Arumugan, R. J. Quinn, J. Redburn, G. Guymer and P. Grimshaw. 2005. Grandisine A and B, novel indolizidine alkaloids with -opioid receptor-binding affinity from the leaves of the human Australian rainforest tree, *Elaeocarpus grandis*. J. Org. Chem. 70: 1889–1892.

Carvalho, F. C., M. L. Bruschi, R. C. Evangelista and M. P. D. Gremião. 2010. Mucoadhesive drug delivery systems. Bras. J. Pharm. Sci. 46: 381–387.

Chernyak M. Canadian NHP Market: Headed In The Right Direction2012; (November). Available from: http://www.nutraceuticalsworld.com/issues/2012-11/view_features/canadian-nhp-market-headed-in-the-right-direction/. Discovery, Development, and Regulation of Natural Products (PDF Download Available). Available from: https://www.researchgate.net/publication/239730566_Discovery_Development_and_Regulation_of_Natural_Products [accessed Mar 27, 2017].

Chin, Y. -W., M. J. Balunas, H. B. Chai and A. D. Kinghorn. 2006. Drug discovery from natural sources. AAPS J. 8: 239–253.

Ciociola, A. A., L. B. Cohen and P. Kulkarni. 2014. How drugs are developed and approved by the FDA: Current process and future directions. Am. J. Gastroenterol. 109: 620–623.

Colegate, S. M. and R. J. Molyneux. 2008. Bioactive Natural Products: Detection, Isolation and Structure Determination. CRC Press; Boca Raton, FL, USA.

Cragg, G. M. and D. J. Newman. 2005. Biodiversity: A continuing source of novel drug leads. Pure Appl. Chem. 77: 7–24.

Dailey, A. and Q. V. Vuong. 2015. Effect of extraction solvents on recovery of bioactive compounds and antioxidant properties from macadamia (*Macadamia tetraphylla*) skin waste. Cogent Food & Agriculture 1: 1115646.

Deleu, D., Y. Hanssens and M. G. Northway. 2004. Subcutaneous apomorphine: An evidence-based review of its use in Parkinson's disease. Drugs Aging 21: 687–709.

Denny, A. and J. Buttriss. 2007. Plant Foods and Health: Focus on Plant Bioactives. Institute of Food Research, Norwich Research Park, Norwich, Norfolk, UK.

Dewick, P. M. 2002. Medicinal Natural Products: A Biosynthentic Approach. 2nd ed., John Wiley and Son, West Sussex, UK, p. 520.

Dias, D. A., S. Urban and U. Roessner. 2012. A historical overview of natural products in drug discovery. Metabolites 2: 303–336.

Directive 1999/31/EC on the Landfill of Waste. Available online: https://www.google.it/#q=Council+Directive+1999%2F31%2FEC+on+the+Landfill+of+Waste (accessed on 3 June 2016).

Directive 2008/98/EC on Waste and Repealing Certain Directives. Available online: http://eur-lex.europa.eu/legal-content/EN/TXT/?uri=CELEX:32008L0098 (accessed on 3 June 2016).

European Commission. Preparatory Study on Food Waste across EU 27. Technical Report-2010-054. Available online: http://ec.europa.eu/environment/eussd/pdf/bio_foodwaste_report.pdf (accessed on 3 June 2016).

Evens, A. M., M. Hutchings and V. Diehl. 2008. Treatment of Hodgkin's lymphoma: The past, present and future. Nat. Clin. Pract. Oncol. 5: 543–556.

Falkenberg, S. S., S. O. Mikalsen, H. Joensen, J. Stagsted and H. H. Nielsen. 2014. Extraction and characterization of candidate bioactive compounds in different tissues from salmon (*Salmo salar*). Int. J. Appl. Res. Nat. Prod. 7: 11–25.

Farha, S., E. Chatterjee, S. G. A. Manuel, S. A. Reddy and R. D. Kale. 2012. Isolation and characterization of bioactive compounds from fruit wastes. Dyn. Biochem. Process Biotechnol. Mol. Biol. 6: 92–94.

Farris, S., L. Introzzi, J. M. Fuentes-Alventosa, N. Santo, R. Rocca and L. Piergiovanni. 2012. Self-assembled pullulan-silica oxygen barrier hybrid coatings for food packaging applications. J. Agric. Food Chem. 60: 782–790.

Faulkner, D. J. 1988. Marine natural products. Nat. Prod. Rep. 20: 269–309.

Figueira, A., J. Janick and J. N. BeMiller. 1994. Partial characterization of cacao pod and stem gums. Carbohydr. Polym. 24: 133–138.

Figuerola, F., M. L. Hurtado, A. M. Estevez, I. Chiffelle and F. Asenjo. 2005. Fiber concentrates from apple pomace and citrus peel as potential fiber sources for food enrichment. Food Chem. 91: 395–401.

Furiga, A., A. Lonvaud-Funel and C. Badet. 2009. *In vitro* study of antioxidant capacity and antibacterial activity on oral anaerobes of a grape seed extract. Food Chem. 113: 1037–1040.

Gapor, M. T. 1995. Palm vitamin E–A value-added tocotrienols-rich fraction (TRF) from palm oil. Palm Oil Dev. Palm Oil Res. Inst. Malays. 22: 7–17.

Gapor, M. T. and H. A. Rahman. 2000. Squalene in oils and fats. Palm Oil Dev. Malays. Palm Oil Board 32: 36–40.

Genovese, S., M. Curini and F. Epifano. 2009. Chemistry and biological activity of azoprenylated secondary metabolites. Phytochem. 70: 1082–1091.

Goodman, J. W. and V. Walsh. 2001. The Story of Taxol: Nature and Politics in the Pursuit of an Anti-cancer Drug. Cambridge University Press.

Gottlieb, O. R. 1990. Phytochemicals: Differentiation and function. Phytochem. 29: 1715–1724.

Guaadaoui, A., S. Benaicha, N. Elmajdoub, M. Bellaoui and A. Hamal. 2014. What is a bioactive compound? A combined definition for a preliminary consensus. Int. J. Nutr. Food Sci. 3: 174–179.

Harborne, J. B. 1989. Methods in plant biochemistry. *In*: P. M. Dey and J. B. Harborne (eds.). Plant Phenolics. Academic Press, London, UK.

Harborne, J. B., H. Baxter and G. P. Moss. 1999. Phytochemical Dictionary: Handbook of Bioactive Compounds from Plants (2nd ed.). Taylor & Francis, London, UK.

Harvey, A. L., R. A. Edrada-Ebel and R. J. Quinn. 2015. The re-emergence of natural products for drug discovery in the genomics era. Nat. Rev. Drug Discov. 14: 111–129.

Hollman, P. C. H. and M. B. Katan. 1999. Dietary flavonoids: Intake, health effects and bioavailability. Food Chem. Toxicol. 37: 937–942.

http://www.ema.europa.eu/ema/index.jsp?curl=pages/regulation/general/general_content_000208. jsp. Herbal medicinal products.

Jahurul, M. H. A., I. S. M. Zaidul, K. Ghafoor, F. Y. Al-Juhaimi, K. -L. Nyam, N. A. N. Norulaini, F. Sahena and A. K. Mohd Omar. 2015. Mango (*Mangifera indica* L.) by-products and their valuable components: A review. Food Chem. 183: 173–180.

Javed, S., A. Javaid, Z. Mahmood, A. Javaid and F. Nasim. 2011. Biocidal activity of citrus peel essential oils against some food spoilage bacteria. J. Med. Plant Res. 3697–3701.

Jayathilakan, K., K. Sultana, K. Radhakrishna and A. S. Bawa. 2012. Utilization of by-products and waste materials from meat, poultry and fish processing industries: A review. J. Food Sci. Technol. 49: 278–293.

Jeffreys, D. 2004. Aspirin: The Remarkable Story of a Wonder drug. Bloomsbury Publishing, London.

Ji, H. F., X. J. Li and H. Y. Zhang. 2009. Natural products and drug discovery. Can thousands of years of ancient medical knowledge lead us to new and powerful drug combinations in the fight against cancer and dementia? EMBO Rep. 10: 194–200.

Jimenez-Garcia, S. N., M. A. Vazquez-Cruz, R. G. Guevara-Gonzalez, I. Torres-Pacheco, A. Cruz-Hernandez and A. A. Feregrino-Perez. 2013. Current approaches for enhanced expression of secondary metabolites as bioactive compounds in plants for agronomic and human health purposes—A review. Pol. J. Food Nutr. Sci. 63: 67–78.

Joseph, B. and P. A. M. Priya. 2010. Bioactive compounds in essential oil and its effects of antimicrobial, cytotoxic activity from the *Psidium guajava* (L.) leaf. J. Adv. Biotechnol. 9: 10–14.

Jwanny, E. W., S. T. El-Sayed, A. M. Salem, N. A. Mabrouk and A. N. Shehata. 2012. Fractionation, identification and biological activities of Egyptian citrus peel extracts. Australian J. Bas. Appl. Sci. 6: 34–40.

Kaneria, M., Y. Baravalia, Y. Vaghasiya and S. Chanda. 2009. Determination of antibacterial and antioxidant potential of some medicinal plants from Saurashtra region, India. Indian J. Pharm. Sci. 71: 406–412.

Kawada, T., L. W. Leong, A. S. Ong, A. G. M. Top, N. Tsuchiya and H. Watanabe. 1993. Production of High Concentration Tocopherols and Tocotrienols from Palm Oil By-products. U.S. Patent 5,190,618, 2 March 1993.

Kenny, O. M., N. P. Brunton, M. B. Hossain, D. K. Rai, S. G. Collins, A. R. Maguire, P. W. Jones and N. M. O'Brien. 2013. Anti-inflammatory properties of potato glycoalkaloids in stimulated Jurkat and Raw 264.7 macrophages. Life Sci. 92: 775–782.

Khadem, S. and R. J. Marles. 2012. Chromone and flavonoid alkaloids: Occurrence and bioactivity. Molecules 17: 191–206.

Kinghorn, A. D., Y. W. Chin and S. M. Swanson. 2009. Discovery of natural product anticancer agents from biodiverse organisms. Curr. Opin. Drug Discov. Devel. 12: 189–196.

Kinghorn, A. D., L. Pan, J. N. Fletcher and H. Chai. 2011. The relevance of higher plants in lead compound discovery programs. J. Nat. Prod. 74: 1539–1555.

Kumar, S. and A. Sagar. 2007. Microbial associates of *Hippophae rhamnoides* (Seabucktorn). Plant Pathol. J. 6: 299–305.

Lahlou, M. 2013. The success of natural products in drug discovery. Pharmacol. Pharm. 4: 17–31.

Laroze, L., C. Soto and M. E. Zúñiga. 2010. Phenolic antioxidants extraction from raspberry wastes assisted by-enzymes. Electr. J. Biotechnol. 13: 1–4.

Leal, M. C., J. Puga, J. Serodio, N. C. Gomes and R. Calado. 2012. Trends in the discovery of new marine natural products from invertebrates over the last two decades—Where and what are we bioprospecting? PLoS One 7: e30580.

Liu, W. K., S. X. Xu and C. T. Che. 2000. Anti-proliferative effect of ginseng saponins on human prostate cancer cell line. Life Sci. 67: 1297–1306.

Maisuthisakul, P., R. Pongsawatmanit and M. H. Gordon. 2006. Antioxidant properties of Teaw (*Cratoxylum formosum* Dyer) extract in soybean oil and emulsions. J. Agric. Food Chem. 54: 2719–2725.

Maisuthisakul, P., S. Pasuk and P. Ritthiruangdej. 2008. Relationship between antioxidant properties and chemical composition of some Thai plants. J. Food Comp. Anal. 21: 229–240.

Maplestone, R. A., M. J. Stone and D. H. Williams. 1992. The evolutionary role of secondary metabolites—A review. Gene 115: 151–157.

Martin, J. G. P., E. Porto, C. B.Corrêa, S. M. de Alencar, E. M. da Gloria, I. S. R. Cabral and L. M. de Aquino. 2012. Antimicrobial potential and chemical composition of agro-industrial wastes. J. Nat. Prod. 5: 27–36.

McConnell, O., R. E. Longley and F. E. Koehn. 1994. Chapter 5. *In*: V. P. Gullo (ed.). The Discovery of Marine Natural Products with Therapeutic Potential. Butterworth-Heinemann; Boston, MA, USA.

Milner, S. E., N. P. Brunton, P. W. Jones, N. M. O' Brien, S. G. Collins and A. R. Maguire. 2011. Bioactivities of glycoalkaloids and their aglycones from solanum species. J. Agric. Food Chem. 59: 3454–3484.

Mishra, B. B. and V. K. Tiwari. 2011. Natural products in drug discovery, clinical evaluations and investigations. Research Signpost, 1–62.

Mithen, R. 2006. Sulphur-containing compounds. *In*: A. Crozier, M. N. Clifford and H. Ashihara (eds.). Plant Secondary Metabolites: Occurrence, Structure and Role in the Human Diet. Blackwell Publishing, Oxford.

Molinski, T. F., D. S. Dalisay, S. L. Lievens and J. P. Saludes. 2009. Drug development from marine natural products. Nat. Rev. Drug Discov. 8: 69–85.

Montaser, R. and H. Luesch. 2011. Marine natural products: A new wave of drugs? Future Med. Chem. 3: 1475–89.

Mudaliyar, P., L. Sharma and C. Kulkarni. 2012. Food waste management—lactic acid production by lactobacillus species. I.J.A.B.R. 2: 34–38.

Nascimento, T. A., V. Calado and C. W. P. Carvalho. 2012. Development and characterization of flexible film based on starch and passion fruit mesocarp flour with nanoparticles. Food. Res. Int. 49: 588–595.

Nazzaro, M., M. V. Mottola, F. La Carac, G. Del Monacoc, R. P. Aquino and M. G. Volpe. 2012. Extraction and characterization of biomolecules from agricultural wastes. Chem. Eng. Trans. 27: 331–336.

Nohynek, L. J., H. L. Alakomi, M. P. Kähkönen, M. Heinonen, I. M. Helander, K. M. O. Caldentey and R. H. P. Pimiä. 2006. Berry Phenolics: Antimicrobial properties and mechanisms of action against severe human pathogens. Nutr. Cancer 54: 18–32.

Ogunleye, D. S. and S. F. Ibitoye. 2003. Studies of antimicrobial activity and chemical constituents of *Ximenia americana*. Trop. J. Pharm. Res. 2: 239–241.

Ortholand, J. Y. and A. Ganesan. 2004. Natural products and combinatorial chemistry: Back to the future. Curr. Opin. Chem. Biol. 8: 271–80.

Paiva, P. N. G., F. S. Gomes, T. H. Napoleão, R. A. Sá, M. T. S. Correia and C. B. B. Coelho. 2010. Antimicrobial activity of secondary metabolites and lictins from plants. Res. Technol. Edu. Top. Appl. Microb. Biotechnol. 396–406.

Parsaeimehr, A., E. Sargsyan and A. Vardanyan. 2011. Expression of secondary metabolites in plants and their useful perspective in animal health. ABAH Bioflux 3: 115–124.

Paulsen, B. S. 2010. Highlights through the history of plant medicine. pp. 18–29. *In*: A. Bernhoft (ed.). Bioactive Compounds in Plants—Benefits and Risks for Man and Animals. The Norwegian Academy of Science and Letters, Oslo.

Porter, L. J. 1989. Tannins. pp. 389–419. *In*: J. B. Harborne (ed.). Methods in Plant Biochemistry: Vol. 1. Plant Phenolics. Academic Press, London, UK.

Purvis, W. 2000. *Lichens*. Natural History Museum, London/Smithsonian Institution; Washington D.C., USA, p. 112.

Radwan, H., F. K. Alanazi, E. I. Taha, H. A. Dardir and I. M. Moussa. 2010. Development of a new medium containing date syrup for production of bleomycin by *Streptomyces mobaraensis* ATCC 15003 using response surface methodology. Afr. J. Biotechnol. 9: 5450–5459.

Rekha, M. R. and S. Chandra. 2007. Pullulan as a promising biomaterial for biomedical applications: A perspective. Trends Biomater. Artif. Organs. 20: 116–121.

Rhodes, P. L., J. W. Mitchell, M. W. Wilson and L. D. Melton. 2006. Antilisterial activity of grape juice and grape extracts derived from *Vitis vinifera* variety Ribier. Int. J. Microb. 107: 281–286.

Rouhi, A. M. 2003. Moving beyond natural products. Chem. Eng. News. 13: 77–91.

Samuel, Y. K. C. 2006. Crude gum from cocoa of Malaysian origin: Part I: Rheological Properties. Malaysian Cocoa J. 2: 28–31.

San-Martin, A., R. Negrete and J. Rovirosa. 1991. Insecticide and acaricide activities of polyhalogenated monoterpenes from Chilean *Plocamium cartilagineum*. Phytochem. 30: 2165–2169.

Sánchez-Moreno, C. 2002. Compuestos polifenólicos: Estructura y classificación: presencia en alimentos y consumo: biodisponibilidad y metabolismo. Alimentaria 329: 19–28.

Santos, T. M., M. S. M. F. Souza, C. A. Cacere, M. F. Rosa, J. P. S. Morais, A. M. B. Pinto and H. M. C. Azeredo. 2014. Fish gelatin films as affected by cellulose whiskers and sonication. Food Hydrocoll. 41: 113–118.

Sarker, S. D., Z. Latif and A. I. Gray. 2006. Natural Product Isolation. *In*: D. Satyajit (ed.). Methods in Biotechnology: Natural Product Isolation. Human Press Inc; Totowa, NJ, USA.

Sáyago-Ayerdi, S. G., S. Arranz, J. Serrano and I. Goñi. 2007. Dietary fiber content and associated antioxidant compounds in roselle flower (*Hibiscus sabdariffa* L.) beverage. J. Agric. Food Chem. 55: 7886–7890.

Seabra, I. J., M. E. M. Braga, M. T. P. Batista and H. C. De Sousa. 2010a. Fractioned high–pressure extraction of anthocyanins from elderberry (*Sambucus nigra* L.) pomace. Food Bioprocess Tech. 3: 674–683.

Seabra, I. J., M. E. M. Braga, M. T. P. Batista and H. C. De Sousa. 2010b. Effect of solvent (CO_2/ethanol/H_2O) on the fractionated enhanced solvent extraction of anthocyanins from elderberry pomace. J. Supercrit. Fluids 54: 145–152.

Silva, Larissa Morais Ribeiro da, Evania Altina Teixeira de Figueiredo, Nagila Maria Pontes Silva Ricardo, Icaro Gusmao Pinto Vieira, Raimundo Wilane de Figueiredo, Isabella Montenegro Brasil and Carmen L. Gomes. 2014. Quantification of bioactive compounds in pulps and by-products of tropical fruits from Brazil. Food Chemistry 143: 398–404.

Sousa, C., C. Gabriel, F. Cerqueira, M. C. Manso and A. F. Vinha. 2015. Coffee industrial waste as a natural source of bioactive compounds with antibacterial and antifungal activities. *In*: A. Méndez-Vilas (ed.). The Battle Against Microbial Pathogens: Basic Science, Technological Advances and Educational Programs. Formatex Research Center, Badajoz, Spain.

Spatafora, C. and C. Tringali. 2012. Valorization of vegetable waste: Identification of bioactive compounds and their chemo-enzymatic optimization. The Open Agriculture Journal, 2012, 6: 9–16.

Steele, T. W. J., C. L. Huang, E. Widjaja, F. Y. C. Boey, J. S. C. Loo and S. S. Venkatraman. 2011. The effect of polyethylene glycol structure on paclitaxel drug release and mechanical properties of PLGA thin films. Acta Biomater. 7: 1973–1983.

Strobel, G. and B. Daisy. 2003. Bioprospecting for microbial endophytes and their natural products. Microbiol. Mol. Biol. Rev. 67: 491–502.

Suleria, H. A. R., P. Masci, G. Gobe and S. Osborne. 2016. Current and potential uses of bioactive molecules from marine processing waste. J. Sci. Food Agric. 96: 1064–1067.

Tan, Y. A., R. Sambanthamurthi, K. Sundram and M. B. Wahid. 2007. Valorisation of palm by-products as functional components. Eur. J. Lipid Sci. Technol. 109: 380–393.

Teixeira, A., N. Baenas, R. Dominguez-Perles, A. Barros, E. Rosa, D. A. Moreno and C. Garcia-Viguera. 2014. Natural bioactive compounds from winery by-products as health promoters: A review. Int. J. Mol. Sci. 15: 15638–15678.

Tg, S., C. Rha, R. Sambanthurthi, A. J. Sinskey, Y. A. Tan, P. Manickam Kalyana Sundram and M. B. Wahid. 2013. Compositions Comprising Shikimic Acid Obtained from Oil Palm Based Materials and Method of Producing Thereof. EP Patent 2,582,654, 24 April 2013.

The American Heritage Medical Dictionary. 2007. Houghton Mifflin Company, Boston MA.

Tiaz, L. and E. Zeiger. 2006. Secondary metabolites and plant defense. pp. 283–308. *In*: Plant Physiology, 4th ed. Sinauer Associates, Inc., Sunderland, Massachusetts.

Tringali, C., G. Oriente, M. Piattelli, C. Geraci, G. Nicolosi and E. Breitmaier. 1988. Crenuladial, an antimicrobial diterpenoid from the brown alga Dilophus ligulatus. Can. J. Chem. 66: 2799–2802.

van Poppel, G., D. T. Verhoeven, H. Verhagen and R. A. Goldbohm. 1999. Brassica vegetables and cancer prevention. Epidemiology and mechanisms. Adv. Exp. Med. Biol. 472: 159–68.

Veldhuizen, E. J. A., J. L. M. V. Bokhoven, C. Zweijtzer, S. A. Burt and H. P. Haagsman. 2006. Structural requirements for the antimicrobial activity of carvacrol. J. Agric. Food Chem. 54: 1874–1879.

www.fftc.agnet.org/library.php?func=view&id=20120103110652&type_id=1. Improved Utilization of Fishery By-Products as Potential Nutraceuticals and Functional Foods. (accessed on 7 June 2016).

Index

A

accelerated solvent extraction 216, 218, 240
acid hydrolysis 131, 203, 262, 332
acidic extraction 202
aconitum alkaloids 78
alkali hydrolysis 131
alternative physical treatments 219
ammonia pretreatment 132
amylolytic enzymes 322–325
Anthocyanidins 8–10, 28, 37, 45, 133, 137,
    200, 215, 221, 255, 257, 278, 280, 284,
    387, 388
anti-cancer 147, 389, 395
anti-inflammatory activity 38, 65, 223
antibacterial effects 40, 223, 224
antibacterial properties 40, 366
antioxidant activity 28–31, 37, 64, 65, 135, 138,
    145, 147, 161, 165, 179, 182, 184, 200,
    223, 224, 237, 242, 245, 255, 263, 277,
    284, 293, 295, 300, 350, 361, 364–366,
    370, 388, 392
antioxidant compounds 200
antiviral properties 41
Apigenin 8, 31–35, 37, 38, 45, 140, 146, 236,
    241, 278, 388
apple peel 14–16, 40, 255, 282, 327, 328
Artemisinin 73, 390

B

bacterial pathogens 39
Berberine 6, 60, 61, 65, 67–70, 73, 76, 78
bio-oxidation of terpenes 172
bioactive compounds from fungi 390
bioactive compounds from marine sources 390
bioactive compounds from plants 133, 136,
    261, 390
bioactive compounds in pharmaceutical
    products 384
bioalcohol 318, 329
Bioavailability 42–44, 74–76, 143, 145, 148,
    177, 246, 278, 374, 383
bioflavors 171, 172, 175, 176

biogas 239, 318, 330, 333
biological pretreatment 130, 132, 133
biotechnological production of carotenoids 170
biovanillin production 175
by-products from the agrofood industry 314,
    315, 335
by-products of fruits and vegetables 257, 259,
    329

C

carcinogenic activity 45
cardiovascular disease 2, 15, 28, 33, 68, 160,
    216, 237, 342, 364, 366, 372
castanea sativa by-products 361
cereals and legumes 315, 316
chemical pretreatment 131
chikungunya virus 73
chitosan 148, 179, 180–182, 184, 224, 245,
    352, 395
Chlorophyll 14, 91, 146, 284, 343, 345, 352
coffee by-products 360, 371
collection of isolated compounds 111
column and reaction conditions 96
column liquid chromatography 93
commodity chemicals 334
conventional extraction technologies 239
cosmetic interest 359, 367
curcumin 31, 32, 76
cytotoxicity 31, 45, 71, 73, 74, 76, 197, 373

D

Diabetes 17, 28, 35, 36, 46, 68–70, 196, 197,
    206, 230, 342, 363, 366
dietary fibre 283
dispersive-solid-phase extraction 221

E

egg products 317
electron capture detectors 101
encapsulation 148, 149, 159, 177–180, 183,
    184, 245–247, 283, 353, 374
enzymatic and alkali extraction 202, 203

enzyme-assisted extraction 140, 262, 332
enzymes and flavouring agents 255
enzymes used in bioconversion of food by-
        products 331
evaporative light scattering detectors 111
exotic fruit juice manufacturing 276
extraction method for dietary fiber 202
extraction of phenolic compounds from rice
        bran 205
extraction of phytic acids from rice bran 206
extraction of rice bran protein 203
extraction techniques 5, 133, 136, 138, 149,
        165, 169, 170, 177, 216, 217, 220, 221,
        239, 240, 241, 243, 247, 260, 286, 293
extrusion methods 184

F

fermentation methods 141, 149
fermentation technology 319, 320, 322, 323,
        329
ferulic acid 29, 38, 44, 142, 146, 161–167, 175,
        184, 200, 201, 257, 279, 280, 295, 349
flame ionisation detectors 99–101
flavanones 8–10, 28, 29, 40, 43, 45, 133, 178,
        215, 255, 278, 292, 387, 388
flavones 8, 9, 28, 32, 33, 39, 45, 133, 134, 215,
        236, 241, 255, 277, 278, 298, 346, 387,
        288, 393
flavonoids 7–9, 12, 14, 28–34, 36, 38, 41–43,
        45, 129, 133–135, 137, 138, 140, 144, 145,
        147, 180, 200, 201, 214, 215, 221, 236,
        241, 243, 255, 277, 278, 281, 282, 288,
        291, 293, 322, 332, 343, 345, 360–362,
        365, 366, 387, 388, 393, 394
flavonols 8, 9, 14, 28, 33, 41, 45, 133, 134, 214,
        215, 221, 222, 241, 255, 257, 277, 278,
        282, 288, 289, 298, 346, 387, 388
flavor active compound 160, 173
food and feed 167, 331, 333
food by-products and their effects on skin 366
food by-products in final cosmetic products 372
food waste 2–4, 13–15, 20, 101, 112, 123, 127,
        159, 160, 273, 283, 315, 334, 343, 353,
        359, 375, 391, 392, 396
freeze-drying 148, 183, 184, 221, 284
fruit by-products 256
functional food development 143
functional foods 2, 133, 134, 137, 143–145,
        149, 183, 213, 218, 219, 230, 240, 242,
        247, 272, 277, 299–301, 333
functional foods from residues/by-products 145

G

γ-oryzanol (steryl ferulate) 200
γ-oryzanol extraction and purification 205

gas chromatography 78, 93, 96, 99, 100, 102,
        263
gas chromatography/mass spectrometry 100,
        102
genotoxic activity 44, 45
gingerol 31, 347
glucosinolates 12, 61, 145, 388, 389
grape berry 13, 15, 17
grape pomace 14, 137, 214, 256–258, 260, 261,
        276, 279, 282, 285, 294, 300, 324, 332,
        349, 350, 362, 363, 394
grape stems and canes 214

H

high performance liquid chromatography 78,
        106, 262, 299
high voltage electric discharge 220
high voltage electrical discharge (HVED) 293
high-performance TLC 87
horticultural wastes 252, 253
HPLC columns 107, 109
HPLC detectors 110, 111
human pathogenic fungi 41
hydrodistillation 133, 134, 149, 260, 261

I

Immune system 37, 66, 71, 145
immunomodulatory effects 37, 71, 72
indole alkaloids 63, 68, 390
industrial enzymes 322, 323, 331
Infectious diseases 39, 42, 46, 72
Inflammation 17, 30, 32, 34, 36, 37, 60, 65, 66,
        129, 147, 223, 370, 384
instant controlled pressure drop-assisted
        extraction 140
Isoflavones 8–10, 28, 30, 43, 45, 133, 255, 363,
        364, 366, 369, 387, 388
isoquinoline alkaloids 63–65, 67, 73

K

keamferol 31, 32

L

licorice 36, 40, 76, 144
lignans 8, 11, 12, 29, 30, 37, 43, 141, 215, 236,
        255, 277, 278, 387
lignocellulolytic enzymes 322, 323, 325
lignocellulosic biomass 129–132, 332–334
lipid oxidation 202, 244, 342–344, 346,
        348–351, 353, 354
liquid chromatography/mass spectrometry 100,
        104
liquid scintillation counting 88
lyophilized microcapsules 183

M

maceration 133, 140, 149, 214, 260, 261
malaria 39, 42, 72, 73
mass spectrometers 100, 103–105
meat and derivatives 315, 316
medicago by-products 363
membrane separation processes 239, 243, 247
microtubule-binding natural products 66
microwave assisted extraction 5, 134, 138, 144, 169, 176, 219, 262, 286, 321
mobile phase 87–91, 93, 94, 96, 99, 104–112, 287–292, 298
modern-medicine 1
myricetin 8, 31, 32, 34, 36, 38, 41, 222, 223, 278, 290, 395, 388

N

natural colouring agents 353
NMR spectrum 117–121
non-UV active compounds 91
nuclear magnetic resonance (NMR) 116
nutraceutical applications 13, 242, 244–246

O

obesity 36, 37, 46, 65, 70, 196, 197, 206, 223, 230, 342
olive by-products 231, 235–239, 242, 361, 370, 371
olive mill waste water (OMWW) 231, 361
olive pomace 40, 171, 179, 181, 231, 233, 235–237, 239–241, 243, 245, 247, 330, 361
olives 13, 15, 17, 18, 33–35, 40, 129, 146, 147, 171, 173, 176, 179, 181, 230–247, 255, 257, 260, 261, 264, 330, 347, 348, 350, 361, 369–371, 394
one-dimensional NMR 120
organic acids 110, 132, 161, 234, 252, 284, 285, 318, 326–329, 334, 335, 361, 363
Organosolv pretreatment 132
oxidative stress 28, 30, 31, 33, 68, 167, 215, 222, 223, 369, 370

P

pectin 140, 145, 147, 148, 178, 180, 181, 183, 184, 196, 215, 216, 234, 256, 258, 260, 263, 276, 278–283, 294, 296, 297, 299, 324, 325, 351, 352, 392
pectinases 140, 145, 260, 261, 322, 324–326, 332, 334
phenolic acids 8, 10–12, 14, 27–29, 37, 43, 133, 138, 140, 145, 161, 200, 201, 205, 206,

214, 215, 221, 222, 236, 241, 255, 262, 277, 278, 281, 282, 298, 332, 362, 366, 387, 388, 393
phenolic acids in rice bran 201
phosphorus compounds in rice wastes 201
photo-oxidation mechanism 344
photoionisation detector 101
physical pretreatment 130
piperidine and pyridine 63
plant alkaloids 5, 66, 78
plant derived phenolic compounds 31
plant-derived natural lycopene 168
potential food losses 3
pressurized liquid extraction 5, 135, 149, 218, 240, 247, 285
pretreatment of materials 129
pulsed electric field extraction 139, 149
pulsed ohmic heating 220
pyrrolizidine (Senecio) alkaloids 64

Q

quantitative analysis 87, 88, 93, 102, 106
QuEChERS 221, 222
Quercetin 8, 14, 29, 31–38, 45, 145–146, 222, 223, 257, 262, 264, 278–281, 289–292, 295, 322, 347, 349, 361–363, 368–370, 388
quercus by-products 365
quinoline alkaloids 63, 65
quinolizidine (lupin) alkaloids 64

R

reactive oxygen species (ROS) 28, 65, 222, 369
refractive index detector 110
resin chromatography 239, 243, 247
rice bran dietary fiber and diabetes 196
rice bran dietary fiber and obesity 197
rice bran protein 198, 199, 203 205
rice bran dietary fiber and gastrointestinal health 197

S

sample injection 94, 95
sample introduction 106
sample preparation 94, 118, 218, 221, 260, 283, 284, 287, 288, 290, 292
secondary plant metabolites 1, 3, 4, 7
selecting a detector 100
separation processes for bioflavors 176
side effects and toxicity 44, 46, 76
solid state fermentation (SSF) 128, 141, 171, 260, 321
solid waste from fruit juice 279

solid-liquid extraction 5, 217, 262
solid-phase extraction 170, 177, 221, 394
solvent elution 109
solvent extraction 134, 135, 149, 165, 168, 169,
    176, 205, 206, 216–218, 239, 240, 262,
    284–287, 293, 332
sources and types of alkaloids 61
Soxhlet extractor 133, 168
spectrophotometric detection 112
spray drying 148, 178–183, 245–247
stationary phase 86, 87, 89, 90, 93–100,
    106–109
steroid alkaloids 64
sterols 19, 143, 161, 200, 280, 283, 350, 387,
    389, 393
stilbenes 8, 11, 29, 33, 34, 42, 43, 147, 214,
    215, 219, 221, 222, 255, 257, 277, 278,
    292, 387
storage by-products 234–236, 239
subcritical water extraction 136, 285
supercritical fluid extraction 136, 149, 205,
    217, 240, 241, 247, 285, 289, 332, 394

T

tannase 140, 260, 261, 322, 324, 326
tannins 12, 28–30, 39 41, 42, 44, 71, 133, 134,
    141, 142, 144, 215, 216, 219, 221, 232,
    234, 255, 260, 277, 278, 280–282, 288,
    289, 298, 317, 326, 360, 362, 366, 387,
    388
terpenoid alkaloids 64
terpenoids 4, 12, 64, 134, 138, 144, 167, 172,
    370, 387, 389

thermal conductivity detector 100
thin layer radiochromatography 88
thin-layer chromatography 86, 166
tomato 5, 6, 13, 18, 19, 43, 129, 137, 146,
    168–170, 182, 183, 252, 254, 255, 257,
    259, 322, 332, 364, 365, 368, 369, 372
tomato by-products 364, 372
traditional medicine 1, 70, 235, 390
tripterygium wilfordii 66
tropane alkaloids 62, 63, 77, 78
twigs and olive leaves 231
two-dimensional (2D) NMR 121
two-phase centrifugation system 241

U

ultrasonic assisted extraction 139, 149, 296
ultraviolet (UV) detector 110
utilisation of bioactive compounds 127, 195,
    213, 230, 252, 272, 342, 346, 348, 350,
    351, 353, 391

V

valorization processes 318
vegetable by-products 252, 256, 258
viewing compounds 91
vincosamide 68

W

waste from fruit juice industry 273, 279
wine industry 14, 129, 213, 214, 216
winemaking wastes 213, 224, 225